WITHDRAWN

D0301817

Second Edition

Introduction to COMPOSITE MATERIALS DESIGN

Second Edition

Introduction to COMPOSITE MATERIALS DESIGN

Ever J. Barbero

CRC Press is an imprint of the
Taylor & Francis Group, an **informa** business

Cover Photo: © Gilles Martin-Raget / BMW ORACLE Racing

CRC Press
Taylor & Francis Group
6000 Broken Sound Parkway NW, Suite 300
Boca Raton, FL 33487-2742

Printed in the United States of America on acid-free paper
10 9 8 7 6 5 4

International Standard Book Number: 978-1-4200-7915-9 (Hardback)

Library of Congress Cataloging-in-Publication Data

Barbero, Ever J.
Introduction to composite materials design / author, Ever J. Barbero. -- 2nd ed.
p. cm.
Includes bibliographical references and index.
ISBN 978-1-4200-7915-9 (hardcover : alk. paper)
1. Composite materials--Mechanical properties. 2. Manufacturing processes. I. Title.

TA418.9.C6B37 2011
620.1'1892--dc22 2010018871

Visit the Taylor & Francis Web site at
http://www.taylorandfrancis.com

and the CRC Press Web site at
http://www.crcpress.com

Dedicado a las instituciones en que cursé mis estudios, a saber,
Colegio Dean Funes, Escuela Nacional de Educación Técnica
Ambrosio Olmos, Universidad Nacional de Río Cuarto, y
Virginia Polytechnic Institute and State University

Contents

Acknowledgment

I wish to thank my colleagues and students for their valuable suggestions and contributions to this textbook. Thanks to Adi Adumitroiae and Julio C. Massa for their participation in the development of Chapters 9 and 10, respectively. Thanks to Fabrizio Greco and Paolo Lonetti for their initial writing of, and major contributions to Chapter 13. Thanks to Jim Gauchel for his review and contributions to the section on matrix materials, to Daneesh McIntosh for her contribution to the section on nanomaterials, and to Daniel Cortes, Carlos Dávila, Pere Maimi, Sergio Oller, and Girolamo Sgambitterra for providing insightful suggestions for various sections, including those related to damage and failure. Thanks to the reviewers of the first edition, Matt Fox, Jim Harris, Kevin Kelly, Pizhong Qiao, Hani Salim, Marco Savoia, Malek Turk, Youqi Wang, Ed Wen, and to the reviewers of the second edition, Fritz Campo, An Chen, Joaquin Gutierrez, Xavier Martinez, Pizhong Qiao, Sandro Rivas, and Eduardo Sosa. Furthermore, thanks to the many students at West Virginia University and elsewhere around the world that studied from the first edition and pointed out sections that needed rewriting. My gratitude to my wife, Ana Maria, who typed the entire manuscript for the first edition, and had the patience to let me go through the monumental task of preparing this second edition, and to my children, Margaret and Daniel, who maintain that this is not really a second edition but rather an entirely new book, or so it seems to them in proportion to the time I have spent on it.

Preface to the Second Edition

It has been over ten years since the first edition appeared. In the meantime, utilization of composites has increased in almost every market. Boeing's 787 main technological advance is based on widespread incorporation of composites, accounting for about 50% of the aircraft. Its use allows the plane to be lighter and consequently more fuel efficient. It also allows higher moisture content in the cabin, thus increasing passenger comfort. Cost and production time still hamper utilization of composites in the automotive sector, but as in all other industries, there is a relentless transformation from using conventional to composite materials in more and more applications.

Increased utilization of composites requires that more and more engineers be able to design and fabricate composite structures. As a result, practising engineers and students are equally interested in acquiring the necessary knowledge. This second edition incorporates the advances in knowledge and design methods that have taken place over the last ten years, yet it maintains the distinguishing features of the first edition. Like the first edition, it remains a textbook for senior-level undergraduate students in the engineering disciplines and for self-studying, practicing engineers. Therefore, the discussion is based on math and mechanics of materials background that is common by the senior year, avoiding tensor analysis and other mathematical constructs typical of graduate school.

Seventy eight fully developed examples are distributed throughout the textbook to illustrate the application of the analysis techniques and design methodology presented, making this textbook ideally suited for self-study. All examples use material property data and information available in thirty five tables included in the textbook. One hundred and sixty eight illustrations, including twelve carpet plots, aid in the explanation of concepts and methodologies. Additional information and software is provided in the Web site [1]. Finally, one hundred and sixty seven exercises at the end of chapters will challenge the reader and provide opportunity for testing the level of proficiency achieved while studying.

Experience from instructors all over the world has confirmed that Chapters 1 to 7 can be taught in a 1-semester undergraduate course, assigning Chapters 2 and 3 for independent reading. Those seven chapters remain in the same order in this second edition. Since they have been expanded to accommodate new information, a number of sections have been marked with an (*) to indicate topics that could be skipped on an introductory course, at the instructor's discretion.

Experience from self-taught practitioners all over the world demanded to maintain the ten chapters from the first edition, including composite beams, plates, and shells, as well as to add new topics such as fabric-reinforced composites and external strengthening of concrete. Therefore, all ten chapters from the first edition have been revised, updated, and expanded. Three additional chapters have been added, expanding significantly the coverage of analysis and design of practical composite applications. From the material in Chapters 8 to 13, an instructor/reader can pick topics for special projects or tailor a follow-up graduate course. One such course, *structural composites design,* is now taught at several universities in the United States as an advanced-undergraduate/introductory-graduate course.

Two new topics, design for reliability and fracture mechanics, are now introduced in Chapter 1 and applied throughout the book. The composite property tables in Chapter 1 have been expanded in order to support an expanded set of examples throughout the book.

Chapter 2 is thoroughly revised and updated, including new information on modern fibers, carbon nanotubes, and fiber forms such as textiles. A new section on fabric-reinforcement serves as introduction for the new Chapter 9 on Fabric Reinforced Composites. More material properties for fibers and matrices are given in the tables at the end of Chapter 2, to support the revised and expanded set of examples throughout the book.

Chapters 3, 5, and 6 sustained the least changes. Chapter 3 was revised and updated with a new section on Vacuum Assisted Resin Transfer Molding (VARTM).

Major advances in prediction of unidirectional-lamina properties were incorporated in Chapter 4, which, as a result, it is heavily updated and expanded. For example, prediction of fracture toughness in modes I and II of the unidirectional lamina are now included, and they serve as background for the discussion about in-situ strength values in Chapter 7. The more complex sections have been rewritten in an attempt to help the student and the instructor make faster progress through complex material. In each section, a short summary describes the main concepts and introduces practical formulas for design. This is followed by (*)-labeled sections for further reading, provided the time allows for it. The sections on prediction of longitudinal compressive strength, transverse tensile strength, transverse compressive strength, and in-plane shear strength have been re-written in this way. This layout allows for in-depth coverage that can be assigned for independent study or be left for later study. In this way, new topics are added, such as Mohr-Coulomb theory, as well as mode I and mode II fracture toughness of composites.

Over the last ten years, the most advances have occurred on the understanding of material failure. Consequently, Chapter 7 has been thoroughly revised to include the most advanced prediction and design methodologies. As in the first edition, Chapter 7 remains focused on design and can be the ending chapter for an undergraduate course, perhaps followed by a capstone design project. However, it now transitions smoothly into Chapter 8, thus providing the transition point to a graduate course on *structural composites design.*

Chapters 8 to 13 cover applied composites design topics without resorting to finite element analysis, which is left for other textbooks used for more advanced graduate courses [2]. Chapters 8 to 13 are designed for a new course aimed simultaneously at the advanced-undergraduate and introductory-graduate levels, but selected topics can be used to tailor the introductory course for particular audiences, such as civil engineering, materials engineering, and so on.

Chapter 8 includes the methodology used to perform damage mechanics analysis of laminated composites accounting for the main damage modes: longitudinal tension, longitudinal compression, transverse tension, in-plane shear, and transverse compression. The methodology allows for the prediction of damage initiation, evolution, stiffness reduction, stress redistribution among laminae, and ultimate laminate failure.

Chapter 9 includes an in-depth description of fabric-reinforced composites, including textile and nontextile composites. The methodology for analysis of textile-reinforced composites includes the prediction of damage initiation, evolution, stiffness reduction, and laminate failure.

Chapters 10, 11, and 12 are revised versions of similarly-titled chapters in the first edition. The chapters have been revised to include design for reliability and to correct a few typos on the first edition.

Finally, Chapter 13 is a new chapter dealing with external strengthening of reinforced-concrete beams, columns, and structural members subjected to both axial and bending loads. External strengthening has emerged as the most promising and popular application of composite materials (called FRP) in the civil engineering sector. Therefore, this chapter offers an opportunity to tailor a course on composites for civil engineering students or to inform students from other disciplines about this new market.

In preparing this second edition, all examples have been revised. The number of examples has grown from 50 in the first edition to 78 in this one. Also, the exercises at the end of chapters have been revised. The number of exercises has grown from 115 in the first edition to 167 in this one. I trust that many students, practicing engineers, and instructors will find this edition to be even more useful than the first one.

Ever J. Barbero, 2010

MATLAB® is a registered trademark of The MathWorks, Inc. For product information, please contact: The MathWorks, Inc. 3 Apple Hill Drive, Natick, MA 01760-2098 USA Tel: 508 647 7000 Fax; 508 647 7001 E-mail: info@mathworks.com Web: www.mathworks.com

Preface to the First Edition

This book deals with the design of structures made of composite materials, also called *composites*. With composites, the material and the structure are designed concurrently. That is, the designer can vary structural parameters, such as geometry, and at the same time vary the material properties by changing the fiber orientation, fiber content, etc. To take advantage of the design flexibility composites offer, it is necessary to understand material selection, fabrication, material behavior, and structural analysis. This book provides the main tools used for the preliminary design of composites. It covers all design aspects, including fiber and matrix selection, fabrication processes, prediction of material properties, and structural analysis of beams, plates, shells, and other structures. The subject is presented in a concise form so that most of the material can be covered in a one-semester undergraduate course.

This book is intended for senior-level engineering students, and no prior knowledge of composites is required. Most textbooks on composites are designed for graduate courses; they concentrate on materials behavior, leaving structural analysis and design to be covered by other, unspecified, graduate courses. In this book, structural analysis and design concepts from earlier courses, such as mechanics of materials, are used to illustrate the design of composite beams, plates, and shells.

Modern analysis and design methodology have been incorporated throughout the book, rather than adding a myriad of research-oriented material at the end of the book. The objective was to update the material that it is actually taught in a typical senior technical-elective course rather than adding reference material that is seldom taught. In addition, design content is included explicitly to provide the reader with practical design knowledge, thus better preparing the student for the workplace. Among the improvements, it is worth mentioning the following: A chapter on materials and a chapter on processing, which emphasize the advantages and disadvantages of various materials and processes, while explaining materials science and process-engineering topics with structural-engineering terminology. In Chapter 4, proven micromechanical formulas are given for all the properties required in the design, as well as reference to the American Society of Testing Materials (ASTM) standards used for testing. In Chapter 6, shear-deformable lamination theory is presented in lieu of the obsolete classical lamination theory. In Chapter 7, the truncated-maximum-strain-criterion, widely accepted in the aerospace industry, is explained in detail. Chapters 10, 11, and 12 present simple, yet powerful methods for the preliminary design of composite beams, plates, stiffened panels, and shells. The material in these later chapters does not require, for the most part, any background beyond that provided by the typical engineering curricula in aerospace, civil, or mechanical engineering.

Design content is distributed throughout the book in the form of special design-oriented sections and examples. The presentation emphasizes concepts rather than mathematical derivations. The objective is to motivate students who are interested in designing useful products with composites rather than performing research. Every final equation in every section is useful in the design process. Most of the equations needed for design are programmed into the accompanying software to eliminate the need for tedious computations on part of the students. The software, entitled Computer Aided Design Environment for Composites (CADEC), is a windows application with an intuitive, web-browser-like graphical user interface, including a help system fully cross-referenced to the book. Examples are used to illustrate aspects of the design process. Suggested exercises at the end of each chapter are designed to test the understanding of the material presented.

Composites design involves synthesis of information about materials, manufacturing processes, and stress-analysis to create a useful product. An overview of the design process as well as composites terminology are introduced in Chapter 1, followed by a description of materials and manufacturing processes in Chapters 2 and 3. Composites design also involves stress- and deformation-analysis to predict how the proposed structure/material combination will behave under load. Since a one-semester course could be spent on analysis alone, an effort has been made in this book to simplify the presentation of analysis methods, leaving time for design topics.

Composites design can be accomplished following one of various methodologies outlined in Chapter 1 and developed throughout Chapters 4 to 7. The instructor can choose from the various design options described in Section 1.2 to strike a balance between simplicity and generality. Self-study readers are encouraged to read through Chapters 4 to 7, with the exception of those sections marked with a star (*), which can be studied afterward. The book is thoroughly cross-referenced to allow the reader to consult related material as needed. Since the constituent materials (fiber and matrix) as well as the manufacturing process influence the design of a composite structure, the designer should understand the characteristics and limitations of various materials as well as manufacturing processes used in the fabrication of composites, which are described in Chapters 2 and 3.

Prediction of composite properties from fiber and matrix data is presented in Chapter 4. Although composite properties could be obtained experimentally, the material in Chapter 4 is still recommended as the basis for understanding how fiber-reinforced materials work. Only those formulas useful in design are presented, avoiding lengthy derivations or complex analytical techniques of limited practical use. Some of the more complex formulas are presented without derivations. Derivations are included only to enhance the conceptual understanding of the behavior of composites.

Unlike traditional materials, such as aluminum, composite properties vary with the orientation, having higher stiffness and strength along the fiber direction. Therefore, the transformations required to analyze composite structures along arbitrary directions not coinciding with the fiber direction are presented in Chapter 5. Furthermore, composites are seldom used with all the fibers oriented in only one direction. Instead, laminates are created by stacking laminae with fibers in various orientations to efficiently carry the loads. The analysis of such laminates is presented in Chapter 6, with numerous design examples.

While stress-analysis and deformation-controlled design is covered in Chapter 6, failure prediction is presented in Chapter 7. Chapter 7 includes modern material, such as the truncated-maximum-strain criterion, which is widely used in the aerospace industry. Also, a powerful preliminary-design tool, called *carpet plots*, is presented in Chapters 6 and 7 and used in examples throughout the book.

The instructor or reader can choose material from Chapters 10 to 12 to tailor the course to his/her specific preferences. Chapter 10 includes a simple preliminary-design procedure (Section 10.1) that summarizes and complements the beam-design examples presented throughout Chapters 4 to 7. In Section 10.2, a novel methodology for the analysis of thin-walled composite sections is presented, which can be used as part of an advanced course or to tailor the course for aerospace or civil engineering students. All of the thin-walled beam equations are programmed into CADEC to eliminate the need for tedious computations or programming by the students.

Chapter 11 is intended to provide reference material for the preliminary design of plates and stiffened panels. Rigorous analysis of plate problems has not been included because it requires the solution of boundary value problems stated in terms of partial differential equations, which cannot be tackled with the customary background of undergraduate students. Similarly, Chapter 12 can be used to tailor a course for those interested in the particular aspects of composite shells. Again, complex analytical or numerical procedures have not been included in favor of a simple, yet powerful membrane-analysis that does not require advanced analytical or computational skills.

Introduction to Composite Materials Design contains more topics than can be taught in one semester, but instructors can tailor the course to various audiences. Flexibility is built primarily into Chapters 2, 3, 10, 11, and 12. Chapters 2 and 3 can be covered in depth or just assigned for reading, depending on the emphasis given in a particular curriculum. Chapters 10, 11, and 12 begin with simple approximate methods that can be taught quickly, and they evolve into more sophisticated methods of analysis that can be taught selectively depending on the audience or be left for future reading. Video references given in Chapter 1 provide an efficient introduction to the course when hands-on experience with composite manufacturing is not feasible.

An effort has been made to integrate the material in this book into the undergraduate curriculum of aerospace, civil, and mechanical engineering students. This has been done by presenting stress-analysis and structural-design in a similar fashion as covered in traditional, mandatory courses, such as mechanics of materials, mechanical design, etc. Integration into the existing curriculum allows the students to assimilate the course content efficiently because they are able to relate it to their previously acquired knowledge. Furthermore, design content is provided by special design-oriented sections of the textbook, as well as by design examples. Both, integration in the curriculum and design content are strongly recommended by engineering educators worldwide, as documented in engineering accreditation criteria, such as those of Accreditation Board of Engineering Technology (ABET).

The market for composites is growing steadily, including commodity type applications in the automotive, civil infrastructure, and other emerging markets. Because of the growing use of composites in such varied industries, many practicing engineers feel the need to design with these new materials. This book attempts to reach both the senior-level engineering student as well as the practicing engineer who has no prior training in composites. Practicality and design are emphasized in the book, not only in the numerous examples but also in the material's explanation. Structural design is explained using elementary concepts of mechanics of materials, with examples (beams, pressure vessels, etc.) that resemble those studied in introductory courses. I expect that many students, practicing engineers, and instructors will find this to be a useful text on composites.

Ever J. Barbero, 1998

List of Figures

List of Tables

List of Symbols

overline $\overline{(\ )}$	*Transformed,* usually to laminate coordinates
widetilde $\widetilde{(\ )}$	*Undamaged* (virgin), or *effective* quantity
$\widehat{(\ )}$	Average
α	Load factor. Also, fiber misalignment
α_σ	Standard deviation of fiber misalignment
α_0	Angle of the fracture plane
α_1, α_2	Longitudinal and transverse coefficient of thermal expansion (CTE)
α_A, α_T	Axial and transverse CTE of fibers
$[\alpha]$	Membrane compliance of a laminate
α, β, γ	Thickness ratio used with carpet plots
$[\alpha]$	In-plane compliance of a laminate
$[\beta]$	Bending-extension compliance of a laminate
$[\delta]$	Bending compliance of a laminate
θ_k	Orientation of lamina k in a laminate
β_1, β_2	Longitudinal and transverse coefficient of moisture expansion
δ_b, δ_s	Bending and shear deflections of a beam
ϵ_{1t}	Ultimate longitudinal tensile strain
ϵ_{2t}	Ultimate transverse tensile strain
ϵ_{1c}	Ultimate longitudinal compressive strain
ϵ_{2c}	Ultimate transverse compressive strain
ϵ_{fu}	Ultimate fiber strain (tensile)
ϵ_{mu}	Ultimate matrix strain (tensile)
$\boldsymbol{\epsilon}$	Strain tensor
ε_{ij}	Strain components in tensor notation
ϵ_α	Strain components in contracted notation
ϵ_α^e	Elastic strain
ϵ_α^p	Plastic strain
$\epsilon_x^0, \epsilon_y^0, \gamma_{xy}^0$	Strain components at the midsurface of a shell
$\widetilde{\epsilon}$	Effective strain in contracted notation ($\epsilon_6 = \gamma_6$)
$\widetilde{\varepsilon}$	Effective strain in tensor notation ($\epsilon_6 = \gamma_6/2$)
γ_{6u}	Ultimate shear strain
γ_{xy}^0	In-plane shear strain
$\kappa_x^0, \kappa_y^0, \kappa_{xy}^0$	Curvatures of the midsurface of a shell

λ	Lame constant. Also, crack density
μ	Mean value of a distribution
η	Eigenvalues
η_L, η_T	Coefficients of influence, longitudinal, transverse
η_2, η_4, η_6	Stress partitioning parameters
$\eta_i = E_i/E$	Modular ratio in the transformed section method
ϕ	Resistance factor. Also, angle of internal friction
$\phi(z)$	Standard PDF
ϕ_x, ϕ_y	Rotations of the normal to the midsurface of a shell
ϖ	Standard deviation
ρ	Density
ρ_f, ρ_m, ρ_c	Density of fiber, matrix, and composite
ψ	Load combination factor
$\boldsymbol{\sigma}$	Stress tensor
σ_0	Weibull scale parameter
σ_{ij}	Stress components in tensor notation
σ_α	Stress components in contracted notation
$\widetilde{\sigma}$	Effective stress
τ_L, τ_T	Longitudinal and transverse shear stress
ν	Poisson's ratio
ν_{12}	In-plane Poisson's ratio
ν_{23}, ν_{13}	Intralaminar Poisson's ratios
ν_{xy}	Laminate Poisson ratio x-y
ν_A	Axial Poisson's ratio of fibers
Γ	Gamma function
$\Lambda_{22}^0, \Lambda_{44}^0$	Dvorak parameters
$\mathbf{\Omega} = \sqrt{\mathbf{I\text{-}D}}$	Integrity tensor
$2a_0$	Representative crack size
g	Damage activation function
d_f	Degradation factor
k_f	Fiber stress concentration factor
m	Weibull shape parameter
n_1, n_2, n_3	Components of the vector normal to a surface
$p(z)$	Probability density function (PDF)
$r_1(\varphi), r_2(\varphi)$	Radii of curvature of a shell
t_k	Thickness of lamina k in a laminate
t_1, t_2, t_3	Projection components of a vector on the coordinate axes $1, 2, 3$
$t_t = 4a_0$	Transition thickness
u, v, w	Components of the displacement along the directions x, y, z
u_0, v_0, w_0	Components of the displacement at the midsurface of a shell
w	Fabric weight per unit area
z	Standard variable
$[A]$	Membrane stiffness of a laminate with components A_{ij} ; $i, j = 1, 2, 6$
$[B]$	Bending-extension coupling stiffness of a laminate B_{ij}

C_{ij}	3D stiffness matrix
C_{DF}	Cumulative distribution function (CDF)
C_F, C_σ	Strength and load coefficient of variance (COV)
CTE	Coefficient of thermal expansion
D	Damage tensor
$[D]$	Bending stiffness of a laminate with components D_{ij} ; $i, j = 1, 2, 6$
E_x, E_y, G_{xy}	Laminate moduli
E	Young's modulus
E_1, E_2	Longitudinal and transverse moduli
E_A, E_T	Axial and transverse moduli of fibers
F	Resistance (material strength)
F_{1t}	Longitudinal tensile strength
F_{2t}	Transverse tensile strength
F_6	In-plane shear strength
F_{1c}	Longitudinal compressive strength
F_{2c}	Transverse compressive strength
F_4	Intralaminar shear strength
F_{ft}	Apparent fiber tensile strength
F_{mt}	Apparent matrix tensile strength
F_{csm-t}	Tensile strength of a random-reinforced lamina
F_x^b, F_y^b, F_{xy}^b	Flexural strength
G	Shear modulus
G_{12}	In-plane shear modulus
G_{23}	Transverse shear modulus
G_A, G_T	Axial and transverse shear modulus of fibers
G_{Ic}, G_{IIc}	Fracture toughness mode I and II
$[H]$	Transverse shear stiffness of a laminate H_{ij} ; $i, j = 4...5$
HDT	Heat distortion temperature
I	Moment of inertia of the cross-section of a beam
I_F	Failure Index
M	Bending moment applied to a beam
M_x, M_y, M_{xy}	Bending moments per unit length at the midsurface of a shell
M_x^T, M_y^T, M_{xy}^T	Thermal moments per unit length
N_x, N_y, N_{xy}	Membrane forces per unit length at the midsurface of a shell
N^T, ϵ^T	Membrane thermal force and strain caused by thermal expansion
N_x^T, N_y^T, N_{xy}^T	Thermal forces per unit length
Q	Reduced stiffness matrix in lamina coordinates x_1, x_2, x_3
Q^*	Intralaminar reduced stiffness matrix
$\overline{Q}$	Reduced stiffness matrix in laminate coordinates X, Y, Z
$\widetilde{Q}$	Undamaged reduced stiffness matrix in lamina coordinates
$\overline{\widetilde{Q}}$	Undamaged reduced stiffness matrix in laminate coordinates
Q^{CSM}	Reduced stiffness of a random-reinforced lamina
R	Strength ratio, safety factor
$[R]$	Reuter matrix

$R_e = 1 - C_{DF}$ Reliability
S_{ij} 3D compliance matrix
S^* Intralaminar components of the compliance matrix S
SFT Stress-free temperature [$°C$]
T_g Glass transition temperature
$T = T(\theta)$ Stress transformation matrix from laminate to lamina coordinates
$T^{-1} = T(-\theta)$ Stress transformation matrix from lamina to laminate coordinates
V_x, V_y Transverse shear forces per unit length at the midsurface of a shell
V_f, V_m Fiber and matrix volume fraction
V_v Void content (volume fraction)
W_f, W_m Fiber and matrix weight fraction

Symbols Related to Fabric-reinforced Composites

θ_f, θ_w Undulation angle of the fill and gap tows, respectively
Θ_f, Θ_w Coordinate transformation matrix for fill and gap
a_f, a_w Width of the fill and gap tows, respectively
g_f, g_w Width of the gap along the fill and gap directions, respectively
h_f, h_w, h_m Thickness of the fill and gap tows, and matrix region, respectively
n_g Harness
n_s Number of subcells between consecutive interlacings
n_i Number of subcells in the interlacing region
$z_f(x), z_w(y)$ Undulation of the fill and gap tows, respectively
A_{fill}, A_{warp} Cross-section area of fill and warp tows
F_{fa} Apparent tensile strength of the fiber
F_{mta}, F_{msa} Apparent tensile and shear strength of the matrix
L_{fill}, L_{warp} Developed length of fill and warp tows
T_f, T_w Stress transformation matrix for fill and gap
V_f^o Overall fiber volume fraction in a fabric-reinforced composite
V_{meso} Volume fraction of composite tow in a fabric-reinforced composite
V_f^f, V_f^w Volume fraction of fiber in the fill and warp tows
w Weight per unit area of fabric

Symbols Related to Beams

β Rate of twist
ϕ Angle of twist
λ^2 Dimensionless buckling load
ω_s Sectorial area
$\overline{\omega}$ Principal sectorial area
η_c, ζ_c Mechanical shear center
$\Gamma_{s''}$ Area enclosed by the contour
e_b, e_q Position of the neutral surface of bending and torsion
q Shear flow

s, r	Coordinates along the contour and normal to it
$\overline{y}_c, \overline{z}_c$	Mechanical shear center
$\overline{z}_G, \overline{z}_\rho, \overline{z}_M$	Geometric, mass, and mechanical center of gravity
(EA)	Axial stiffness
$(EI_{y_G}), (EI_{z_G})$	Mechanical moment of inertia
$(EI_{y_G z_G})$	Mechanical product of inertia
$(EI_\eta), (EI_\zeta)$	Bending stiffness with respect to principal axis of bending
$(E\omega)$	Mechanical sectorial static moment
(EI_ω)	Mechanical sectorial moment of inertia
(GA)	Shear stiffness
(GJ_R)	Torsional stiffness
$(EQ_{\omega\zeta})$	Mechanical sectorial linear moment
$(EQ_\zeta(s))$	Mechanical static moment
K	Coefficient of restraint
N_{xs}^i	Shear flow in segment i
L_e	Effective length of a column
T	Torque
$Z = I/c$	Section modulus

Symbols Related to Strengthening of Reinforced Concrete

α, α_i	Load factor, partial load factors
α_c, β_c	Stress-block parameters for confined section
β_1	Stress-block parameter for unconfined section
ε_{bi}	Initial strain at the soffit
ε_c	Strain level in the concrete
ε_{cu}	Ultimate axial strain of unconfined concrete
ε_{ccu}	Ultimate axial compressive strain of confined concrete
ε_f	Strain level in FRP
ε_{fd}	FRP debonding strain
$\varepsilon_{fu}, \varepsilon_{fe}$	FRP allowable and effective tensile strain
ε_{fu}^*	FRP ultimate strain
ε_s	Strain level in the steel reinforcement
ε_y	Steel yield strain
κ_a	FRP efficiency factor in determination of f'_{cc}
κ_b	FRP efficiency factor in determination of ε_{ccu}
κ_v	Bond reduction factor
κ_ε	FRP efficiency factor
ϕF	Factored capacity
ϕ	Strength reduction factor (resistance factor)
$\phi(P_n, M_n)$	Factored (load, moment) capacity
ϕ_{ecc}	Eccentricity factor
ρ_f	FRP reinforcement ratio
ρ_g	Longitudinal steel reinforcement ratio

ψ	Load combination factor
ψ_f	FRP strength reduction factor
c	Position of the neutral axis
c_b	Position of the neutral axis, BSC
b_f	Width of FRP laminae
b, h	Width and height of the beam
d	Depth of tensile steel
d_{fv}	Depth of FRP shear strengthening
$f_{c,s}$	Compressive stress in concrete at service condition
f'_c	Concrete compressive strength, unconfined
f'_{cc}	Concrete compressive strength, confined
$f_{f,s}$	Stress in the FRP at service condition
f_l	Confining pressure
f^*_{fu}	FRP tensile strength
f_{fu}, f_{fe}	FRP allowable and effective tensile strength
$f_{s,s}$	Stress in the steel reinforcement at service condition
f_y	Steel yield strength
h, b	Height and width of the cross-section
n	Number of plies of FRP
n_f, n_s	Number of FRP strips and steel bars in shear
r_c	Radius of edges of a prismatic cross-section confined with FRP
s_f, s_s	FRP and steel spacing in shear
t_f	Ply thickness of FRP
w_f	Width of discontinuous shear FRP
A_c	Area of concrete in compression
A_e	Area of effectively confined concrete
A_f	Area of FRP
A_g	Gross area of the concrete section
A_{fb}	Area of FRP, BSC
A_{si}	Area of i-th rebar
$A_s(A'_s)$	Tensile (compressive) steel reinforcement area
A_{st}	Sum of compressive and tensile steel reinforcement areas
A_{sv}	FRP and steel shear area
C	Axial compressive force in the concrete
C_E	Environmental exposure coefficient
D	Column diameter
E_c	Modulus of concrete
E_f	Modulus of FRP
E_s	Modulus of steel
$(EI)_{cr}$	Bending stiffness of the cracked section
F	Nominal capacity (strength)
L	Load (applied load, moment, or stress)
L_e	Active bond length of FRP
M_n	Nominal moment capacity

M_u	Required moment capacity
P_n	Nominal axial compressive capacity (strength)
P_u	Required axial strength
S_{DL}	Stress resultant of the dead load
S_{LL}	Stress resultant of the live load
T_s, T_f	Tensile force in steel and FRP
U	Required capacity
V_n	Nominal shear capacity
V_u	Required shear capacity
V_c, V_f, V_s	Nominal shear strength of concrete (C), FRP (F), and steel stirrups (S)

List of Examples

Chapter 1

Introduction

Materials have such an influence on our lives that the historical periods of humankind have been dominated, and named, after materials. Over the last forty years, composite materials, plastics, and ceramics have been the dominant emerging materials. The volume and number of applications of composite materials has grown steadily, penetrating and conquering new markets relentlessly. Most of us are all familiar with fiberglass boats and graphite sporting goods; possible applications of composite materials are limited only by the imagination of the individual. The main objective of this book is to help the reader develop the skills required to use composites for a given application, to select the best material, and to design the part. A brief description of composites is presented in Section 1.1 and thoroughly developed in the rest of this book.

There is no better way to gain an initial feeling for composite materials than actually observing a composite part being fabricated. Laboratory experience or a video about composite fabrication is most advantageous at this stage. Materials and supplies to set up for simple composites fabrication as well as videos that teach how to do it are available from various vendors [3–11]. Alternatively, a visit to a fabrication facility will provide the most enlightening initial experience. After that, the serious work of designing a part and understanding why and how it performs can be done with the aid of the methods and examples presented in this book, complemented with further reading and experience. There are several ways to tackle the problems of analysis and design of composites. These are briefly introduced in Section 1.2 and thoroughly developed in the rest of this book.

1.1 Basic Concepts

A composite material is formed by the combination of two or more distinct materials to form a new material with enhanced properties. For example, rocks are combined with cement to make concrete, which is as strong as the rocks it contains but can be shaped more easily than carving rock. While the enhanced properties of concrete are strength and ease of fabrication, most physical, chemical, and processing-related properties can be enhanced by a suitable combination of materials. The most common composites are those made with strong fibers held together in a binder. Particles or flakes are also used as reinforcements but they are not as effective as fibers.

The oldest composites are natural: Wood consists of cellulose fibers in a lignin matrix and human bone can be described as fiber-like osteons embedded in an interstitial bone matrix. While manmade composites date back to the use of straw-reinforced clay for bricks and pottery, modern composites use metal, ceramic, or polymer binders reinforced with a variety of fibers or particles. For example, fiberglass boats are made of a polyester resin reinforced with glass fibers. Sometimes, composites use more than one type of reinforcement material, in which case they are called *hybrids*. For example consider reinforced concrete, a particle-reinforced composite (concrete) that is further fiber-reinforced with steel rods. Sometimes, different materials are layered to form an enhanced product, as in the case of sandwich construction where a light core material is sandwiched between two faces of stiff and strong materials. Composite materials can be classified in various ways, the main factors being the following:

- Reinforcement
  - Continuous long fibers
    - Unidirectional fiber orientation
    - Bidirectional fiber orientation (woven, stitched mat, etc.)
    - Random orientation (continuous strand mat)
  - Discontinuous fibers
    - Random orientation (e.g., chopped strand mat)
    - Preferential orientation (e.g., oriented strand board)
  - Particles and whiskers
    - Random orientation
    - Preferential orientation
- Laminate configuration
  - Unidirectional lamina: a single lamina (also called *layer* or *ply*), or several laminae (plural) with the same material and orientation in all laminae.
  - Laminate: several laminae stacked and bonded together, where at least some laminae have different orientation or material.
  - Bulk composites: for which laminae cannot be identified, including bulk molding compound composites, particle-reinforced composites, and so on.
- Hybrid structure
  - Different material in various laminae (e.g., bimetallics)
  - Different reinforcement in a lamina (e.g., intermingled boron and carbon fibers)

Fiber reinforcement is preferred because most materials are much stronger in fiber form than in their bulk form. This is attributed to the sharp reduction in the number of defects in the fibers compared to those in bulk form. The high strength of polymeric fibers, such as aramid, is attributed to the alignment of the polymer chains along the fiber as opposed to the randomly entangled arrangement in the bulk polymer. Crystalline materials, such as graphite, also align along the fiber length, increasing the strength. Whiskers, which are elongated single crystals, are extremely strong because the dislocation density of a single crystal is lower than in the poly crystalline bulk material.

The main factors that drive the use of composites are weight reduction, corrosion resistance, and part-count reduction. Other advantages that motivate some applications include electromagnetic transparency, wear resistance, enhanced fatigue life, thermal/acoustical insulation, low thermal expansion, low or high thermal conductivity, etc.

Weight reduction provides one of the more important motivations for use of composites in transportation in general and aerospace applications in particular. Composites are lightweight because both the fibers and the polymers used as matrices have low density. More significantly, fibers have higher values of strength/weight and stiffness/weight ratios than most materials, as shown in Figure 1.1.

However strong, fibers cannot be used alone (except for cables) because fibers cannot sustain compression or transverse loads. A binder or matrix is thus required to hold the fibers together. The matrix also protects the fibers from environmental attack. Therefore, the matrix is crucial for the corrosion resistance of the composite. Because of the excellent resistance to environmental and chemical attack of polymer matrices and most fibers, composites have conquered large markets in the chemical industries, displacing conventional materials such as steel, reinforced concrete, and aluminum. This trend is expanding into the infrastructure construction and repair markets as the resistance to environmental degradation of composites is exploited [12].

Since polymers can be molded into complex shapes, a composite part may replace many metallic parts that would otherwise have to be assembled to achieve the same function. Part-count reduction often translates into production, assembly, and inventory savings that more than compensate for higher material cost.

Since the fibers cannot be used alone and the strength and stiffness of polymer are negligible when compared to the fibers, the mechanical properties of composites are lower than the properties of the fibers. Still, composites are stiffer and stronger than most conventional materials when viewed on a per unit weight basis, as shown in Figure 1.1 by the data of unidirectional composites. The reduction from fiber to composite properties is proportional to the amount of matrix used. This effect will be thoroughly investigated in Chapter 4.

Since the fibers do not contribute to the strength transversely to the fiber direction, and the strength of the matrix is very low, it becomes necessary to add laminae with various orientations to face all the applied loads. One way to achieve this is to create a laminate, as shown in Figure 1.2, stacking laminae with vari-

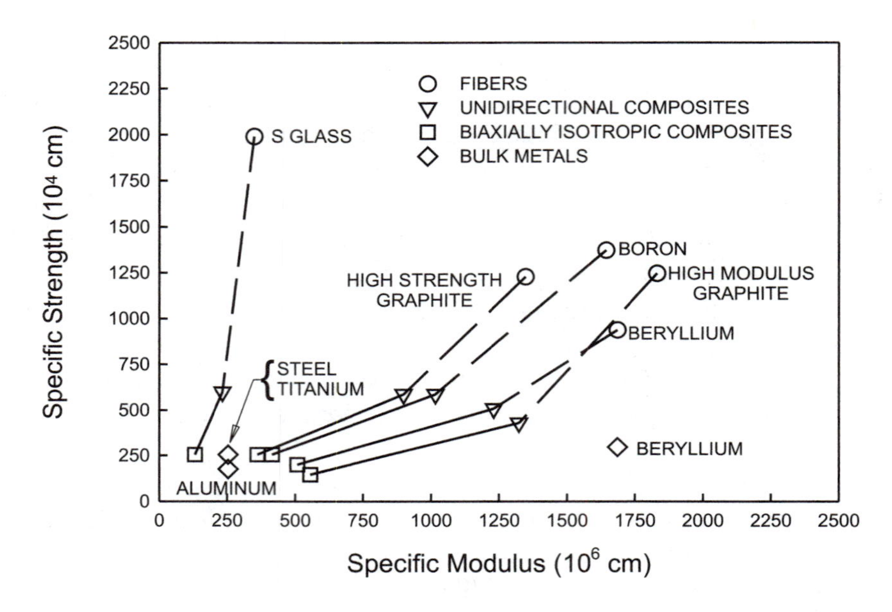

Figure 1.1: Comparison of specific strength and stiffness of composite materials and metals [13].

ous orientations. Although such a laminate can handle multidirectional loads, the strength and stiffness per unit weight of the laminate on a given direction are lower than the corresponding values for a unidirectional composite. This effect, which is thoroughly investigated in Chapters 5 and 6, is illustrated in Figure 1.1 by the data of biaxial composites.

Another solution to handle multidirectional loads is to incorporate bidirectional reinforcements, such a woven cloth, in each lamina. The composite may even be reinforced in such a way that the properties are the same in every direction. This can be accomplished by using random reinforcing material such as continuous or chopped strand mat, or by using short fibers dispersed in the matrix. These types of reinforcements are also used to lower the cost but the properties of such composites are lower than those of continuous fiber composites.

Filler particles, such as calcium carbonate, are commonly used mixed with the matrix for weight and/or cost reduction, to delay flame and reduce smoke, and to reduce the degradation of the polymer by the ultraviolet (UV) radiation present in sunlight. However, filler particles are not considered reinforcements; their effect is accounted for by modifying the properties of the matrix.

Hybrids are used for many reasons. In one case, glass-reinforced or aramid-reinforced laminae are placed on the surface of a carbon-reinforced laminate. The carbon fibers provide stiffness and strength and the glass fibers provide protection against impact from flying objects or projectiles. In another example, a boron-reinforced lamina is sandwiched between carbon-reinforced laminae. Boron fibers

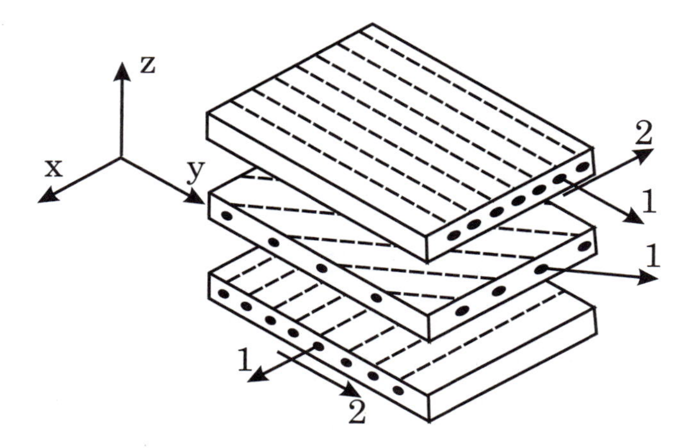

Figure 1.2: Assembly of three laminae into a laminate.

provide high compressive strength but they are very expensive and difficult to handle. Therefore, the carbon-reinforced faces provide high tensile strength and simplify the fabrication while reducing the overall cost. Finally, the most common hybrid is sandwich construction. A lightweight core, such as foam or honeycomb, is sandwiched between two strong and stiff faces. The core separates the two faces so that the moment of inertia of the faces is large, resulting in high bending stiffness, while the core adds little to the weight and cost of the product.

1.2 The Design Process

Design is the process that involves all the decisions necessary to fabricate, operate, maintain, and dispose of a product. Design begins by recognizing a need. Satisfying this need, whatever it is, becomes the problem of the designer. The designer, with the concurrence of the user and other parties involved (marketing, etc.), defines the problem in engineering as well as in layman's terms, so that everyone involved understands the problem. Performance criteria are defined at this stage, meaning that any solution proposed later will have to satisfy them to be considered an acceptable solution.

Definition of the problem leads to statements about possible solutions and specifications of the various ingredients that may participate in the solution to the problem. Synthesis is the selection of the optimum solution among the many combinations proposed. Synthesis, and design in general, relies on analysis to predict the behavior of the product before one is actually fabricated.

Analysis uses mathematical models to construct an abstract representation of the reality from which the designer can extract information about the likely behavior of the real product. The optimized solution is then evaluated against the performance criteria set forth in the definition of the problem. The performance criteria become the stick by which the performance, or optimality, of any proposed solution is measured. In optimization jargon, performance criteria become either objective functions or constraints for the solution. An iterative process takes place as depicted in Figure 1.3 [14].

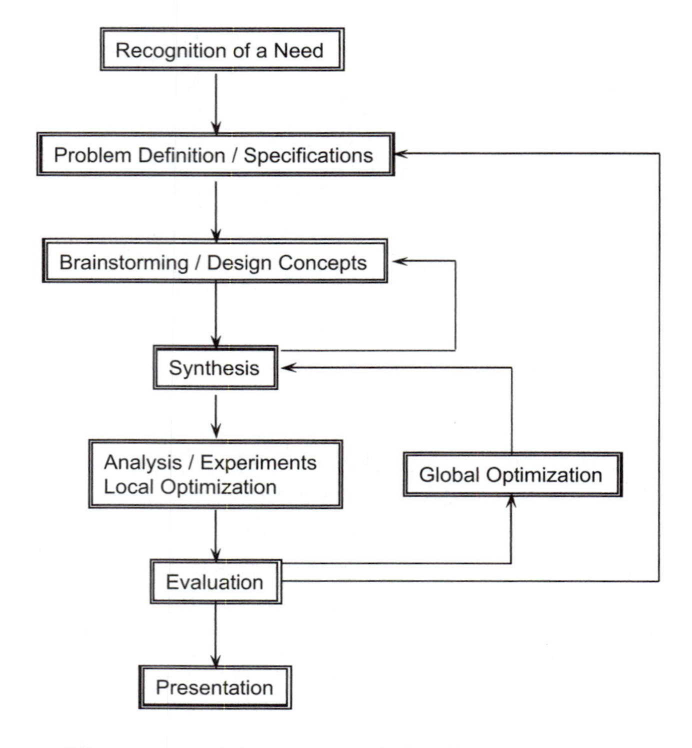

Figure 1.3: Schematics of the design process.

Example 1.1 *Choose a material to carry the loads* $\sigma_x = 400\ MPa$, $\sigma_y = 100\ MPa$, $\sigma_{xy} = 200\ MPa$. *This example illustrates a popular methodology used for structural analysis and design with isotropic materials.*

Solution to Example 1.1 *Principal stress design consists of transforming the applied state of stress* (σ_x,σ_y,σ_{xy}) *into two principal stresses* (σ_I, σ_{II}), *which can be computed using Mohr's circle, or*

$$\sigma_I = \frac{\sigma_x + \sigma_y}{2} + \sqrt{\left(\frac{\sigma_x - \sigma_y}{2}\right)^2 + \sigma_{xy}^2}$$

$$\sigma_{II} = \frac{\sigma_x + \sigma_y}{2} - \sqrt{\left(\frac{\sigma_x - \sigma_y}{2}\right)^2 + \sigma_{xy}^2} \tag{1.1}$$

The principal stresses are oriented at an angle (Figure 1.4, see also [14])

$$\theta = \frac{1}{2}\tan^{-1}\left(\frac{2\sigma_{xy}}{\sigma_x - \sigma_y}\right) \tag{1.2}$$

with respect to the x, y, coordinate system. The principal axes (I, II), are oriented along the principal stresses.

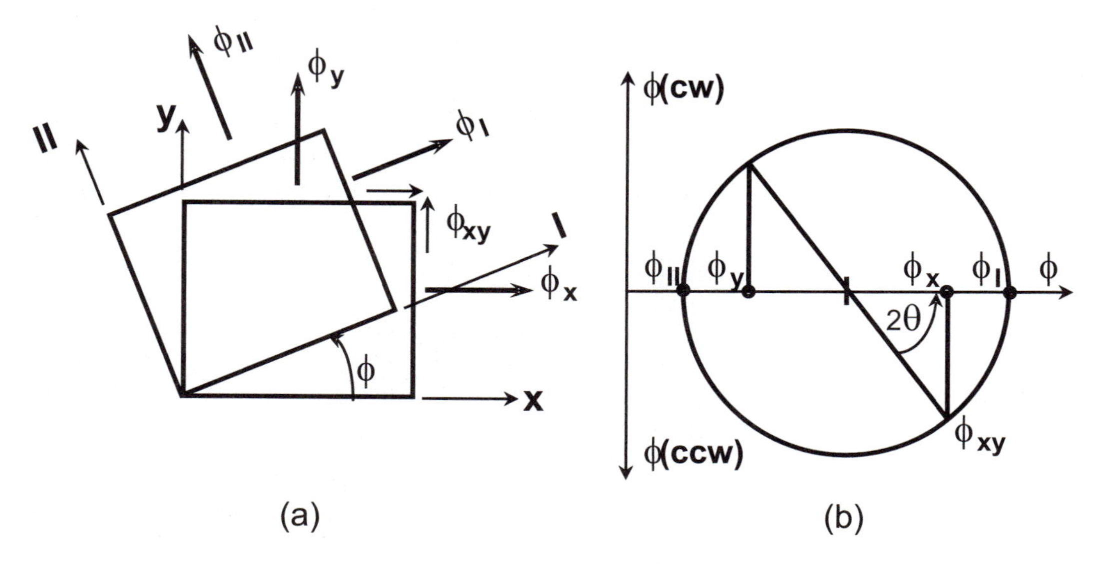

Figure 1.4: Mohr circle for transformation of stresses.

Using this methodology, compute the principal stresses (1.1) and the orientation of the maximum principal stress σ_I with respect to the x-axis (1.2). To simplify the computations introduce the quantities p, q, and R

$$p = \frac{\sigma_x + \sigma_y}{2} = 250 \ MPa$$
$$q = \frac{\sigma_x - \sigma_y}{2} = 150 \ MPa$$
$$R = \sqrt{q^2 + \sigma_{xy}^2} = 250 \ MPa$$

which can be easily identified in the Mohr's circle as the average stress, one-half the difference of the two normal stresses, and the radius of the circle, respectively. Then, the principal stress and their orientation are

$$\sigma_I = p + R = 500 \ MPa$$
$$\sigma_{II} = p - R = 0$$
$$\theta = \frac{1}{2} \tan^{-1}\left(\frac{\sigma_{xy}}{q}\right) = 26.57^\circ$$

This means that the load of $500 \ MPa$ *has to be carried in the direction* $\theta = 26.57^\circ$. *This load can be easily carried by a unidirectional lamina with the fibers oriented at* 26.57°. *From Tables 1.3–1.4, select a material that can carry* $500 \ MPa$. *E-glass–epoxy has a tensile strength of* $1020 \ MPa$, *and assuming a resistance factor* $\phi = 0.7$ *(see section 1.4), yields a load factor* $\alpha = 1.428$; *that is, the nominal load can be exceeded by* $\approx 142\%$.
The advantage of this procedure is that the shear stress is zero in principal axes. Then, the largest of σ_I and σ_{II} is compared to the strength of the material. Principal stress design is

very popular for the design of metal structures because the strength of isotropic materials does not depend on the orientation. But principal stress design is of limited application to composites design. The limitations of this method for composites design are illustrated in Examples 5.1, 5.4, and 5.5.

Example 1.2 *In Example 1.1 (p. 6), what happens if the sign of the applied shear stress is reversed, to* $\sigma_{xy} = -200\ MPa$ *after the structure was built with fiber orientation of* $26.57°$*?*

Solution to Example 1.2 *Recompute the principal stresses and the orientation as* $\sigma_I = 500, \sigma_{II} = 0$, *and* $\phi = -26.57$. *Now the principal stress would be applied at* $53.13°$ *from the previously selected fiber orientation. Although there is only one principal stress different from zero, it cannot be compared to any strength property (tensile or compressive, along or transverse to the fibers) because* σ_I *is not in the fiber direction or at* $90°$ *with it. This is the main problem when applying principal stress design to composites. In the case of metals, the strength is independent of the direction. Therefore, it is always possible to compare the principal stress with the strength of the material, regardless of the direction of the principal stress. For composites, the best option is to compute the stresses along the fiber direction* σ_1, *transverse to it* σ_2, *and the shear stress in the material coordinates* σ_{12}. *This is done in Section 5.4. Then, these values can be compared with the longitudinal, transverse, and shear strength of the material (see Example 5.5, p. 158).*

1.3 Composites Design Methods

Composite materials are formed by the combination of two or more materials to achieve properties (physical, chemical, etc.) that are superior to those of the constituents. The main component of composite materials are the matrix and the reinforcement. The most common reinforcements are fibers, which provide most of the stiffness and strength. The matrix binds the fibers together providing load transfer between fibers and between the composite and the external loads and supports.

The design of a structural component using composites involves concurrently material design and structural design (e.g., the geometry). Unlike conventional materials (e.g., steel), the properties of the composite material can be designed simultaneously with the structural aspects. Composite properties (e.g., stiffness, thermal expansion, etc.) can be varied continuously over a broad range of values, under the control of the designer.

Properties of fibers and matrices are reported separately in Chapter 2. These properties can be combined using micromechanics formulas presented in Chapter 4 to generate properties for any combination of fibers and matrix. Although micromechanics can predict the stiffness of a material very well, it is not so accurate at predicting strength. Therefore, experimental data of strength is very valuable in design. For this reason, manufacturers and handbooks tend to report composite properties rather than fiber and matrix properties separately. The problem is that the reported properties correspond to a myriad of different reinforcements and processing techniques, which makes comparison among products very difficult. Typical properties of unidirectional composites are listed in Tables 1.3–1.4, compiled from manufacturer's literature, handbooks, and other sources [15–25].

The design of composites can be done using composite properties, such as those given in Tables 1.3–1.4, as long as experimental data are available for all types of fiber/matrix combinations to be used in the laminate. In this case, Chapter 4 can be skipped, and the analysis proceeds directly from Chapter 5. However, Chapter 4 provides a clear view of how a composite material works and it should not be completely disregarded by the reader. While using experimental composite properties eliminates the need for micromechanics modeling, it requires a large investment in generating the experimental data. Furthermore, a change of matrix material later in the design process invalidates all the basic material data used and requires a new experimental program for the new matrix. Most of the time, experimental material properties are not available for the fiber/matrix/process combination of interest. Then, fiber and matrix properties, which are readily available from the material supplier, or can be measured with a few tests, are used in micromechanics (Chapter 4) to predict lamina properties. This is often done when new materials are being evaluated or in small companies that do not have the resources to generate their own experimental data. Then, the accuracy of micromechanics results can be evaluated by doing a few selected tests. The amount of testing will be determined by the magnitude of the project and the availability of resources, and in some cases testing may be deferred until the prototype stage.

Once the properties of individual laminae are known, the properties of a laminate can be obtained by combining the properties of the laminae that form the laminate (see Figure 1.1) as explained in Chapter 6. However, the design may start directly with experimental values of laminate properties, such as those shown in Table 1.5 compiled from [26] (see also Table 2.8). These laminate properties can be used to perform a preliminary design of a structure, as explained in Chapters 6 to 12. Note however that the effect of changing the matrix or the manufacturing process is unknown and any such change would require to repeat the whole experimental program. The cost of experimentation is likely to limit the number of different laminates for which data is available. When laminate properties are not available from an experimental program, they can be generated using micromechanics and macromechanics.

1.4 Design for Reliability

The uncertainties associated with external loads, material strength, and fabrication/construction tolerances can be accounted for in design by probabilistic methods. The *reliability* R_e is a measure of the ability of a *system*[1] to perform a required function under stated conditions for a specified period of time.

[1]In this book, system is a general term used to refer to the material, component, member, structure, and so on. Its specific meaning will be apparent from the context.

In (classical) allowable strength design, material properties, loads, and dimensions are implicitly assumed to be deterministic; that is, their values have no variability whatsoever.[2] However, material properties, such as moduli E and strength F, are obtained by experimentation, which clearly have variability. Assuming the data is normally distributed (see section 1.4.1), each set of data **E** and **F**, is represented by a mean (μ_E, μ_F) and a standard deviation (ϖ_E, ϖ_F).

The loads applied to the system also have variability, which translates into variability of the calculated stress at a point within the system. If the loads are normally distributed, the stress at a point is represented by a mean μ_σ and a standard deviation ϖ_σ. Furthermore, the dimensions also vary within specified fabrication tolerances.

Often it is convenient to use the coefficient of variance (COV) defined as $C = \varpi/\mu$, in lieu of standard deviation ϖ, one advantage being that COV are dimensionless, while the standard deviations are not.

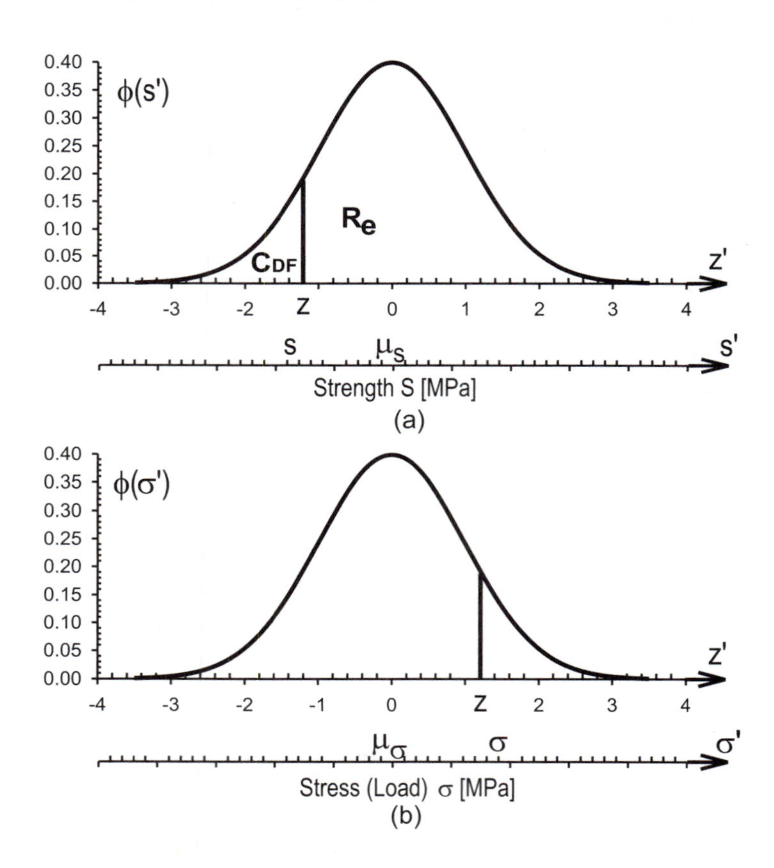

Figure 1.5: Standard normal probability density function $\phi(z)$. The second axis indicates (a) strength s', (b) stress σ'. C_{DF}: probability of failure, R_e: reliability.

[2]The classical approach was followed in the first edition of this book.

1.4.1 Stochastic Representation

A quantity that displays variability is represented by a stochastic variable $\mathbf{X}$, which can be thought as an infinite set of different values. The probability of finding a particular value x in that set is represented by a probability density function $p(x)$. Material properties, loads, and dimensions often display probabilities that are well represented by the normal (Gaussian) probability density function (PDF)

$$p(x) = \frac{1}{\varpi_x \sqrt{2\pi}} \exp\left[-\frac{1}{2}\left(\frac{x-\mu_x}{\varpi_x}\right)^2\right] \tag{1.3}$$

where μ_x, ϖ_x, are the mean and standard deviation, respectively.[3]

The change of variable

$$z = \frac{x - \mu_x}{\varpi_x} \tag{1.4}$$

yields a simplified equation for the *normal PDF*

$$p(z) = \frac{1}{\varpi_x} \phi(z) \tag{1.5}$$

in terms of the *standard normal PDF*

$$\phi(z) = \frac{1}{\sqrt{2\pi}} \exp\left(-\frac{z^2}{2}\right) \tag{1.6}$$

The standard normal $\phi(z)$, shown in Figure 1.5a, has zero mean, $\mu = 0$ and unit standard deviation, $\varpi = 1$. The PDF gives the probability $p(x)$ that and event x occurs. Using strength F to illustrate the concept, $p(F)$ is the probability that a specimen of material fails with a strength F. The probability $C_{DF}(x)$ that an event occurs for any $x' < x$ is the sum (integral) of all the probabilities $p(x')$ for $x' < x$, called *cumulative distribution function* (CDF), and given by

$$C_{DF}(x) = \int_{-\infty}^{z} p(z')dz' \tag{1.7}$$

and z computed with (1.4). Equation (1.7) cannot be integrated in closed form but can be integrated numerically. For example, in MATLAB®, `cdf('norm',-1.645,0,1)` yields the CDF at $x = -1.645$ for a PDF with $\mu = 0$ and $\varpi = 1$.

While $C_{DF}(x)$ represents the probability of failure, the reliability $R_e(x)$ represents the probability that a system will not fail if $x' > x$ (Figure 1.5a). Since the

[3]Fiber strength data is represented better by a Weibull PDF (Appendix B).

total area under the probability *density* function (PDF), by definition as a *density*, is unitary, we have

$$R_e(x) = 1 - C_{DF}(z) \tag{1.8}$$

again, with z computed with (1.4).

To avoid performing a numerical integration for each particular case, $C_{DF}(z)$ has been tabulated for $z > 0$, as shown in Table 1.1, but it is more convenient to calculate it numerically, for example using MATLAB `cdf('norm',z,0,1)`. Since the PDF (Figure 1.5a) is symmetric with respect to $z = 0$, then $R_e(-z) = C_{DF}(z)$. Therefore, values of reliability can be read from the same table. Since one is interested in high values of reliability, i.e., $R > 0.5$, then $z < 0$. The reliability values 90%, 95%, 97.5%, 99%, 99.5%, and 99.9% are boldfaced in Table 1.1 because of their importance in design. The corresponding values of z are read by adding the value in the top row to the value on the left column. For example, $z = 1.9 + 0.0600$ yields a CFD $F = 0.9750$. Therefore, $z = -1.9600$ yields a reliability $R_e = 0.9750$.

Example 1.3 *Consider the experimental observations of strength reported in Table 1.2 (p. 23). Sort the data into 7 bins of equal width ΔS to cover the entire range of the data. Build a histogram reporting the number of observations per bin. Compute the mean, standard deviation, and coefficient of variance. Plot the histogram. Plot the normal probability density function.*

Solution to Example 1.3 *The data* **F** *is sorted, then grouped into bins of equal width $\Delta F = 10$ in Table 1.2 (Histogram) from the minimum strength $F_{min} = 30\ MPa$ to the maximum $F_{max} = 100\ MPa$. The data is represented by a normal distribution with $\mu_F = 65.4, \varpi_F = 14.47, C_F = 0.22$. The resulting histogram and PDF are shown in Figure 1.6. Hint: in MATLAB* `pdf('norm',x,m,v)` *evaluates the normal PDF $p(x)$ with mean* `m` *and standard deviation* `v`*. Then,* `[Hist,Centers]=hist(F,7)` *returns the frequency and centers, and* `bar(Centers,Hist)` *plots the histogram.*

1.4.2 Reliability-based Design

Let's assume that the strength data **F** is normally distributed.[4] Furthermore, let's assume that the applied load **L** varies also with a normal distribution, producing a state of stress $\boldsymbol{\sigma}$ that is normally distributed. Then, every combination of particular values F, σ, yields a margin of safety $m = F - \sigma$. The set of all possible margins of safety is denoted by a stochastic variable

$$\mathbf{m} = \mathbf{F} - \boldsymbol{\sigma} \tag{1.9}$$

[4] If the number of data points available is too low to support the use of a normal distribution, the approach is more complex [27], but the idea remains the same.

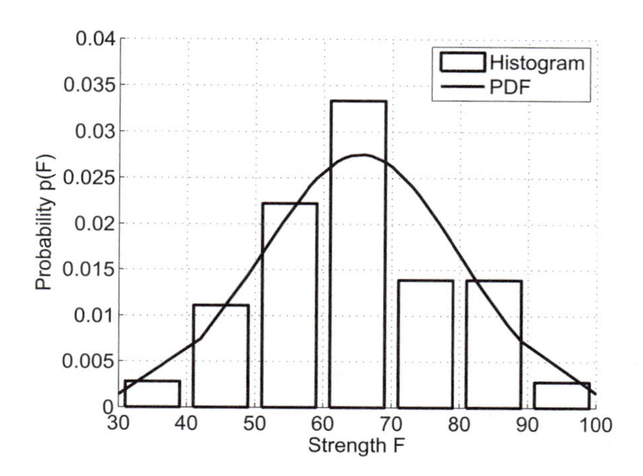

Figure 1.6: Histogram and associated PDF for the data in Table 1.2, Example 1.3.

which is also normally distributed. Therefore, **m** is represented by a mean μ_m and a standard deviation ϖ_m, which can be calculated as

$$\begin{aligned}\mu_m &= \mu_F - \mu_\sigma \\ \varpi_m &= \sqrt{\varpi_F^2 + \varpi_\sigma^2}\end{aligned} \tag{1.10}$$

and has a reliability $R_e(m) = 1 - C_{DF}(m)$ just like any other normal distribution. However, in design, one is interested only in situations that have a positive margin of safety, i.e., $m' > m = 0$. In this case, x in (1.4) stands for the margin of safety. Then, setting $x = 0$, and substituting (1.10) in (1.4), yields the reliability $R_e(z) = 1 - C_{DF}(z)$ with

$$z = \frac{0 - \mu_m}{\varpi_m} = \frac{\mu_\sigma - \mu_F}{\sqrt{\varpi_F^2 + \varpi_\sigma^2}} \tag{1.11}$$

In other words, if $\mu_F, \mu_\sigma, \varpi_F, \varpi_\sigma$, are known, a design having a strength ratio[5] $R = \mu_F/\mu_\sigma$ has a reliability $R_e(z)$ with z calculated with (1.11), then substituted in (1.8).

So far, the reliability R_e has been calculated in terms of given strength **F** and stress $\boldsymbol{\sigma}$. However, the stress can be controlled by design, that is, by changing the geometry, configuration, and/or material. Therefore, it is possible to control the reliability through design. In other words, we are interested in finding out what value of strength ratio would yield a desired value of reliability. For example, the

[5]For R, the term **strength ratio** (Section 7.1.1) is preferred to **safety factor** because R alone does not represent a safe design; for that one needs to know the reliability R_e as well.

designer decides, or the code of practice requires, that the system be designed for a 95% reliability. A value $R_e = 0.95$ corresponds to $z = -1.645$. Then, having set a value for z, it is possible to extract a strength ratio R from (1.11). In (1.11), take the square of both sides, then substitute COVs for standard deviations, i.e., $C_F = \varpi_F/\mu_F, C_\sigma = \varpi_\sigma/\mu_\sigma$ and solve for μ_F/μ_σ to get [14, (5.42)]

$$R = \frac{1 \pm \sqrt{1 - (1 - z^2 C_F^2)(1 - z^2 C_\sigma^2)}}{1 - z^2 C_F^2} \tag{1.12}$$

with the plus sign for $R_e > 0.5$, minus otherwise[6] Therefore, given the COV of load and strength, the designer can immediately determine from (1.12) the safety factor needed; with the safety factor understood in classical terms, i.e., in terms of mean values of stress and strength. Reliability-based design is illustrated with examples throughout this book (e.g., Example 7.11, etc.).

Example 1.4 *Consider the set of measured strength values* $\mathbf{F} = [72, 82]$ MPa *and the set of expected loads, or rather stresses induced by expected loads* $\boldsymbol{\sigma} = [60, 80]$ MPa. *Calculate the reliability of the system.*

Solution to Example 1.4 *The margin of safety* $\mathbf{m}$, *given by (1.9) is calculated considering all the pairs of stress and strength, as shown in the table.*

F	σ	m	Success/Failure
72	60	12	✓
72	80	-8	×
82	60	22	✓
82	80	2	✓

From $\mathbf{m}$, *one concludes that the reliability is 75% and the probability of failure is 25%.*

Example 1.5 *Consider the experimental observations of strength reported in Example 1.3, p. 12. Assume that material is subject to a state of stress with COV 0.05.*

- *What is the required safety factor for 99% reliability?*
- *What is the maximum value of mean stress that can be applied to the material?*

Solution to Example 1.5 *For* $R_e = 0.99$, *Table (1.1) or MATLAB* `z=fzero(@(z) cdf('norm',z,0,1)-(1-.99),0);` *yields* $z = -2.326$. *From Example 1.3 (p. 12), the COV of the material strength is* $C_F = 0.22$, *then, (1.12) yields*

$$R = \frac{1 + \sqrt{1 - (1 + 2.326^2 \times 0.22^2)(1 + 2.326^2 \times 0.05^2)}}{1 + 2.326^2 \times 0.22^2} = 2.03$$

[6]Of course, designers are interested in reliability higher than 50%, so the plus sign would apply.

and the maximum mean stress that can be applied is

$$\mu_\sigma = R\ \mu_F = 2.03 \times 65.5 = 132.97\ MPa.$$

1.4.3 Load and Resistance Factor Design

Load and Resistance Factor Design (LRFD) is a simpler and more intuitive design methodology that builds upon the *design for reliability* discussed in Section 1.4. LRFD splits the safety factor (1.12) into a resistance factor and a load factor. The resistance factor ϕ accounts for the variability of the strength. The load factor α accounts for the variability of the loads. Using two factors, a design slightly more conservative than using (1.12) is obtained. But more importantly, it forces the designer to elucidate, separately, the roles of the variability of strength and variability of loads.

According to LRFD, there are two complementary aspects to be considered separately: the resistances of the structure and the loads applied to it. Therefore, the design must satisfy an inequality such as the following[7] [28, 29]

$$\phi\ F > \alpha_D\ L_D + \psi\ \sum_{j=1}^{N_L} \alpha_j L_j \tag{1.13}$$

where

$$\begin{aligned} F &= \text{Material Resistance (Strength)} \\ \phi &= \text{Resistance Factor} \\ L_D &= \text{Dead Load} \\ L_j &= \text{Other Loads} \\ \alpha_D &= \text{Dead Load Factor} \\ \alpha_j &= \text{Live Load Factors} \\ \psi &= \text{Load Combination Factor} \\ N_L &= \text{Number of Other Loads} \end{aligned} \tag{1.14}$$

where *other loads* includes live, pressure, thermal, acceleration, dynamic loads, and any other loads that are not permanently applied to the structure. Permanent loads, such as gravity loads, are grouped into the *dead load* term.

The resistance factor takes into account the variability of the strength of the material; therefore it is always less than one. The load factor takes into account the variability of the applied loads; therefore, it is always greater than one. The load combination factor ψ is included to take into account the likelihood that all of the live loads be present simultaneously. If only one live load is considered ($N_L = 1$),

[7] Variations exist among different codes but all conform more or less to the form presented here.

then $\psi = 1$. If there is a dead load, such as a gravity load, it is always applied, and thus it should not be reduced by a load combination factor.

Most of the design codes, notably civil engineering building and construction codes, *specify* the *values* to be used for load and resistance factors for most anticipated situations so that the designer does not have to deal with the actual variability of the data. Instead, the body that writes the code considers the typical variability of the data and decides the actual values of the load and resistance factors to be used for each anticipated situation. However, many applications and materials are either not covered by design codes or the applicable codes do not anticipate the situations for which new materials can be used. In any case, the load and resistance factors can be calculated in terms of the variability of the loads and the material strength data as shown in the next sections. Besides, an understanding of the rational process used for determination of load and resistance factors is useful for their correct application to design, even if the values used are codified.

1.4.4 Determination of Resistance Factors

First, let's assume in this section only, that the applied stress $x = \alpha L$ is deterministic, i.e., it has no variability. Given a set of strength data represented by mean μ_F, and standard deviation ϖ_F, and a required value of reliability R_e, one can easily find the value of z for that reliability (Figure 1.5a). Note that the value of z is negative when $R_e > 0.5$. Then, from (1.4), once can easily compute the minimum (deterministic) value of strength x that corresponds to z, and express it as the product of a resistance factor ϕ times the mean strength $\mu_m = F$, as follows

$$F_{min} = \mu_F + z\ \varpi_F = \phi\ \mu_F \tag{1.15}$$

from which the resistance factor (also called *strength reduction factor*) is calculated easily for a given reliability R_e, thus z, in terms of the known COV of the strength data, as follows

$$\phi = \frac{\mu_F + z\ \varpi_F}{\mu_F} = 1 + z\ C_F < 1 \tag{1.16}$$

which is less than 1 because $z < 0$. The use of resistance factors for design is illustrated throughout this book (e.g., Examples 7.7, etc.) Usually, the number of data points available is too low to support the use of a normal distribution, and a more complex approach is required for determination of the strength reduction factor [27].

Example 1.6 *Calculate the nominal resistance F and the resistance factor ϕ from the data in Table 1.2 in Example (1.3) for a 95% reliability.*

Solution to Example 1.6 *The mean and standard deviation are calculated in Example (1.3). The nominal resistance is* $F = \mu_F = 65.5\ MPa$. *The standard deviation is* $\varpi_F = 14.34\ MPa$. *The COV is* $C_F = \varpi_F/\mu_F = 14.34/65.5 = 0.22$. *For 95% reliability, Table (1.1) or MATLAB* `z=fzero(@(z) cdf('norm',z,0,1)-(1-.95),0);` *yields* $z = -1.645$. *Therefore, the resistance factor is*

$$\phi = 1 - 1.645 \times 0.22 = 0.64$$

1.4.5 Determination of Load Factors

In this section only, let's assume that the strength $x = \phi F$ is deterministic, i.e., it has no variability. For the loads, the reliability $R_e(x)$ is given by the area to the left of x, and the probability of failure $C_{DF}(z)$ by the area to the right of x (Figure 1.5b). This is because, when the strength x is deterministic, any stress $\sigma' > x$ causes failure, while any stress $\sigma' < x$ does not. For $R_e > 0.5$, $z > 0$ in Figure 1.5b, and the reliability $R_e(z)$ is given by the area to the left of z. But such area is the same as $R_e(-z)$ on Figure 1.5a. Therefore, one can replace $-z$ for z in (1.16) to obtain the load factor as

$$\alpha = \frac{\mu_\sigma - z\,\varpi_\sigma}{\mu_\sigma} = 1 - z\,C_\sigma > 1 \tag{1.17}$$

which is greater than 1 because $z < 0$. The use of load factors for design is illustrated throughout this book (e.g., Examples 6.8, etc.)

Example 1.7 *Calculate the load factor* α *for a 95% reliability if the COV of the load is* 0.10.

Solution to Example 1.7 *For 95% reliability, Table 1.1 or MATLAB* `z=fzero(@(z) cdf('norm',z,0,1)-(1-.95),0);` *yields* $z = -1.645$. *Therefore, the load factor is*

$$\alpha = 1 - (-1.645) \times 0.10 = 1.1645$$

Note: When the load and resistance factors are calculated as explained in this section, LRFD uncouples the variability of material failure from the variability of the loads by assigning a resistance factor ϕ to the former and a combined-effective load factor α to the later. To illustrate this concept, consider the case when only dead load is active in (1.13). Then, using the notation of the previous section, $F = \mu_F, L = \mu_\sigma$ and (1.13) becomes

$$\phi\ \mu_F > \alpha\ \mu_\sigma \tag{1.18}$$

Noting that $\alpha > 0, \phi < 0$, the safety factor is $R = \alpha/\phi > 0$ as expected. Now, taking into account (1.16) and (1.17), the safety factor in LRFD is

$$R = \frac{\alpha}{\phi} = \frac{1 - zC_\sigma}{1 + zC_F} \tag{1.19}$$

Numerical comparison between values of R calculated with (1.12) and values calculated with (1.19), for the particular case of $C_F = C_\sigma$ and $R_e = 95\%$ shows that LRFD is conservative by about 10% at COV=0.1.

1.4.6 Limit States Design

The *limit states design method* is a probabilistic design method that seeks to design a structure able to resist load combinations associated to *ultimate limit states* and *serviceability limit states*, which are defined as follows:

1. *Ultimate limit states*: the structure is subjected to the *most severe loading* conditions. The design should *anticipate damage* of the structural elements, but *avoid collapse*, partial or complete, of the structure. The nonlinearity of the material behavior is included in the analysis in order to account for damage (Chapters 8 and 13).

2. *Serviceability limit states*: the structure is subjected to *in-service loading* conditions. The performance of structure is checked to avoid *deformations, vibrations, and damage* that would impair the *usability* of the structure.

Explicit, separate verification of the design for *ultimate* and *serviceability* is similar to the classical *design for strength* and *design for stiffness*, where stiffness is a catch all for deflections, vibrations, and so on, while strength includes buckling and all other modes that are catastrophic in nature. Examples of *design for strength* and *design for stiffness* are presented throughout the book and specifically addressed in the first twelve chapters. The *limit states design method* is described and utilized in Chapter 13.

1.5 Fracture Mechanics

Strength of materials courses emphasize *strength design* where the calculated stress σ is compared to a material property, the *strength* F and the design is supposed to be adequate if $F > R\sigma$, where R is a safety factor. This approach enjoys popularity because of tradition (it was the first engineering approach) and because it is simple. However, it suffers from at least two shortcomings. The first is that it does not take into account the often unknown variability of the applied loads and unknown variability of material properties. The second is that strength design does not give good results for brittle materials where not only the applied load and strength of the material play a role but also the existing flaw sizes play a prominent role. The first shortcoming is addressed by employing *design for reliability* (see Section 1.4). The second shortcoming is addressed by using *fracture mechanics* instead of strength for the design.

In the fracture mechanics approach, it is recognized that no matter how careful the manufacturing process is, there will always be defects. Whether these defects grow into catastrophic cracks or not depends of the size of the defects, the state of stress to which they are subjected, and the toughness of the material. An energy approach to fracture mechanics is used in this book.

The energy approach is based on thermodynamics, and as such it states that a crack will grow only if the total energy in the system after crack growth is less than the energy before growth. The net energy is the difference between the potential energy of the system minus the energy absorbed by the material while creating the crack. The potential energy is composed of elastic strain energy U minus the work done by the external loads W, i.e., $\Pi = U - W$.

The energy absorbed to create a crack in a perfectly elastic material goes into surface energy γ_e of the newly formed surfaces. In an elastoplastic material, plastic dissipation γ_p also absorbs energy during crack growth. In general, all the dissipative phenomena occurring during crack growth absorb a combined 2γ $[J/m^2]$ of energy per unit area of new crack A, where the surface area created is twice the crack area A.

The change in potential energy during an increment of crack area A is called *energy release rate* (ERR)

$$G = -\frac{d\Pi}{dA} \tag{1.20}$$

The energy absorbed per unit crack area created is $G_c = 2\gamma$. Therefore, the fracture criterion states that the crack will not grow as long as

$$G \leq G_c \tag{1.21}$$

In this book, in what it is a deviation of classical fracture mechanics nomenclature [30], the critical ERR G_c is called *fracture toughness* to convey the idea that the material property represents the toughness of the material. The critical ERR G_c is a material property independent of size and geometry of the specimen used to determine its value [30]. The methodology used to experimentally measure fracture toughness of composites is described, for example in [31]. Such material property can then be used to predict fracture in structures of varied size and geometry.

The example below illustrates the fact that fracture is controlled by three factors: the toughness of the material, the applied load, *and* the size of the flaw.

Example 1.8 *Consider a beam of width B that is peeled of a substrate by the action of a fixed load "P", creating a crack of length "a", as shown in Figure 1.7. Calculate the energy release rate G_I in mode I.*

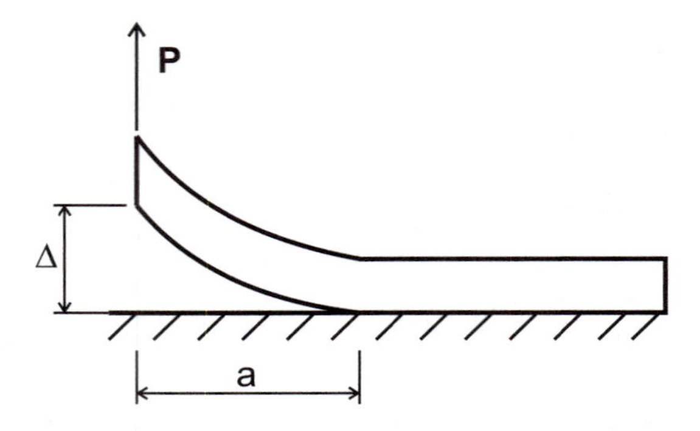

Figure 1.7: Beam peeling off from a substrate under the action of a constant applied load P.

Solution to Example 1.8 *The work done by external forces is*

$$W = P\Delta$$

and the strain energy is

$$U = \int_0^{\Delta} P d\Delta = \frac{P\Delta}{2}$$

Therefore, the potential energy is

$$\Pi = -\frac{P\Delta}{2}$$

Recalling (1.20), the ERR at constant load P is

$$G_I = \frac{P}{2B}\frac{d\Delta}{da}\bigg|_P$$

Defining the compliance $S = \Delta/P$, we have

$$G_I = \frac{P^2}{2B}\frac{dS}{da}$$

For a cantilever beam of length a, the tip deflection is

$$\Delta = \frac{Pa^3}{3EI}$$

where E is the modulus and $I = bh^3/12$ is the moment of inertia. Therefore, the compliance is

$$S = \frac{\Delta}{P} = \frac{a^3}{3EI}$$

yielding and ERR for this system

$$G_I = \frac{P^2 a^2}{2BEI}$$

For a given material with fracture toughness G_c, the crack will not grow as long as $G \le G_c$, or

$$\frac{P^2 a^2}{2BEI} \le G_c$$

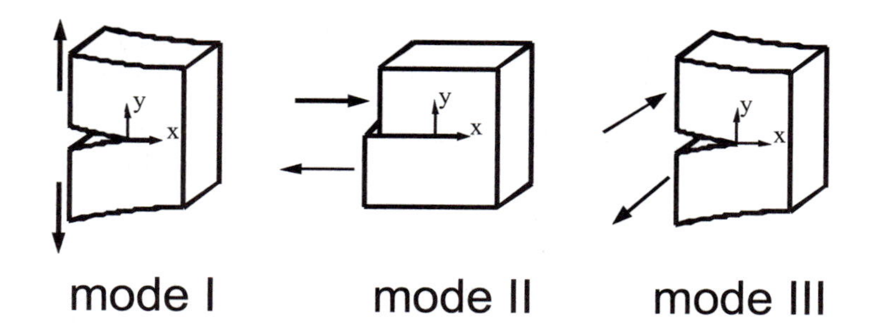

Figure 1.8: Modes of fracture. Mode I: opening or peeling, mode II: in-plane shear, mode III: out-of-plane shear.

Unlike strength design, where a stress is compared to a strength value, in fracture mechanics there are three factors at play: the fracture toughness G_c, the applied load, and the size of the flaw. In this example, the applied load "P" and flaw size "a" are equally important, with both affecting the likelihood of crack growth quadratically.

Besides the typical *opening* mode illustrated in Figure 1.7, a crack may be subject to in-plane shear (mode II) and out-of-plane shear (mode III), as illustrated in Figure 1.8. The concept of fracture mechanics is used primarily in Chapters 7 and 8.

Exercises

Exercise 1.1 *A cylindrical pipe with diameter $d = 1.0\ m$ is subject to internal pressure $p = 5.0\ MPa$ but it is free to expand along the length, so that only hoop stress $\sigma = pd/2t$ is developed in the wall. Use Tables 1.3–1.4, a resistance factor $\phi = 0.25$ and a load factor $\alpha = 1.0$, i.e., the internal pressure has no variability. Do not consider those materials for which the required data is not available.*

(a) What is the optimum orientation for the fibers?, and why?

(b) What is the required wall thickness and material type that yields the minimum thickness?

(c) What is the required wall thickness and material type for minimum weight?

(d) What is the required wall thickness and material type for minimum cost? Assume the following costs (\$/kg) for the materials in Table 1.3, from left to right: 1.60, 3.20, 1.50, 5.80, 2.0, 2.0, and for the materials in Table 1.4, from left to right: 10.80, 10.80, 10.80, 15.00, 15.00, 15.00, 20.00, 20.00.

Exercise 1.2 *Compare the weight per unit length of the minimum weight pipe of Exercise 1.1 with a pipe made out of 6061-T6 aluminum with yield strength 270 MPa and density $\rho = 2.70\ g/cc$. Use the same load and resistance factors.*

Exercise 1.3 *A component of a composite rocket motor case is subject to a state of stress $\sigma_x = 10\,MPa$, $\sigma_y = 5\,MPa$, $\sigma_{xy} = 2.5\,MPa$. Use the material data for carbon–epoxy (AS4/3501-6) in Table 1.4. Using principal stress design, compute:*

(a) the optimum fiber orientation

(b) the load factor α that can be tolerated for the lamina with optimum orientation, considering all possible modes of failure and a resistance factor $\phi = 1.0$.

(c) the resistance factor ϕ that can be tolerated for the lamina with optimum orientation, considering all possible modes of failure and a load factor $\alpha = 1.0$.

Exercise 1.4 *Consider a material with measured strength values* **F** *given in Table 1.2. Assume the expected stress values to be one-half of the values in Table 1.2. Calculate the safety factor and the reliability of the system.*

Exercise 1.5 *For a fixed, given load, the stress can be increased or decreased by changing the geometry of the system. Calculate the % change needed on the stress in Exercise 1.4 to achieve a reliability of 99.5%.*

Exercise 1.6 *(a) Calculate the load and resistance factors using the data of Exercise 1.4 and a reliability of 99%. (b) Calculate the resulting safety factor.*

Exercise 1.7 *(a) Calculate the load and resistance factors for a reliability of 99% when the strength has a mean value of 65.4 MPa and COV 22%; the stress has a mean value 32.7 MPA and COV 11%. (b) Calculate the resulting safety factor and compare it to the safety factor in Exercise 1.6 and comment on your finding.*

Table 1.1: Standard normal cumulative distribution function $C_{DF}(z)$ and reliability $R_e(-z)$.

z	0	0.026	0.045	0.06	0.076	0.082
0	0.5	0.5104	0.5179	0.5239	0.5303	0.5327
1	0.8413	0.8476	0.852	0.8554	0.859	0.8604
1.2	0.8849	0.8899	0.8934	0.8962	0.899	**0.900**
1.6	0.9452	0.948	**0.950**	0.9515	0.9531	0.9537
1.9	0.9713	0.9729	0.9741	**0.975**	0.9759	0.9763
2.3	0.9893	**0.990**	0.9905	0.9909	0.9912	0.9914
2.5	0.9938	0.9942	0.9945	0.9948	**0.995**	0.9951
3	0.9987	0.9988	0.9988	0.9989	0.999	**0.999**

Table 1.2: Strength data (in MPa) for Example 1.3.

Strength data

66	100	54	64	51	53	49	77	64	80	87	63
69	83	69	59	67	64	81	70	58	82	74	51
58	42	49	30	64	58	49	73	89	67	62	80

Sorted strength data

30	42	49	49	49	51	51	53	54	58	58	58
59	62	63	64	64	64	64	66	67	67	69	69
70	73	74	77	80	80	81	82	83	87	89	100

Centers of bins located at

35	45	55	65	75	85	95

Frequency of observations in each bin

1	4	8	12	5	5	1

Table 1.3: Typical properties of unidirectional composites, obtained from product literature and other sources, with some values estimated for a broad class of materials. The designer is responsible for obtaining actual values.

		E-glass –Epoxy	S-Glass –Epoxy	E-glass –Polyester	Kevlar 49™ –Epoxy	E-glass –LY556 [25]	E-glass –MY750 [25]	E-glass –Epoxy [32]
Density [g/cc]	ρ	2.076	1.993	1.85	1.380	-	-	-
Longitudinal Modulus [GPa]	E_1	45	55	37.9	75.8	53.48	45.6	44.7
Transverse Modulus [GPa]	E_2	12	16	11.3	5.5	17.70	16.2	12.7
In-plane Shear Modulus [GPa]	G_{12}	5.5	7.6	3.3	2.07	5.83	5.83	5.8
In-plane Poisson's ratio	ν_{12}	0.19	0.28	0.3	0.34	0.278	0.278	0.297
Transverse Poisson's ratio	ν_{12}	0.31	-	-	-	0.4	0.398	0.410
Longitudinal Tensile Strength [MPa]	F_{1t}	1020	1620	903	1380	1140	1280	-
Transverse Tensile Strength [MPa]	F_{2t}	40	40	40	34.5	35	40	43.4
Longit. Compressive Strength [MPa]	F_{1c}	620	690	357	586.0	570	800	-
Transv. Compressive Strength [MPa]	F_{2c}	140	140	68	138.0	138	145	-
In-plane Shear Strength [MPa]	F_6	60	60	40	44.1	61	73	135.8
Interlaminar Shear Strength [MPa]	F_4	-	-	-	150	-	-	-
Ult. Longit. Tensile Strain [%]	ϵ_{1t}	1.29	1.68	1.64	1.45	2.132	2.807	-
Ult. Transv. Tensile Strain [%]	ϵ_{2t}	-	-	-	-	.197	.246	-
Ult. Longit. Comp. Strain [%]	ϵ_{1c}	-	-	-	-	1.065	1.754	-
Ult. Transv. Comp. Strain [%]	ϵ_{2c}	-	-	-	-	.644	1.2	-
Ultimate Shear Strain [%]	γ_{6u}	-	-	-	-	3.8	4.0	-
Fracture Toughness Mode I [J/m^2]	G_{Ic}	-	-	334	-	165	165	360
Fracture Toughness Mode II [J/m^2]	G_{IIc}	-	-	456	-	-	-	1400
Longitudinal CTE [$10^{-6}/^\circ$ C]	α_1	3.7	3.5	6.5	-2.0	8.6	8.6	-
Transverse CTE [$10^{-6}/^\circ$ C]	α_2	30	32	22	60	26.4	26.4	-
Stress-free Temperature [$^\circ$ C]	SFT	126	126	25	126	120	120	-
Longitudinal Moisture Expansion	β_1	0	0	0	0.01	-	-	-
Transverse Moisture Expansion	β_2	0.2	0.2	0.2	0.2	-	-	-
Fiber Volume Fraction	V_f	0.6	0.6	0.5	0.6	0.6	0.6	-
Void Content	V_v	-	-	0.02	-	-	-	-
Fiber Misalignment [deg]	α_σ	2.97	2.98	3.53	1.34	-	-	-
Ply Thickness [mm]	t_k	0.144	0.144	0.144	0.144	0.6	0.6	0.144

Table 1.4: Typical properties of unidirectional (carbon-fiber) composites, obtained from product literature and other sources, with some values estimated for broad class of materials. The designer is responsible for obtaining actual values.

		AS4 –3501-6	AS4 –3501-6 [25]	T300 –914-C [25]	T800 –3900-2	IM7 –8551-7	IM7 –8552	AS4 –APC2[a]	Avimid –K-III[b]
Density [g/cc]	ρ	1.58	-	-	-	-	1.55	1.6	-
Longitudinal Modulus [GPa]	E_1	142	126	138	155.8	151	171.4	138	110
Transverse Modulus [GPa]	E_2	10.3	11.0	11.0	8.89	9.0	9.08	10.2	8.3
In-plane Shear Modulus [GPa]	G_{12}	7.2	6.6	5.5	5.14	5.6	5.29	5.7	5.7
In-plane Poisson's ratio	ν_{12}	0.27	0.28	0.28	0.3	0.3	0.32	0.3	-
Transverse Poisson's ratio	ν_{23}	-	0.4	0.4	0.064	0.184	0.4[c]	-	-
Longitudinal Tensile Strength [MPa]	F_{1t}	1830	1950	1500	2698	-	2326.2	2070	-
Transverse Tensile Strength [MPa]	F_{2t}	57	48	27	-	-	62.3	86	37
Longit. Compressive Strength [MPa]	F_{1c}	1096	1480	900	1691	1200	1200.1	1360	1000
Transv. Compressive Strength [MPa]	F_{2c}	228	200	200	-	-	199.8	-	-
In-plane Shear Strength [MPa]	F_6	71	79	100	186	71	92.3–120	186	63
Interlaminar Shear Strength [MPa]	F_4	-	-	-	-	-	137.2	150	-
Ult. Longit. Tens. Strain [%]	ϵ_{1t}	1.29	1.38	1.087	1.68	1.64	1.62	1.45	-
Ult. Transv. Tens. Strain [%]	ϵ_{2t}	-	0.436	0.245	-	-	-	-	-
Ult. Longit. Comp. Strain [%]	ϵ_{1c}	-	1.175	0.652	-	-	-	-	-
Ult. Transv. Comp. Strain [%]	ϵ_{2c}	-	2.0	0.245	-	-	-	-	-
Ultimate Shear Strain [%]	γ_{6u}	-	2.0	4.0	-	-	3.22	-	-
Fracture Toughness Mode I [J/m^2]	G_{Ic}	-	220	220	-	-	277	-	-
Fracture Toughness Mode II [J/m^2]	G_{IIc}	-	-	-	-	-	788	-	-
Longitudinal CTE [$10^{-6}/^\circ$ C]	α_1	-0.9	-1.0	-1.0	-	-	-5.5	0.5	-
Transverse CTE [$10^{-6}/^\circ$ C]	α_2	27	26	26	-	-	25.8	30	-
Stress-free Temperature [$^\circ$ C]	SFT	-	177	120	-	-	176	-	-
Longitudinal Moisture Expansion	β_1	0.0	-	-	0.0095	-	-	-	-
Transverse Moisture Expansion	β_2	0.2	-	-	0.321	-	-	-	-
Fiber Volume Fraction	V_f	0.6	0.6	0.6	-	0.573	0.591	0.61	-
Void Content	V_v	-	-	-	-	0.001	-	-	0.005
Fiber Misalignment [deg]	α_σ	1.70	-	-	1.75	1.27	1.59	2.74	1.52
Ply Thickness [mm]	t_k	0.125	0.8	0.8	0.125	0.125	0.125	0.125	0.125

[a]PEEK matrix. [b]Polyimide matrix; remainder are all epoxy-matrix composites. G_{23}=2.8–5.9 MPa (range reported in the source).

Table 1.5: Laminate properties of a general-purpose polyester resin reinforced with E-glass stitched fabric [33].[a]

		Orientation 0°		Orientation 45°				
Denomination	Property	Modulus E_x [MPa]	Strength F_x [MPa]	Modulus E_{45} [MPa]	Strength F_{45} [MPa]	V_f [%]	Laminate Thickness [mm]	Density [g/cc]
M1500	Tensile		128	6752	128			
[CSM]	Compressive	7303	179	7303	179			
	Flexural	6960	212	6960	212			
	Intralaminar		24.8		24.8	17.1	1.041	1.4223
CM1808	Tensile	13780	201					
[CSM/0/90]	Compressive	11713	187					
	Flexural	13091	310					
	Intralaminar		20.3			26.6	1.219	1.5458
TVM 3408	Tensile	15502	229	14469	214			
[CSM/ 45/0]	Compressive	17914	262	17914	250			
	Flexural	16536	386	15158	351			
	Intralaminar		24.1		22.7	34.2	1.727	1.6446
XM 2408	Tensile	10680	98	15158	236			
[CSM/±45]	Compressive	15158	228	22392	262			
	Flexural	10335	222	16536	400			
	Intralaminar		24.4		28.8	37.0	1.422	1.6810
UM1608	Tensile	12746	214					
[CSM/0]	Compressive	13091	228					
	Flexural	13091	310					
	Intralaminar		25.5			29.9	1.143	1.5887

[a]See Table 2.8 for definition of fiber architecture. Matrix density $\rho_m = 1.2$ g/cc.

Chapter 2

Materials

Composite materials are formed by the combination of two or more materials to achieve properties (physical, chemical, etc.) that are superior to those of its constituents. The main components of composite materials, or composites, are fibers and matrix. The fibers provide most of the stiffness and strength. The matrix binds the fibers together thus providing load transfer between fibers and between the composite and the external loads and supports. Also, it protects the fibers from environmental attack. Other substances are used to improve specific properties. For example, fillers are used to reduce the cost and improve processability and dimensional stability [34].

The design of a structural component using composites involves simultaneous material and structural design. Unlike conventional materials (e.g., steel), the properties of the composite material can be designed simultaneously with the structural aspects. Composite properties (e.g., stiffness, thermal expansion, etc.) can be varied continuously over a broad range of values, under the control of the designer. The objective of this chapter is to describe the constituents used in the fabrication of the composite material. The capabilities and limitations of various processing techniques used to fabricate the material and the parts are presented in Chapter 3.

A brief review of the most common types of materials used in the fabrication of composites is presented in this chapter, with emphasis on properties, advantages, disadvantages, and cost. No attempt is made to explain how fibers and polymers are produced; excellent material science books and handbooks cover that subject [35–37]. Comprehensive lists of material suppliers can be found in specialized publications [38, 39], trade associations [40, 41], or the World Wide Web. Another important aspect of design and fabrication is to have standard test methods to verify the material properties used in design [42]. While testing of composites is a very active area of research, only references to standard test methods are made in this book.

2.1 Fiber Reinforcements

Fibers are used in composites because they are lightweight, stiff, and strong. Fibers are stronger than the bulk material from which they are made. This is because of the preferential orientation of molecules along the fiber direction and because of the reduced number of defects present in a fiber as opposed to the bulk material. Whereas the tensile strength of bulk E-glass is low (about 1.5 GPa [43]), the same material reaches 3.5 GPa in fiber form, mainly because of the reduction in the number and size of surface defects. Typical properties of fibers are presented in Tables 2.1–2.2.[1]

Tensile strength of fibers are usually reported in the literature as the mean value of a number of tests on individual fibers or as the tensile strength of fiber bundles, which may or may not be impregnated with a resin and cured into a composite fiber bundle. It is often difficult to discern how the data has been obtained. Furthermore, fibers do not have a single value of strength but rather have individual fiber strengths that follow a Weibull distribution (Appendix B). The cumulative probability of a fiber failing with a strength less than or equal to σ is given by

$$F(\sigma) = 1 - exp\left[-\left(\frac{\sigma}{\sigma_0}\right)^m\right] \tag{2.1}$$

where σ_0 is the *Weibull scale parameter* and m is the *Weibull shape parameter*. The shape parameter controls how dispersed the distribution of strength is. The scale parameter is related to the average strength, as shown in Appendix B. Average fiber strength values are reported in Tables 2.1–2.2. Values of Weibull shape parameter m are reported in Tables 2.3–2.4. Sometimes the distribution of strength is bimodal due to two failure mechanisms; one due to intrinsic defects and the other due to extrinsic defects [44]. In such a case, a bimodal Weibull is used and two scale parameters m_1 and m_2 are reported in Table 2.3–2.4 [45, 46].

Fibers are used as continuous reinforcements in unidirectional composites by aligning a large number of them in a thin plate or shell, called *lamina* or *ply*. A unidirectional lamina has maximum stiffness and strength along the fiber direction and minimum properties in a direction perpendicular to the fibers. When the same properties are desired in every direction on the plane of the lamina, randomly oriented fibers are used.

The creep properties of the composite are dominated by the matrix (see Section 2.8). By choosing fibers with very low creep compliance (e.g., carbon or glass), polymer matrix composites (PMC) can be made as creep resistant as necessary. Composites reinforced with chopped fibers, whiskers (elongated single crystals), and particles may experience noticeable creep, even at room temperature, because they do not have continuous fibers to suppress it. This is the main reason for using continuous fibers for structural applications.

---

[1]Properties were obtained from product literature and other sources and some values were estimated for a broad class of materials; the designer is responsible for obtaining actual values.

2.2 Fiber Types

A wide variety of fibers can be used as reinforcements in structural applications. Fibers can be classified by their length—short, long, or continuous fibers; according with their strength and/or stiffness—low (LM), medium (MM), high (HM), and ultrahigh modulus (UHM); or according to their chemical composition—organic and inorganic. The most common inorganic fibers used in composites are glass, carbon, boron, ceramic, mineral, and metallic. The organic fibers used in composites are polymeric fibers. Choosing a fiber type involves a tradeoff among mechanical and environmental properties, and cost.

2.2.1 Glass Fibers

Glass fibers are processed from bulk glass—an amorphous substance fabricated from a blend of sand, limestone, and other oxidic compounds. Hence, the main chemical constituent of glass fibers (46–75%) is silica (S_iO_2). Controlling the chemical composition and the manufacturing process, a wide variety of glass fibers for different types of applications can be obtained, but they exhibit the typical glass properties of hardness, corrosion resistance, and inertness. Furthermore, they are flexible, lightweight, and inexpensive. These properties make glass fibers the most common type of fiber used in low-cost industrial applications.

The high strength of glass fibers is attributed to the low number and size of defects on the surface of the fiber. All glass fibers have similar stiffness but different strength values and different resistance to environmental degradation. E-glass fibers (E for electrical) are used where high tensile strength and good chemical resistance is required. E-glass is the preferred structural reinforcement because of the combination of mechanical performance, corrosion resistance, and low cost (about $1.60/kg). S-glass^{TM} [47] and S-2-glass^{TM} (S for strength) have the highest strength but they are of limited application because they cost three to four times more than E-glass. For this reason, low cost carbon fibers are now considered as an alternative to S-glass and S-2-glass. C-glass (C for corrosion) is used for corrosion-resistant applications. D-glass (D for dielectric) is used for electrical applications such as the core reinforcement of high voltage ceramic insulators. A-glass and AR-glass (alkaline resistant) are used for lightweight surfacing veils or mats. R-glass fiber is the European counterpart of the American high performance S-glass. Their high modulus and strength, as well as very good temperature, humidity, and fatigue stability makes them popular for filament winding and sheet molding compounds for high performance industrial applications [38].

The maximum strength measured in single fiber tests (ASTM D3379) may yield up to 3.5 GPa for E-glass and 4.8 GPa for S-glass but these values cannot be realized in a composite. Damage produced during the various stages of processing reduce the fiber strength to about 1.75 GPa for E-glass and 2.10 GPa for S-glass (up to 50% reduction; strongly dependent on processing conditions and type of loads). The reduction of fiber strength in the composite with respect to the strength of the virgin reinforcement is also caused by residual stresses and secondary loads (shear and transverse to fiber direction), among other factors [48]. Typical reduction of fiber strength in filament wound composites are listed in Table 2.5 on page 66.

The corrosion resistance of fiber depends on the composition of the fiber, the corrosive solution, and the exposure time. Tensile strength of glass fibers decreases at elevated temperature, with S- and R-glass having better strength retention at elevated temperatures, as shown in Table 2.6. The operating temperature limitations of polymer matrix composites is controlled by the matrix, which can resist up to 275°C depending of the matrix type, and to a lesser extent by the reduction in strength fibers at high temperatures. Fiber tensile strength also decreases with chemical corrosion, since glass fibers are susceptible to chemicals such as alkalis, while humidity enables those chemicals to reach the surface of the fiber. Chemical corrosion and dissolution influence the growth and propagation of surface micro-cracks that are inherent to glass fibers. S and R-glass again show a better stability in humid environments. Also, tensile strength reduces with time under sustained loads. This effect is called *static fatigue* or *stress corrosion* [49, 50]. Mainly to account for this effect, a stress ratio (similar to a safety factor) of 3.5 is used in the design of glass reinforced composites for pressure vessels subject to permanent load [51]. Data for commercially available glass fibers are given in Table 2.1 (see also [43]). Glass fiber diameters range from 9.5 to 24.77 microns, and they are designated with a letter code (see p. 2.62 in reference [15]).

2.2.2 Silica and Quartz Fibers

Silica fibers and *quartz fibers* [52,53], are distinguished from classical glass fibers by their high concentration of silica (S_iO_2), 96–98% for silica fibers and 99.95–99.97% for quartz fibers (*Refrasil* fibers produced by Hitco Materials, and *Astroquartz II* fibers by J.P. Stevens). Silica and quartz fibers costs up to 25–50% more than glass fibers, with a corresponding gain in physical and mechanical properties. They feature comparable or better stiffness, strength than glass fibers, and thermal stability that glass fibers, with long-term working temperatures up to 900°C for silica fibers and up to 1050°C for quartz fibers. They also display good thermal and electrical insulation properties, and very good stability under different chemicals, being virtually insensitive to humidity. Their properties make them attractive for high-temperature, high chemical corrosion applications. Also, quartz fibers have better radio-frequency transparency, which is needed for antenna applications. The typical diameter of Astroquartz^{TM} II fibers is 9 microns. By further processing and diameter reduction, very high strength values can be achieved. An application of this type of reinforcement combined with polymer matrices is in ablative structures providing thermal protection for atmospheric reentry and rocket engines nozzles. Ablative insulation involves heat dissipation by vaporizing a thin sacrificial layer of polymer reinforced with quartz fibers. Use of these these fibers requires costly tooling due to their high toughness.

2.2.3 Carbon Fibers

Carbon fibers, also called *graphite fibers*, are lightweight and strong fibers with excellent chemical resistance. They dominate the aerospace market. The mechanical properties of carbon fibers are determined by the atomic configuration of carbon chains and their connections, which are similar to the graphite crystal structure. The strength of carbon fiber is controlled by orienting the carbon atomic structures with their strongest atomic connections along the carbon fiber direction.

The properties of carbon fibers depend on the raw material, the process used for its manufacture, and the specific manufacturing process used. Two main raw materials, or precursors, are used: *polyacrylonitrile* (PAN) and *pitch*. Pitch fibers are less expensive but have lower strength than PAN fibers. The tensile strength of pitch fibers is about one-half and their compressive strength is about one-third of that of PAN fibers, due to intrinsic structure that makes them more sensible to surface defects. PAN fibers dominate the high performance market for aerospace applications because they can be made with a variety of stiffness and strength values, as it can be seen in Table 2.1 (see also [54]).

Unlike glass fibers, carbon fibers are available with a broad range of stiffness values (Table 2.1). The stiffness is controlled by the thermal treatment of both PAN and pitch-based carbon fibers. Thermal treatment influences both carbon content and orientation of the strongest carbon links along the fiber direction, thus influencing fiber stiffness. Due to the broad range of stiffness and strength of carbon fibers, classification of carbon fibers in terms of modulus and tenacity (strength) ranges is sometimes used [52] as shown in Table 2.7.

With reference to Table 2.1, IM6 and HMS4 are intermediate modulus fibers; P55 is an high modulus fiber; UHM and M50 (PAN based fibers) and P100 (pitch based fiber) are ultrahigh modulus fibers; T300 [55], UHM, and AS4D [56] are high strength fibers; and IM6 is an super high strength fiber. Very high modulus is achieved in P100 by using a pitch precursor in *liquid crystal phase* that allows for higher orientation of the carbon structures in the final fiber.

One motivation for using high modulus fibers is to fabricate composites that imitate steel or aluminum, so that a metallic part of a structure can be replaced by a composite part of lesser weight. For example, a composite with 50% fiber volume fraction of M50 fibers would have approximately the same stiffness of steel and one-fourth of the weight. However, substitution with composites in a structure that was designed with metals almost invariably leads to an inefficient design of the composite part. A more valid motivation for using high modulus fibers is to maximize the stiffness-to-weight ratio of structures such as space telescopes, where both weight and deformations are critical.

The maximum operating temperature of carbon fibers varies from 315°C to 537°C [15] but it may be further limited by the operating temperature of the matrix as is the case with polymer matrix composites. Being stiffer than glass fibers, carbon fibers provide better fatigue characteristics to the composite by reducing the amount of strain in the polymer matrix for a given load [57]. Also, the stress corrosion (static fatigue) phenomenon is less marked for carbon fibers. This is the main reason for using a lower stress ratio (safety factor) of 2.25 for carbon fiber, instead of 3.5 for fiberglass, in the design of pressure vessels subject to permanent loading [51]. Carbon fibers are good electrical conductors.

Some of the limitations of carbon fibers are their low shock resistance (due to their high rigidity, high fragility), as well as susceptibility to chemical attack in the presence of oxygen and another oxidizing compounds at temperatures higher than 400°C. Galvanic corrosion will take place if carbon fiber composites are in electrical contact with metals, due to their good electrical resistance. Therefore, an insulating barrier needs to be created between carbon fiber composites and metal parts in the same structure. This is usually accomplished by adding a lamina of glass-mat reinforced composite (about 0.5 mm thick).

The major limiting factor for the application of carbon fibers is the cost. Generally speaking, the cost of carbon fibers can be justified when weight savings offer a large payoff, such as in aerospace applications, or when high temperature performance, corrosion resistance, improved fatigue strength, or the long-term retention of strength are essential for the intended application. Carbon fibers are lighter and stiffer than glass fibers but it is difficult to show an economic advantage when carbon fiber composites are used to substitute a conventional material such as steel in a structure that is not weight critical. Further cost savings can be achieved by part integration and reduced installation and maintenance cost, but these are usually independent of the type of fiber. The high cost of carbon fibers is better justified when the material is used as an enabling material rather than a substitution material. An enabling material allows the structure to accomplish a mission that is not possible with other materials. However, the bulk of applications most likely to be encountered are substitution applications, where the designer proposes to use composites to accomplish basically the same mission that is currently achieved with a conventional material. In the latter case, initial product cost becomes of paramount importance.

One of the latest research in the field of carbon technologies has lead to a new kind of carbon fibers, called *carbon nanotubes* (CNTs), one of the most fascinating materials ever discovered. The arrangements of carbon atoms in CNTs is controlled so that a perfectly cylindrical atomic structure of about 1 nm diameter and about 1 μm long, with the strong carbon bonds orientated along the cylinder axis. Because of this controlled nanostructure, the inherent defects that appear in any material and that are mostly responsible for the strength limitations are virtually eliminated. The advantage of this high atomical orientation makes CNTs the strongest and most stiff fibers so far—their modulus being about 1000 GPa and their strength about 30 GPa. To take advantage of their superior stiffness and strength, nanotubes are used as a filling agent in a polymeric matrix, and the compound is extruded to obtain CNT filled fibers, the fiber itself being a nanocomposite system. During this process, the alignment of the CNTs is controlled so that their axis is aligned as much as possible with the fiber length. Of course neither the original stiffness or the original strength of the CNT can be achieved in the nanocomposite fiber, but notable improvements can be made. Furthermore, superior heat transfer and electrical properties are obtained. So far, high cost makes CNTs hardly accessible for industrial applications.

2.2.4 Carbon Nanotubes

Carbon nanotubes (CNT) are comprised entirely of carbon atoms that lie in a nearly defect free graphitic plane rolled into a seamless cylinder. The absence of defects and the strength of the carbon-carbon bonds are responsible for the incomparable stiffness and strength of CNT, up to 1.5 Tera-Pascal (TPa) modulus and 200 Giga-Pascal (GPa) tensile strength.

Two structural forms of carbon nanotubes exist, single-walled (SWNT) and multi-walled (MWNT). MWNTs were discovered first, by Iijima in 1991. They are comprised of multiple concentric open-ended single-walled nanotubes forming a single structure with multiple walls, with a constant interlayer separation of 3.5 Amstrong (3.5 Å = 0.35 nm), diameters in the range 2–100 nm and lengths up to tens of microns [58,59]. SWNTs have a diameter of about 1 nm and consist of a single graphite sheet rolled into a seamless cylinder with caps (hemispheric Buckminister fullerene molecules) at both ends and can have lengths up to centimeters [59, 60]. The near perfect atomic configuration of the walls of a single-walled carbon nanotube requires that nearly 10 to 20 primary carbon-carbon bonds be broken to completely cleave the structure. This affords carbon nanotubes great stiffness and strength.

The future of carbon nanotubes is to create lighter-weight, mechanically superior composites that take advantage of the superior properties of nanotubes. Nanocomposites produced by incorporating nanotubes into a matrix have the potential to surpass the performance of conventional fiber-reinforced composites while being economically feasible because substantially lower loading of nanotubes may be sufficient to yield superior performance. However, the cost of single-walled carbon nanotubes still hovers near $500/gram and more research and development is still required to produce large quantities of high quality nanotubes and their composites and to harness the full potential of these novel fibers. Research is being done to understand how to use stable configurations of individual, well-dispersed nanotubes to improve the mechanical properties of a nanocomposite system. Additionally, research is being done to identify the mechanisms responsible for their increased performance. Specifically, researchers are interested in elucidating whether nanotubes act as stress concentration points in a matrix, thus promoting the prolongation of crazing or microcracking, or act as physical bridges to cracks in the matrix, thus preventing crack propagation, or simply carry a higher load due to their higher stiffness and strength.

Various measurement techniques have been used to measure the mechanical properties of both MWNTs and SWNTs ranging from the amplitudes of their intrinsic thermal vibrations to cantilever type bending methods to direct stress-strain measurements [61–65]. Some of the first measurements on CNTs were performed via a direct stress-strain method [66]. Properties of MWNTs produced by the arc discharge method were measured, yielding modulus values in the range 0.27–0.95 TPa, with fracture occurring at strains of up to 12%, and strength values in the range 11–63 GPa. In these experiments, failure occurred at the outer tube with the inner walls shearing outward in a telescopic manner. Measurement of the stress-strain response of small bundles of SWNTs by a similar direct method yielded modulus values in the range 0.32–1.47 TPa and strength values in the range 10–52 GPa, with failure occurring at a strain of about 5.3%. SWNTs failures initiated at the perimeter of small nanotube bundles, with the rest of the tubes slipping apart instead of fracturing [67]. Theoretically, it is predicted that individual nanotubes have elastic moduli approaching $1\,TPa$ and tensile strengths in the range 20–200 GPa [68–70]. This far outmatches traditional macroscopic materials such as high strength steel, which has a modulus of about 200 GPa and tensile strength of about 1–2 GPa [67].

The impressive mechanical properties of these novel carbon-based materials render them a natural candidate for incorporating into thermoplastic or thermoset matrices to create nanocomposites, where the nanotubes act as the reinforcing fiber in the matrices. If CNTs behaved like continuous fibers, the composite performance could be predicted by the rule of mixtures (Equation (4.24) in Section 4.2.1), and the enhancement of tensile modulus and strength of the composite system would be enormous, even for low values of fiber volume fraction. However, this idealistic situation rarely applies because incorporating well-dispersed, aligned, individual carbon nanotubes, in large quantities in a matrix is difficult. Suitably dispersed, surfactant-free, purified single-walled nanotubes often exist as small bundles or ropes of tubes held together via Van der Waals bonds [71]. These small ropes or bundles of nanotubes, which remain bundled after incorporation into a matrix, tend to slip when a tensile force is applied, thus limiting the performance of the nanocomposite.

Due to SWNT's small dimensions, their surface area is very large compared to their volume, with aspect ratios of about 1000. In principle, such a large area should enhance load transfer to the surrounding matrix. However, the high aspect ratio does not always facilitate interaction with the matrix as would be expected of a perfectly dispersed system of individual nanotubes, but can in fact promote re-agglomeration of nanotubes, even when attempts are made to initially disperse them in the matrix. Since agglomeration and imperfect dispersion results in limited interfacial stress transfer between the matrix and the stronger CNTs, the expected enhancements in mechanical properties of the overall system are not achieved. Often, higher than ideal loading of nanotubes are used in an attempt to compensate for the shortcomings in expected performance, but higher loadings may actually result in decreased strength and modulus of the nanocomposite [71].

In order to overcome bundling of nanotubes, chemical functionalizations are used to pull apart the nanotube ropes and promote interfacial bonding with the surrounding matrix [72,73]. However, this is not without difficulty as the naturally chemically inert nature of the graphic surface of nanotubes has to be overcome. In other cases, neat ropes of nanotubes are spun into fibers in super acidic solutions [74], or arrays of nanotubes are grown on surfaces [75], all with the goal of identifying usable individual nanotubes or more consistent ropes of nanotubes for various applications. If individual nanotubes are not pre-treated, then dispersion techniques are employed during processing to create the nanocomposite. These methods include polymer infiltration of nanotube thin film networks, melt processing and melt spinning of nanocomposite fibers via solution based methods, covalent functionalization and polymer grafting to nanotube sidewalls, and in situ polymerization in the presence of carbon nanotubes [76–80].

2.2.5 Organic Fibers

Among a large variety of organic fibers, *polymeric fibers* are very attractive for high-stiffness, high-strength structural applications, mainly because of their very low density, that results in very high values of the *specific* strength and stiffness values. Polymeric fibers comprise many different products, depending on the base polymer used as a precursor. The manufacturing process used to make polymeric fibers comprises two main phases: transformation of the solid bulk polymer to liquid phase by either melting or solution, followed by extrusion and cooling. *Liquid Crystal Polymer* (LCP) fibers refers to the aforementioned manufacturing process and the fact that some polymers feature anisotropic crystalline behavior in their liquid phase, forming stronger chemical connections along certain directions. These orientated strong connections are then preserved in the fiber fabrication process, resulting in strong fibers along their axis, with anisotropic behavior. In this way, LCP fibers are able to feature similar strength as carbon fibers.

The best-known polymer fibers are the ones processed from polyamide, called *aramid fibers* (aromatic polyamide), produced by DuPont [81], Teijin [82], and Akzo Nobel [83] under the trade names *Kevlar*^{TM}, *Technora*^{TM}, and *Twaron*^{TM}, respectively. Aramid fibers are LCP fibers. They have high energy absorption during failure, which makes them ideal for impact and ballistic protection. Because of their low density, they offer high tensile strength-to-weight ratio, and high modulus-to-weight ratio, which makes them attractive for aircraft and body armors. Since aramid fibers are made of a polymer material, they have similar characteristics to the polymer matrices discussed in Section 2.5. They have low compressive strength; they creep, absorb moisture, and are sensitive to ultraviolet (UV) light. Also, their mechanical properties vary with temperature, with the tensile strength at 177°C reducing to 75–80% of their room temperature value [38]. Various types of aramid fibers offer different material properties, *Kevlar 49* being the most commonly used (Table 2.2). *Technora* fibers overcome some of the weaknesses of aramid fibers, having improved thermal stability; they can be used up to 200–250°C but loss of stiffness and with up to 50% loss of strength.

The aforementioned advantages and disadvantages of aramid fibers are shared by all polymeric fibers. PBO based fibers are represented by *Zylon*^{TM} [84] produced by Toyobo, which feature higher strength and stiffness than aramid and excellent tenacity. Polyethylene fibers, the most known being *Spectra*^{TM} [85] produced by Honeywell International Inc., have lower moisture absorption, lower density, and stiffness–strength properties comparable to aramid fibers, but they also have lower maximum operating temperature (about 120°C). They are attractive for construction of radomes because they are highly transparent to electromagnetic waves. Potential applications for structural components is limited because of their low maximum operating temperature. Polyester polymeric fibers such as *Vectran*^{TM} [86], produced by Kuraray Co., have outstanding properties as shown in Table 2.2 but they lose about 50% of their strength at about 150°C.

2.2.6 Boron Fibers

High stiffness, high strength, and low density are common to *boron fibers* (Table 2.1 [87]. They are most notably used as reinforcement in aerospace and sporting goods. The strength values are controlled by the statistical distribution of flaws during the manufacturing process. The mechanical properties are preserved at high temperatures (typically, the tensile strength at 500°C is about 60% from the initial strength value at room temperature). They also feature high toughness, high fatigue strength, and a very good compressive behavior. They are fragile, with low impact tenacity. Boron fibers are produced by chemical vapor deposition on a tungsten wire. Because of slow production rate, boron fibers are among the most costly of all the fibers presently made, thus motivating substitution of boron with carbon fibers whenever possible.

2.2.7 Ceramic Fibers

Ceramic fibers are used for high temperature applications. Silicon Carbide (SiC) fibers are produced like boron fibers but on a carbon substrate. They have their greatest application as reinforcement for metal matrices, primarily titanium, but they have also been used in combination with temperature-resistant polymeric matrices. Like boron fibers, SiC fibers are characterized by high stiffness and high strength (Table 2.1) [87] and most notably they exhibit higher temperature capability as they keep their initial tensile strength up to about 1300°C. Because of limited use and low production volumes, the cost of these fibers is high.

Alumina fibers are based on alumina oxide Al_2O_3. Unlike other ceramic fibers, they are not produced by chemical deposition, and thus have high volume availability and relatively low cost (200 to 1100/kg depending on fiber type, tow size, and quantity). Properties of *Nextel*^{TM} fibers [88] are given in Table 2.1. First used to reinforce diesel pistons, ceramic fibers add strength and hardness to metal matrix composites at high temperatures, where degradation of metal properties usually occurs.

2.2.8 Basalt Fibers

Basalt fibers [52] can be classified as mineral fibers. This new product uses volcanic basalt rock, melted and extruded into fibers. Basalt fibers feature better mechanical properties than glass fibers, and less costly than the carbon fibers. They display excellent thermal stability, high strength and stiffness, good chemical stability, good corrosion resistance, and good matrix adherence, not being affected by any kind of radiation. They are potentially applicable for thermal protection and structural applications.

2.2.9 Metallic Fibers

Metallic fibers [52] include fibers made from a variety of base metals and alloys. By processing the raw material into fibers of small dimensions of the order of a μm, flaws inherent in the bulk material are virtually eliminated and enhanced properties are achieved. The fiber strength is directly related to the fiber diameter, that is in turn related to the fiber manufacturing cost. The fiber material can be selected for the required application, such as lightweight aluminum fibers, strong steel fibers, or stiff tungsten fibers. Other advantages of these fibers are their excellent electrical and thermal conductivity. Metallic fibers are typically used as filler nets for polymer matrices, conferring electrical conductivity, electromagnetical interference protection or lighting strike protection to the final structural composite material. Two examples of metallic fibers, high carbon steel fibers and tungsten fibers, are presented in Table 2.1. Nonmetallic fibers covered by a thin metallic coating, combine the advantages of both components, the substrate fiber and the metal coating.

A variety of new fibers are continuously introduced or improved but acceptance of new fiber products in the market for structural applications is slow because of the large number of characterization tests required to assure safe and effective application under a variety of conditions. While high strength-to-weight and stiffness-to-weight ratios are advantageous for aerospace applications, other characteristics of the composites that can be produced with these fibers are likely to be the limiting factors in their application. Environmental degradation, maximum operating temperature, transverse and shear strength are some of the limiting factors that need to be investigated for any new material.

2.3 Fiber–Matrix Compatibility

Most fibers are covered with a substance called *sizing*. Fibers sustain considerable damage during processing as they rub against each other and the equipment. The sizing helps protect the fibers by serving as a lubricant and antistatic agent, and helping a bundle of fibers stick together as a unit. Furthermore, the sizing incorporates a coupling agent to promote bonding with the matrix in the composite. Therefore, the same fiber may have a different sizing depending on the type of polymer that is being used as matrix. The intralaminar strength of the composite (see Section 4.4.10) provides an indication of the quality of the fiber–matrix bond. For this reason, the intralaminar shear strength tests ASTM D2344, D4475, and D3914 are used in industry to evaluate the fiber–matrix compatibility, which largely depends on the sizing of the fiber. For glass fibers, the sizing also serves to protect the fiber from moisture attack during the service life of the composite. Water reaching the surface of the fiber may leach some of the chemical components from the glass, leaving a porous surface. The pores then become defects that effectively reduce the strength of the fiber.

2.4 Fiber Forms

Fibers provide strength and stiffness while having low density, constrain the creep of the matrix, reduce thermal expansion, and they can be used to either increase or decrease other physical properties such as thermal conductivity, electric conductivity, and radiation transparency. The matrix provides stress transfer among fibers and protects them from chemical attack and abrasion.

In any product, a key role is played not only by physical properties, but also by cost. Fiber reinforcements are a major component of the tradeoff between performance and cost. High performance can be achieved by a judicious design and manufacturing of the product, but more than one solution often exist; the final choice governed by cost. A large variety of fibers forms exist in order to satisfy the requirements of various processing methods. Fiber reinforcements may be obtained as continuous or discontinuous fibers, unidirectional aligned fibers or textile products. Among textile products is it possible to use woven, knitted, and braided fabrics, to name a few. The reinforcement form is a design parameter and it affects both the quality and cost of the composite material.

Most fibers can be obtained in the form of prepreg tape in which the fibers are held together by an epoxy resin and a fiberglass backing (Section 3.2). Because the making of the prepreg tape introduces additional labor, the cost of prepreg tape is usually 75–100% higher than the cost of the fibers. As explained in Chapter 3, the fabrication methods using prepreg tape are slow and labor intensive. Therefore, most of the new applications of composites use fibers in their simplest, unprocessed forms. For example, pultrusion and filament winding use roving or tow and resin to produce the final product without intermediate operations. Woven or stitched fabrics facilitate the fabrication of laminates in Resin Transfer Molding and other processes, while adding only 20–40% to the cost of the reinforcement.

2.4.1 Continuous and Discontinuous Fibers

A first classification for fiber forms can be made considering that composites are reinforced either with continuous fibers, discontinuous fibers, or particles. When maximum achievable stiffness and strength are desired, *continuous fibers* are used. In continuous fiber reinforced composites, the load is carried mostly by the fibers oriented along the load direction. Continuous fibers are packaged in various forms that will be discussed in the following sections. The composite material manufacturing process using continuous fibers is costly and slow. This is in part due to higher costs of the continuous fiber system itself, and in part due to the specific methods of using the continuous fibers for composite materials manufacturing, such as hand lay-up, prepreg lay-up, vacuum bagging processing, resin transfer molding, and so on. An alternative at continuous fiber reinforcement, leading to reduced manufacturing cost, is to use *discontinuous fibers.* Using this kind of reinforcement, not only the price of the fiber material is less, but also less expensive methods to fabricate the final part are possible, such as injection molding and so on. Discontinuous fibers are short fibers obtained by chopping continuous fibers, or produced directly as short

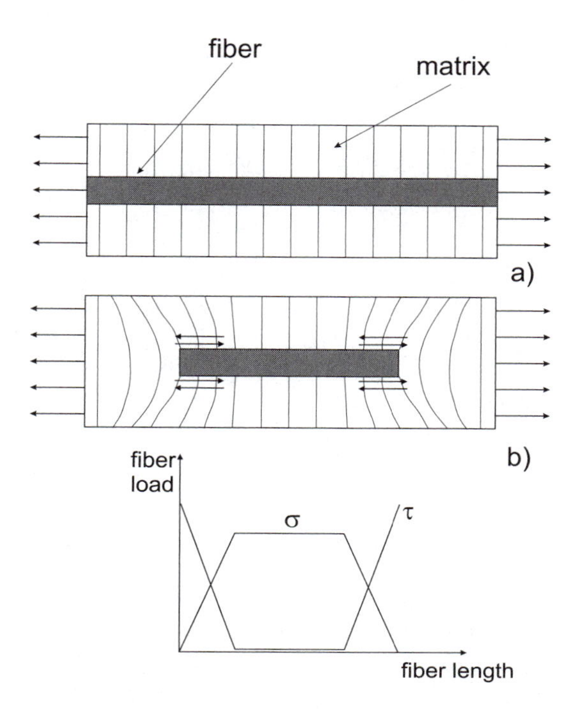

Figure 2.1: Load transfer process in (a) continuous fiber reinforcement and (b) short fiber reinforcement.

fibers to reduce cost. The mechanism of load transfer using discontinuous fibers is different to that of continuous fibers reinforcement (Figure 2.1). The quality of the fiber-matrix interface plays a crucial role since the load is transferred from fiber to fiber through the matrix. It can be seen in Figure 2.1 that, because of end fiber effect, the fiber does not use all its length for load carrying, and the resulting reinforcing effect is lower compared with continuous fibers. In order for the short fiber to achieve its maximum load carrying capacity, it has to have a minimum length called *critical length* L_c. The maximum axial load carrying capacity of the fiber is reached only at a certain distance from the fiber ends, where the shear effects are dominant. The critical fiber length is not an intrinsic fiber quality, but it depends on both the fiber length to diameter ratio and the fiber-matrix system. In this way, if a short fiber length is satisfactory for reinforcing a certain polymer matrix, it can be inadequate for reinforcing another matrix. For the most common glass or carbon fibers with polymer matrices, the critical length can vary from a fraction of one millimeter to a few millimeters [53]. Due to their easy processing and low cost, short fiber reinforcement has been studied and continuously improved. Composite materials based on this reinforcement type may reach 50% strength and 90% stiffness [53] of continuous reinforcement with the same fiber material. The quality of these composites can be improved by enhancing the fiber-matrix interface and ensuring a preferential orientation of the short fibers along the load direction.

Whiskers are a type of very high performance short fibers. These are elongated, controlled growth, monocrystals. Because of the monocrystal structure, the manufacturing defects inherent to all bulk materials are virtually eliminated and very high strength is obtained in the direction of the strongest atomic connections of the crystal. The alignment of the crystal structure with the longest dimension of the whisker also results in strong anisotropic properties for the whisker.

Since the orientation of short fibers in the composite cannot be controlled easily, they are assumed to be randomly oriented unless special provisions are taken to control the fiber orientation. Composites made with short fibers arranged randomly have nearly isotropic properties in the plane of the laminate. Perhaps the main disadvantage for structural applications is that short fiber composites do not reduce the creep of polymer matrices as effectively as continuous reinforcement. Furthermore, short fiber composites usually have lower strengths than continuous fiber composites.

2.4.2 1D Textiles: Strand, Tow, End, Yarn, and Roving

Besides continuous and discontinuous reinforcements, textile reinforcements provide alternatives that may increase performance while reducing manufacturing cost. For example, the resin transfer molding process is particularly suited for incorporating woven fabric reinforcement. The textile industry and the structural composite materials industry thus collaborate in this area. The individual fiber is the precursor for the textile tread, and the thread is in turn the precursor for the reinforcing fabric, which is the precursor for the structural composite. The design process starts with thread design and manufacturing, with the thread properties playing an important role in the final composite material properties. For example, an increase in thread twist reduces its modulus but it can improve the fatigue behavior of the composite.

The textile design and manufacturing process leads to the following classification of textile forms (textile structures):

- 1D textile, or individual thread. These are called *strand, tow, end, yarn,* or *roving,* and defined later in this section.

- 2D textile, or fabric. It uses the 1D textile, laying threads on various patterns on a surface. The resulting composite material is a 2D structure, such as a plate or a shell.

- 3D textile. It uses the 1D textile, arranging threads in complicated 3D forms by special textile processing methods. The resulting composite material features a 3D solid behavior.

A *strand*, or *end*, is an untwisted bundle of continuous filaments (fibers) used as a unit, called *tow* if made of carbon fibers. All fibers in a strand are produced at the same time, from a single furnace, and spun together.

A *yarn* or *thread* is a twisted strand.

A *roving* is a collection of parallel strands, twisted or untwisted; each strand being a separate bundle of fibers as noted before. A roving is produced by winding together the number of strands needed to achieve the desired weight per unit length (TEX in g/km) or length per unit weight (Yield in yd/lb). Also, *yarn number* and *yarn count* are expressions of linear density of the yarn. TEX is inversely proportional to YIELD, as in

$$\text{TEX [g/km]} = \frac{496,238}{\text{YIELD [yd/lb]}} \tag{2.2}$$

Another expression of linear density is the *Denier*, which is weight in grams of 9000 m of yarn. Since all these names refer to the same yarn characteristic, direct or inverse relationships exist between them, corrected by the system of units used by each of them. The higher the TEX, the larger the cross-sectional area occupied by the fibers

$$A\ [\text{cm}^2] = 10^{-5}\frac{\text{TEX [g/km]}}{\rho_f\ [\text{g/cm}^3]} \tag{2.3}$$

where the brackets [ ] indicate the units of each quantity in the equation and ρ_f is the density of the fibers. The area of a roving is useful for the computation of fiber volume fraction of composites (see Section 4.1.1).

The term *tow* is commonly used with carbon fibers and their linear density is usually expressed as a count, or K-number, which represents the number of fibers in a tow in thousands. The K-number is related to TEX through the fiber diameter and the density as follows

$$\text{TEX} = \frac{\pi d_f^2}{4}\rho_f K \tag{2.4}$$

with TEX in [g/km], K in thousands, ρ_f in [g/cc], and d_f in microns (1 micron $= 10^{-6}$ m). However, this equation does not give accurate TEX results if the K number or the fiber diameter are only approximate. For accurate prediction of the weight or volume fraction in the composite, it is advisable to request the TEX value directly from the manufacturer or to measure it [36]. Also, it must be noted that a variability of $\pm 6\%$ in TEX is not unusual, both for glass and carbon fiber tows and rovings, even within the same material lot.

A *direct-draw* roving is wound into a spool directly from the furnace with the required numbers of strands to achieve the desired yield. An *indirect-draw* roving is formed in a secondary operation to increase the yield, by rewinding several direct draw rovings into one. In each roving, the fibers wound on the outside of the spool are longer than the ones on the inside. When the roving is unwound to fabricate the composite, not all the fibers have the same length, which creates microcatenary. An indirect-draw roving has in general more microcatenary than a direct-draw roving because several direct draw rovings are rewound to obtain the higher yield. To measure the microcatenary, a length of roving (about 25 m) is stretched between two supports and the fibers are loosely separated by hand. The shorter fibers will be stretched and the longer fibers will hang in a catenary shape (similar to power lines) because of gravity. Microcatenary results in misalignment in the composite which is adverse to the compressive strength of the material [89].

Rovings are used directly in pultrusion and filament winding without intermediate operations, thus reducing the cost of the final part. Common fiberglass rovings have yields between 56 and 250 yd/lb (TEX between 8861 and 1985 g/km). Carbon fibers are commonly packaged in $3K$ to $36K$ tows (TEX 198 to 2290).

2.4.3 2D Textiles: Fabrics

A large variety of textile fabric forms can be used as reinforcements for composite materials. The simplest, the least elaborated and least expensive ones are the *nonwoven fabrics*. A nonwoven fabric is usually called a *mat*, which is made by randomly oriented chopped fibers such as chopped strand mat, randomly oriented short fibers, or swirled tows or rovings. The later is called *continuous strand mat* or simply CSM and it is made by continuous tows or rovings swirled on a table and pressed and/or loosely held together with a very small amount of binder (adhesive). Sometimes the mat fibers are better fixed by stitching them to a backing surface or among themselves without a backing. Stitched nonwoven fabrics can be made into very heavy fabrics, thus reducing the time and cost of composite processing, provided they can be adequately infiltrated with resin.

A *veil* is a thin mat used as a surfacing lamina to improve corrosion resistance of the composite (see Section 2.9) taking advantage of the higher matrix content held up in the veil. Since the veil layer is smooth, without the texture of a woven fabric, veil also is used to improve the appearance of the part by hiding the texture of the reinforcement underneath. Veils and mats have fibers oriented randomly in every direction, leading to isotropic properties in the composite. Despite their low costs, the mechanical properties of the composite material using this kind of reinforcement are also low.

A *woven fabric* is a two-dimensional reinforcement obtained by interlacing of yarns (Figure 2.2) in the weaving machine. Fabric reinforcements offer mechanical properties somewhat lower than unidirectional continuous fibers (see Chapter 9). Usually the woven fabric is processed by yarns interlacing along two orthogonal directions: the yarn along the weaving direction is called *warp* yarn, and the interlacing yarn perpendicular to the weaving direction is called *fill* or *weft* yarn.

Balanced reinforcing properties can be obtained in both fill and warp directions, using the same yarn type in both weaving axes. This weaving architecture is called *biaxial woven fabric* (Figure 2.3) and it is the most common one used as structural reinforcement. A wide variety of weaving patterns are possible in the case of biaxial woven fabrics structures, the most common ones being plain weave, twill weave, and satin weave (Figure 2.3 a, b, and c, respectively). Different weaving patterns assure different properties and of the textile structure and of the final composite material.

Drape is the ability of the fabric to conform to complex multicurved surfaces, which is needed for the manufacturing of complex parts. Drape is highly influenced by the weaving pattern, with plain, twill, and satin weave having increasingly higher drape.

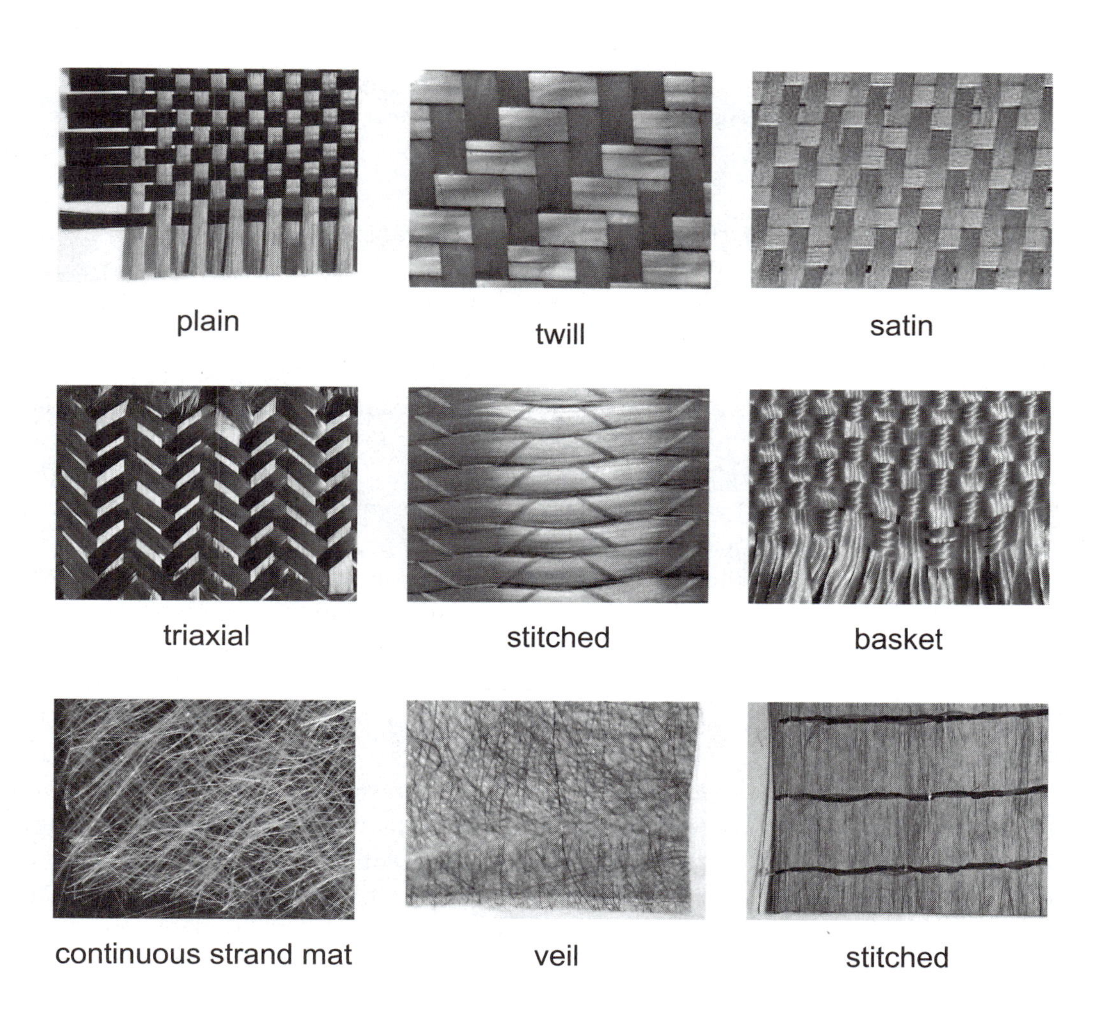

Figure 2.2: Fiber forms.

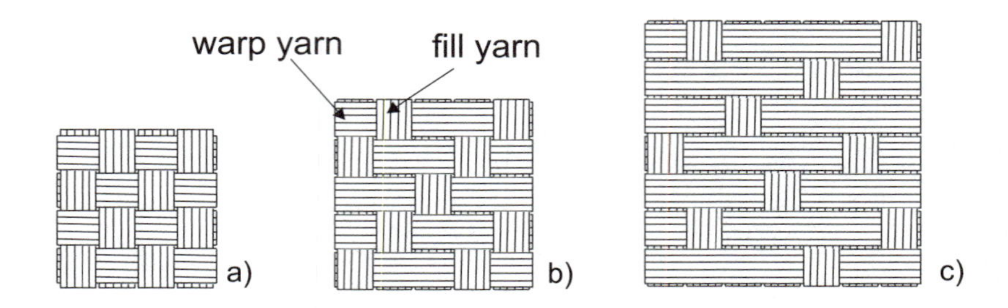

Figure 2.3: Weaving patterns for biaxial woven fabrics: (a) plain, (b) twill, and (c) satin weave.

Crimp is a measure of the yarn undulation or waviness in a woven fabric. A higher crimp represents a higher yarn undulation. Usually the crimp is expressed as *percentage crimp* or *crimp amplitude.* Percentage crimp is defined as the difference between the length of the straightened yarn and the distance between the ends of the woven yarn, in percent. Crimp amplitude (units of length) refers to the deflection of the yarn axis with respect to the middle surface of the fabric. The crimp level is influenced by the yarn thickness and by the gap between yarns. For fabric reinforced composites (see Chapter 9), crimp influences negatively the mechanical properties of the composite material, by affecting the overall fiber volume fraction and the yarn orientation in the composite.

The design of laminated composite materials using woven fabric laminae is similar to that of designing using unidirectional laminae. Laminates with balanced off-angle laminae (e.g., $\pm\theta$ with the same amount of $+\theta$ and $-\theta$ strands) are the most common. These produce specially orthotropic composites (see Section 6.3.8) that have several advantages for manufacturing, design, and performance, as shown later in this book.

Besides popular biaxial woven fabrics such as plain weave, twill, and satin, other woven fabrics with unique features exist. If reinforcement is desired in only one direction, but maintaining the ease of handling and drape properties of woven fabrics, the solution is called *uniaxial woven fabric* (Figure 2.4). In this case, normal yarns are used along the warp direction but only thin yarns that can be made from another low cost, low quality material are used along the fill direction, their role being to keep the warp yarns together. In this way the mechanical reinforcing effect is obtained only along one axis (the warp axis). By contrast, if a multiaxial reinforcement effect is desired (including bias directions), the solution is a *triaxial woven fabric* (Figure 2.4), which is obtained by interlacing three sets of yarns at different angles. Quasi-isotropic material properties can be obtained in this way, but this also implies special weaving technologies that impact the product cost.

Yet more types of textiles used for reinforcements exist. *Knit fabrics* are made, not by interlacing, but by interlooping yarns (Figure 2.4). *Braid fabrics* are obtained by simultaneously winding three or more yarns around one axis, resulting in a long, narrow, 2D textile structure.

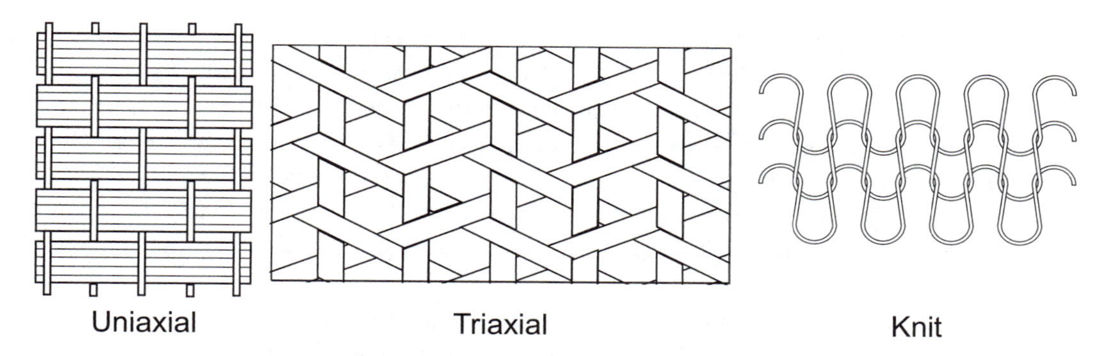

Figure 2.4: Uniaxial, triaxial, and knit woven fabrics.

All the 2D textiles (mat, fabric, knit, braid) are described by and their fiber architecture and their weight per unit of surface area in [g/m^2]. The fiber architecture refers to the orientation and weight of each component of the fabric, as illustrated in Table 2.8. Detailed description and analysis of textile-reinforced composites is presented in Chapter 9.

By special manufacturing methods developed by the textile industry, it is possible to obtain a new category of 3D textile architecture called *3D textiles*. These are infiltrated with a resin to obtain composite materials with reinforcing fibers in all three directions. The resulting composite material can be processed as a solid 3D bulk material that is then machined to shape or processed directly into the final shape of the desired product. 3D textiles can be made as 3D weaves or 3D braids.

2.5 Matrix Materials

The matrix has many functions. It holds the fibers together, thus transferring the load through the interface to the reinforcing fibers and to the composite from external sources. It carries some of the loads, particularly transverse stress (Section 4.4.4), intralaminar shear stress (Section 4.4.10), and bearing stress. Some properties of the composite, such as transverse stiffness and strength, are matrix dominated. Therefore, they affect the matrix selection more than fiber-dominated properties. The properties of the matrix determines the allowable service conditions for the composite including temperature range, chemical resistance, abrasion resistance, and weathering capability. Also, the matrix plays a leading role in regards to heat and electrical conductivity the composite as well as dominating the external characteristics such as appearance. Matrix materials can be polymers, metals, or ceramics. Polymer matrices are the most common because of their ease of fabrication of very complex parts with low tooling cost and low capital investment manufacturing costs. Many commercial composite applications were initially designed using unreinforced plastics.

Unreinforced plastic designs provided many functional improvements versus natural materials such as wood, and conventional materials such as metals. Through part integration, low cost tooling and unlimited freedom for aesthetic design, unreinforced plastics replaced their predecessors in many nonstructural applications. However, under continuous loading they suffer from creep problems. Reinforcing plastics with fibers or fiber systems virtually eliminates the creep problem, thus making polymer matrix composites suitable for structural applications. The continuous and virtually unlimited variation of mechanical properties under the designer's control creates opportunities for optimization not available with isotropic material systems.

Since polymer matrix composites (PMC) are the most commonly used type of composite material, they are emphasized in this section. The rest of the book, however, applies equally to any type of composite. For an in-depth discussion of other matrix materials see [15, 35, 90, 91].

Matrix dominated properties depend strongly on the maximum/minumum temperature capability of the matrix. The performance of the composite may vary drastically when the operating temperature approaches or exceeds the operating temperature limits of the matrix. Thermoset polymer matrices such as polyester resins and epoxy resins lose modulus and thus the ability to transfer load when they are exposed to temperatures above or near their primary or glass transition temperatures. Aluminum softens and melts under high heat. Ceramic matrices may self-destruct from internal shrinkage stresses when exposed to cryogenic conditions.

The matrix of polymer matrix composites (PMC) is a polymer that can be either a thermoset or a thermoplastic. Various terms are used in the marketplace to refer to PMC. Advanced composite materials (ACM) usually refers to more expensive and high performance composites targeted to aerospace applications, mostly carbon–epoxy and carbon–thermoplastic systems. Fiber reinforced plastics (FRP) usually refers to less expensive materials targeted to consumer goods and mass markets, including fiberglass–polyester systems. However, the distinction is no longer clear since many new materials marketed as FRP systems also have high performance, particularly performance/cost ratios, that exceed those of some materials labeled as ACM.

Other designations attempt to differentiate the fiber type used. For example GRP and GFRP both refer to glass fiber reinforced plastics, and CFRP refers to carbon-reinforced plastics. Sometimes the term RP (reinforced plastic) is used when the term composite may be confusing. This is the case in the area of civil construction market [92], where composite construction means structures made of steel structural shapes and reinforced concrete used together, notably for bridge construction. The term composites will be used through this book to refer to all types of fiber-reinforced materials.

The first step of the design process for any product is to determine the functional and structural requirements of the product, which may be a part, an assembly, or a complex system of structural components. The short- and long-term loading, support conditions, and other structural aspects constitute the *structural requirements* of the product. Cost, production rate, and service environment (e.g., operating temperature, aesthetics, and weathering) constitute the *functional requirements* of the part. Functional requirements are as important or more important for matrix selection than the structural requirements because the reinforcement affords the designer great flexibility to satisfy the latter. Expending sufficient effort on determining the functional requirements of an application can focus materials selection on the total optimization of the product throughout its life cycle, including recycle and/or disposal.

The designer must keep an open mind when selecting materials. For example, it is oversimplistic to think that fiberglass reinforced polyester matrix composites are required for all low cost parts and all aerospace applications should be high performance matrices reinforced with carbon or boron fibers. Also, the designed must recognize that there are many customers for the product as it passes from manufacturing through assembly, marketing, sales, and final use. The material chosen must satisfy all of these many customers. A well-thought out set of functional requirements gives the designer the information needed to carry out a successful materials selection process. It will help reduce the selection to a few candidate matrices and a few candidate processing methods from the vast possibilities available.

The functional requirements encompass the life cycle of the product and highlights the capabilities required during each of the segments of the application's existence. It creates a picture of the ideal composite materials for the specific application. Comparing the requirements for the matrix of this ideal composite to the performance capabilities of existing matrices helps optimize the matrix selection process by focusing on the needs of the application, rather than the desires of the designer or manufacturer.

The selection of a matrix for a PMC parallels alloy selection in steel and aluminum parts. All steel is not equal, nor does every aluminum alloy and heat treatment give the same processing and environmental performance. Their microstructure and alloying components must be matched to the end use and manufacturing process. With the vast assortment of PMCs available, the need for a defined selection process becomes a critical component of the design process.

The functional requirements guide the selection of a matrix by balancing the functional and structural properties required by the application. Matrix dominated parameters such as corrosion resistance and flammability of the polymer matrix are given special attention in the selection process. The corrosion resistance of the resulting composite, which is normally dominated by the matrix, is matched to the application's needs. Even though carbon fibers are chemically inert under most common conditions, if the matrix degrades, the integrity of the composite will be severely compromised. Some properties such as transverse stiffness (see sections 4.2.2 and 4.2.4) and strength (see Section 4.4.4) are dominated by the matrix, with the fibers having little influence. Furthermore, not every resin can be processed by every processing technique (Chapter 3). Quite the contrary, resins are available in families of products, each member of the family targeting a specific processing technique. Selection of the matrix can determine the processing technique used. Conversely, production rate and cost requirements may dictate a processing method, which then highly influences the matrix selection.

Selecting candidate processing methods must be done at the beginning of the design process for several reasons. First, processing methods impose constraints on the design process by limiting parameters like the fiber volume fraction and fiber orientations as explained in Chapter 3. Second, the production rate and cost are controlled by the processing technique. Third, every production technique imposes some limit on part size and complexity, which in turn are related to the resin system used. For example, the lower the viscosity of the resin, the larger and thicker a part can be made. In summary, once the designer selects a few candidate matrix systems that meet the functional requirements of the application, one may then concentrate on optimizing the system's mechanical, corrosion, flammability, and processing capabilities at the minimum cost possible.

2.6 Thermoset Matrices

A thermoset matrix is formed by the irreversible chemical transformation of a resin system into a cross-linked polymer matrix. In general, the polymer is called a *resin system* during processing and matrix after the polymer has cured. Thermoset resins have low viscosity, which allows for excellent impregnation of the fiber reinforcement and high processing speeds. Thermoset resins are the most common resin systems used in composite systems because of their ease in processing and wide range of performance. Shelf life is the time the unmixed resin system can be stored without degradation. Refrigerated storage is usually recommended to achieve the nominal shelf life of the product. Pot life or gel time is the time the mixed resin can be handled before the viscosity grows to a point where processing is no longer possible. The gel time can be determined by ASTM D2471 for resins and ASTM D3532 for carbon–epoxy prepreg.

Depending on the choice of initiator or curing system, catalyst, and the reactivity of the resin, cure cycles can vary from minutes to hours, and they can take place at room temperature or at elevated temperature. The reactions can be exothermic or endothermic depending on the resin. Gelation rate depends on the type of reaction used to cure the resin system. Gelation in radical cure system is normally rapid. In condensation cured systems, transition to the gel state is still relatively quick, but occurs at a much higher overall degree of cure. Gelation occurs when the resin has reached a point at which the viscosity has increased so much that the resin barely moves when probed with a sharp instrument (ASTM D2471). During cure, the mixture releases or absorbs heat, thickens, solidifies, and shrinks. The volumetric shrinkage upon curing varies depending on the type and composition of the resin and curative. Volumetric shrinkage should not be confused with thermal shrinkage due to temperature change. Volumetric shrinkage is caused by the change in free volume that occurs when the resin cures. It is directed toward the center of mass of the section. Since the fiber reinforcement does not shrink, internal stress can be induced, causing cracking and fiber misalignment, as well as dimensional inaccuracy and surface roughness.

Volumetric shrinkage is of critical importance to the designer because of its effect on the final dimension of the part and therefore on the dimensions of the tooling from which the part is manufactured. A typical ladder rail die will have its flange portions angled at 91–92 degrees so that the part produced by the die will have web to flange angles of 90 degrees. To make a solid rod of exactly 1/2 inch diameter requires adjusting for the thermal expansion of the die during heating, the volumetric shrinkage of the part during cure and the thermal shrinkage of the part as it cools from the reaction temperature back to room temperature.

The most common thermoset resins are: polyesters, vinyl ester, epoxy, and phenolic. All resins provide higher thermal insulation than most commonly used construction materials. Vinyl esters can have higher elongation to failure, which allows more load to be transferred to the reinforcement and improved corrosion resistance versus polyesters. Polyesters have a wide range of properties that can be matched to differing applications and process requirements. They tend to have moderate physical properties, but can be very cost effective because of their low initial cost and low conversion costs. Epoxy resins are considered high performance resins in the sense that their elongation to failure and higher service temperature are superior to that of most other commonly used thermoset resins. Phenolic resins are normally used in application were smoke generation and longer-term flame spread is required by the customer or by code. Phenolic resins are more difficult to process than other common thermoset resins, so a functional reason normally drives their selection.

2.6.1 Polyester Resins

Polyester resins are high value, low cost resins with high performance/cost ratio. Polyester resins are low viscosity, clear liquids based on unsaturated polyesters, which are dissolved in a reactive monomer, such as styrene. The addition of heat and a free radical initiator system, such as organic peroxides, will result in a cross-linking reaction between the unsaturated polymer and unsaturated monomer, converting the low-viscosity solution into a three-dimensional thermoset polymer. The ratio of saturated to unsaturated acid used to form the backbone of the polyester resin controls the cross-link density of polymer network. The distance between unsaturation sites and the molecular stiffness of the backbone between the sites as well as the stiffness of the monomer controls the rigidity of the matrix product.

Cross-linking can also be accomplished at room temperature through the use of peroxides and suitable accelerators and promoters. The monomer type plays an important role in the thermal performance of a polyester resin. A polyester resin in vinyl toluene, for example, shows superior thermal performance (see Section 2.10) when compared to the same resin in styrene [93].

Polyester resins can be formulated to have good UV resistance to structural degradation and can be used in many outdoor applications. They can survive exposure to the elements for periods exceeding 30 years, although some discoloration and loss of strength may occur [93, 94]. The onset of surface degradation is marked by a yellow discoloration that becomes progressively darker as erosion occurs. In cases where filler is used as part of the matrix formulation, as erosion occurs the filler is exposed at the surface. This process is known as chalking because the particles are similar to chalk dust on a blackboard. In translucent systems, UV radiation causes yellowing of the composite as a whole, although the color is usually more intense on the surface. Superior durability, color retention, and resistance to fiber erosion can be obtained when styrene is supplemented with a more resistant monomer such as methyl methacrylate (MMA). The gloss retention of styrene/MMA monomer blends is usually better than for either monomer alone. The refractive index of MMA is also lower than for styrene, allowing the formulation of polyester resins with refractive index more closely matched to that of glass fibers. This, combined with improved UV resistance, has resulted in the use of MMA polyesters to prepare glass-reinforced, transparent building panels that can be used in greenhouses, skylights, and other applications [93].

In many applications, polyester resins are required to have some degree of resistance to burning, which can be accomplished by using either a filler or a specially formulated flame-retardant polyester resin, depending on the degree of resistance required. Incorporating halogen into a polyester resin has been found to be an effective way of improving flame retardancy [93]. However, in applications were smoke density and toxicity are issues, halogen containing matrices may be outlawed by code.

Polyester resins have been used in applications requiring resistance to chemical attack. Numerous applications for corrosion-resistant tanks, pipes, ducts, and liners can be found in the chemical process and the paper industries [95]. Different polyester resin classifications are used in corrosion resistant applications depending upon the specific chemical environment. Chlorendic resins are recommended for strong acid environments, especially at elevated temperatures, while bisphenol-A-fumarate is better in strong basic solutions. Using glass fiber does not improve and may even reduce the corrosion resistance of polyester resins. This is especially true in strong acids such as hydrofluoric acid or strong caustic environment because these chemicals can attack and dissolve the glass fiber.

Representative data of mechanical properties of resins are given in Table 2.9. These properties correspond to unreinforced cast samples of resin, also called *clear castings* or *neat resin samples*. The isophthalic resin shows higher tensile and flexural properties than the orthophthalic resin. The BPA fumarate and the chlorendic are much more rigid, and in clear castings this introduces brittle behavior with low tensile elongation and strength. Using any fiber reinforcement dramatically improves the mechanical properties of all resins. Properties of unreinforced resin derived from casting do not necessarily represent the properties of the same resins manufactured in composite form by a particular manufacturing process because the interface chemistry between the reinforcement and the resin can influence both the resin network next to the reinforcement as well as in the bulk, away from the reinforcement. Therefore, it is always preferable to compare resins in composite form.

Isophthalic resins have moderate corrosion resistance at temperatures below 82°C. They can be flexible or rigid, the latter having higher chemical resistance because of their lower free volume that inhibits diffusion of the corrosion media into the polymer and reduced sites for chemical attack. They have high resistance to water; good resistance to gasoline, oil, weak acids and alkali, but low resistance to peroxides and hypochlorites. Their elongation to failure is moderate, limiting their ability to transfer load to the reinforcing fibers. They offer low fire retardancy by themselves. However, fiber-rich surfaces may provide significant fire retardancy. Furthermore, brominated versions of isophthalic resin formulations are available to improve fire retardancy. Typical properties are given in Table 2.9 [93].

Orthophthalic resins, also called *general purpose polyesters,* are used for structural applications that require lower functional performance than other polyesters because they have low corrosion resistance and weathering stability. As with any generalization about a large family of resins, there are some orthophalitic systems that outperform some specific isophthalic systems.

Chlorendic resins are used for high temperature applications, up to 176°C. They are highly resistant to acids, peroxides, and hypochlorites. They have good resistance to solvents, but low resistance to alkaline environments. Their strength is moderate and their flame retardancy can be high, up to Class I (see Section 2.10) for special formulations [93].

Bisphenol-A-fumarates are better than isophthalic in severe acid and alkaline environments up to 121°C, with moderate strength. They provide low flame retardancy (although fire retardant formulations are available) and moderate resistance to solvents, peroxides, and hypochlorites [93].

2.6.2 Vinyl Ester Resins

Vinyl ester resins have higher elongation and corrosion properties than polyesters, providing a transition in properties and cost to the high-performance epoxy resins, but maintaining the processing versatility of polyesters. There is a great variety of vinyl ester resins available for applications up to 121°C. They are highly resistant to acids, alkalis, solvents, hypochlorites, and peroxides. Brominated versions have high flame retardancy. Typical properties are given in Table 2.9 [93, 96]. The cost of vinyl ester resins is between that of polyesters and epoxies.

2.6.3 Epoxy Resins

Epoxy resins are widely used because of their versatility, high mechanical properties, and high corrosion resistance. Epoxies shrink less than other materials (1.2–4% by volume), which helps explain their excellent bond characteristics when used as adhesives. Most epoxies can be formulated to be less affected by water and heat than other polymer matrices. Epoxy resins are also favored for their simple cure process that can be achieved at any temperature between 5 to 150°C, depending on the type of curative and accelerators used.

One of the main fields of application of epoxies is the aircraft industry. Epoxies are used as adhesives for aircraft honeycomb structures and as laminating resins for airframe and missile applications, for filament wound structures, and for tooling.

They are very useful as body solders and caulking compounds for the building and repair of plastic and metal boats and automobiles. In building and highway construction, they are used as caulking and sealant compounds (when high chemical resistance is required), and as concrete topping compounds [97].

Epoxy resins are also used as casting compounds for the fabrication of short-run and prototype molds, stamping dies, patterns, and tooling. Finally, epoxies have a wide range of uses in the electrical business because of their excellent electrical insulation. They can be used as potting and encapsulating compounds, impregnating resins, and varnishes for electric and electronic equipment. The cost of epoxy resins is proportional to the performance of the resin that varies over a broad range, but they are usually more expensive than vinyl esters.

The typical mechanical properties of some epoxy resins (unreinforced) are given in Table 2.9. The Epon 9310/9360 system is targeted for pultrusion processing. The Epon 9310 resin system was cured with 33 phr[2] of 9360 curing agent [98]. Note the detrimental influence of temperature on modulus and strength, which is true for polymers in general. The Epon 9420/9470 is a system used in RTM, filament winding, pultrusion, and for prepreg. The 9420 resin system was cured with (a) 24.4 phr and (b) 32.4 phr of 9470 curing agent. Note in Table 2.9 that higher glass transition temperature (T_g, see Section 2.8) results in a more brittle matrix with lower elongation and mechanical properties. That means that there is always a tradeoff between high temperature application and mechanical properties when tailoring a resin system.

Moisture absorption is another consideration because of the detrimental effects that moisture has on mechanical properties of the matrix and the composite. For example, Epon HPT 1072/1062-M (100/53.2 phr) is a high temperature resin with a low water gain, about 1.4% per weight (Table 2.9).

Epoxy matrices can be used at service temperatures up to 175°C. To increase the toughness of the resin and the composite, the basic thermoset epoxy resins are toughened with additives, including the addition of thermoplastics. Service temperatures for toughened epoxies are lower than for brittle epoxies, up to 125°C. The operating temperature is always below the glass transition temperature (T_g, see Section 2.8), which is high for brittle epoxies (up to 247°C) and lower for toughened epoxies (between 76 and 185°C) [15].

2.6.4 Phenolic Resins

Phenolic resins have low flammability and low smoke production, compared to other low cost resins. Furthermore, they have good dimensional stability under temperature fluctuations and good adhesive properties. Phenolic resins are attractive for aircraft, mass transit vehicles, and as interior construction materials where outgassing due to fire must be extremely low. Phenolics are being used in sheet molding

[2]phr = parts per hundred in weight. For example, 33 phr of curing agent and 0.65 phr of accelerator means for every 100 parts of resin, 33 parts of curing agent and 0.65 parts of acelerator are added, all on a weight basis.

compound (SMC), pultrusion [99, 100], and filament winding. Processing of phenolic resins is quite different from other thermoset resins that they tend to substitute, especially polyester resins, but successful production has been achieved at good production rates and mechanical properties. Cost of phenolic resins is competitive with polyesters [101].

Other resins used to achieve low flammability and toxicity, such as bismaleimide and polystyrylpyridine, along with various flammability tests are reported in [102]. Room and high temperature data for a bismaleimide resin, a 70/30 mix of compimide 796 and TM-123 from Shell Chemical Co., is shown in Table 2.9.

2.7 Thermoplastic Matrices

A thermoplastic polymer does not undergo any chemical transformation during processing. Instead, the polymer is softened from the solid state to be processed, and it returns to a solid after processing is completed. Thermoplastics have high viscosity at processing temperatures, which makes them more difficult to process. The high shear stresses needed to make thermoplastics flow cause damage to the fibers resulting in a reduction of fiber length in the order of 10 to 100 times. Therefore, one of the major goals in the production of new thermoplastic materials and processes is to reduce viscous effects in the fluid. Since impregnation is impaired by high viscosity, special care must be taken to assure contact between the fibers and the polymer. For example, intermingled fiber and polymer strands are used to form a tow that then is processed by melting the polymer with heat. These intermediate operations and slower processing rates add cost to the final product. Thermoplastics do not require refrigerated storage and they have virtually unlimited shelf and pot life. Also, thermoplastic composites can be repaired because the transition to the softened stage can be accomplished any number of times by application of heat.

Polyether ether ketone (PEEK) is a common thermoplastic matrix for high performance applications. It has very high fracture toughness, which is important for damage tolerance of composites (Table 2.10). PEEK is a semicrystalline thermoplastic with very low water absorption (about 0.5% by weight) at room temperature, much lower than for most epoxies.

Polyphenylene sulfide (PPS) is a semicrystalline thermoplastic with an excellent chemical resistance. Polysulfone (PSUL) is an amorphous thermoplastic with very high elongation to failure and excellent stability under hot and wet conditions.

Four types of polyimides are listed in Table Table 2.10. Polyetherimide (PEI) and polyamide-imide (PAI) are amorphous thermoplastics with high glass transition temperatures. The two other thermoplastics in Table Table 2.10, K-III and LARC-TPI, are prepolymers in a solvent solution that are used to coat fibers. The coatings are then cured at about 300°C becoming amorphous polymers with high glass transition temperatures. After cure, the polymer can be heated and reshaped as a regular thermoplastic with good high temperature performance.

2.8 Creep, Temperature, and Moisture

Unlike metals, polymers creep at room temperature. Also, the mechanical properties of polymers depend strongly on temperature and moisture. Data for a large number of polymers and elastomers is reported in [103, 104]. Creep can be defined as the continuous growth of deformation with time under constant load. Creep is just one manifestation of the viscoelastic behavior of polymers, which is evidenced also as stress relaxation and as load rate effects [105].

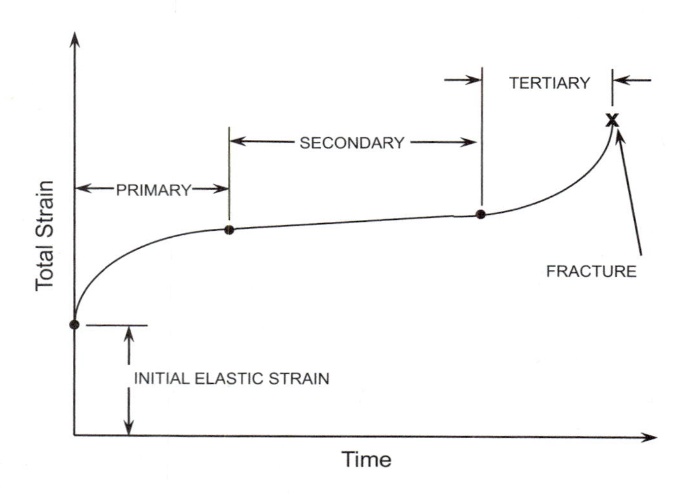

Figure 2.5: Typical results from a creep test.

A simple creep experiment consists of loading a sample with a constant tensile load (σ constant) and recording the elongation over time (ASTM D2990). Figure 2.5 is representative of such a test. There is an instantaneous elastic deformation, followed by the primary creep region where deformations grow fast. The primary creep region occurs over a short period of time. For this reason, it is of interest during processing and assembly. Anybody who has assembled wooden furniture has been instructed to retighten the bolts after a couple of weeks. This is to take out the slack caused by stress relaxation in the wood that occurs shortly after assembly. The secondary creep region is characterized by a constant slope, called *creep rate,* and it extends over a long period of time. This is the region of interest in design because it encompasses the range of time for which the structure will be in operation. If the creep rate is not a function of the stress level, the material is called *linear viscoelastic,* otherwise *nonlinear viscoelastic* (or *viscoplastic* [106]). The tertiary creep region occurs usually for high stress values and long times. It is characterized by the simultaneous accumulation of creep strain and material damage [107]. The tertiary region may not occur at all for certain materials or below certain stress levels.

A simple relaxation test consists of applying a constant tensile deformation (ΔL constant) and measuring the stress over time. Figure 2.6 is representative of such a test. The initial load is the elastic load necessary to cause the instantaneous strain.

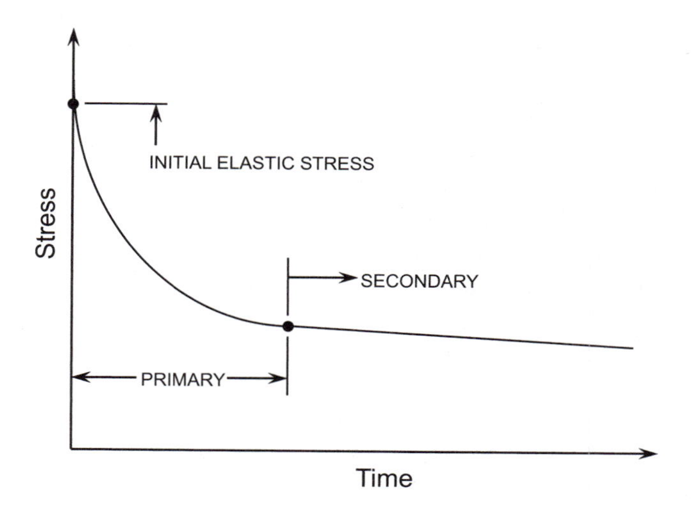

Figure 2.6: Typical results from a stress relaxation test.

After that, the sample relaxes and the load decreases.

The viscoelastic behavior of polymers is also evidenced by load-rate effects. If the load is not applied suddenly but over certain time, the stress–strain plot (see Figure 2.7) changes as a function of the loading rate. The slower the application of the load, the larger the strain. This is because longer times allow the material to accumulate more creep strain. The initial portion of the stress-strain plots is linear, while the behavior is viscoelastic. The plot becomes nonlinear at high stress levels, indicating viscoplastic behavior [106]. One of the main reasons for using fiber reinforcements is to limit the creep of polymer matrices so that structures can take loads over long periods of time. The viscoelastic behavior of composites can be predicted from the creep behavior of the matrix and the reinforcement using micromechanics [108–110], then used for predictions of time-dependent behavior of structural members [111].

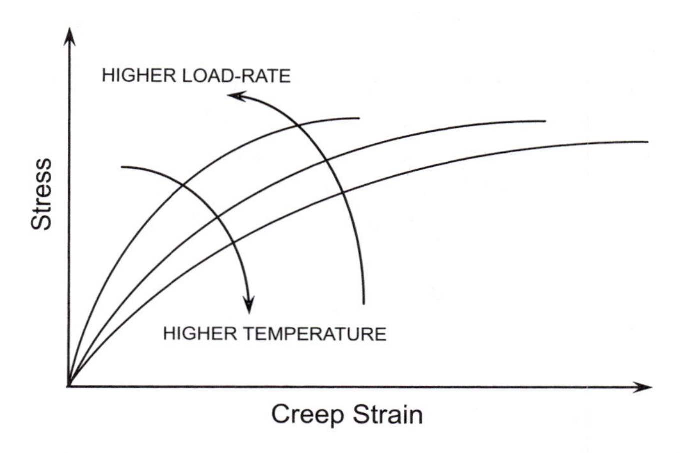

Figure 2.7: Typical results for various values of loading rate and temperature.

Polymers are aging materials. They become stiffer and more brittle with time. These are manifestations of physical aging. When the polymer is quenched rapidly from its processing temperature down to room temperature, the polymer is frozen in a nonequilibrium state containing more free volume that afforded by the equilibrium state at room temperature. As time elapses, the material state, particularly its free volume, drifts towards the equilibrium state. During this process, the material becomes stiffer and more brittle. The effects of physical aging can be characterized experimentally and the results used to predict the aging of the polymer [112–115].

Temperature has a remarkable effect on the behavior of polymers. Thermoset polymers have an amorphous structure and they show a distinctive glass transition temperature T_g, at which the material goes from a glassy (rigid) behavior to a rubbery behavior. Structural polymer matrix composites always operate at temperatures below T_g. On the other hand, thermoplastics may be amorphous or semicrystalline, exhibiting also a melting temperature T_m of the crystalline phase in addition to the glass transition temperature T_g of the amorphous phase. Polymers show a marked reduction of the modulus E_m with temperature, both above and below T_g (see Table 2.9).

Even well below the T_g, an increase in temperature accelerates the creep process; that is, creep accumulates faster at higher temperatures. Therefore, temperature has been used as an acceleration factor to predict the creep at times longer than the time span of the experiments. This is the basis for the time-temperature superposition principle (TTSP) described in [116, 117] and [118, ASTM D2990]. TTSP works great to predict the temperature effects on creep during relatively short time spans encountered in processing of thermoplastics and similar situations. However, the classical TTSP fails at long time spans typical of structural applications (years) when the effects of physical aging are noticeable. The combined effects of temperature and physical aging can be characterized experimentally and the results used to predict the long-term creep of the polymers [113–115].

If the polymer is exposed to a moist or wet environment, it will absorb moisture through a diffusion process [119, 120]. Moisture lowers the modulus, strength, and T_g of the polymer. The effect of moisture can be modeled also with the superposition principle [121]. Based on the preceding discussion, it is clear that it is very important to report the loading rate, temperature, and humidity along with the results of any experiment with polymer matrix composites. It is also important to evaluate the effects of operating temperature and moisture on the properties of the composite for design purposes. The combined effects of temperature and moisture can be characterized experimentally and the results used to predict the creep of polymer matrix composites [121].

The glass transition temperature T_g of polymers reduces as a result of moisture absorption. The following empirical equation [122] models data for some epoxy resins

$$T_{gw} = (1 - 0.1m + 0.005m^2)T_{gd} \tag{2.5}$$

where m is the moisture content by weight, T_{gd} and T_{gw} glass transition temperatures of the dry and wet polymer matrix, respectively.

In general, all polymer properties, including strength and stiffness, reduce with temperature and moisture content (see Table 2.9). The retention ratio of a given property is defined as the quotient of the property at a given temperature and moisture content divided by the reference value at room temperature and dry condition. The retention ratio of some epoxy resins can be modeled by the empirical formula [122][3]

$$\text{retention ratio} = \left[\frac{T_{gw} - T}{T_{gd} - T_0}\right]^{0.5} \tag{2.6}$$

The glass transition temperature T_g can be measured using differential thermal analysis (DTA) or differential scanning calorimetry (DSC), according to ASTM D3418 and E1356. Another measure of the softening characteristics of polymers and polymer matrix composites is their *heat distortion temperature* (HDT), which can be measured according to ASTM D648. A sample is loaded under three-point bending with a constant maximum tensile stress of either 1.82 MPa or 0.455 MPa depending on the material. The temperature is increased at a rate of 2°C/min until the deflection reaches 0.25 mm. The temperature at this point is recorded as the HDT. Although DSC equipment is very common in laboratories today, and T_g provides a better characterization of the polymer, the HDT is broadly used and reported in the literature.

2.9 Corrosion Resistance

Chemical degradation of a polymer matrix composite (PMC) may occur as a result of matrix degradation, fiber degradation, or fiber–matrix interface degradation. The polymer may degrade because of absorption of solvents, (including water), oxidation, UV radiation, and thermal degradation. Alkalis, fluorides, hot water, and hydrochloric acid may attack glass fibers if these are absorbed into the matrix. The absorption will be accelerated if the matrix contains crazes or cracks. Therefore, the chemical resistance of a PMC largely depends on the strength of the matrix (to resist matrix cracking) and the loads applied to the material. The stress level accelerates alkali-induced degradation of glass-reinforced composites because it tends to open the matrix cracks, facilitating alkalis to reach and attack the glass fiber. The fiber–matrix interface may be also attacked, resulting in lower resistance to delamination and lower intralaminar strength (Section 4.4.10).

Chemical attack is defined by four variables: extent of degradation, time, temperature, and concentration of the attacking chemical. Chemical degradation is largely influenced by the operating temperature. To a certain extent, the time-temperature superposition [116] may be applied to predict long term or elevated temperature behavior. But, beyond a certain threshold temperature, the mechanisms of chemical degradation or those of substance ingress may change and the

[3] While these empirical equations show the correct trend for typical epoxy resins, they may not be accurate for other polymers and they should be used with caution.

degradation may proceed at a much higher rate. For this reason, recommended maximum application temperatures listed in matrix selection guides [93, 94] should not be exceeded without comprehensive material testing.

Incomplete cure of thermoset polymers is extremely detrimental to the corrosion resistance of the material. Therefore, the processing method plays an important role in corrosion resistance. Higher degree of cure can be achieved at higher temperatures and allowing more time for the reaction to complete, but high temperature and longer processing time increase processing costs. The corrosion resistance of candidate thermosets must be compared at similar level of cure (or processing conditions). Data from different manufacturing processes (e.g., autoclave curing vs. room temperature curing) are not directly comparable.

The degree of cure achieved during processing can be monitored online with the process using dielectric probes, or checked afterwards. Simple tests such as Barcol hardness (ASTM D2583) and *acetone test* are routinely used for qualitative checking for incomplete cure. In the acetone test, the surface is rubbed with acetone; if it feels soft or tacky the cure is not complete. Differential scanning calorimetry (DSC) is a quantitative test of degree of cure. A DSC test is basically an instrumented postcure test of a sample of the material. The DSC equipment performs a temperature sweep of the sample and records any heat of reaction that may originate from the sample. If the initial cure was complete there will be no heat of reaction. Otherwise the area under the DSC curve represents the residual heat of reaction H_R released by the sample, indicating the need for a postcure. A DSC of a virgin resin sample gives the total heat of reaction H_T (see ASTM D3418 and D5028). Then, it is possible to measure the degree of cure $\alpha, (0 < \alpha < 1)$ of the composite sample, as $\alpha = 1 - H_R/H_T$.

Chemical resistance tests (e.g., ASTM C581) use a standard construction that simulates the chemical barrier used in tanks and other FRP components in contact to aggressive chemical substances. They can be useful as a quick check on the effects of new fluids and mixtures. However, before final matrix selection, testing of samples produced by the chosen manufacturing process with laminate construction that is similar to the actual production component should be performed to verify their ability to meet their chemical resistance requirements.

Resin selection guides from resin manufacturers (e.g., [93, 99]) can be used as part of the material selection process during preliminary design. At this stage, cost considerations and alternative structural configurations should also be considered in order to meet the functional requirements of the product.

2.10 Flammability

Organic materials, including polymers, will burn in the presence of a flame. When polymer composites are heated the polymer begins to dissociate chemically. The temperature at which this decomposition occurs and the type of fumes produced depends on the structure of the polymer. Several methods exist to assess the behavior of Polymer Matrix Composites at high temperature.

The standard ASTM E84 can be used to rate materials for their burning characteristics. The material is placed in a forced air tunnel and a controlled flame is used to ignite a fire on the surface of the sample. The time to ignition and the distance that the induced flame advances in the direction of the air stream are used to obtain a flame spread classification number (FSC) between 0 and 100. Based on the FSC number, materials are classified: class I for materials with FSC 25 or lower, class II for FSC larger than 25 but lower than 75, and class III for FSC larger than 75.

The oxygen index (ASTM D2863) is the minimum concentration of oxygen that will just support flaming combustion. ASTM E906 can be used to determine the release rate of heat and visible smoke when the material is exposed to radiant heat. Another measure of surface flammability under radiant heat is provided by ASTM E162. In addition, the optical density of the smoke generated can be measured by ASTM E662. Other tests and specifications are also used to assess various aspects of flammability [100, 102]. A design for fire performance following British standards is described in [123].

Different resins and their composites will start decomposing at elevated temperatures. Isophthalic resins can be used up to 98°C, some vinyl esters up to 135°C, bisphenol-A-fumarate and chlorendic resins up to 140°C. The maximum operating temperature of a fiber-reinforced composite may be higher or lower than that of the resin. Fibers with high thermal conductivity help distribute and dissipate the heat away from the heat source, thus improving the temperature rating of the material. Low thermal conductivity of some fibers is detrimental because higher temperatures may be reached around the heat source.

It is very important to take into account that all mechanical properties of polymers change with temperature even well below the maximum operating temperatures (Section 2.8). Also, polymers and PMC have low thermal conductivity (Section 4.3.3). This may be an advantage because exposure to high temperatures for short periods of time will not raise the temperature in the core of a thick laminate. On the other hand, low heat dissipation means that prolonged exposure to high temperature may not be well tolerated.

Local Building Codes and Standards may supersede individual ASTM standard tests. Many applications have designated tests that must be performed on actual parts or mock ups of actually parts. UL industries performs many of these tests as part of industry's standard practice.

Exercises

Exercise 2.1 *Define and explain the following terms: strand, end, yarn, and roving.*

Exercise 2.2 *Does the use of glass reinforcement help the corrosion resistance of polyester resin?*

Exercise 2.3 *Within the scope of the information provided in this chapter, select a resin to be used in a strongly acid environment.*

Exercise 2.4 *Within the scope of the information provided in this chapter, select a resin to be used in a strongly basic environment.*

Exercise 2.5 *Name two types of fibers and give the main advantage of each. The advantages have to be different so that the two types of fibers differ from each other.*

Exercise 2.6 *Name two matrices and give the main advantage of each. The advantages have to be different so that the two matrices differ from each other.*

Exercise 2.7 *Compute the cross-sectional area occupied by a 36K tow of carbon fiber, with a fiber diameter of 7 microns.*

Exercise 2.8 *Compute the cross-sectional area occupied by a roving of fiberglass with a Yield of 56 yards/lb. How many 112 Yield roving are needed to obtain the same cross-sectional area.*

Exercise 2.9 *Compute the thickness of fiber in a CSM mat made out of* $800\ g/m^2$ *of E-glass fibers.*

Exercise 2.10 *A cylindrical pipe with diameter* $d = 40\,cm$ *is subject to internal pressure* $p = 20\,MPa$ *with a variability given by a COV 20%. The pipe is free to expand along the length so that no axial stress develops in its wall, only hoop stress. The pipe is hoop wound and it is assumed that the fibers carry all the load. Therefore, the stress in the fibers is* $\sigma_f = pd/2t_f$ *where* t_f *is the thickness occupied by the fibers only. Assume that a matrix, which does not contribute at all to the strength of the material, occupies a thickness* $t_m = t_f$. *The material properties of the fibers in Tables 2.1–2.2. Assume a variability of strength given by a COV of 35% and matrix density* $\rho_m = 1.2\ g/cc$. *Designing for a reliability of 99%,*

(a) *What is the required wall thickness* ($t = t_f + t_m$) *and fiber type that yields the minimum thickness (use tables in this chapter only)?*

(b) *What is the required wall thickness and fiber type for minimum weight?*

(c) *What is the required wall thickness and fiber type for minimum cost? Use costs in $/kg for the composites given in Exercise 1.1, p. 21.*

Exercise 2.11 *Select the appropriate resin to fabricate fiber-reinforced pipe depending of the substance it will carry. Consider each case separately.*

(a) *a solution of sulfuric acid at room temperature*

(b) *an alkaline solution at room temperature*

(c) *hot water*

(d) *hot gasses, up to* $140°C$.

Exercise 2.12 *For each of the following criteria, select the appropriate resin that meets the criterion (taken separately).*

(a) *high fracture toughness*

(b) *very low flammability and low smoke production under heat and fire*

(c) *low water absorption*

(d) *maximum possible operating temperature*

(e) *good resistance to UV and appropriate for translucent panels*

(f) *low shrinkage during curing of the part for good dimensional tolerances in the product*

Exercise 2.13 *Name three parameters that must be reported along with experimental data of stiffness and strength of PMC.*

Exercise 2.14 *Do PMC operate at temperatures (a) below, (b) at, or (c) above the glass transition temperature?*

Exercise 2.15 *What class of polymers have a melting temperature?*

Exercise 2.16 *Compare the strength and stiffness data at* 149° *C (dry) of Epon 9310/9360 with an extrapolation of the room temperature data using the retention ratio.*

Exercise 2.17 *Compare the flexural modulus and strength at* 249° *C (dry) of clear cast bismaleimide 792/TM-123 with an extrapolation of the room temperature data using the retention ratio.*

Exercise 2.18 *What is the detrimental effect of an incompletely cured thermoset matrix on the life of a PMC?*

Exercise 2.19 *Explain how the matrix influences the extent of chemical attack on the fibers. Explain the effect of the applied load.*

Exercise 2.20 *What is the effect of temperature and moisture on the mechanical properties of a polymer? Use empirical formulas to plot the estimated stiffness and strength variation of Epon 9420/9470(A) from room temperature to high temperature. Based on the data in Table 2.9, what is a reasonable maximum operating temperature for this resin?*

Table 2.1: Typical properties of inorganic fibers.[a]

Fiber	Modulus [GPa]	Tensile Strength [GPa][b]	Elongation [%]	Poisson's Ratio	Longit. Thermal Expan.	Transv. Thermal Expan.	Density [g/cc]	Thermal Conduct. [W/m/°C]	Max. Oper. Temp.	Resistivity μOhm m
Glass fibers										
E-glass	72.35	3.45	4.4	0.22	5.04–5.4	-	2.5–2.59	1.05	550	-
S-glass	85.00	4.80	5.3	0.22	1.6–2.9	-	2.46–2.49	1.05	650	-
R-glass	86.00	4.40	5.2	0.15–0.26	3.3–4	-	2.55	1	-	$2\ 10^6$
C-glass	69.00	3.31	4.8	-	6.3	-	2.56	1.05	600	-
D-glass	55.00	2.50	4.7	-	3.06	-	2.14	-	477	-
Carbon fibers										
Torayca T300 (PAN)	230	3.53	1.5	0.2	-0.6	7-12	1.75	3.06	-	18
Torayca T1000 (PAN)	294	7.06	2.0	-	-	-	1.8	-	-	-
Torayca M50 (PAN)	490	2.45	0.5	-	-	-	1.91	54.43	-	8
Hexcel AS2 (PAN)	227	2.756	1.3	-	-	-	1.8	8.1–9.3	-	15–18
Hexcel AS4-D (PAN)	241	4.134	1.6	-	-0.9	-	1.77	8.1–9.3	-	15–18
Hexel IM6 (PAN)	279	5.740	1.9	-	-	-	1.76	8.1–9.3	-	15–18
Hexel IM7 (PAN)	276	5.310	1.7	-	-	-	1.78	8.1–9.3	-	15–18
Hexel HMS4 (PAN)	317	2.343	0.8	-	-	-	1.8	64–70	-	9–10
Hexcel UHM (PAN)	441	3.445	0.8	-	-	-	1.85	6.5	-	120
Thornel P55 (pitch)	379	1.9	0.5	-	-1.3	-	2	120	-	8.5
Thornel P100 (pitch)	758	2.41	0.32	-	-1.45	-	2.16	520	-	2.5
Thornel K1100 (pitch)	965	3.1	-	-	-	-	2.2	900–1000	-	-
Boron, Ceramic and Mineral fibers										
Boron[c]	400	2.7–3.7	0.79	0.2	4.5	0.2	2.57	38	315[d]	-
SCS-6	427	2.4–4	0.6	0.2	4–4.8	-	3	10	-	-
Nextel 720	260	2.1	-	-	6	-	3.4	-	1200[d]	-
Astroquartz	69	3.45	5	-	0.54	-	2.2	2	1050	10^{10}
Basalt	89-125	4.1–4.8	3.15	-	5.5–8.0	-	2.6–2.8	-	700	-
Metallic fibers										
Steel	210	4.0	-	-	11.8	-	7.9	-	-	-
Tungsten	350-410	3.2–4.2	-	-	4.5	-	19.3	-	-	-

[a]The designer is responsible for obtaining actual values. [b]See Table 2.5. [c]Compressive strength 6.9 GPa. [d]Long term.

Table 2.2: Typical properties of organic (polymeric) fibers, obtained from product literature and other sources, with some values estimated for a broad class of materials. The designer is responsible for obtaining actual values.

Fiber	Modulus [GPa]	Tensile Strength [GPa][a]	Compress. Strength [GPa]	Elon-gation [%]	Poisson's Ratio	Longit. Thermal Expan.	Transv. Thermal Expan.	Density [g/cc]	Thermal Conduct. [W/m/°C]	Max. Oper. Temp.	Resistivity μOhm m
Kevlar 29	62	3.792	-	-	-	-	-	1.44	-	-	-
Kevlar 49	131	3.62	0.72	2.8	0.35	-2	59	1.45	0.04	160[b]	-
Kevlar 149	179	3.62	0.69	1.9	-	-	-	1.47	-	-	-
Technora H	70	3	0.6	4.4	0.35	-6	59	1.39	-	160[b]	-
Spectra 900	79	2.6	-	3.6	-	-	-	0.97	-	-	-
Spectra 1000	113	3.25	-	2.9	-	-	-	0.97	-	-	-
Spectra 2000	124	3.34	-	3.0	-	-	-	0.97	-	-	-
Vectran NT	52	1.1	-	-	-	-	-	1.4	-	-	-
Vectran HT	75	3.2	-	3.8	-	-4.8	-	1.41	1.5	-	-
Vectran UM	103	3.0	-	2.8	-	-	-	1.4	-	-	-
Zylon HM	270	5.8	0.561	2.5	-	-6	-	1.56	-	-	-

[a]See Table 2.5. [b]Long term.

Table 2.3: Weibull shape parameters m of carbon fibers. The designer is responsible for obtaining actual values.

Fiber name	Condition			Weibull Modulus			Material	Ref.
	Temp [$°C$]	Treatment	Time	m	m1	m2		
HTA5131	RT				6.3 ± 0.1	4.1 ± 0.6	PAN-based carbon	
HTA5131	1800				9.1 ± 0.9	2.1 ± 0.6	PAN-based carbon	
HTA5131	2100				8.6 ± 0.8	2.2 ± 0.9	PAN-based carbon	[45]
HTA5131	2400				10.3 ± 0.4	1.9 ± 0.3	PAN-based carbon	
T1000GB				5.9			PAN-based carbon	
K13D				4.2			Pitch-based carbon	
XN-05				7.9			Pitch-based carbon	
T300				7			PAN-based carbon	[124]
M40B				6.8			PAN-based carbon	
M60JB				5.8			PAN-based carbon	
XN-90				5			Pitch-based carbon	
HT (high strength)				5.89			PAN-based carbon	
HT-O (Plasma-oxidized)				7.741			PAN-based carbon	
UHM (ultrahigh modulus)				3.85			Pitch-based carbon	[125]
UHM-O (Plasma-oxidized)				3.424			Pitch-based carbon	
P120J untreated			-	3.85			carbon	
P120J-3 min-75 W		Oxidated	3 min	3.42			carbon	
P120J-10 min-75 W		Oxidated	10 min	4.55			carbon	
P120J-3 min-100 W		Oxidated	3 min	3.65			carbon	
P120J-3 min-150 W		Oxidated	3 min	3.88			carbon	
P120J-10 min-150 W		Oxidated	10 min	3.75			carbon	[126]
P75S untreated			-	6.45			carbon	
Ribbon untreated			-	4.04			carbon	
Ribbon-3 min-75 W		Oxidated	3 min	4.77			carbon	
C230 untreated			-	5.89			carbon	
C230-3 min-75 W		Oxidated	3 min	7.74			carbon	
T300	RT			6.9			PAN-based carbon	[127]

Table 2.4: Weibull shape parameter m of various fibers and composites. The designer is responsible for obtaining actual values.

Fiber name	Condition			Weibull Modulus			Material	Ref.
	Temp [$^{\circ}C$]	Treatment	Time	m	m1	m2		
E-glass P122 1200 Tex				6.2	175	5.5	E-glass fiber	
E-glass P122 2400 Tex				3.8	97	4	E-glass fiber	[46]
Kevlar 49	RT	H2O	7 d	14.01			aramid	
E-glass	RT	H2O	7 d	6.7			glass	[128]
Carbon	RT	H2O	7 d	4.5			carbon	
GFRP (glass fiber-reinforced polyester)				5			GFRP	[129]
GF-1		untreated		3			glass	
GF-2		Silane		5.2			glass	
CF-1		untreated		4.7			carbon	[130]
CF-2		Silane		3.7			carbon	
OP-1		untreated		5.7			optical	
E-glass 357D-AA				3.301			E-glass	
E-glass P 227				2.827			E-glass	[131]
VARTM (Section 3.7)				19.7			E-glass–vinyl ester	[132]
Graphite				1.71			graphite	
Fiber of C/C composite				2.58			carbon	[133]
Matrix of C/C composite				2.6			carbon	
Fiber of C/C composite				3.41			carbon	[134]

Table 2.5: Fiber strength reduction [135].

Fiber	Strength reduction (%)
E-glass	25–50
S-2 glass	24
Kevlar 49	31
Kevlar 149	14
Carbon ASW-4	17
Carbon T-700	22
Carbon IM-6	21
Carbon T-40	21

Table 2.6: Temperature effect on strength of glass fibers.

Temperature °C	E-glass GPa	S-glass GPa
23	3.45	4.8
371	2.62	4.5
538	1.72	2.5

Table 2.7: Classification of carbon fibers according to modulus.

Denomination	Acronym	Modulus GPa
low modulus	LM	below 200
standard modulus	SM	200–250
intermediate modulus	IM	250–350
high modulus	HM	350–450
ultrahigh modulus	UHM	above 450
high tenacity	HT	above 3
super high tenacity	SHT	above 4.5

Table 2.8: Fiber architecture of various stitched fabrics [33].

Denomination	Chopped Strand Mat	0/90 (balanced)	±45 (balanced)	0(warp)	90(weft)
	g/m^2	g/m^2	g/m^2	g/m^2	g/m^2
M1500	450	-	-	-	-
C24	-	800	-	-	-
CM1808	225	600	-	-	-
CM1810	300	600	-	-	-
Q30	-	-	440	405	170
QM5620	600	1100	800	-	-
TH27	-	-	450	-	450
TVM3408	225	-	609	541	-
UM1810	300	-	-	600	-
UM1608	240	-	-	533	-
XM2408	225	-	800	-	-

Table 2.9: Typical properties of thermoset matrices. Typical data from product literature and other sources, some values estimated for a broad class of materials, the designer is responsible for obtaining actual values.

Thermosets	Tensile Modulus [GPa]	Tensile Strength [MPa]	Compressive Strength [MPa]	Shear Strength [MPa][a]	Tensile Elongation [%]	Flexural Modulus [GPa]	Flexural Strength [MPa]	CTE $\left[\frac{10^{-6}}{°C}\right]$	HDT[b] [°C]	ν	Tg [°C]	ρ [g/cc]
Polyester												
Orthophthalic	3.4	55.2	-	-	2.1	6.9	220.7	-	79.4	0.38	-	-
Isophthalic	3.4	75.9	117.2	75.9	3.3	7.6	241.4	30	90.6	0.38	-	-
BPA Fumarate	2.8	41.4	103.5	-	1.4	9	158.6	-	129.4	0.38	-	-
Chlorendic	3.4	20.7	103.5	-	-	9.7	193.1	-	140.6	0.38	-	-
Vinyl Ester												
Derakane 411-45	3.4	82.7	117.1	82.7	5-6	3.1	124	-	104	0.38	-	-
Epoxy												
8551-7	4.089	99.2	-	-	4.4	-	-	-	-	-	157	1.272
8552	4.667	100.0	-	-	1.7	-	-	-	-	-	200	1.301
9310/9360 @23°C	3.12	75.8	-	-	4.0	-	-	54	-	0.38	185	1.2
9310/9360 @149°C	1.4	26.2	-	-	5.2	-	-	-	-	-	185	1.2
9420/9470(A) @23°C	2.66	57.2	-	-	3.1	-	-	-	-	-	195	1.162
9420/9470(B) @23°C	2.83	77.2	-	-	5.2	-	-	-	-	-	155	1.158
HPT1072/1062-M												
@23°C	3.383	-	-	-	-	3.383	131	-	-	-	239	-
Bismaleimide												
796/TM-123 @24°C	3.582	-	-	-	-	3.582	132	-	-	-	260	-
796/TM-123 @249°C	-	-	-	-	-	2.48	90	-	-	-	260	-

[a]Back-calculated from composite data. [b]See the list of symbols on p. xxxiii.

Table 2.10: Typical properties of thermoplastic matrices. Typical data from product literature and other sources, some values estimated for a broad class of materials, the designer is responsible for obtaining actual values.

Thermoplastics	Tensile Modulus [GPa]	Tensile Strength [MPa]	Tensile Elongation [%]	ν	CTE $\left[\frac{10^{-6}}{°C}\right]$	Tg [°C]	Tm[a] [°C]	Process Temp. [°C]	HDT [°C]	G_{IC} [KJ/m^2]	ρ [g/cc]
PEEK	3.24	100[b]	50	0.4	47	143	343	400	160	4.03	1.32
PPS	3.3	82.7	5	0.37	49	90	290	343	135	-	1.36
PSUL	2.48	70.3[b]	75	0.37	56	190	-	300	174	2.45	1.24
PEI	3	105[b]	60	0.37	56	217	-	343	200	2.8	1.27
PAI	2.756	89.57	30	0.37	36	243	-	300	274	3.5	1.4
K-III	3.76	102	14	0.365	-	250	-	-	-	1.9	1.31
LARC-TPI	3.72	119.2	5	0.36	35	264	325	343	-	1[c]	1.37

[a]See the list of symbols on p. xxxiii. [b]Yield point. [c]Extrapolated value.

Chapter 3

Manufacturing Processes

The choice of manufacturing process depends on the type of matrix and fibers, the temperature required to form the part and to cure the matrix, and the cost effectiveness of the process. Often, the manufacturing process is the initial consideration in the design of a composite structure. This is because of cost, production volume, production rate, and adequacy of a manufacturing process to produce the type of structure desired. Each manufacturing process imposes particular limitations on the structural design. Therefore, the designer needs to understand the advantages, limitations, costs, production rates and volumes, and typical uses of various manufacturing processes. In the design of a composite structure, the material is designed concurrently with the structure. Because of this freedom, high performance structures can be designed, provided the designer understands how the material is going to be produced. The basic characteristics of manufacturing processes that are relevant to the structural designer are presented in this chapter. Hand lay-up, prepreg lay-up, bag molding, autoclave processing, compression molding, resin transfer molding (RTM), vacuum assisted resin transfer molding (VARTM), pultrusion, and filament winding are described.

Processing of polymer matrix composites involves the following unit operations:

1. Fiber placement along the required orientations;
2. Impregnation of the fibers with the resin;
3. Consolidation of the impregnated fibers to remove excess resin, air, and volatile substances;
4. Cure or solidification of the polymer;
5. Extraction from the mold; and
6. Finishing operations, such as trimming.

Various manufacturing processes differ in the way these operations are performed, but they are all executed. Some of the operations may be combined into a simple step to save time. For example, fiber placement, impregnation, and consolidation are simultaneous in filament winding. Some operations may be performed ahead of time, such as the impregnation of a prepreg that is subsequently used in hand lay-up. The variations in the way the material is processed have significant impact on the cost, production rate, quality, and performance of the final product. Each processing method has inherent advantages and limitations that affect the structural and material design.

3.1 Hand Lay-up

The hand lay-up technique, also called *wet lay-up*, is the simplest and most widely used manufacturing process. Basically, it involves manual placement of the dry reinforcements in the mold and subsequent application of the resin (Figure 3.1). Then, the wet composite is rolled using hand rollers to facilitate uniform resin distribution and removal of air pockets. This process is repeated until the desired thickness is reached. The layered structure is then cured. The emission of volatiles, such as styrene, is high as in any other open mold method. The hand lay-up process may be divided into four basic steps: mold preparation, gel coating, lay-up, and curing.

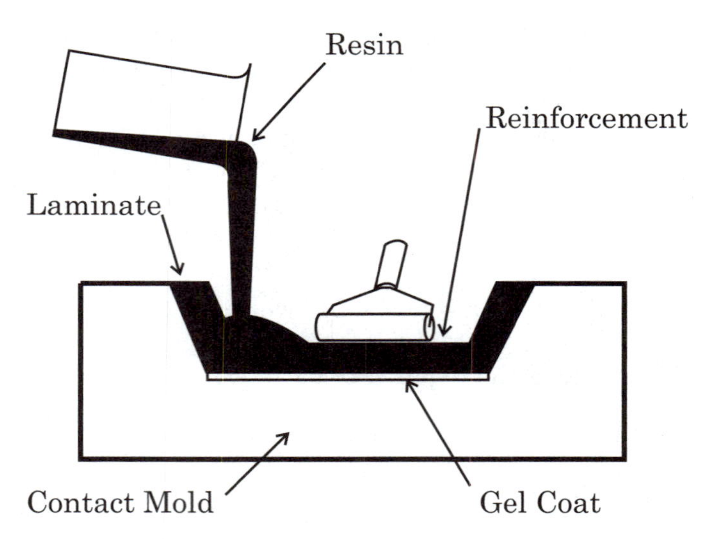

Figure 3.1: Hand lay-up.

The mold preparation is one of the most critical steps in the lay-up process. The mold may be made of wood, plaster, plastics, composites, or metals depending on the number of parts, cure temperature, pressure, etc. Permanent molds, used for long runs, are made of metals. Molds made of composites are mostly used for low volume production since they do not respond well to repeated use. The mold may be male or female type, depending on which surface needs to be smooth. A coating of release agent is applied to the mold to facilitate the removal of the finished part.

Table 3.1: Advantages and disadvantages of the hand lay-up process.

Advantages	**Disadvantages**
Large parts with complex geometries can be produced	Only one surface of the molded part is smooth
Minimal equipment investment	Quality depends on the skill of workers
Minimal tooling cost	Labor intensive
Void content under 1%	Low production rate
Sandwich construction is possible	High emission of volatiles
Inserts and structural reinforcements can be easily accommodated	Product uniformity is difficult to maintain
Parts requiring excellent finish can be easily manufactured	Long curing times at room temperature
Curing ovens are not necessary	

The release agents in common use are wax, polyvinyl alcohol, silicones, and release papers. The choice of the release agent depends on the type of material to be molded and the degree of luster desired on the finished product.

A gel coat is applied after the preparation of the mold to produce a good surface appearance of the part being molded. The coating is normally a polyester, mineral-filled, pigmented, nonreinforced lamina. This resin lamina is applied to the mold before the reinforcements. Thus, the gel surface becomes the outer surface of the laminate when molding is complete. This surface forms a protective lamina through which the fibrous reinforcements do not penetrate and the product may require no subsequent finishing operations.

The final steps involve material preparation, fiber placement, and curing. The fiber is applied in the form of chopped strand mat, cloth, or woven roving (Section 2.4). Premeasured resins and catalysts are mixed together thoroughly. The resin mixture is then applied to the fibers. Serrated hand rollers are used to compact the material against the mold to ensure complete air removal. Curing is usually accomplished at room temperature and the final molded part is removed by pulling it from the mold. Some of the advantages and disadvantages of the hand lay-up process are listed in Table 3.1 and typical applications are listed in Table 3.2.

The production rates and costs of the hand lay-up technique vary widely and depend on the fibers and matrix used, size of the part to be manufactured, and the process used. The cost of tooling depends on the number of parts to be made because higher quality molds are needed for larger runs. A new mold is constructed for every new item. The cost of equipment depends on the production rate because of the need to set up several lines working at high speed for high production rates. The cost per part is minimized by choosing the appropriate mold construction and adjusting production rates to the available equipment. Finally, the cost per part is affected by the quality required because of the need for either semiskilled or skilled workers. Cost per unit weight may be up to \$20/kg for high quality aircraft parts using glass fibers.

Table 3.2: Some applications of hand lay-up.

Applications	Products
Marine	boats, boat hulls, ducts, pools, tanks, furniture
Aircraft	rocket motor nozzles, and other aircraft parts
Structural	furnace filters, structural supports, flat and corrugated sheets, corrosion duct work, housings, pipe
Consumer	bicycle parts, truck parts

Partial automation of the hand lay-up process is accomplished by the spray-up process, which differs from the hand lay-up in the method of application of the resin and reinforcement. Continuous fiber is chopped and sprayed on the mold simultaneously with resin using a chopper gun. Spray-up is an old process, mostly used for producing parts of constant thickness: truck body parts, small boats, shower units, and custom automotive parts. Since the operator entirely controls the deposition of the spray on the mold, the quality of the product depends entirely upon operator skill. The physical properties of the product obtained are inferior and not uniform from part to part. The costs of the spray chopper guns and other special tooling add to the total cost. Complete automation of the hand lay-up technique has been tried but it was found to be very costly and inefficient, involving high equipment and tooling costs.

3.2 Prepreg Lay-up

Prepreg is a preimpregnated fiber-reinforced material where the resin is partially cured or thickened. The fibers are arranged in a unidirectional tape, a woven fabric, or random chopped fiber sheets. The basic difference between prepreg lay-up and the conventional hand lay-up is that when using prepreg the impregnation of the fibers is made prior to molding.

Prepregs are widely used for making high performance aerospace parts and complex geometries. Most prepregs are made from epoxy resin systems and reinforcements usually include glass, carbon, and aramid fibers. In most of the prepreg systems the resin content is higher than desired in the final part. The removal of this excess resin assists in removing the entrapped air and volatiles that may produce voids in the final part if not removed. This is necessary because for each 1% of voids there is 7% reduction in the intralaminar shear strength and significant reductions in the compressive strength occuring for void content above 2% [136, 137]. Lower resin content also reduces the weight and cost without affecting the strength. New prepregs are made with near net-resin contents to avoid the removal of excess resin that has become costly. The near net-resin content of the resin means that the amount of resin in the prepreg is maintained near to the resin content desired in the final part. These prepregs are made using a hot melt impregnation method that minimizes the volatiles present in the prepreg.

The prepregs are usually supplied in rolls of convenient widths (30–60 cm). They

Table 3.3: Advantages and disadvantages of using prepregs.

Advantages	Disadvantages
High fiber volume fractions	Slow and labor intensive
Uniform fiber distribution	More expensive curing equipment
Simplified manufacturing	Added cost of making prepreg

are normally cut to fit in the mold and laid up lamina by lamina until the desired thickness is reached. Since the resin is partially cured, prepregs have a limited shelf life, which is extended by storing them in freezers. An autoclave or vacuum is usually required to assist in consolidating and curing parts laminated with prepregs (Section 3.3). Some of the advantages and disadvantages of using prepreg are listed in Table 3.3

Prepregs using thermoplastic matrices are available with most of the fibers in most of fiber forms. They need to be heated to achieve tack[1] and drape[2] since they are usually stiff at room temperature. Processing of thermoplastics differs from the processing of thermoset resins. The temperatures and pressures used are usually higher for thermoplastics because the viscosity of thermoplastics needs to be lowered by heat during processing. Then, curing is replaced by simply cooling down to room temperature.

3.3 Bag Molding

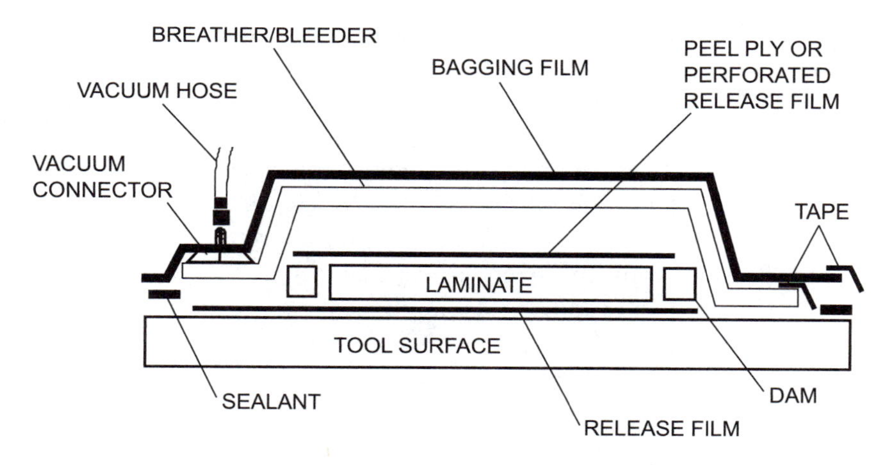

Figure 3.2: Vacuum bagging.

A uniform pressure applied to the laminate before it is cured improves consolidation of the fibers and removal of the excess resin, air, and volatiles from the matrix. Pressure is applied with the aid of a flexible diaphragm or bag (Figure 3.2).

[1]Tack is the ability of the prepreg to stick temporarily to the mold.

[2]Drape is the ability of a fabric to conform to a complex multicurved surface. See Section 2.4.3.

The laminae are laid up in a mold and resin is spread. Release film or a release agent is used on both sides of the laminate to prevent it from sticking to the mold or to the breather. Sometimes, a peel-ply is used to leave an imprint or pattern on the surface to enhance adhesive bonding (secondary bonding) at a later time. The breather/bleeder combination helps distribute the vacuum and channels the volatiles and excess resin to the vacuum port. The laminate is then covered with a flexible bag, which is perfectly sealed to the tool. Then, vacuum is applied and the part is cured with heat and pressure. By applying vacuum under the bag, the atmospheric pressure acts uniformly over the laminate. The vacuum helps withdraw excess volatile compounds, such as residual solvent, trapped air, or low molecular weight components of the resin. After the cycle, the materials become an integrated molded part shaped to the desired configuration. There are three basic methods of applying pressure to the laminate: pressure bag, vacuum bag, and autoclave processing, the latter two being the most popular methods.

Vacuum bags allow for the production of large, high quality, lower cost composite parts. The main advantages of vacuum bagging are that the vacuuming and curing equipment can be used for a variety of parts. The size of part that can be made by bag molding is limited only by the curing equipment, including the size of the curing oven or autoclave. However, the quality of the part is dependent on the worker's skill.

Bagging procedures for thermoplastics are similar to thermoset materials except that the bagging materials must be able to withstand the high temperatures. Materials like Kapton vacaloy are used because they can withstand up to 370°C. Most of the thermoplastics melt in the 260–370°C range, as compared to 120–180°C maximum operating temperature in case of thermosets. The molding of thermoplastics must also take into account the high temperatures required to obtain good flow and compaction. The higher viscosities of thermoplastics make the consolidation process very complicated and high pressures may be necessary. Thermoplastics are used to obtain improved properties in hot and wet environments and higher impact resistance. The high cost of production of thermoplastics are sometimes compensated because of production advantages, such as no need for refrigeration, reduced shipping costs, etc. Specific applications involving high impact resistance and improved mechanical properties include aerospace and medical equipment (where performance and quality are critical but cost is not such a concern), food processing, camera and watch cases, electrical and electronic equipment, etc.

3.4 Autoclave Processing

Autoclaves are pressure vessels that contain compressed gas during the processing of the composite. They are used for the production of high quality, complex parts. The method is good for large parts and moderate production quantities.

Autoclave processing of composites is an extension of the vacuum bag technique, providing higher pressure than available with a vacuum and giving greater compression and void elimination. The composite part is laid up and enclosed in a vacuum bag. Full or partial vacuum is drawn within the bag, and gas pressure greater than atmospheric is applied on the exterior of the bag. The part temperature is then raised to initiate cure of the polymer. Higher temperature also reduces the viscosity of the polymer, helping wetting of the reinforcement and consolidation of the composite.

Augmented pressure exerts mechanical forces on the unconsolidated composite, increases the efficiency of transport of volatiles to the vacuum ports, and causes increased wetting and flow of the resin. The volume of the trapped air and released volatile is reduced proportionally to the applied pressure vacuum at a given temperature. Therefore, porosity and voids are minimized. Also, the transport of volatile materials in the molten polymer to the vacuum ports is more efficient at elevated pressure.

The majority of autoclaves for composite manufacturing are cylindrical pressure vessels with domed ends, one of which is the door or entrance. The autoclaves are usually mounted horizontally on the factory floor to provide easy access to the interior. Diameter is the limiting factor in size. Large-diameter autoclaves require extremely thick walls and become very expensive. Most research and development autoclaves are about 1 m in diameter, while production autoclaves run from about 1 to 8 m in diameter.

Tools for autoclave use are usually simple male or female forms. The surface of the part facing the tool surface is more precise and has better surface finishing than the bag surface. The choice of materials for tooling is usually based on the temperatures to be encountered during curing and the production quantities required. Plaster and wood masters may be used to produce prototype parts, provided a low temperature cure is employed. For production parts, polyester, epoxy, aluminum, steel, and cast epoxy tooling are used. Large epoxy tools must be reinforced to give dimensional stability to the mold. Production tools frequently contain built-in heating elements and sensors, vacuum ports around the outside of the lay-up, and provisions for stiffening and handling.

Autoclave operations consume large amounts of energy and materials, including industrial gases (nitrogen is used for pressurization as an alternative to high-pressure compressors) and bagging materials. Autoclave operations are also labor intensive and time-consuming. The curing cycle and consolidation of a part in an autoclave is long and intensive. In the case of very thick parts, the curing cycle may have to be repeated several times to complete curing. Matched tool molding in presses is usually more economical in the case of large production runs of relatively small composite parts. For large parts exceeding the size of available presses, autoclave processing is the only choice, and for the intermediate production runs of the aerospace industry, autoclave processing is usually the most economical choice.

The production rate depends on many factors including the tools used, the size of the part, and the number of laminae. First the part has to be formed by the hand lay-up process which is time-consuming. For vacuum bagging, a pump is used to create a vacuum of about 100 KPa, at a rate of 60 to 400 liters per minute. Therefore, the size of the part determines the vacuuming time. The production rate of autoclaves is also determined by the cure profile that includes heat up, curing, and cooling time. The size of the tool used also effects the production rate because the tool has to be heated too. Small parts can be cured in about three to five hours and large parts can take as long as twelve to sixteen hours to cure.

3.5 Compression Molding

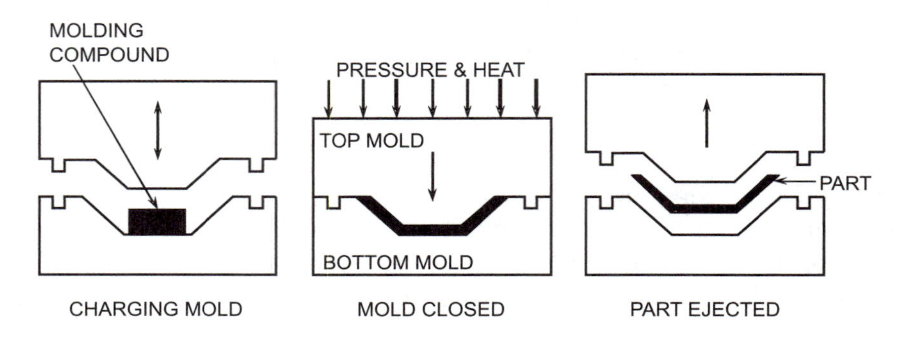

Figure 3.3: Compression molding.

The process of compression molding employs matched male and female metal dies to form the mold. A measured amount of loading compound (fiber plus resin) is charged into the mold (Figure 3.3). A hydraulic press, using heat and a relatively high pressure, is used to cure the fibers and resin by closing the male and female halves of the mold. After the material is cured, the pressure is released and the part is ejected from the mold. Post molding operations include the removal of flash and postcuring.

Compression molding is a simple and well-established process. It minimizes part setup costs, nearly eliminates material waste, reduces secondary finishing, and requires minimal labor. The process can be easily automated allowing high volume production with very good part to part uniformity. Small parts can be produced as fast as 15 parts per minute and larger pieces (e.g., automobile bumpers) at a rate up to 24 parts per hour.

The press is the most important and expensive piece of equipment in the compression molding process. Compression molding presses are usually vertical acting with movable male and female dies constructed of tool steel. The surfaces of the mold are polished and sometimes chrome-plated to increase surface hardness. Mold press sizes range from 100 ton to 4000 ton producing parts from under 1 kg to over 75 kg. Optional equipment includes preheaters and preformers that allow faster molding cycles and reduce entrapped air in the final part.

Compression molding does not allow a high content of continuous fibers. Therefore the parts are not suitable for primary structures, although they are used for some secondary structures. These compression molding components are also used when high stiffness is required, by designing ribs and flanges into the part.

The most common loading compounds used in compression molding are BMC (bulk molding compound) and SMC (sheet molding compound). Chopped fibers, fiber preforms, and prepreg are also used.

BMC is a dough-like mixture with fiber content, ranging from 20% to 50%. A combination of fillers (wood flower, minerals, cellulose, etc.) are mixed with the resins in a blade-type mixer. The reinforcement may be glass, cellulose, cotton, or other fibrous materials. This batch material is placed in a mold at 150 to 200°C and molded at about 3 to 4 MPa.

SMC has longer fibers and a higher fiber content than BMC. It can be molded into thick and thin sections maintaining maximum fiber integrity. SMC sheets contain the resin, fillers, catalyst, and preimpregnated reinforcements, which are cut into suitable size sheets or charges to be placed in the hot mold. The mold temperature is generally 150° to 200°C and it is molded at 7 to 14 MPa of pressure.

3.6 Resin Transfer Molding

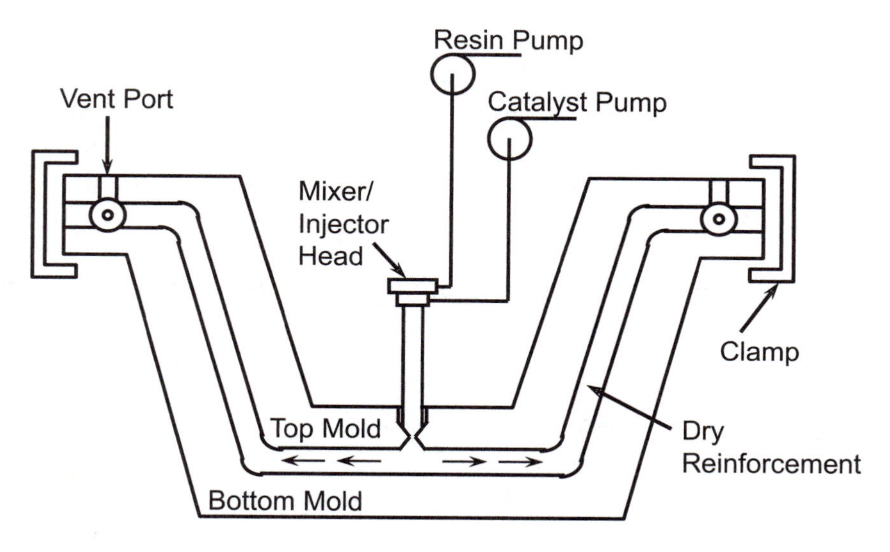

Figure 3.4: Resin Transfer Molding (RTM).

Resin transfer molding (RTM) uses a mold, with inlets to introduce the resin and outlets that allow air to escape (Figure 3.4). The fiber reinforcement is placed dry in the mold, and the mold is closed. Liquid resin is pumped into the mold through the inlet, soaking the fibers and filling the mold cavity. When the mold is full, the resin supply is removed, the mold inlets and outlets are sealed, and heat is applied to cure the resin. After the resin is completely cured, the mold is opened and the resulting composite part is removed.

The RTM process can produce large continuous fiber-reinforced composites with complicated shapes and relatively short cycle times. The process is differentiated from other molding processes in that all the reinforcement is placed dry in the mold before any resin is applied. This allows for better control over the orientation of the fibers, thus improving material properties. Also, the process is cleaner, with less volatile organic compounds (VOC) released, and less prone to defects. There are several variations of the basic RTM process.

In *preform molding*, the mold material is commonly steel. A preform is placed in the mold and a measured quantity of resin is pumped into the open tooling. The tool is then closed and compressed causing the resin to flow and wet out the preform. Cycle time for large parts of uniform thickness is often 3 minutes or less.

A *preform* is usually made by spraying chopped reinforcement onto a perforated screen and applying a vacuum until the binder, which is sprayed along with the reinforcement to hold the fibers together, has time to cure. Binder is the curing agent used to hold the fiber mesh or fiber preform together before the actual resin is transferred through the system. Binders are usually 75% water based by weight with a corresponding bonding agent that will hold the fiber preform together. The type of binder used depends on the type of fiber and resin to be used in the process, so that the binder does not create an interfacial problem between the fiber and matrix.

Structural reaction injection molding (SRIM), uses a preform that is placed in the mold before resin is introduced. Since the resin and the initiator are highly reactive, they are stored in two separate tanks. Resin and initiator from the two tanks are injected into a mixing chamber, and the mixture pumped into the mold, where it cures rapidly. Cycle times of one minute are possible for production of relatively small parts.

In *flexible resin transfer molding* (FRTM), rather than using a rigid mold, two elastomeric diaphragms are used to contain the dry fibers. Resin is drawn through the flexible mold by means of a vacuum applied at the mold outlet. After the fibers have been impregnated with resin, the flexible mold is then shaped to form the desired part. Since the reinforcements are placed on a flat diaphragm, the time and labor associated to traditional RTM lay-up on a contoured mold are virtually eliminated [138].

Physical properties of molded RTM components tend to be very consistent. This is at the cost of a relatively high waste factor around the part perimeter. Corners and edges tend to be resin rich making it difficult to achieve uniform fiber content through the part. When low-cost materials are used in mold construction, mold pressures must be low, resulting in slow fill times and limited fiber contents. Poor temperature resistance of the mold, coupled with poor heat transfer, restrict the resin chemistry to slow cure times with minimum exothermic to prevent resin degradation or tool damage. Reinforcements are generally cut and placed in the mold by hand for each molding, which increases cycle time considerably for complex parts. A major limitation is the fact that the mold design is very critical and requires great skill. Mold design, particularly inlet and outlet design and location, is increasingly performed using software packages to simulate the flow of resin through the mold cavity [139].

The cost and time to build a preform for resin transfer molding are significant.

The absence of reinforcement at part edges may be a limitation if ribs and bosses are required in a design. Ribs and bosses must be loaded individually in the tool cavity, and maintaining reinforcement at the part edge while avoiding resin richness at corners of the part can be difficult. Scrap losses also may be more costly as component integration increases, and replacement cost can be significant if a large component is defective. Producing mold-in holes are usually difficult in RTM. Routinely, parts are limited to about 12 mm thickness because of the difficulty in resin transfer through large mediums. Tolerance for RTM are difficult to keep but some applications can keep tolerance as close as ± 0.2 mm. Reinforcement movement during resin injection is sometimes a problem causing leakage and nonuniform resin transfer.

Production rates strictly depend on the size and type of part being produced. The RTM production area can be configured to operate a variety of molds simultaneously. An RTM machine can process 45 kg/min and 2–8 parts/hr while it may take 2 hrs to produce a similar part by spray-up (Section 3.1) [140]. An example of a high volume part is a balsa-core sandwich panel for wet electrostatic precipitators. Each panel is 3.65 m by 3.0 m with thickness ranging from 3.8 to 45.7 cm. The weight of the panel is 272 kg and the production time is one week [140].

The production costs of RTM are lower than processes using prepreg because the cost of raw materials (fiber and resin) are significantly lower. In some cases the cost of RTM has been found to be as low as 1/3 of using prepreg and vacuum bag process to produce the same part. For a particular example using the same part, the production cost of RTM was estimated at 80% of the cost for hand lay-up [141], including equipment depreciation, material costs, scrap rate, and labor costs. Similar savings are reported in [142]. Equipment costs for RTM may be as low as 35% of the cost for compression molding and autoclave processes. In general, the equipment cost for RTM is usually lower than other composite processes except for hand lay-up. However, RTM has a fairly high labor cost compared to SMC and Injection Molding.

Vacuum assisted resin injection molding, also called *vacuum assisted resin transfer molding* (VARTM) has become very popular during the last decade. The process is explained in detail in the next section.

3.7 Vacuum Assisted Resin Transfer Molding

In Vacuum Assisted Resin Transfer Molding (VARTM), the vacuum is applied to the outlet of the mold and the resin is drawn into the mold by vacuum only. A schematic of the process is shown in Figure 3.5. The pressure applied to the impregnated reinforcement is due to the pressure differential between the vacuum applied and the atmospheric pressure action on the exposed surface of the vacuum bag. Since vacuum is applied instead of pressure, half of the mold may be replaced by a vacuum bag. Resin flow can be assisted by microgrooves built into the mold or into a distribution medium placed beneath the vacuum bag. Some aspects of this technology are patented and commercialized under the name SCRIMP™ [143]. Since the pressure differential is much lower than the pressure used in conventional

RTM, and curing is most commonly done at ambient temperature, the cost of the mold can be reduced substantially; heavy steel molds needed in pressurized RTM can be replaced by lighter molds made of wood, epoxy, or light-gauge steel. Cycle times for this process range from minutes to hours for large complex parts. Typical applications of VARTM include large, complex parts such as boat hulls and so on.

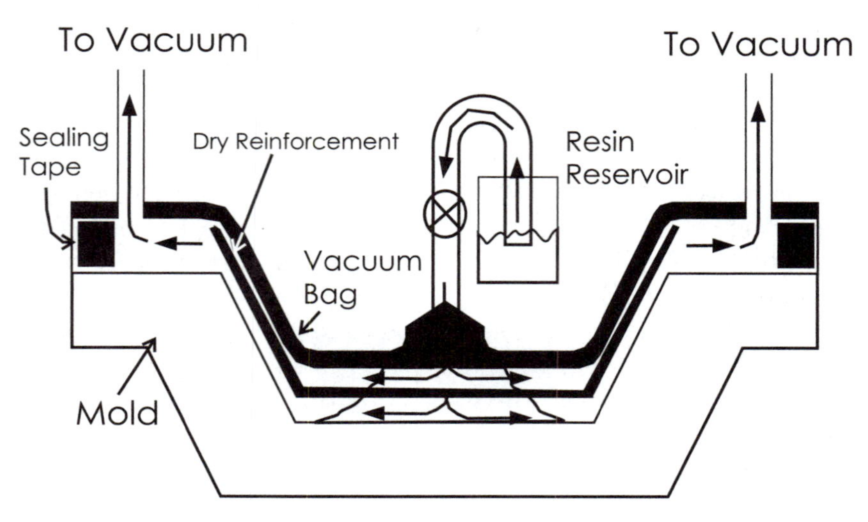

Figure 3.5: Vacuum Assisted Resin Transfer Molding (VARTM).

A typical sequence of operations in the VARTM process is as follows:

- lay-up all laminae of dry fabric or a preform on the tool.
- Optional. Cover with a highly permeability resin distribution medium on top of the dry fabric to help spread the resin quickly over the entire part. In this way, flow through the thickness is aided by gravity.
- Cover with a vacuum bag.
- Place one or more resin inlet ports strategically so that resin drawn by the vacuum pump thoroughly impregnates the entire lay-up, leaving no dry spots.
- Evacuate the entrapped air with a vacuum pump.
- Continue to apply vacuum to draw liquid resin from an external reservoir into the lay-up. At this stage the resin should impregnate all of the fabric, leaving no dry spots.

Epoxy, polyester, vinyl ester, and phenolic resins can all be used for VARTM, and bismaleimides can be used if the tooling (mold, vacuum bag, etc.) can sustain the elevated process temperature required. Since resin has to flow through the thickness of the fabric lay-up, woven fabrics cannot be too crimped,[3] not too tight.

[3]Crimp is a measure of the yarn undulation or waviness in a woven fabric. See Section 2.4.3.

Rather loose, drapeable[4] fabrics work better. Stitched materials are ideal for this process because the tows are sufficiently loose to allow the resin to seep through. Honeycomb core cannot be used because the cells would fill with resin. Foams need to be sturdy enough to resist the pressure (equal to the vacuum applied).

High fiber volume fraction can be obtained, as high as 60%, with very low void content. Very low volatile organic compounds (VOC) are released to the atmosphere the resin is enclosed at all times, which is a significant advantage when VARTM is compared to wet lay-up methods. Faster lay-up can be achieved with lower labor cost because of the ease of lay-up of dry fabrics when compared with wet lay-up. Final parts made by this process have good dimension tolerances, which are comparable to RTM and prepreg manufacturing, and better than hand and wet lay-up.

Since VARTM is normally used for large parts, any unimpregnated area may result in very expensive scrap. The user of VARTM must use trial and error and or simulation software [144] to learn how to to prevent unimpregnated areas. Furthermore, the tool has to be sturdy enough to withstand the weight of fabric plus resin plus tooling (vacuum bag, etc.).

3.8 Pultrusion

Pultrusion is a continuous manufacturing process used to manufacture constant cross-section shapes of any length. Pultrusion is a low cost process because it achieves direct conversion of continuous fibers and resin into a finished part. The fibers are continuously impregnated and pulled through a heated die, where they are shaped and cured.

In the simplest pultrusion line (Figure 3.6), roving and mat are dispensed from the creel and mat racks and guided through preforming guides. The preforming guides position the reinforcements in the appropriate locations in the cross-section of the product, as specified by the design. The reinforcements enter dry into the injection chamber where they are wetted by resin supplied under pressure. Frequently, the injection chamber is an integral part of the die. The cross-section of the die gives the final shape to the product. As the wet reinforcement travels through the die, curing takes place aided by the heat supplied by the in-line heaters. As it cures, the composite shrinks and separates from the walls of the die, leaving the die as a finished product. The cured part is then pulled by reciprocating pullers, synchronized to provide a constant speed. The product is therefore continuously produced in virtually infinite length. A moving cutoff saw clamps itself to the moving product whenever a preset length of product is available, thus cutting the part without stopping the process. The cutting length is chosen as to facilitate transportation of the pultruded product.

The reciprocating pullers may be substituted by a caterpillar puller where two rubber belts clamp the part while continuously pulling it by friction. A resin bath can be used to wet the reinforcements before entering the die. In this case the

[4]Drape is the ability of a fabric to conform to a complex multi-curved surface. See Section 2.4.3.

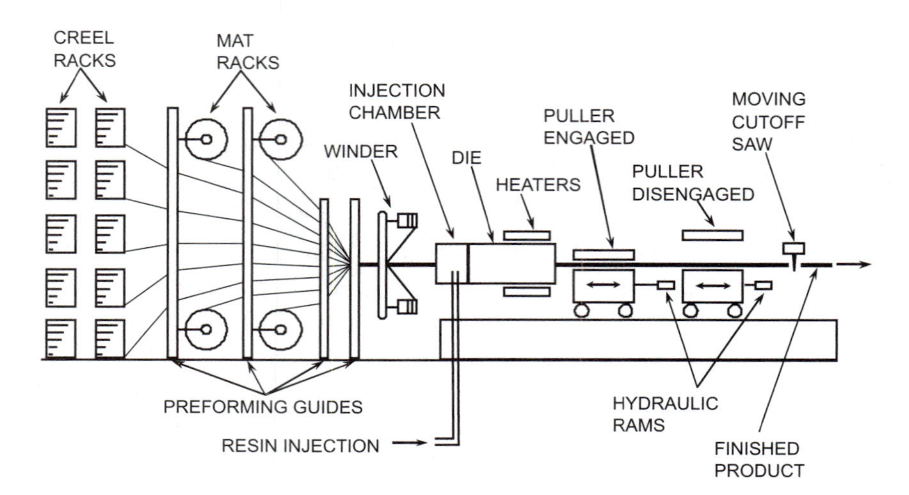

Figure 3.6: Pultrusion.

injection chamber and associated pressurization tank are not used. However, using an injection chamber has the advantage of virtually eliminating the volatile content (VOC) that otherwise would emanate from an open resin bath. Styrene emissions are drastically reduced, as low as 1% of an open bath operation. Also, the time for start-up and shutdown of the pultrusion line can be significantly reduced. Using an injection chamber, the dry fibers can be positioned with accuracy at the inlet of the die that translates in better uniformity of the material. Finally, the injection pressure can be adjusted and continuously monitored to improve wet-out of the fiber lay-up.

Both closed sections (e.g., box-beam) and open sections (e.g., I-beam) can be produced, but it is easier to produce closed sections. Closed sections, such as a box beam, are fabricated using a mandrel cantilevered behind the entrance to the die. The pultrusion line also can be fitted with a rotating winder to apply reinforcements at an angle (usually $\pm\theta$) around the product. This is commonly used to fabricate pipe and drive-shafts. Other additions to the line consist of fiber preheaters and radio-frequency (RF) heaters, especially for thick parts or when a thermoplastic matrix is used. Thermoplastics can be fed in the form of fibers commingled with the reinforcements. Another option is to feed the line with prepreg but this adds one more operation, the manufacturing of prepreg, and thus additional cost.

Operational costs of pultrusion are low. The major cost of the process is in equipment, the chrome plated dies, and the design and tune-up of the guiding system. For these reasons, pultrusion is ideally suited for high volume applications. Any type of fibers can be used but glass dominates the market because of cost. Special formulations of various resins have been developed for pultrusion, including polyesters, vinyl esters, epoxies, and phenolics. Even thermoplastics have been pultruded, but polyesters and vinyl esters dominate the market because of cost. The main fiber form used is roving and continuous strand mat (CSM), the later used to provide transverse strength and to facilitate production. However, more elaborate fiber forms, like stitched bidirectional materials, are also used. Except for the case of thermoplastics and specialty applications, no preimpregnated fibers are used, contributing to a lower cost.

Pultrusion has some constraints regarding fiber orientation and content. A minimum of roving or longitudinal fibers must be used to be able to pull the product. Bidirectional mats must be stitched to prevent warping of the fibers near the edges. Fiber volume fraction can seldom exceed 45%, with 30% being a more typical value. Fillers are commonly added to reduce the cost of the resin and to keep the solid content at a constant value, thus facilitating processing. A minimum solid content (fiber plus filler) is necessary for successful pultrusion without scaling and sloughing (about 70–75% by weight). The required fiber volume may be lowered if the difference is replaced by fillers, within certain limits. The thickness of solid pultruded walls are limited to about 12 mm when using standard conduction heaters because of curing limitations and intralaminar cracking. Furthermore, the void content may be difficult to control in thick sections.

If the die is not heated, the composite leaves the die uncured allowing for additional operations on the uncured part. In this case, curing can be accomplished in a tunnel-oven downstream from the die. In the tunnel-oven process, the rovings and mats are drawn through a resin bath and then through a sizing brush, which removes entrapped air and excess resin. The wet rovings are then molded into the desired size and shape by a die (unheated). The pultruded part is then pulled through a tunnel oven and cured. The cured part is pulled with a caterpillar or two alternating pullers. The tunnel-oven process has become less used today as it has limitations in the shapes it can produce and a need for a secondary finishing. However, it is used for producing reinforcing bars for concrete. After the uncured rod comes out of the die, it is wrapped at an angle with roving, or even coated with sand. Then, it is cured in the tunnel-oven. In this case, the roughness of the surface is an advantage, allowing the bar to achieve the desired bond with concrete.

Only straight, constant cross-sections can be produced with the die-cure method. Pulforming, however, can be used for curved shapes, like leaf springs. In step-molding or pulforming, rovings and mats are drawn through an impregnating bath. The material is then pulled into a mold, the mold is closed, and heat is applied. Once this portion of the product is cured, it is drawn out of the mold, and the next section is pulled into the mold. An advantage of this system is that it allows for the production of a nonuniform cross-section. However, the process is slow and it is difficult to produce large cross-sections.

Pultrusion rates of production vary widely depending upon the type of machine being used and the type of cross-section being produced. On the average, a standard beam cross-section can be produced at about 2 meters per minute, while a panel production line can produce approximately 20 m^2 per minute.

3.9 Filament Winding

Most shapes generated through this process are surfaces of revolution, such as pipes, cylinders, and spheres. In filament winding, continuous reinforcements, such as roving, are wound onto a mandrel until the surface is covered and the required thickness is achieved. The process uses raw materials, fiber and resin, in a fairly automated process with low labor, thus contributing to a low production cost. The preprogrammed rotation of the mandrel and horizontal movement of the delivery eye produce the helical pattern depicted in Figure 3.7, which is the simplest mode of operation of an helical winding machine.

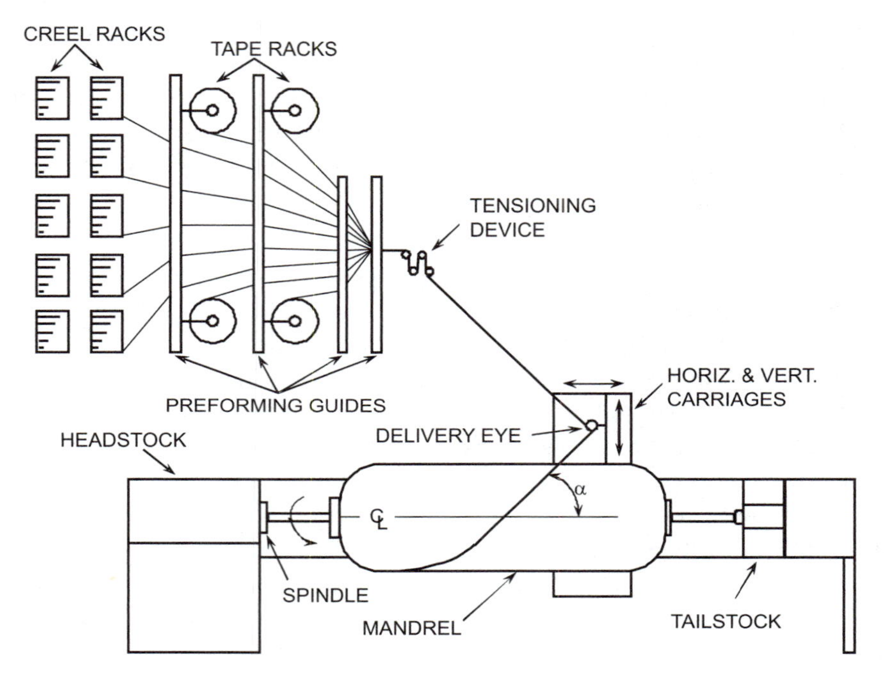

Figure 3.7: Filament winding.

There are two basic types of winding machines: helical and polar. The helical winding machine is similar to a lathe. The mandrel rotates continuously while the delivery eye moves back and forth. The rotational speed of the mandrel and the linear speed of the delivery eye can be adjusted to produce any fiber orientation between 5° and 90°, the latter called *hoop winding*. Several back-and-forth travels of the carriage are needed to complete a lamina covering the mandrel. Such a lamina is always a two-ply balanced laminate at $\pm\theta$. The fiber reinforcements are delivered from creel and tape racks, and through a tensioning device or brake that can be adjusted to control the tension in the reinforcement. Next, the reinforcement goes through a resin bath where it picks up resin. Then, the wet reinforcement is delivered through the delivery eye that is mounted on a carriage. In addition to the spindle rotation, the carriage and delivery eye can move in a number of ways designed to help place the reinforcement along complicated contours. A helical winder with three possible movements, called *axes*, is depicted in Figure 3.7, but machines with up to six axes are available. A six axes machine independently controls its spindle rotation, horizontal carriage feed, radial carriage position, delivery eye angle and yaw, and vertical carriage feed. Winders employing fewer axes are used for simple parts such as golf shafts and larger number of axes are used for more complex components such as windmill blades.

A helical winder naturally produces a geodesic path, that is the path followed by a string under tension on the surface of the mandrel. An example of such a path used for winding a cylindrical vessel is shown in Figure 3.7. For more complex shapes, the winder can be programmed to deviate from the geodesic path. In this case, the roving tends to slip back into the geodesic path. The difference between the geodesic and the set path is the slip angle, which is limited by processing conditions. A string free to slip, stretched between two points on the convex side of any surface, follows a geodesic path. If the shape of the surface can be designed so that the geodesic path coincides with the resultant of the hoop and meridional forces (N_x and N_y in Chapter 12), the shape is called a *geodesic dome.* The design of such a shape is used for the end domes of pressure vessels, and it is described in [48].

Polar winders are used to produce spherical vessels or cylindrical vessels with length/diameter ratio less than 2.0. A polar winder is mechanically simpler, thus less expensive, and faster than a helical winder. It consists of an arm that rotates around the mandrel delivering the roving into a planar path. The mandrel is stepped slowly so that the arm covers its surface. Except for the perfect sphere, the planar path always has a slip angle with respect to the geodesic path that limits the applicability of polar winding to nearly spherical shapes.

After winding the part is moved to a gas fired or electric oven, thus freeing the winder for winding another part. The need for continuous tension of the fiber around the mandrel virtually prevents the manufacturing of shapes with negative curvature ($r_2 < 0$ in Figure 12.5), unless special fixtures are used. Small radii of curvature are also a problem because fiber breakage and sudden changes in curvature tend to create resin rich zones. The need for a mandrel and for its removal after the composite is cured also limits the shapes that can be wound. In general filament winding finds most of its applications in surfaces of revolution.

Several types of mandrels have been developed to facilitate removal. The easiest alternative used for some pressure vessels is to use a metallic liner as a mandrel and leave the liner as an integral part of the end product. This is sometimes required to prevent the leakage of gasses by diffusion through the composite wall. Collapsible mandrels are made of segments that can be disassembled after the part is cured. These are the most expensive mandrels, and thus they are used for large volume productions. A soluble sand mandrel is made of sand and polyvinyl alcohol. The mixture is cast in two or more parts, that when assembled, give the desired shape. Once the composite is cured, the mandrel is dissolved by injecting hot water. Plaster molds are used only for prototypes or low runs of large parts because they are labor intensive and damage to the part may result during removal.

Besides using wet reinforcements, it is possible to use prepreg or wet rerolled material, but these options invariably add more operations and cost to the product. Using wet reinforcements, fiber placement, impregnation, and consolidation are achieved simultaneously. The wet reinforcement is placed on the mandrel under tension, thus compacting the material previously wound. The maximum tension that can be used is a function of the fiber strength and the feed rate being used. The consolidation is not as good as that obtained with an autoclave resulting in higher void content and somewhat lower mechanical properties. However, not needing an autoclave is advantageous because it reduces the cost through lower capital expense and lower processing time. Furthermore, large parts that would not fit in any available autoclave can be fabricated by filament winding.

The maximum thickness that can be wound is limited by fiber slippage and wrinkling under the pressure of new laminae on top. When the thickness is large, it may be necessary to stop winding and let the part cure partially, until the resin gels, before adding more laminae. This slows the process resulting in additional cost. Therefore, as with virtually all processes, relatively thin laminates are preferred from a production point of view.

The major limitations of filament winding are size restrictions, geometric possibilities, the orientation of the fibers, and the surface finish of the final product. Void content may be high since no vacuum or autoclave is used and the resin cures at low temperature.

Production rates for filament winding processes vary greatly because the size of the part and the mandrel type dictate the amount of time needed to setup and remove a part from the winding machine. If setup and removal time are not considered, production rate is dictated by the feed rate at which fibers are wound onto the mandrel. Feed rates vary according to the strength of the fiber used, typically 0.6–1.2 m/s for production using a wet fiber setup.

Exercises

Exercise 3.1 *Explain polar winding and its limitations.*

Exercise 3.2 *What are the differences between SMC and BMC?*

Exercise 3.3 *With reference to autoclave and vacuum bagging: (a) What is a release film?, (b) What is a peel-ply?, and (c) What is a breather?*

Exercise 3.4 *Name four manufacturing processes. For each give its main advantage and disadvantage.*

Exercise 3.5 *What manufacturing process would you recommend to produce a cylindrical tube for each of the following configurations of the reinforcement:*

(a) only unidirectional fibers along the length of the tube

(b) only hoop fibers

(c) a combination of unidirectional and ±45° (angle measured with respect to the axis of the tube)

(d) *only chopped or continuous strand mat reinforcement*

(e) *a short length of tube fabricated as a prototype with arbitrary fiber orientations*

Exercise 3.6 *What manufacturing process would you recommend to produce a large quantity of the following parts while minimizing cost?*

(a) *automotive door panel where structural performance is not critical but surface finish must be excellent*

(b) *aircraft door panel where structural performance is critical but surface finish is not important*

(c) *a few prototype samples of both (a) and (b)*

Exercise 3.7 *Same as exercise 3.6 but you are required to use a thermoplastic matrix. What additional considerations must be accounted for in this case?*

Chapter 4

Micromechanics

Micromechanics is the study of composite materials taking into account the interaction of the constituent materials in detail. Micromechanics allows the designer to represent an heterogeneous material (see Section 4.1.3) as an equivalent homogeneous material, usually anisotropic (see Section 4.1.4 and Figure 4.1).

Micromechanics can be used to predict stiffness (with great success) and strength (with lesser success). There are several approaches, increasingly complex, to derive micromechanics formulas. To fix ideas, the simplest and most intuitive approach (the mechanics of materials approach) is introduced first. Then, more accurate formulas and comparisons with experimental data are presented.

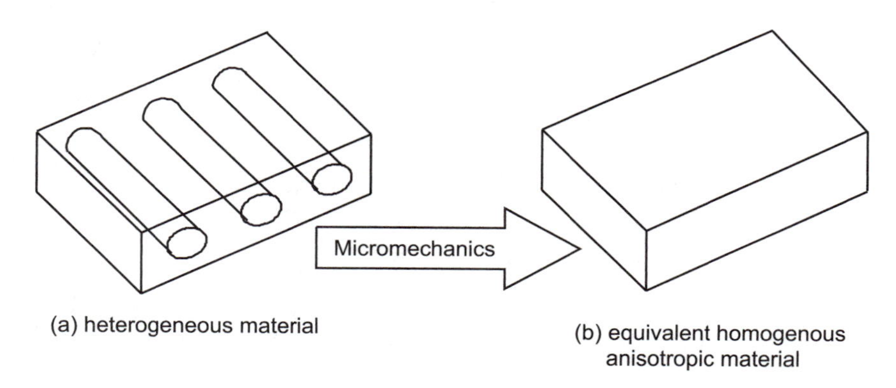

Figure 4.1: Micromechanics process.

Composite materials are the first class of materials that are designed concurrently with the structure. When designing with metals, the only choices are the type of alloy and the thermal treatment, followed by the design of the geometry. Composite materials are produced as combinations of various fibers with various matrices. The designer can change fibers, matrices, the relative amount of each constituent, and the geometry of the part, simultaneously. The relative amount of fiber and matrix is expressed in terms of fiber and matrix volume fractions, which can be varied over a broad range during manufacturing.

4.1 Basic Concepts

Basic concepts and definitions used in the rest of this chapter are described in this section.

4.1.1 Volume and Mass Fractions

The properties of a composite are controlled by the relative volume of fiber and matrix used. The fiber volume fraction is defined as

$$V_f = \frac{\text{volume of fiber}}{\text{total volume}} \tag{4.1}$$

The amount of matrix by volume is the matrix volume fraction, defined as

$$V_m = \frac{\text{volume of matrix}}{\text{total volume}} \tag{4.2}$$

Since the total volume is the sum of the fiber volume plus the matrix volume

$$V_f + V_m = 1 \tag{4.3}$$

The amount of fiber by weight in the composite is the fiber weight fraction

$$W_f = \frac{\text{weight of fiber}}{\text{total weight}} \tag{4.4}$$

The amount of matrix by weight (or mass) is the matrix weight fraction

$$W_m = \frac{\text{weight of matrix}}{\text{total weight}} \tag{4.5}$$

Since the total weight is the sum of the fiber weight plus the matrix weight

$$W_f + W_m = 1 \tag{4.6}$$

The mass of any material (fiber, matrix, and composite) is equal to the product of the density times the volume. Therefore, the density of the composite can be computed as

$$\rho_c = \rho_f V_f + \rho_m V_m \tag{4.7}$$

The volume of any material (fiber, matrix, and composite) can be written as the mass divided by the density. Therefore, the density of the composite can be computed also as

$$\frac{1}{\rho_c} = \frac{W_f}{\rho_f} + \frac{W_m}{\rho_m} \tag{4.8}$$

In design of composite structures, volume fractions are used because they enter directly into the computations of stiffness, etc. But during processing, weight fractions are used because it is much easier to weigh components to be mixed than to measure their volumes. The conversion from one to another is simple if the density of the composite has been computed. In this case, by writing the definition of weight fraction in terms of density times volume, the relationship between volume and weight fraction becomes

$$W_f = \frac{\rho_f}{\rho_c} V_f$$
$$W_m = \frac{\rho_m}{\rho_c} V_m \tag{4.9}$$

When more than two constituents enter in the composition of the composite material, the previous formulas become

$$\rho_c = \sum_{i=1}^{n} \rho_i V_i$$
$$\frac{1}{\rho_c} = \sum_{i=1}^{n} \frac{W_i}{\rho_i}$$
$$W_i = \frac{\rho_i}{\rho_c} V_i \tag{4.10}$$

where n is the number of constituents. The computed composite density may differ from the experimental value $\rho_{c,\mathrm{exp}}$ because of voids. The volume fraction of voids V_v can be measured experimentally using [118, ASTM D2734], or it can be estimated as

$$V_v = \frac{\text{void volume}}{\text{total volume}} = \frac{\rho_c - \rho_{c,\mathrm{exp}}}{\rho_c} \tag{4.11}$$

where ρ_c is computed with (4.7) and $\rho_{c,\mathrm{exp}}$ is measured experimentally [118, ASTM D792]. The fiber weight fraction can be measured by weighting a composite sample, then removing the resin and weighing the fibers. The resin can be removed by matrix digestion [118, ASTM D3171], solvent extraction [118, ASTM C613], or by burning the resin in an oven.

It is also possible to determine the fiber volume or weight fraction analytically if the fiber architecture is known. For a unidirectional material

$$V_f = \frac{\eta \; \mathrm{TEX}}{10,000 \; \rho_f \; t_c} \tag{4.12}$$

where η is the number of tows per unit width (tows per cm) perpendicular to the fiber direction, TEX is the weight of the tow (in g/km), ρ_f is the density of the fibers in g/cc (cc = cm^3), and t_c is the thickness of the composite lamina in mm.

For a continuous strand mat (CSM), textile, or chopped strand mat

$$V_f = \frac{w}{1,000 \rho_f t_c} \tag{4.13}$$

where w is the weight in grams of a square meter of mat [g/m^2], ρ_f is the density of the fibers in g/cc and t_c is the thickness of the composite lamina in mm. Conversions to the U.S. customary system are given in the Appendix.

Example 4.1 *Compute the fiber volume fraction of a unidirectional lamina with 5 tows per cm, each tow is E-glass of 113 Yield, and the final lamina thickness is* 2.0 *mm.*

Solution to Example 4.1 *Using the conversion factors from the Appendix*

$$TEX\left[\frac{g}{km}\right] = \frac{496,238}{Yield\left[\frac{yd}{lb}\right]} = \frac{496,238}{113} = 4391\ g/km$$

From Table 2.1, the density of E-glass is 2.5 *g/cc. Then, using (4.12)*

$$V_f = \frac{(5)4391}{10,000(2.5)2.0} = 0.439 = 43.9\%$$

Example 4.2 *Compute the fiber volume fraction of a laminate* 12.7 *mm thick built with 22 laminae of double bias [±45] stitched fabric (Hexcel DB243) having* 23.7 *oz/sqyd nominal weight of E-glass. Neglect any CSM backing material.*

Solution to Example 4.2 *First convert the weight per unit area to metric units*

$$w = 23.7\frac{oz}{yd^2}\frac{28.35g/oz}{(0.9144m/yd)^2} = 23.7(33.9) = 803.5\ g/m^2$$

Then, using (4.13)

$$V_f = \frac{803.5(22)}{1000(2.5)12.7} = 0.556 = 55.6\%$$

4.1.2 Representative Volume Element

The fiber volume fraction V_f can take any value over a broad range allowed by the particular manufacturing process (e.g., by hand lay-up between 0% and 60% by volume). To test for the properties (mechanical, thermal, etc.) of all these possible material combinations is impossible; thus, the importance of micromechanics becomes apparent.

Micromechanics is used to predict the properties of the composite material based on the known (tested) properties of the constituents (fiber and matrix). Micromechanics analyzes the material considering the properties of the matrix, the fiber, and the geometry of the microstructure. This is done once and for all, so that the designer is not concerned with the microstructure during the structural design process.

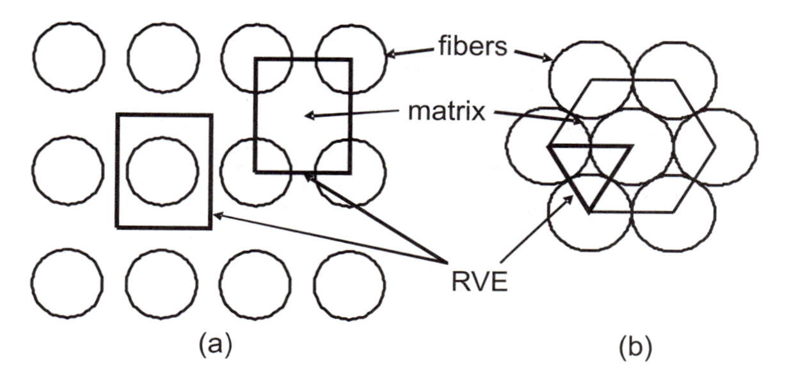

Figure 4.2: Typical representative volume elements (RVE) for (a) rectangular packing array, and (b) hexagonal packing array.

To avoid considering all the fibers included in a composite during the derivation of the equations, micromechanics uses the concept of a representative volume element (RVE). An RVE (Figure 4.2) is the smallest portion of the material that contains all of the peculiarities of the material; therefore, it is representative of the material as a whole. The stresses and strains are nonuniform over the RVE because the composite is a heterogeneous material (see Section 4.1.3). However, the volume occupied by the RVE can be replaced by an equivalent homogeneous material (see Section 4.1.4) without affecting the state of stresses around the RVE (see Figure 4.3). In fact, the state of stresses in the rest of the structure will not change as long as the designer looks at a scale larger than the dimensions of the RVE. The fiber spacing and the lamina-thickness are typical RVE dimensions (Figure 4.2).

4.1.3 Heterogeneous Material

Heterogeneous materials have properties (mechanical, etc.) that vary from point to point. Consider a cross-section of a tree, where each growth ring is different from the rest. The lighter rings (summer wood) are softer and the darker rings (winter wood) are stiffer. In contrast, a homogeneous material (e.g., steel) has the same properties everywhere.

4.1.4 Anisotropic Material

Isotropic materials (e.g., aluminum) have the same properties in any direction. Anisotropic materials have properties (mechanical, etc.) that vary with the orientation. Anisotropic materials may be homogeneous but the properties change depending on the orientation along which the property is measured. A typical example is the modulus of elasticity of wood, which is higher along the length of the tree and lower across the growth rings. Although wood is a heterogeneous material (each growth ring is different from the rest, see Section 4.1.3), when looking at a large piece of wood, the material is assumed homogeneous. That is, for simplicity,

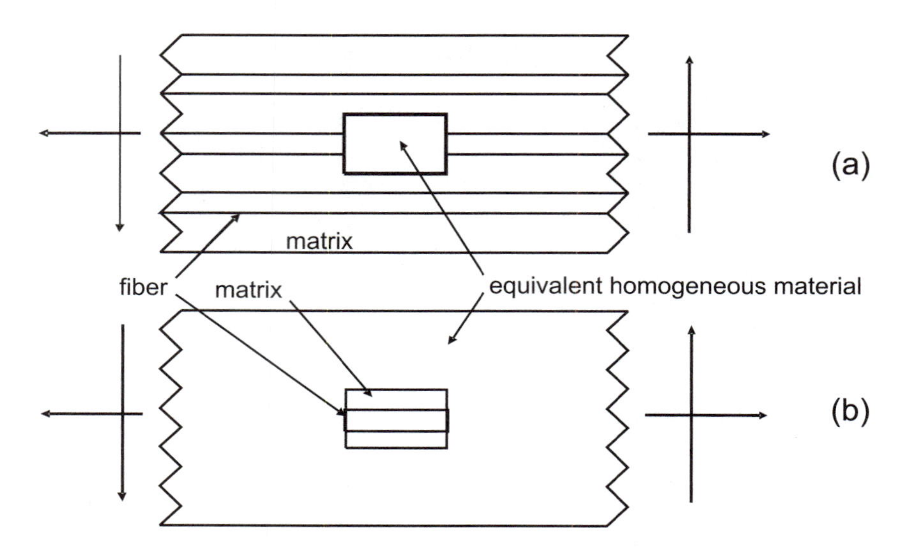

Figure 4.3: (a) RVE replaced by equivalent homogenous material, (b) all structure but the RVE replaced by equivalent material.

the peculiarities of the growth rings are ignored. However, the differences in the modulus along various directions must be recognized.

The stiffness of an isotropic material is completely described by two properties, for example the modulus of elasticity E and Poisson's ratio ν. In contrast, up to 21 properties may be required to describe anisotropic materials ([145, p. 224]).

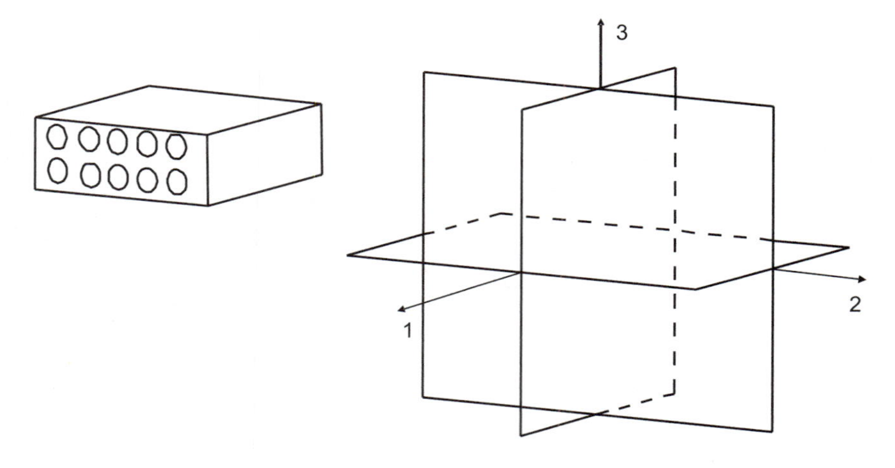

Figure 4.4: An orthotropic material has three planes of symmetry.

4.1.5 Orthotropic Material

An orthotropic material has three planes of symmetry (Figure 4.4) that coincide with the coordinate planes. A unidirectional fiber reinforced composite may be considered to be orthotropic. One plane of symmetry is perpendicular to the fiber direction, and the other two can be any pair of planes parallel to the fiber direction and orthogonal among themselves. Only nine constants are required to describe an orthotropic material.

4.1.6 Transversely Isotropic Material

A transversely isotropic material has one axis of symmetry. For example, the fiber direction of a unidirectional fiber reinforced composite can be considered an axis of symmetry if the fibers are randomly distributed in the cross-section with a uniform probability density function. In this case, any plane containing the fiber direction is a plane of symmetry. A transversely isotropic material is described by five constants.

4.1.7 Isotropic Material

The most common materials of industrial use are isotropic, like aluminum, steel, etc. Isotropic materials have an infinite number of planes of symmetry, meaning that the properties are independent of the orientation. Only two constants are needed to describe the elastic behavior of isotropic materials. Since the Young modulus E and the Poisson's ratio ν can be obtained experimentally from a single test, they are used often to describe isotropic materials. However, other elastic properties can be used as well. For example, the shear modulus of isotropic materials is related to E and ν by

$$G = \frac{E}{2(1+\nu)} \tag{4.14}$$

Therefore, the elastic behavior of isotropic material can be represented by the pair (E, G). Yet another valid combination, which is popular in theoretical elasticity treatises, is the pair (λ, G), where λ is Lame's constant, the latter related to previously defined properties as

$$\lambda = \frac{E}{(1+\nu)(1-2\nu)} \tag{4.15}$$

4.2 Stiffness

In the mechanics of materials approach, both fibers and matrix are assumed to be isotropic (see Section 4.1.4). The stiffness of an isotropic material is completely represented by two properties: the modulus of elasticity E and the Poisson's ratio ν. Using micromechanics, the combination of two isotropic materials (fiber and matrix) is represented as an equivalent, homogeneous, anisotropic material (see Section 4.1.4). The stiffness of the equivalent material is represented by five elastic properties:

E_1 : modulus of elasticity in the fiber direction
E_2 : modulus of elasticity in the direction transverse to the fibers

G_{12} : in-plane shear modulus
G_{23} : out of plane shear modulus
ν_{12} : in-plane Poisson's ratio

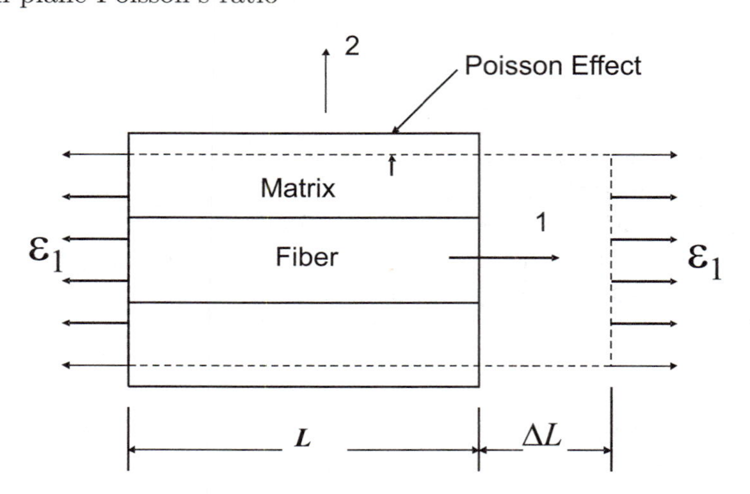

Figure 4.5: RVE subject to longitudinal uniform strain.

4.2.1 Longitudinal Modulus

The longitudinal modulus, or modulus of elasticity in the fiber direction can be predicted very well by (4.24), called the *rule of mixtures* (ROM) formula. The main assumption in this formulation is that the strains in the direction of the fibers are the same in the matrix and the fiber. This implies that the fiber–matrix bond is perfect. When the material is stretched along the fiber direction, the matrix and the fibers will elongate the same, as shown in Figure 4.5. This basic assumption is needed to be able to replace the heterogeneous material in the RVE by a homogeneous one (Figure 4.3) while satisfying compatibility of displacements with the rest of the body. Or vice-versa, replace everything but the RVE by an equivalent material while satisfying compatibility.

By definition of strains

$$\epsilon_1 = \frac{\Delta L}{L} \tag{4.16}$$

Since both fiber and matrix are isotropic and elastic, their stress–strain law is

$$\begin{aligned}\sigma_f &= E_f \epsilon_1 \\ \sigma_m &= E_m \epsilon_1\end{aligned} \tag{4.17}$$

The average stress σ_1 acts on the entire cross-section of the RVE with area

$$A = A_f + A_m \tag{4.18}$$

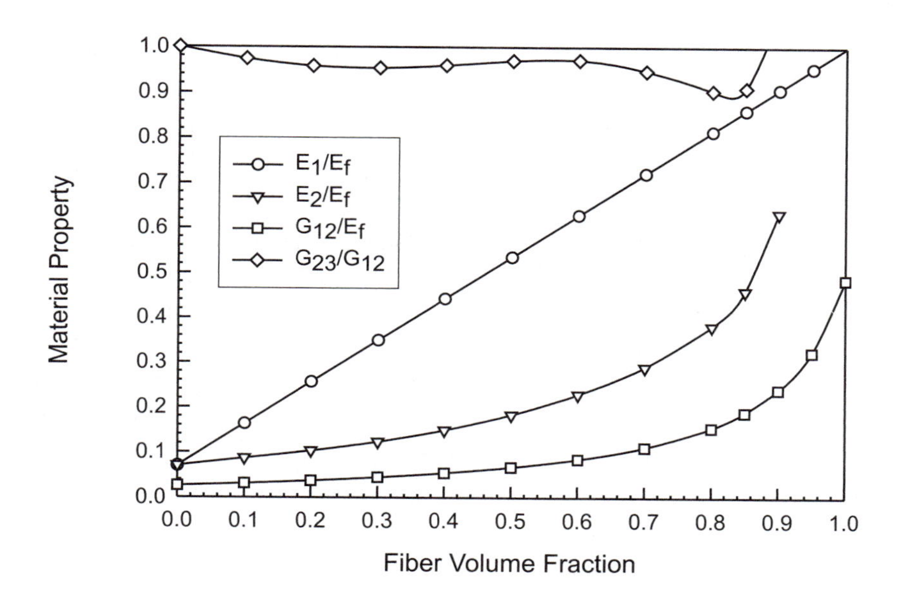

Figure 4.6: Micromechanics predictions as a function of the fiber volume fraction (Section 4.2.8).

The total load applied is

$$P = \sigma_1 A = \sigma_f A_f + \sigma_m A_m \tag{4.19}$$

Then

$$\sigma_1 = \epsilon_1 (E_f V_f + E_m V_m) \tag{4.20}$$

where

$$V_f = A_f / A \text{ and } V_m = A_m / A \tag{4.21}$$

For the equivalent homogeneous material

$$\sigma_1 = E_1 \epsilon_1 \tag{4.22}$$

Then, comparing (4.20) with (4.22) results in

$$E_1 = E_f V_f + E_m V_m \tag{4.23}$$

Finally, using (4.3), the rule of mixture formula is obtained

$$E_1 = E_f V_f + E_m (1 - V_f) \tag{4.24}$$

that can be written as

$$E_1 = E_m + V_f(E_f - E_m) \tag{4.25}$$

According to the rule of mixtures, the property E_1 depends linearly on V_f and the properties of the constituents, as shown in Figure 4.6 for an E-glass composite ($E_f =$ 72.3 GPa, $E_m = 5.05$ GPa, $\nu_f = 0.22, \nu_m = 0.35$). In most cases, the modulus of the fibers is much larger than that of the matrix (see Tables 2.1, 2.2, 2.9, 2.10). Then, the first term dominates, making the contribution of the matrix to the composite longitudinal modulus negligible. This indicates that the longitudinal modulus E_1 is a fiber-dominated property. Also note that V_f cannot reach all the way to 100% as Figure 4.6 may suggest. To begin with, the best packing of cylindrical fibers is the hexagonal array (Figure 4.2) with V_f about 90%. Furthermore, processing conditions usually limits V_f to values much lower than that (e.g., for pultrusion to about 45%).

4.2.2 Transverse Modulus

In the determination of the modulus in the direction transverse to the fibers, the main assumption is that the stress is the same in the fiber and the matrix. This assumption is needed to maintain equilibrium in the transverse direction. Once again, the assumption implies that the fiber–matrix bond is perfect. The RVE, subject to a uniform transverse stress, is shown in Figure 4.7. Note that, for simplicity, a cylindrical fiber has been replaced by a rectangular one. In reality, most micromechanics formulations (except advanced formulations [108, 146]) do not represent the actual geometry of the fiber at all.

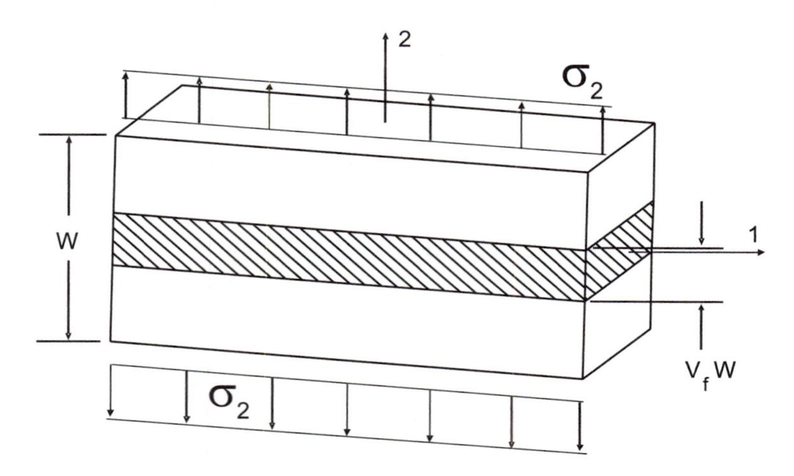

Figure 4.7: RVE subject to transverse uniform stress.

Since fiber and matrix are assumed to be liner elastic materials, the strains in the fiber and the matrix are

$$\epsilon_f = \frac{\sigma_2}{E_f}; \; \epsilon_m = \frac{\sigma_2}{E_m} \tag{4.26}$$

These strains act over a portion of the RVE, ϵ_f over $V_f W$, and ϵ_m over $V_m W$, while the average strain ϵ_2 acts over the entire width W. The total elongation is

$$\epsilon_2 W = \epsilon_f V_f W + \epsilon_m V_m W \tag{4.27}$$

Cancelling W and using Hooke's law for the constituents, which are assumed to be isotropic

$$\epsilon_2 = \frac{\sigma_f}{E_f} V_f + \frac{\sigma_m}{E_m} V_m \tag{4.28}$$

Since the stress is the same in the fiber and the matrix ($\sigma_f = \sigma_m = \sigma_2$)

$$\epsilon_2 = \frac{\sigma_2}{E_f} V_f + \frac{\sigma_2}{E_m} V_m \tag{4.29}$$

Then, comparing with Hooke's law for the equivalent material ($\sigma_2 = E_2 \epsilon_2$), the transverse modulus is given by

$$\frac{1}{E_2} = \frac{V_m}{E_m} + \frac{V_f}{E_f} \tag{4.30}$$

Equation 4.30 is known as the inverse rule of mixtures (IROM). The fibers do not contribute appreciably to the stiffness in the transverse direction unless V_f is very high (see Figure 4.6), in which case the assumptions made in this section are not valid. Therefore, it is said that E_2 is a matrix-dominated property. The prediction obtained with (4.30) is a lower bound, which is not accurate in the majority of cases. A lower bound means that the property is underestimated. The inverse rule of mixture is a simple equation to be used for qualitative evaluation of different candidate materials but not for design calculations.

A better prediction can be obtained with the semi-empirical Halpin-Tsai [147] formula

$$E_2 = E_m \left[\frac{1 + \zeta \eta V_f}{1 - \eta V_f} \right] \tag{4.31}$$
$$\eta = \frac{(E_f/E_m) - 1}{(E_f/E_m) + \zeta}$$

where ζ is an empirical parameter obtained by curve-fitting (4.31) with the results of an analytical solution (too complex to describe here). The value $\zeta = 2$ usually gives good fit for the case of circular or square fibers. For rectangular fibers, a good estimate is $\zeta = 2a/b$, where a and b are the dimensions of the cross-section of the fiber.

The periodic microstructure model (see Section 4.2.8) leads to accurate formulas for E_2 and also G_{12}, G_{23}, ν_{12}, and E_1. Since the PMM formulas are relatively complex, they have been programmed in the accompanying software. Comparison with experimental data is presented in [108,146] and Figure 4.8 for glass–epoxy with $E_f/E_m = 21.19$, $\nu_f = 0.22$, $\nu_m = 0.38$.

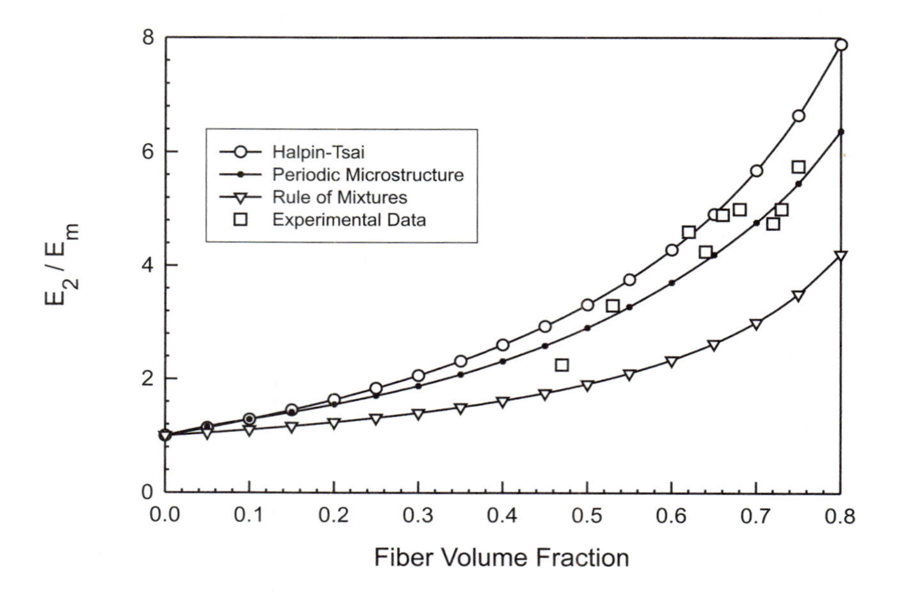

Figure 4.8: Comparison of predicted values of E_2/E_m with experimental data for glass–epoxy.

4.2.3 In-plane Poisson's Ratio

The Poisson's ratio is defined as minus the quotient of the resulting strain over the applied strain, as follows

$$\nu_{ij} = -\frac{\epsilon_j}{\epsilon_i} \tag{4.32}$$

That is, in a test in which load is applied in the i-direction, strain is induced by Poisson's effect on the perpendicular j-direction. The mechanics of materials approach leads to a rule of mixtures equation for the in-plane Poisson's ratio

$$\nu_{12} = \nu_f V_f + \nu_m V_m \tag{4.33}$$

An approximate prediction of Poisson's ratio is usually sufficient in design. Since the Poisson's ratio for the matrix and the fibers are not very different, (4.33) predicts also a similar value for the composite. Poisson's ratios are difficult to measure, mainly that of the fiber. Some fibers like carbon fibers are not even isotropic.

Finally, Poisson effects are usually secondary effects. For all these reasons, (4.33) is generally adequate for design. Since the Poisson's ratio is predicted by a rule of mixtures equation (4.33), the plot as a function of fiber volume fraction is similar to the plot of the longitudinal modulus.

4.2.4 In-plane Shear Modulus

The in-plane shear stress $\sigma_6 = \tau_{12} = \tau_{21}$ deforms the composite as in Figure 4.9a.[1] The strength of material approach leads to an inverse rule of mixtures equation [148] for the in-plane shear modulus

$$\frac{1}{G_{12}} = \frac{V_m}{G_m} + \frac{V_f}{G_f} \tag{4.34}$$

Equation 4.34 predicts that the in-plane shear modulus is a matrix-dominated property in the case of stiff fibers. The inverse rule of mixtures can be rewritten as

$$G_{12} = \frac{G_m}{V_m + V_f G_m / G_f} \tag{4.35}$$

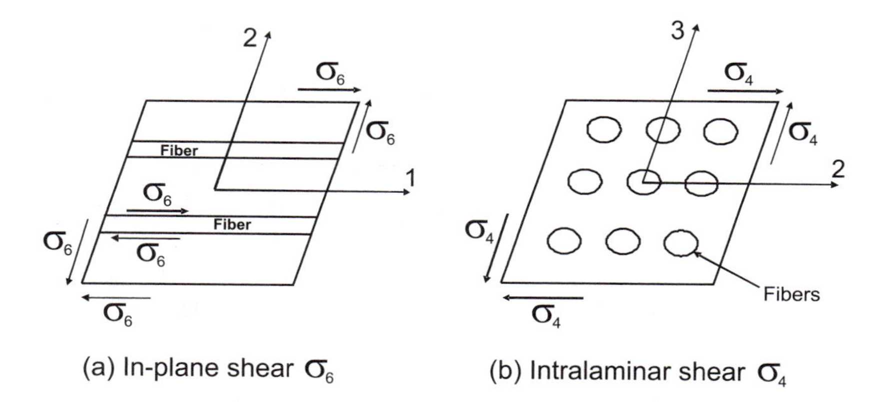

Figure 4.9: Distinction between in-plane and intralaminar shear stress.

If the fibers are much stiffer than the matrix ($G_m \ll G_f$), the in-plane shear modulus can be approximated by the matrix-dominated approximation

$$G_{12} \cong \frac{G_m}{1 - V_f} \tag{4.36}$$

Once again, the inverse rule of mixtures gives a simple but not accurate equation for the prediction of the in-plane shear modulus. The cylindrical assemblage model (CAM, [149]) gives a better approximation

$$G_{12} = G_m \left[\frac{(1 + V_f) + (1 - V_f) G_m / G_f}{(1 - V_f) + (1 + V_f) G_m / G_f} \right] \tag{4.37}$$

[1] See Section 5.1 for definition of the material coordinate system. See also Section 4.4.8.

Again, if ($G_m \ll G_f$), the matrix-dominated approximation is

$$G_{12} \cong G_m \left(\frac{1+V_f}{1-V_f} \right) \tag{4.38}$$

Equation 4.38 deviates slightly from the results of (4.37) as shown in Figure 4.10. The periodic microstructure model (PMM, [108, 146]) yields the following formula

$$G_{12} = G_m \left[1 + \frac{V_f\,(1 - G_m/G_f)}{G_m/G_f + S_3\,(1 - G_m/G_f)} \right] \tag{4.39}$$

where

$$S_3 = 0.49247 - 0.47603 V_f - 0.02748 V_f^2 \tag{4.40}$$

Then, the matrix-dominated PMM formula is

$$G_{12} \cong G_m \left[1 + \frac{V_f}{G_m/G_f + S_3} \right] \tag{4.41}$$

Equation (4.41) gives almost the same results as (4.39), and good agreement with experimental data for most common materials. Experimental data and predicted values of G_{12}/G_m are compared in Figure 4.10 for glass–epoxy with $E_m = 4.0$ GPa, $\nu_m = 0.35$, $E_f = 72.3$ GPa, $\nu_f = 0.22$ (see Table 2.1).

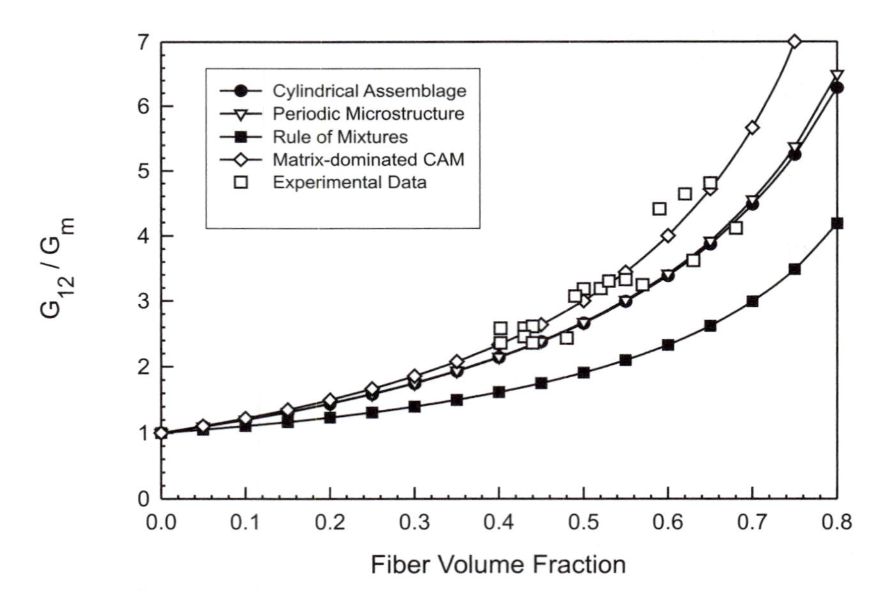

Figure 4.10: Comparison of G_{12}/G_m with experiments for E-glass composite.

If the matrix and the fibers are isotropic (e.g., glass fibers), the shear modulus of the fiber and the matrix can be computed from the known modulus and Poisson's ratio using the following formula, valid for any isotropic material

$$G = \frac{E}{2(1+\nu)} \tag{4.42}$$

Otherwise, transversely isotropic fibers, such as carbon fibers, have axial (A) and transverse (T) properties $E_A, E_T, G_A, \nu_A, \nu_T$, that must be measured independently. Still, G_T can be computed with (4.42) in terms of E_T, ν_T, due to the transverse isotropy of the fiber.

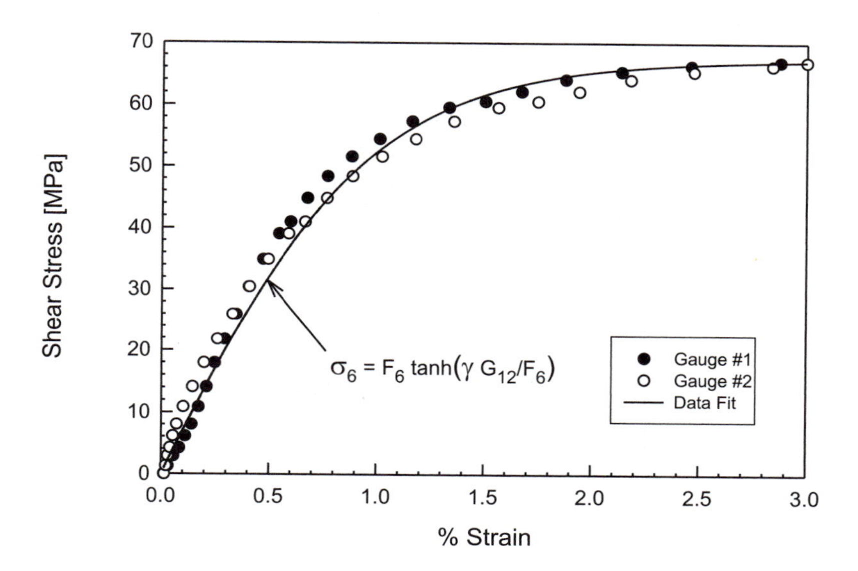

Figure 4.11: In-plane shear stress–strain behavior of unidirectional carbon–epoxy (experimental data according to ASTM D5379).

The shear stress–strain response is nonlinear, as shown in Figure 4.11. Equations (4.35)–(4.42) predict only the initial value of the shear moduli G_{12}; that is, the tangent to the stress–strain plot at the origin. The nonlinear stress–strain response can be described by

$$\sigma_6 = F_6 \tanh\left(\frac{G_{12}}{F_6}\gamma_6\right) \tag{4.43}$$

or by

$$\gamma_6 = \frac{\sigma_6}{G_{12}} + \beta\sigma_6^2 \tag{4.44}$$

where β is adjusted to fit experimental data such as that presented in Figure 4.11. Note that (4.43) predicts the nonlinear stress–strain response in terms of commonly available parameters, namely the in-plane shear stiffness G_{12} and the in-plane shear strength F_6, while (4.44) needs to have β adjusted to experimental data that is not commonly available.

4.2.5 Intralaminar Shear Modulus

The intralaminar stress $\sigma_4 = \tau_{23} = \tau_{32}$ acts across the thickness of the composite as shown in Figure 4.9b.[2] The intralaminar shear modulus can be computed with the semi-empirical stress partitioning parameter technique (Section 4.2.7) as [150]

$$G_{23} = G_m \frac{V_f + \eta_4 (1 - V_f)}{\eta_4 (1 - V_f) + V_f G_m / G_f} \qquad (4.45)$$
$$\eta_4 = \frac{3 - 4\nu_m + G_m/G_f}{4(1 - \nu_m)}$$

The periodic microstructure model [108, 146] accounts for the exact geometry of the fibers and gives similar results to (4.45) for most common composites.

The intralaminar shear stress $\sigma_5 = \tau_{13} = \tau_{31}$ introduces a shear deformation through the thickness of the composite which is similar to that in Figure 4.9a. Therefore, it is usual to assume

$$G_{13} = G_{12} \qquad (4.46)$$

which is exact for a transversely isotropic material (Section 4.1.6) with the axis of symmetry coinciding with the fiber direction.

4.2.6 Restrictions on the Elastic Constants

Several restrictions on the possible values for the various elastic constants presented in this chapter can be derived from elasticity theory. Since the compliance matrix (5.15) must be symmetric, the following must be satisfied,

$$\frac{\nu_{ij}}{E_i} = \frac{\nu_{ji}}{E_j} \quad \text{with } i \neq j \qquad (4.47)$$

which is routinely used to compute the Poisson's ratio ν_{21}. Also, all the diagonal terms of the stiffness matrix (5.19) must be positive. Since all the moduli (E_1, ν_{12}, etc.) are positive, all the diagonal terms of the stiffness matrix are positive if the following conditions are satisfied,

$$0 < \nu_{ij} < \sqrt{\frac{E_i}{E_j}} \quad \text{with } i \neq j \qquad (4.48)$$

$$\Delta = 1 - \nu_{12}\nu_{21} - \nu_{23}\nu_{32} - \nu_{31}\nu_{13} - 2\nu_{21}\nu_{32}\nu_{13} > 0 \qquad (4.49)$$

These and other restrictions [2, Section 1.1.3.1] are used to check the validity of experimental data, as in the following example.

---

[2]See Section 5.1 for the definition of the material coordinate system. See also Section 4.4.8.

Example 4.3 *The elastic properties $E_1 = 19.981$ GPa, and $\nu_{12} = 0.274$ were measured in a longitudinal test (fibers in the direction of loading,) by using two strain gauges, one longitudinal and one transverse. The elastic properties $E_2 = 11.389$ GPa and $\nu_{21} = 0.153$ are measured in a transverse tensile test (fibers perpendicular to loading) in the same way. Check the data for consistency.*

Solution to Example 4.3 *For the test procedure to be valid, all four data values, E_1, E_2, ν_{12}, and ν_{21} must conform to (4.47), within the experimental error. First, compute both sides of (4.47)*

$$\frac{E_1}{\nu_{12}} = \frac{19.981}{0.274} = 72.9\ GPa$$
$$\frac{E_2}{\nu_{21}} = \frac{11.389}{0.153} = 74.4\ GPa$$

The difference is small taking into account that there may be some experimental error. Therefore, the data is considered valid.

4.2.7 Stress Partitioning Parameter (*)[3]

The stress partitioning parameter (SPP) is a typical example of an experimental correction for an otherwise not accurate formula. This approach is used often in the field of composite materials. When an accurate formula is not available, the obvious alternative is to test all the possible volume fractions to be encountered during design and fit the data with a curve fit. This curve fitting approach is too expensive because of the large amount of testing required. A more intelligent approach is to correct the inaccurate formula using limited experimental information. In this semi-empirical method, all the good information that was built in the original formula is preserved while enhancing it with experimental data. The stress partitioning parameter can be used to correct both the transverse modulus E_2 (4.30) and the in-plane shear modulus G_{12} (4.34).

In the derivation of the inverse rule of mixtures equation for E_2 (Section 4.2.2), it was assumed that the stress in the fiber and the matrix were the same

$$\sigma_f = \sigma_m = \sigma_2 \tag{4.50}$$

That is only an approximation brought by modeling the fiber as a rectangle in Figure 4.7, while actually the fiber is cylindrical. To obtain a better approximation for E_2, a cylindrical fiber is considered. For a cylindrical fiber, the stresses σ_f and σ_m are not uniform but they can be represented by their average values. Note that in reality the average stress in the matrix is lower than that in the fiber, because $E_m < E_f$. Without actually computing the real stress field, the stress partitioning parameter η_2 can be defined as the ratio of the *average* stress in the matrix to the *average* stress in the fiber

[3]Sections marked with (*) can be omitted during the first reading but are recommended for further study and reference.

$$\eta_2 = \frac{\hat{\sigma}_m}{\hat{\sigma}_f} \quad ; \quad 0 < \eta_2 \leq 1 \tag{4.51}$$

where the average stresses are defined by averaging the actual stress distribution over the volume

$$\begin{aligned}
\hat{\sigma}_f &= \frac{1}{V_f} \int_{V_f} \sigma_f \, dV \\
\hat{\sigma}_m &= \frac{1}{V_m} \int_{V_m} \sigma_m \, dV \\
\hat{\sigma}_c &= \frac{1}{V_c} \int_{Vc} \sigma \, dV
\end{aligned} \tag{4.52}$$

where a *hat* indicates an average quantity, V_f, V_m, and V_c are the volume of fiber, matrix, and composite, respectively. The average stress in the composite can be decomposed as

$$\hat{\sigma}_c = \frac{V_f}{V_c} \frac{1}{V_f} \int_{V_f} \sigma_f \, dV + \frac{V_m}{V_c} \frac{1}{V_m} \int_{V_m} \sigma_m \, dV \tag{4.53}$$

and recognizing that $\hat{\sigma}_c = \sigma_2$

$$\sigma_2 = V_f \; \hat{\sigma}_f + V_m \; \hat{\sigma}_m \tag{4.54}$$

Using (4.51), the relationship among the averages (4.54) becomes

$$\sigma_2 = (V_f + \eta_2 V_m) \, \hat{\sigma}_f \tag{4.55}$$

Now, rewrite (4.28) using (4.51) and recognizing that in (4.28) σ_f and σ_m are actually average values $\hat{\sigma}_f$ and $\hat{\sigma}_m$

$$\epsilon_2 = \frac{\hat{\sigma}_f}{E_f} V_f + \frac{\eta_2 \hat{\sigma}_f}{E_m} V_m \tag{4.56}$$

Multiply and divide by the parenthesis in (4.55) to get

$$\epsilon_2 = \left(\frac{V_f}{E_f} + \frac{\eta_2 V_m}{E_m} \right) \hat{\sigma}_f \frac{(V_f + \eta_2 V_m)}{(V_f + \eta_2 V_m)} \tag{4.57}$$

Now, recognize σ_2 from (4.55) and divide by it to get the corrected inverse rule of mixture equation for E_2

$$\frac{1}{E_2} = \frac{\epsilon_2}{\sigma_2} = \left(\frac{V_f}{E_f} + \frac{\eta_2 V_m}{E_m} \right) \frac{1}{(V_f + \eta_2 V_m)} \tag{4.58}$$

Next, one experiment is performed at a fixed V_f to experimentally measure E_2. From this experiment, the correction η_2 can be calculated from (4.58), or writing η_2 explicitly as

$$\eta_2 = -\frac{V_f E_m (E_2 - E_f)}{V_m E_f (E_2 - E_m)} \tag{4.59}$$

The next step is to assume that the correction factor η_2 is independent of V_f for the tested material system. Then, (4.58) or

$$E_2 = E_m \frac{V_f + \eta_2 (1 - V_f)}{\eta_2 (1 - V_f) + V_f E_m / E_f} \tag{4.60}$$

can be used to accurately predict the transverse stiffness E_2. Equation (4.60) matches experimental data for values of V_f other than the one at which η_2 was obtained, as long as the fiber, matrix, and process are not changed. Note that in the mechanics of materials derivation (Section 4.2.2) it is assumed that $\sigma_f = \sigma_m$, i.e., $\eta_2 = 1$, and under such assumption, (4.60) reduces to (4.30).

Similarly, performing one experiment at a fixed V_f to experimentally measure G_{12}, the shear-stress partitioning parameter η_6 can be calculated as

$$\eta_6 = -\frac{V_f G_m (G_{12} - G_f)}{V_m G_f (G_{12} - G_m)} \tag{4.61}$$

Then, the following equation

$$G_{12} = G_m \frac{V_f + \eta_6 (1 - V_f)}{\eta_6 (1 - V_f) + V_f G_m / G_f} \tag{4.62}$$

can be used to accurately predict the in-plane shear modulus G_{12}. Finally, the stress partitioning parameter for in-plane shear can be estimated [150] as

$$\eta_6 = \frac{1}{2}(1 + \frac{G_m}{G_f}) \tag{4.63}$$

Substituting (4.63) into (4.62) results in the CAM formula (4.37).

4.2.8 Periodic Microstructure (*)[4]

The periodic microstructure micromechanics (PMM, [108,146]) yields accurate predictions for all the moduli *(plural)* of a composite reinforced by long fibers. The equations for unidirectional composites with isotropic fibers, such as glass, in an elastic matrix are available in [108]. For unidirectional composites with transversely isotropic fibers (e.g., carbon), in a viscoelastic matrix, formulas are given in [146]. The latter can be used for elastic behavior of the matrix, which is a particular case of viscoelastic behavior. Computer programs implementing these equations are available in [1].

Barbero and Luciano [146] used the the Fourier expansion method to obtain the components of the relaxation tensor for a composite with cylindrical fibers and fiber volume fraction V_f. The elastic, transversely isotropic fibers have axial and

[4]Sections marked with (*) can be omitted during the first reading but are recommended for further study and reference.

radial (transverse) moduli E_A, E_T, G_A, G_T, and axial Poisson's ν_A (see discussion following (4.42)). The equations in [146] can be simplified for the particular case of linear elastic matrix behavior as follows

$$C^*_{11} = \lambda_m + 2\,\mu_m - V_f\,\left(-a_4^2 + a_3^2 \right) \left(-\frac{(2\,\mu_m + 2\,\lambda_m - C'_{33} - C'_{23})\,(a_4^2 - a_3^2)}{a_1} + \frac{2\,(a_4 - a_3)\,(\lambda_m - C'_{12})^2}{a_1^2} \right)^{-1}$$

$$C^*_{12} = \lambda_m + V_f \left(\frac{(\lambda_m - C'_{12})\,(a_4 - a_3)}{a_1} \right) \left(\frac{(2\,\mu_m + 2\,\lambda_m - C'_{33} - C'_{23})\,(a_3^2 - a_4^2)}{a_1} + \frac{2\,(a_4 - a_3)\,(\lambda_m - C'_{12})^2}{a_1^2} \right)^{-1}$$

$$C^*_{22} = \lambda_m + 2\,\mu_m - V_f \left(\frac{(2\,\mu_m + 2\,\lambda_m - C'_{33} - C'_{23})\,a_3}{a_1} - \frac{(\lambda_m - C'_{12})^2}{a_1^2} \right) \left(\frac{(2\,\mu_m + 2\,\lambda_m - C'_{33} - C'_{23})\,(a_3^2 - a_4^2)}{a_1} + \frac{2\,(a_4 - a_3)\,(\lambda_m - C'_{12})^2}{a_1^2} \right)^{-1}$$

$$C^*_{23} = \lambda_m + V_f \left(\frac{(2\,\mu_m + 2\,\lambda_m - C'_{33} - C'_{23})\,a_4}{a_1} - \frac{(\lambda_m - C'_{12})^2}{a_1^2} \right) \left(\frac{(2\,\mu_m + 2\,\lambda_m - C'_{33} - C'_{23})\,(a_3^2 - a_4^2)}{a_1} + \frac{2\,(a_4 - a_3)\,(\lambda_m - C'_{12})^2}{a_1^2} \right)^{-1}$$

$$C^*_{44} = \mu_m - V_f \left(\frac{2}{2\,\mu_m - C'_{22} + C'_{23}} - \left(2\,s_3 - \frac{4\,s_7}{2 - 2\,\nu_m} \right) \mu_m^{-1} \right)^{-1}$$

$$C^*_{66} = \mu_m - V_f \left(\left(\mu_m - C'_{66} \right)^{-1} - \frac{s_3}{\mu_m} \right)^{-1} \tag{4.64}$$

where

$$a_1 = 4\,\mu_m^2 - 2\,\mu_m\,C'_{33} + 6\,\lambda_m\,\mu_m - 2\,C'_{11}\,\mu_m - 2\,\mu_m\,C'_{23} + C'_{23}\,C'_{11} + 4\,\lambda_m\,C'_{12} - 2\,{C'_{12}}^2 - \lambda_m\,C'_{33} - 2\,C'_{11}\,\lambda_m + C'_{11}\,C'_{33} - \lambda_m\,C'_{23}$$

$$
\begin{aligned}
a_2 = {} & 8\,\mu_m^3 - 8\,\mu_m^2 C'_{33} + 12\,\mu_m^2 \lambda_m - 4\,\mu_m^2 C'_{11} \\
& - 2\,\mu_m\, C'^{\,2}_{23} + 4\,\mu_m\, \lambda_m\, C'_{23} + 4\,\mu_m\, C'_{11}\, C'_{33} \\
& - 8\,\mu_m\, \lambda_m\, C'_{33} - 4\,\mu_m\, C'^{\,2}_{12} + 2\,\mu_m\, C'^{\,2}_{33} - 4\,\mu_m\, C'_{11}\, \lambda_m + 8\,\mu_m\, \lambda_m\, C'_{12} \\
& + 2\,\lambda_m\, C'_{11}\, C'_{33} + 4\,C'_{12}\, C'_{23}\, \lambda_m - 4\,C'_{12}\, C'_{33}\, \lambda_m - 2\,\lambda_m\, C'_{11}\, C'_{23} \\
& - 2\,C'_{23}\, C'^{\,2}_{12} + C'^{\,2}_{23} C'_{11} + 2\,C'_{33}\, C'^{\,2}_{12} - C'_{11}\, C'^{\,2}_{33} + \lambda_m\, C'^{\,2}_{33} - \lambda_m\, C'^{\,2}_{23}
\end{aligned}
$$

$$
\begin{aligned}
a_3 = {} & \frac{4\,\mu_m^2 + 4\,\lambda_m\,\mu_m - 2\,C'_{11}\,\mu_m - 2\,\mu_m\,C'_{33} - C'_{11}\,\lambda_m - \lambda_m\,C'_{33} - C'^{\,2}_{12}}{a_2} \\
& + \frac{C'_{11}\,C'_{33} + 2\,\lambda_m\,C'_{12}}{a_2} - \frac{s_3 - \dfrac{s_6}{2 - 2\nu_m}}{\mu_m}
\end{aligned}
$$

$$
\begin{aligned}
a_4 = {} & -\frac{-2\,\mu_m\,C'_{23} + 2\,\lambda_m\,\mu_m - \lambda_m\,C'_{23} - C'_{11}\,\lambda_m - C'^{\,2}_{12} + 2\,\lambda_m\,C'_{12} + C'_{11}\,C'_{23}}{a_2} \\
& + \frac{s_7}{\mu_m\,(2 - 2\,\nu_m)}
\end{aligned}
\tag{4.65}
$$

In (4.64) and (4.65), the C' coefficients[5] for the fiber are given in terms of the fiber elastic properties as follows

$$
\begin{aligned}
\Delta &= \frac{1 - 2\nu_A^2 E_T/E_A - \nu_T^2 - 2\nu_A^2 \nu_T E_T/E_A}{E_A E_T^2} \\
C'_{11} &= \frac{1 - \nu_T^2}{E_T^2 \Delta} \\
C'_{22} = C'_{33} &= \frac{1 - \nu_A^2 E_T/E_A}{E_A E_T \Delta} \\
C'_{12} = C'_{13} &= \frac{\nu_A E_T/E_A + \nu_A \nu_T E_T/E_A}{E_T^2 \Delta} \\
C'_{23} &= \frac{\nu_T + \nu_A^2 E_T/E_A}{E_A E_T \Delta} \\
C'_{44} = G_T &= \frac{E_T}{2(1 + \nu_T)} = (C'_{22} - C'_{23})/2 \\
C'_{55} = C'_{66} &= G_A
\end{aligned}
\tag{4.66}
$$

and the *Lamé* constants of the matrix are calculated as follows

$$
\begin{aligned}
\lambda_m &= \frac{E_m \nu_m}{(1 + \nu_m)(1 - 2\nu_m)} \\
\mu_m = G_m &= \frac{E_m}{2(1 + \nu_m)}
\end{aligned}
\tag{4.67}
$$

[5] Coefficients in the stiffness tensor of a transversely isotropic material.

The coefficients s_3, s_6, s_7 account for the geometry of the microstructure, including the geometry of the inclusions and their geometrical arrangement. For cylindrical fibers arranged in a square array we have

$$\begin{aligned} s_3 &= 0.49247 - 0.47603V_f - 0.02748V_f^2 \\ s_6 &= 0.36844 - 0.14944V_f - 0.27152V_f^2 \\ s_7 &= 0.12346 - 0.32035V_f + 0.23517V_f^2 \end{aligned} \tag{4.68}$$

Note that (4.64) yield six independent components of the relaxation tensor. This is because (4.64) represent a composite with microstructure arranged in a square array. If the microstructure is random [2, Figure 1.12], the composite is transversely isotropic (Section 4.1.6) and only five components are independent. When the axis x_1 is the axis of transverse isotropy for the composite, the averaging procedure [2, (6.7)] yields

$$\begin{aligned} C_{11} &= C_{11}^* \\ C_{12} = C_{13} &= C_{12}^* \\ C_{22} = C_{33} &= \frac{3}{4}C_{22}^* + \frac{1}{4}C_{23}^* + \frac{1}{2}C_{44}^* \\ C_{23} &= \frac{1}{4}C_{22}^* + \frac{3}{4}C_{23}^* - \frac{1}{2}C_{44}^* \\ C_{55} = C_{66} &= C_{66}^* \end{aligned} \tag{4.69}$$

Finally, the properties of the tow are obtained from the compliance $[S] = [C]^{-1}$ as

$$\begin{aligned} E_1 &= 1/S_{11} \\ E_2 = E_3 &= 1/S_{22} \\ \nu_{12} = \nu_{13} &= -S_{21}/S_{11} \\ \nu_{23} &= -S_{32}/S_{22} \\ G_{12} = G_{13} &= 1/S_{55} \\ G_{23} &= \frac{E_2}{2(1+\nu_{23})} = 1/S_{44} \end{aligned} \tag{4.70}$$

4.3 Moisture and Thermal Expansion

The change in dimension ΔL of a body as a result of mechanical strain ϵ, change in temperature ΔT, and change in moisture concentration Δm, can be approximated by the linear relationship

$$\Delta L = (\epsilon + \beta \Delta m + \alpha \Delta T)\, L_0 \tag{4.71}$$

where L_0 is the initial length, β is the coefficient of moisture expansion, and α is the coefficient of thermal expansion (see Tables 2.1, 2.2, 2.9 and 2.10).

4.3.1 Thermal Expansion

The thermal expansion coefficient of all resins are positive (about 30 to 100 $10^{-6}/{}^{o}$C), and higher than steel alloys (10 to 20 $10^{-6}/{}^{o}$C). The coefficient of thermal expansion for E-glass is low (5.04 $10^{-6}/{}^{o}$C). Carbon fibers have a negative expansion in the fiber direction (-0.99 $10^{-6}/{}^{o}$C) and a relatively large expansion in the transverse direction (16.7 $10^{-6}/{}^{o}$C). This means that depending on the amount of fibers (fiber volume fraction) it is possible to tailor the coefficient of thermal expansion of the composite to the user's needs. Especially, it is possible to produce a composite material with a very low coefficient of thermal expansion, which is useful when dimensional stability is required (see Example 4.4).

Composite materials have two coefficients of thermal expansion. In the direction of fibers, the thermal expansion behavior is dominated by fibers, and the coefficient is computed [122] as

$$\alpha_1 = \frac{\alpha_A V_f E_A + \alpha_m V_m E_m}{E_1} \tag{4.72}$$

In the direction perpendicular to the fibers, the thermal expansion is dominated by the matrix. In this case, the thermal expansion coefficient is computed [122] as

$$\alpha_2 = \alpha_3 = \alpha_T \sqrt{V_f} + (1 - \sqrt{V_f})(1 + V_f \nu_m \frac{E_A}{E_1})\alpha_m \tag{4.73}$$

where $\alpha_A, \alpha_T, \alpha_m,$ are the transversely isotropic fiber and isotropic matrix thermal expansion coefficients; E_A, E_m, ν_m are the elastic constants of the transversely isotropic fiber and isotropic matrix; subscripts $_{A,T}$, denote the longitudinal and transverse directions of the fiber, respectively; E_1 is the modulus of the lamina in the longitudinal direction, and V_f is the fiber volume fraction. Some of these constants possibly could be function of temperature.

Equations (4.72) and (4.73) consider the case of anisotropic fibers (e.g., carbon fibers). They have been correlated to experimental values and are reported to be quite satisfactory for the prediction of thermal properties. Note that the elastic moduli used in both equations depend on the temperature. Because of this, these formulas cannot be used over a large range of temperatures [150]. Predictions of coefficients of thermal expansion at constant temperature are shown in Figure 4.12.

Since a lamina of random composite is isotropic in the plane of the lamina, the coefficient of thermal expansion α_q of a composite with randomly oriented fibers can be found following the approach described in Section 9.9 [151]

$$\alpha_q = \frac{\alpha_1 + \alpha_2}{2} + \frac{\alpha_1 - \alpha_2}{2} \frac{E_1 - E_2}{E_1 + (1 + 2\nu_{12})E_2} \tag{4.74}$$

where $\alpha_1, \alpha_2, E_1, E_2, \nu_{21}$ are the properties of a fictitious unidirectional composite with the same fiber volume fraction of the random composite, computed with the equations described earlier in this chapter. It must be noted that experimental values of α_q are very scattered and the formula for α_q may underestimate or overestimate

Table 4.1: Coefficients of thermal expansion in Example 4.4.

	$V_f = 0.45$		$V_f = 0.55$	
Fiber	α_1 $[10^{-6}/^oC]$	α_2 $[10^{-6}/^oC]$	α_1 $[10^{-6}/^oC]$	α_2 $[10^{-6}/^oC]$
E-glass	6.74	23.7	6.31	20.4
S-glass	4.16	23.1	3.76	19.5
Kevlar 49^{TM}	-1.02	21.9	-1.33	17.6
Carbon T300	-0.06	22.5	-0.23	18.3

the experimental values. Experimental values can be obtained following standard procedures such as [118, ASTM E289, E831].

Example 4.4 *Select the materials and the processing technique to produce a square tube that will bend the least under a temperature gradient ΔT. The top face of the tube will be exposed to the sun, while the bottom and the sides will be in the shade. The temperature differential between the the top and bottom flange is ΔT. Two fabrication technique are available: hoop filament winding which lays the fibers at 90° with respect to the axis of the tube and pultrusion that aligns the fibers at 0°.*

Solution to Example 4.4 *First, select the resin with the least thermal expansion that is compatible with both processes. Using the tables in Chapter 2, an isophthalic polyester with $\alpha = 60 \cdot 10^{-6}\ /^oC$ is selected. Next, using (4.72) and (4.73), compute α_1 and α_2 using various fibers. The required values of E_1 and E_2 were computed using the ROM. Two values of fiber volume fraction are selected, which can be achieved with both processes. The results are shown in Table 4.1, from which it is clear that the least thermal expansion is in the fiber direction. So the fibers should be aligned along the axis of the tube (pultrusion). Carbon fibers should be used for optimum performance since they yield the lowest thermal expansion or contraction.*

4.3.2 Moisture Expansion

Moisture is absorbed primarily by the polymer matrix with the exception that aramid fibers can also absorb moisture. Once moisture is absorbed, it produces swelling of the polymer. Moisture is absorbed in a Fickian process, and the absorbtion rate is controlled by the diffusivity coefficient. The diffusion of moisture in composite is much slower than thermal conduction. When exposed to a changing environment, the material will reach thermal equilibrium much faster than moisture equilibrium.

The average moisture concentration in the composite is defined as

$$m = \frac{moisture\, weight}{dry\, composite\, weight} \tag{4.75}$$

Then moisture absorbtion can be easily obtained by measuring the weight of samples. However, careful consideration should be given to the void content, since water going to fill up voids does not contribute to the swelling. If the void content is known, and if it can be assumed that all voids will be filled by water, the weight of the water in the voids can be subtracted to get the the weight of water really absorbed by the polymer. The void content per volume V_v can be estimated by (4.11). Then the density ρ of the material can be found as the weight of a sample divided by its volume corrected for void content, $v - v_0(1 - V_v)$, where v_0 is the apparent volume.

While organic fibers (e.g., aramid) absorb water, inorganic fibers (e.g., glass, carbon) do not absorb moisture. In the case of inorganic fibers, the coefficients of moisture expansion in the longitudinal and transverse direction of the composite [122, 150] reduce to

$$\beta_1 = \beta_m(1 - V_f)\frac{E_m}{E_1}$$
$$\beta_2 = \beta_3 = \beta_m(1 - \sqrt{V_f})\left[1 + \frac{\sqrt{V_f}(1 - \sqrt{V_f})E_m}{\sqrt{V_f}E_2 + (1 - \sqrt{V_f})E_m}\right] \tag{4.76}$$

where β_m is the isotropic matrix moisture expansion coefficients, E_m is the modulus of the isotropic matrix, E_1, E_2 are the elastic constants of the unidirectional composite, and V_f is the fiber volume fraction. Predictions using (4.76) are shown in Figure 4.12.

The coefficient of moisture expansion β_q of a composite with randomly orientated fibers can be computed in the same way as the coefficient of thermal expansion in the previous section; that is,

$$\beta_q = \frac{\beta_1 + \beta_2}{2} + \frac{\beta_1 - \beta_2}{2}\frac{E_1 - E_2}{E_1 + (1 + 2\nu_{12})E_2} \tag{4.77}$$

where $\beta_1, \beta_2, E_1, E_2$ and ν_{12} are the properties of a fictitious unidirectional lamina with the same fiber volume fraction of the random composite. This formula has not been thoroughly validated with experimental data.

4.3.3 Transport Properties

Thermal and electrical conductivities, and mass diffusivity, all are computed by the same equations. In this section k_f and k_m indicate either thermal or electrical conductivities, or mass diffusivity, of the fiber and matrix, respectively. In the fiber direction, the conductivity (or diffusivity) can be computed by the rule of mixtures (ROM) as

$$k_1 = k_f V_f + k_m V_m \tag{4.78}$$

and in the transverse direction by the Halpin-Tsai approximation [152]

$$k_2 = k_m \left[\frac{1 + \xi \eta V_f}{1 - \eta V_f} \right]$$
$$\eta = \frac{(k_f/k_m) - 1}{(k_f/k_m) + \xi}$$
$$\xi = \log^{-1} \sqrt{3} \log \frac{a}{b} \tag{4.79}$$

where a/b is the aspect ratio of the cross-sectional dimensions of the fiber, a along the direction of heat (electrical) conduction, and b transverse to it. For circular fibers $a/b = 1$ and $\xi = 1$. These formulas predict well the thermal and electrical conductivities of carbon–epoxy composites for V_f up to 60%. Predictions using these formulas are shown in Figure 4.12 for a composite with $\beta_m = 0.6$, $\alpha_f = 5.4\,10^{-6}/°C$, $\alpha_m = 60\ 10^{-6}/°C$, $\rho_f = 2.5$ g/cc, $\rho_m = 1.2$ g/cc, $\kappa_f = 1.05$ W/m/°C, $\kappa_m = 0.25$ W/m/°C, $E_f = 72.3$ GPa, $E_m = 5.05$ GPa, $\nu_f = 0.22$, $\nu_m = 0.35$, and $\xi = 1$.

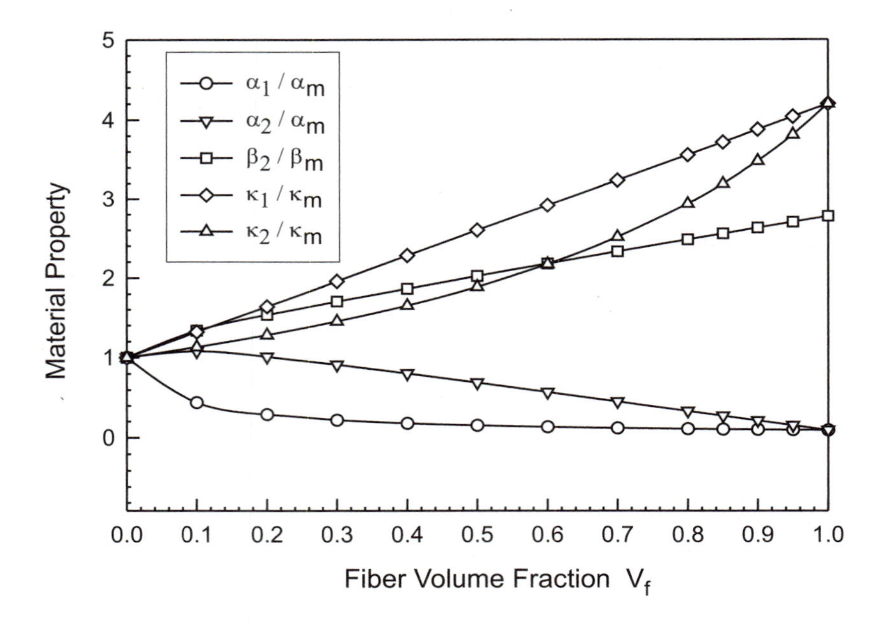

Figure 4.12: Prediction of thermal expansion α, moisture expansion β, and transport properties k.

4.4 Strength

Available micromechanics formulas used to predict strength properties are generally unreliable or require apparent constituent (fiber and matrix) properties that are different from bulk properties. Apparent properties, also called *in-situ* properties, are back-calculated from measured composite properties by using micromechanics formulas. Therefore, the accuracy of apparent properties depends on the quality of the formulas used to back-calculate them. Furthermore, apparent properties thus determined are not valid when any of the constituents changes, or when the processing conditions change, and so on.

The best option is to test actual unidirectional composites for strength. However, it is not always feasible to fabricate and test samples with all the values of fiber volume fraction that may occur in practice. Therefore, tests are performed at one value of fiber volume fraction; then, the actual measured strength values are used to back-calculate the parameters in the micromechanics formulas. Once this is done, the formulas can be used to predict the strength values at other fiber volume fractions as long as the constituents and manufacturing process remain the same. For example, once the longitudinal tensile strength F_{1t} has been measured on samples with fiber volume fraction V_f, the *apparent strength* of the fibers F_{ft} can be back-calculated from (4.82) and subsequently used to predict F_{1t} for other values of V_f, where of course, the values of E_f, E_m are well known.

The reasons for the discrepancy between predictions and experimental data are many, and research is very active in this area. There are two issues of concern. One is how to calculate ultimate strengths (strains) using micromechanics, which is the purpose of this section. The other is how to use the strength values in conjunction with stress analysis, which is the objective of Chapter 7.

4.4.1 Longitudinal Tensile Strength

The simplest model for tensile strength of a continuous fiber reinforced composite is derived by assuming that all the fibers have the same tensile strength. Actually, the tensile strength of fibers is not uniform. Instead, it is described well by a Weibull distribution (Appendix B) with two parameters, the *Weibull characteristic strength* σ_0 and the *Weibull modulus* m, which represent the fiber strength and its dispersion, respectively, as it is explained in Section 8.2. In this section, as a first approximation, it is assumed that all the fibers have the same strength, represented by the apparent strength value F_{ft}.

The second assumption is that both the fibers and the matrix behave linearly up to failure. This is not true for most polymer matrices that exhibit either elastic nonlinear (Figure 4.11) or plastic behavior after a certain elongation. The behavior of polymers is further complicated by their load-rate dependency. That is, polymers are viscoelastic or viscoplastic (see Section 2.8).

The third assumption is that the fibers are brittle with respect to the matrix. As it can be seen in Table 2.1 and 2.9, E-glass has actually more elongation to failure than most resins, invalidating this assumption. However, such large elongation to failure for fibers holds only in ideal conditions, with some fibers breaking at much lesser strain levels.

The fourth assumption is that the fibers are stiffer than the matrix. This is valid for all cases except for ceramic matrix composites (CMC).

Under the assumptions listed, the composite will break when the stress in the fibers reach their apparent strength F_{ft}. After the fibers break, the matrix is unable

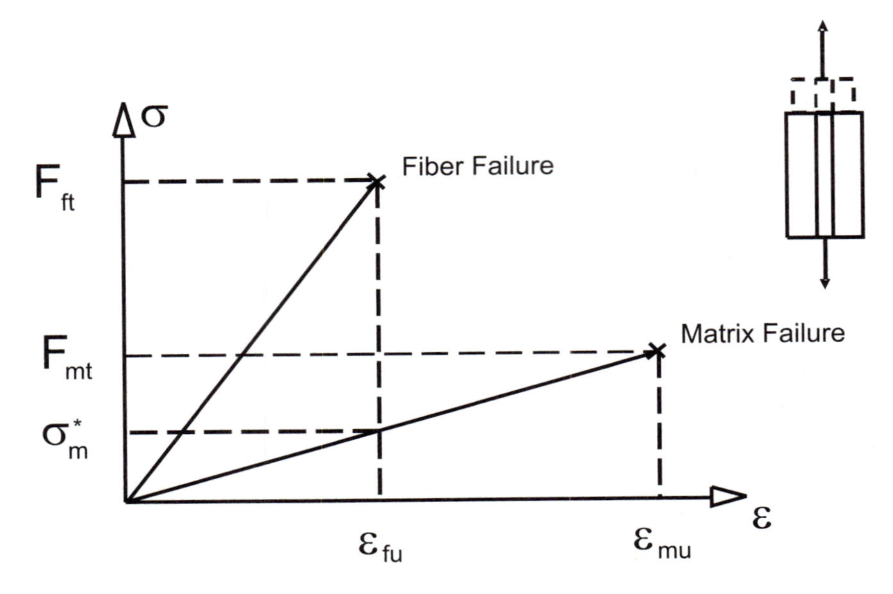

Figure 4.13: Micromechanics of strength assuming uniform fiber strength.

to carry the load. Therefore, the composite elongation to failure ϵ_{1t} is equal to the fiber elongation to failure ϵ_{fu}. At this strain level, the matrix has not failed yet because it is more compliant and can sustain larger strains (Figure 4.13). Under these conditions, it can be assumed that the longitudinal tensile strength is controlled by the fiber strength and represented by

$$F_{1t} = F_{ft} V_f + \sigma_m^* (1 - V_f) \tag{4.80}$$

where the stress in the matrix at failure (see Figure 4.13) is

$$\sigma_m^* = F_{ft} \frac{E_m}{E_f} \tag{4.81}$$

Finally, the tensile strength is predicted by the following equation

$$F_{1t} = F_{ft} \left[V_f + \frac{E_m}{E_f} (1 - V_f) \right] = \frac{F_{ft}}{k_f} \tag{4.82}$$

where $k_f = [V_f + (E_m/E_f)(1 - V_f)]^{-1}$ is the stress concentration factor that magnifies composite stress F_{1t} into fiber stress F_{ft}.

Equation (4.82) assumes that the strain in the matrix and the fibers are the same, which is true if the fiber–matrix bond is perfect. The ultimate strain or stress of the matrix is not realized, because the fibers are more brittle (i.e., fail at a lower strain, Figure 4.13). The underlying assumption is that once the fibers break, the matrix is not capable of sustaining the load and the composite fails. But this is not true for composites with very low fiber volume fraction. If V_f is very low, after the fibers break, the matrix may still be able to carry the load. Then, it is assumed that the broken fibers carry no more load (which is not entirely true [153–155]) resulting in a matrix with a set of holes occupied by broken fibers. For this new material, the tensile strength is

$$F_{1t} = F_{mt}(1 - V_f) \tag{4.83}$$

Equations (4.82) valid for high V_f, and (4.83) valid for low V_f, are plotted in Figure 4.14. The intersection of both curves limits the range of application of each equation. It also denotes the fiber volume fraction at which the lowest strength is achieved, of no practical significance however. To be of any use, composites need to be stronger than the matrix. Therefore, practical composites will have fiber volume fractions above V_{min}, which is found as the intersection of the line F_{mt} constant in Figure 4.14 with (4.80), or

$$V_{\text{min}} = \frac{F_{mt} - \sigma_m^*}{F_{ft} - \sigma_m^*} \tag{4.84}$$

As noted before, fibers do not have uniform strength. However, (4.82) is widely used because of its simplicity. Often, the values of F_{ft} cited in the literature are unreasonably high (e.g., 3.445 GPa for E- or S-glass). These values, which may have been obtained from tests on single fibers in ideal conditions, represent the maximum attainable rather than the typical strength, and therefore, cannot be used in design.

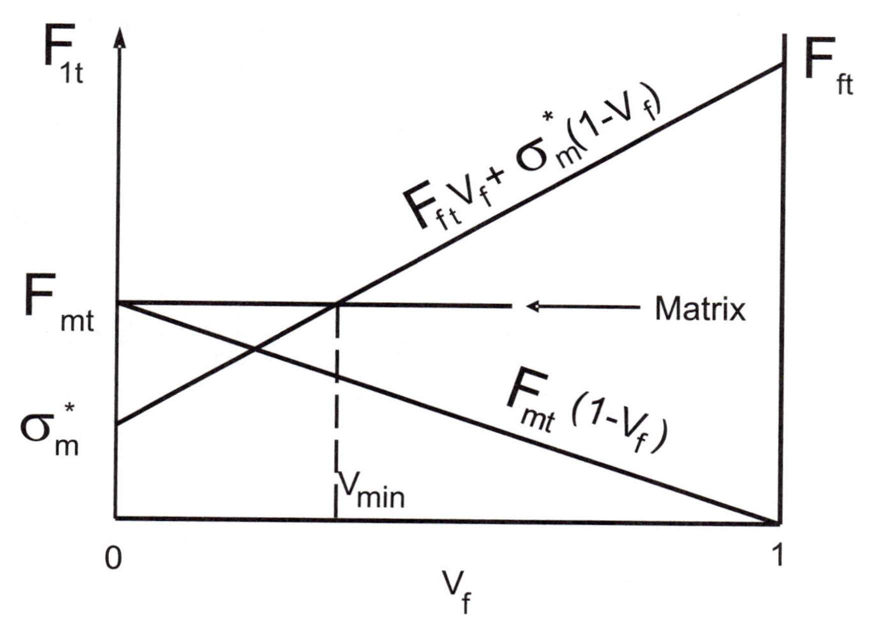

Figure 4.14: Regions of applicability of the two strength formulas.

Since fibers are damaged during processing, it is more advisable, and often done, to use (4.82) with *apparent strength* values F_{ft}. These have been back-calculated, using (4.82), from experimental data (F_{1t}) on composites rather from tests on single fibers. Therefore, the data does represent the fiber material in the real application. If some testing can be afforded, an apparent fiber strength value can be determined for a given V_f. Then, it can be used for other values of V_f as long as the processing method is not changed. Experimental values of tensile strength of composites can be obtained following [118, ASTM D3039].

Example 4.5 *Compute the tensile strength of a Ti-alumina metal matrix composite (MMC) with a fiber volume fraction of 27%. The titanium matrix is Ti 6-4 which at* $427°C$ *has* $E_m = 87.5\ GPa$, $\nu_m = 0.3$. *The alumina fibers have a diameter of* $d_f = 140\ \mu m$, *and at* $427°C$ *have* $E_f = 379\ GPa$, $\nu_f = 0.27$, *and apparent strength* $F_{ft} = 2614\ MPa$.

Solution to Example 4.5 *Using the mechanics of materials approach:*

$$F_{1t} = F_{ft} V_f + \sigma_m^*(1 - V_f)$$
$$\sigma_m^* = F_{ft}\frac{E_m}{E_f} = 2614\frac{87.5}{379} = 603\ MPa$$
$$F_{1t} = 2614(0.27) + 603(1 - 0.27) = 1146\ MPa$$

4.4.2 Longitudinal Compressive Strength

A simple equation used in design to predict the longitudinal compressive strength is given by [156]

$$F_{1c} = G_{12}\ (1 + 4.76\ \chi)^{-0.69} \tag{4.85}$$
$$\chi = \frac{G_{12}\ \alpha_\sigma}{F_6} \tag{4.86}$$

where α_σ is the standard deviation of fiber misalignment, which can be either measured by accepted experimental procedures [157] or calculated from (4.85) in terms of available experimental data for F_{1c} (see Table 1.3). Then, (4.85) can be used to predict values of F_{1c} for materials with other properties such as different shear modulus G_{12} or shear strength F_6 or to estimate the effect of manufacturing improvements that might lead to a reduction of fiber misalignment. Equation (4.85) is derived in Section 4.4.3.

Example 4.6 *Estimate the effect of misalignment on the compressive strength of a carbon–epoxy (AS4–3501) composite fabricated by hand lay-up of prepreg tape, vacuum bagged and oven cured, resulting in a fiber volume fraction* $V_f = 0.6$. *Sample A was fabricated under controlled laboratory conditions, resulting in a standard deviation of misalignment* $\alpha_{\sigma_A} = 1.0°$. *Sample B was fabricated under normal processing conditions, resulting in* $\alpha_{\sigma_B} = 1.7°$.

Solution to Example 4.6 *From Table 2.2, the properties of the carbon AS4-D fiber are:* $E_f = 241$ GPa, $\nu_f = 0.2$ *(assumed equal to that of T300). Since 2.9 does not list epoxy 3501, the following values are assumed:* $E_m = 5.3\ GPa$, $\nu_m = 0.38$.

The compressive strength is given by (4.86) and (4.85). Therefore, the in-plane shear stiffness and strength must be determined first. Assuming both constituents to be isotropic, which is an approximation for carbon fiber, the shear moduli are

$$G_f = \frac{241}{2(1+0.2)} = 100.42\ GPa$$
$$G_m = \frac{5.3}{2(1+0.38)} = 1.92\ GPa$$

Using (4.37), the in-plane shear modulus is

$$G_{12} = 1.92 \left[\frac{(1+0.6)+(1-0.6)1.13/100}{(1-0.6)+(1+0.6)1.13/100} \right] = 7.170\ GPa$$

The shear strength cannot be estimated from fiber and matrix properties alone because it is controlled by the fracture toughness G_{IIc} *of the composite. To a large extent, the latter depends on the flaws in the composite, which are mostly determined by the manufacturing process. Therefore, the shear strength is found in Table 1.4 as* $F_6 = 71\ MPa$. *For sample A, using (4.85–4.86), the compressive strength* F_{1c} *is*

$$\chi = \frac{7,170 \times 1.01 \times \frac{\pi}{180}}{71} = 1.7800$$

$$F_{1c} = 7,170\,(1 + 4.76(1.7800))^{-0.69} = 1,520\ MPa$$

For sample B, $\chi = 2.9961$ *and* $F_{1c} = 1,093\ MPa$. *These values are close to typical experimental data reported in Table 1.4.*

4.4.3 Fiber Microbuckling (*)[6]

The compressive strength of continuous fiber reinforced polymer-matrix composites (PMC) is lower than the tensile strength, about one half or less. The mode of failure is triggered by fiber microbuckling, when individual fibers buckle inside the matrix (Figure 4.15). The buckling process is controlled by fiber misalignment, shear modulus G_{12}, and shear strength F_6 of the composite.

Fiber misalignment measures the waviness of the fibers in the composite. Fiber waviness is always present to some extent, even when great care is taken to align the fibers during processing. Waviness occurs because of several factors. The fibers are wound in spools as soon as they are produced, which induces a natural curvature in the fibers. Then, fibers tend to curl when stretched on a flat mold. Furthermore, many fibers are wound together over a spool to form a tow or roving during fiber production. The fibers wound on the outside of the spool are longer than those wound in the inside. When the tow is stretched, the longer fibers are loose and

[6]Sections marked with (*) can be omitted during the first reading but are recommended for further study and reference.

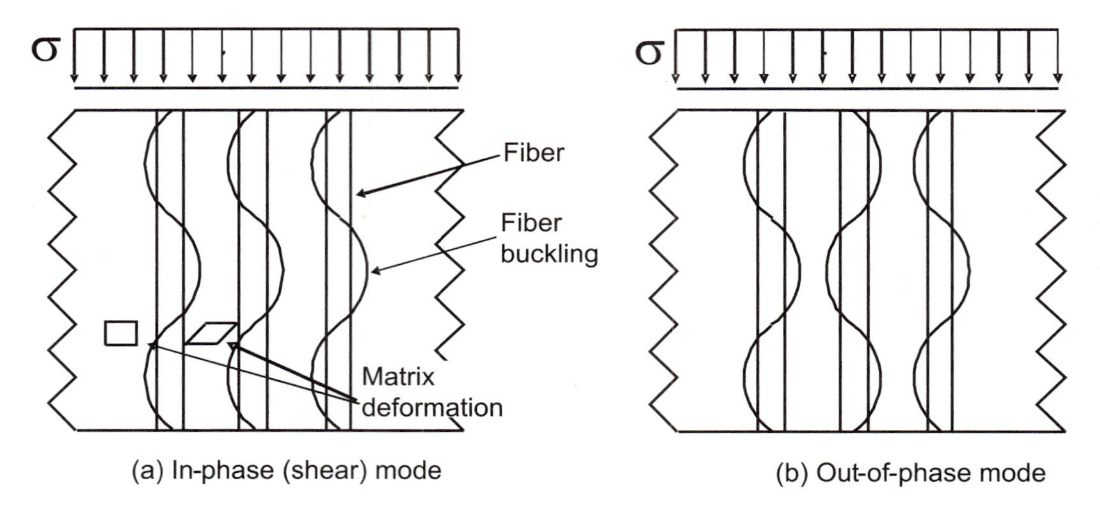

Figure 4.15: Modes of fiber microbuckling.

microcatenary is formed. Microcatenary can be shown by stretching a tow horizontally and letting the longer fibers hang under their own weight. The longer fibers hang in a catenary shape, just as electrical power lines do. In the final composite part, microcatenary appears as fiber misalignment.

Finally, there is the shrinkage of the polymer. During cure, most thermoset polymers shrink 3 to 9% by volume, that is 1 to 3% on any direction. Along the fiber direction, it means that the fiber, which does not shrink during curing, must accommodate the strain in the form of waviness.

The amount of fiber misalignment varies from fiber to fiber and it can be accurately measured by the procedure given in [157]. The measurement technique consists of cutting a sample at an angle ϕ with respect to the fiber orientation as illustrated in Figure 4.16. After polishing the sample and looking at it with a microscope, each fiber shows as an ellipse. The longer dimension of the ellipse b is proportional to the misalignment angle α through the equation

$$\begin{aligned} \sin(\omega) &= \frac{d_f}{b} \\ \alpha &= \omega - \phi \end{aligned} \tag{4.87}$$

where d_f is the fiber diameter. The average, or mean value $\hat{\alpha}$, of all the measured angles will be zero since fibers are misaligned equally at positive and negative angles with respect to the nominal fiber orientation. The experimental data of fiber angle can be fitted accurately with a normal distribution [158]. The standard deviation of the distribution can be computed as

$$\alpha_\sigma = \sqrt{\frac{n \sum_{i=1}^{n} \alpha_i^2 - \left(\sum_{i=1}^{n} \alpha_i\right)^2}{n(n-1)}} \tag{4.88}$$

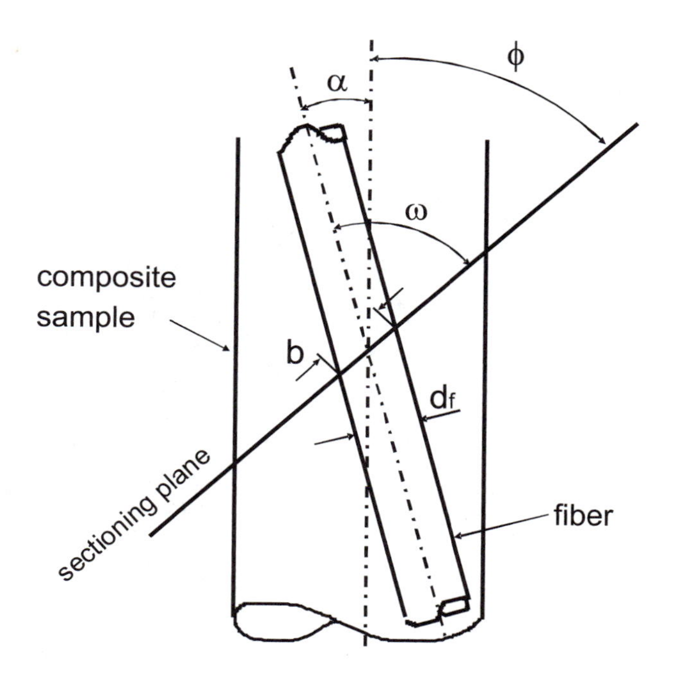

Figure 4.16: Misalignment measuring technique.

where n is the number of observations of misalignment angle α_i. The normal probability density (Appendix A) with zero mean is

$$p(\alpha) = \frac{1}{\alpha_\sigma \sqrt{2\pi}} \exp\left(-\frac{\alpha^2}{2\alpha_\sigma^2}\right) \tag{4.89}$$

shown in Figure 4.17 represents the observed data and indicates the probability of finding a particular value of misalignment α in the cross-section of the composite.

The first formula for compressive strength was proposed by Rosen [159], recognizing the fact that compression failure is triggered by fiber microbuckling. When fibers buckle, they can all buckle in phase (shear mode, Figure 4.15a) or out of phase (extension mode, Figure 4.15b). It can be demonstrated that the shear mode will always occur at a lower compressive stress for polymer matrix composites (PMC) having practical values of fiber volume fraction.

The first approximation to the problem is to assume that the buckling load of the fibers is the limiting factor for compressive strength. To obtain the buckling load, Rosen [159] performed a stability analysis of straight fibers laterally supported by the matrix. Furthermore, the shear stress–strain law of the composite was assumed to be linear, i.e., $\tau = G_{12}\,\gamma_6$. Actually, polymer matrix composites have a nonlinear stress–strain law, as shown in Figure 4.11 and described by (4.43).

Even though fibers are wavy, they are assumed to be straight in Rosen's model because this is the simplest approach to stability problems. The stability analysis is performed on what is called the *perfect system* (straight fibers) [160]. The buckling

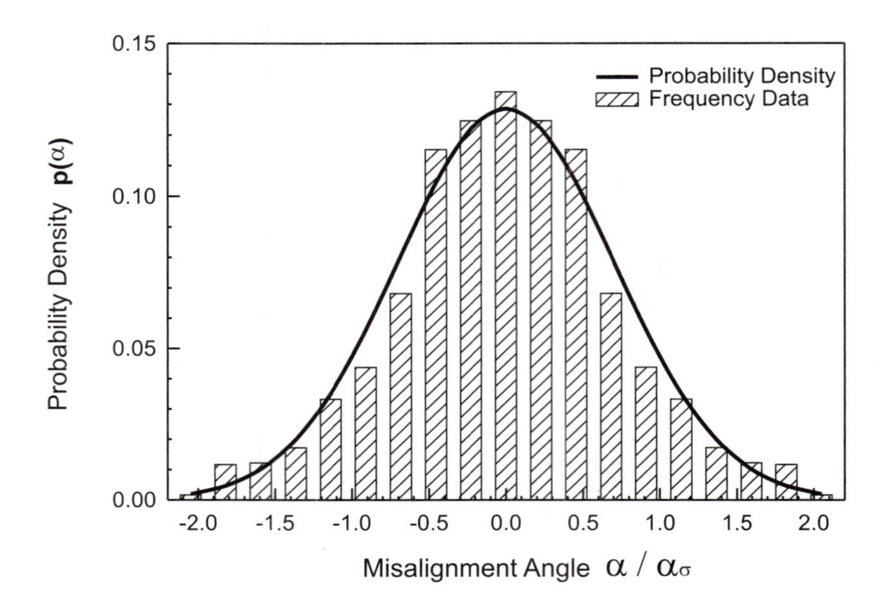

Figure 4.17: Probability density function $p(\alpha)$ describing the number of fibers having a misalignment angle α.

load is found for the perfect system and it is assumed that the load capacity of the imperfect system (wavy fibers), is lower than the buckling load of the perfect system. The buckling stress is then

$$\sigma_{cr} = \frac{G_m}{1 - V_f} \cong G_{12} \tag{4.90}$$

This formula over predicts compressive strength as much as 200%. According to this model, the buckling stress is numerically equal to the matrix-dominated approximation of the composite shear modulus (4.36). An empirical correction of (4.90) can be made by adding a factor k, which is determined from one test at one particular value of the fiber volume fraction, and assuming that the value of the factor k does not vary with the fiber volume fraction

$$\sigma_{cr} = \frac{kG_m}{1 - V_f} \approx kG_{12} \tag{4.91}$$

This approach has been validated for boron–epoxy composites [89]. The main problem with using the empirical correction is that testing must be done, and testing for compressive strength is very difficult. In fact, no single test method can be used for all materials, or even for various thicknesses of samples of the same material. Furthermore, composites from actual production have vastly different values of compressive strength when compared to composites made in the lab, even if the same materials are used. Also, samples cannot be machined out of production parts without introducing significant damage that affects the results. Therefore, there is a strong motivation for deriving a formula that can be related to parameters that can be easily measured on composites manufactured in the shop rather from samples made in the lab.

A better estimate of the buckling stress can be made by taking into account the initial misalignment of the fibers and the nonlinear stress–strain law given by (4.43). The (effective)[7] buckling stress of a fiber bundle in which all the fibers have the same misalignment is given by [156, (9)][8]

$$\begin{aligned}\widetilde{\sigma}(\alpha,\gamma) &= \frac{F_6}{2(\gamma+\alpha)}\cdot\frac{(\sqrt{2}-1)(e^{\sqrt{2}g}-e^{2g})+(\sqrt{2}+1)(e^{(2+\sqrt{2})g}-1)}{1+e^{2g}+e^{\sqrt{2}g}+e^{(2+\sqrt{2})g}}\\ g &= \gamma G_{12}/F_6\end{aligned} \tag{4.92}$$

where G_{12} and F_6 are the in-plane shear modulus (Section 4.2.4) and in-plane shear strength (Section 4.4.8), respectively. Equation (4.92) is plotted in Figure 4.18 [156, Figure 3] for various values of misalignment angle α. The maxima of the curves (4.92) for all misalignment angles, shown in Figure 4.18, represent the imperfection sensitivity curve, also shown in Figure 4.18. The maxima is given by [156, p. 495][9]

$$\begin{aligned}\widetilde{\sigma}(\alpha) &= G_{12}+\frac{16C_2\alpha}{3\pi}-\frac{4\sqrt{2C_2\alpha\,(8C_2\alpha+3\pi G_{12})}}{3\pi}\\ C_2 &= \frac{G_{12}^2}{4F_6}\end{aligned} \tag{4.93}$$

The imperfection sensitivity curve represents the compressive strength $F_{1c}(\alpha)$ of a fiber and surrounding matrix as function of its misalignment α. For negative values of misalignment, it suffices to assume that the function is symmetric, i.e., $F_{1c}(\alpha) = F_{1c}(-\alpha)$.

The stress in the composite is a function of the misalignment α and the shear strain γ, induced when the fibers buckle in phase (Figure 4.15a). The maximum value of (4.92) is the buckling stress $\sigma(\alpha)$ of a material in which fibers have the same misalignment α (Figure 4.19). However, in a real composite not all the fibers have the same value of fiber misalignment. Instead, a normal probability distribution (Figure 4.17) gives the probability of finding a fiber with any angle α.

A formula for the compressive strength of the composite can be derived by taking into account that the fibers with larger value of misalignment will buckle first during the loading process and the stress is redistributed to those that have not buckled.

Once a fiber has buckled, it carries negligible stress, thus overloading those fibers that have lower misalignment. The unbuckled fibers are able to carry the extra load only up to certain point. With reference to Figure 4.19, buckling of the fibers proceeds from large values of misalignment, both positive and negative, towards

[7] See discussion immediately above (4.97).

[8] Note that a *tilde* is used in this book, while an *over line* is used in [2, 156].

[9] Note the redefinition of C_2 with respect to [156, (17)].

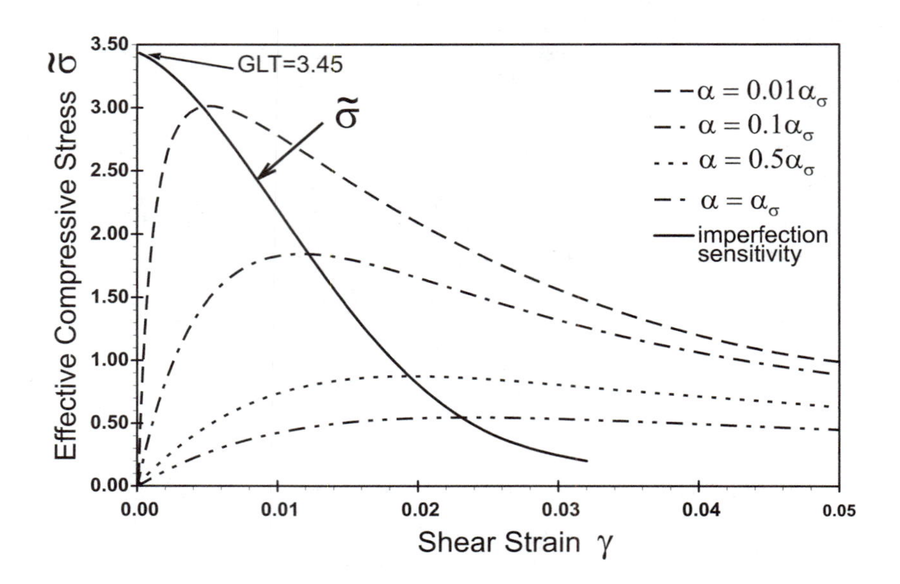

Figure 4.18: Equilibrium states for various values of fiber misalignment α.

lower values of misalignment. The number of unbuckled fibers is proportional to the area under the probability density (Figure 4.17), which is given by

$$A(\alpha) = \int_{-\alpha}^{\alpha} p(\alpha')d\alpha' = F(\alpha) - F(-\alpha) \tag{4.94}$$

where the cumulative distribution function (CDF) is the integral of (4.89)

$$F(\alpha) = \int_{-\infty}^{\alpha} p(\alpha')\ d\alpha' = \frac{1}{2}\left[1 + erf(z)\right] \tag{4.95}$$

where $erf(z)$ is the error function [161] and α_σ is the standard deviation of fiber misalignment given by (4.88) and

$$z = \frac{\alpha}{\alpha_\sigma \sqrt{2}} \tag{4.96}$$

At any given point in the loading process, the applied stress (see Figure 4.19) is equal to the product of the actual (effective) stress $\widetilde{\sigma}$ carried by the unbuckled fibers and surrounding matrix times the area of the unbuckled fibers and surrounding matrix $A(\alpha)$, as follows[10]

[10]The notation of [2, Chapter 8] is used here, so that $\widetilde{\sigma}$ is the effective stress carried by the fibers that remain unbuckled and σ is the applied stress that one would compute as the load over the original area of the sample.

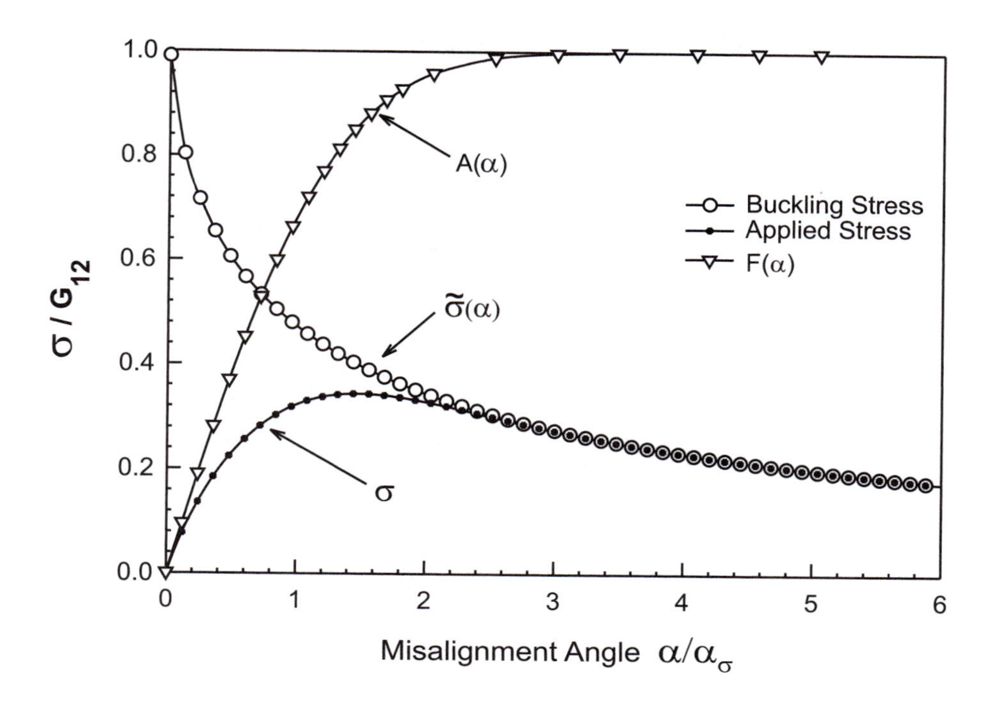

Figure 4.19: Buckling stress $\widetilde{\sigma}$ and applied stress σ (see (4.97)) as a function of fiber misalignment α/α_σ.

$$\sigma = \widetilde{\sigma}(\alpha)A(\alpha) \tag{4.97}$$

where the buckling stress $\widetilde{\sigma}(\alpha)$ is given by (4.93) and it is shown in Figure 4.19 [156, Figures 4–5]. The maximum stress that can be applied is the compressive strength of the material

$$F_{1c} = \max(\sigma) \tag{4.98}$$

Equation (4.98) can be solved explicitly [156, (24)] in terms of the shear modulus G_{12} and a dimensionless number $\chi = G_{12}\alpha_\sigma/F_6$, where α_σ is the standard deviation of fiber misalignment *in radians*, G_{12} is the shear modulus, and F_6 is the in-plane shear strength of the composite. The explicit solution is a rather complex equation, but a very good approximation to it is given by (4.85).

Note that (4.85) is not an empirical equation, but just an approximation to the explicit solution to (4.98). *No experimental data of compressive strength has been used in its derivation.* The constants 4.76 and −0.69 are not adjustable parameters but constants chosen to approximate the explicit solution to (4.98).

Comparison between experimental data and predicted values obtained with (4.85) is shown in Table 4.2. Note that (4.85) and the procedure explained in this section do not apply to Metal Matrix Composites (MMC) or when the misalignment is nearly nonexistent, as it is the case of some boron–fiber composites, because in those cases the mechanism of compression failure is vastly different.

Table 4.2: Comparison of predicted and experimental values of compressive strength for various composite materials.

Material	$\mathbf{V}_f$ [%]	$\mathbf{G}_{12}$ [MPa]	$\mathbf{F}_6$ [MPa]	α_σ [deg]	χ	**Exp.** F_{1c}/G_{12}	**(4.85)** F_{1c}/G_{12}
XAS–914C	60	5133	125	1.01	0.724	0.352	0.357
AS4–PEEK	61	5354	157.5	1.53	0.908	0.310	0.315
AS4–E7K8	60	7923.5	90.95	1.18	1.793	0.213	0.211
Glass–vinyl ester	43	4223	54.77	3.3	4.441	0.124	0.118
Glass–polyester	40.2	3462	40.57	3.45	5.148	0.138	0.107

Although the standard deviation of fiber misalignment α_σ can be measured by optical methods [157], the instrumentation required and effort involved may not be justifiable in certain situations. In these cases, *it is recommended to use experimental data of F_{1c} at one value of fiber volume fraction to back-calculate α_σ in (4.85–4.86). Then, use (4.85) to predict values of F_{1c} for other values of fiber volume fraction.*

Experimental values of compressive strength can be obtained using a number of different fixtures, including those described in [118, ASTM D695, D3410] and [41, SACMA modified D695], among others. An extensive evaluation of most of the existing compression test methods can be found in [162, 163].

4.4.4 Transverse Tensile Strength

Transverse tensile failure of a unidirectional lamina occurs when a transverse crack propagates along the fiber direction, splitting the lamina. Therefore, transverse strength is a fracture mechanics problem. In Section 4.4.5, the following equation is derived to predict the transverse tensile strength F_{2t} of a unidirectional lamina

$$F_{2t} = \sqrt{\frac{G_{Ic}}{1.12^2 \pi (t_t/4) \Lambda_{22}^0}} \tag{4.99}$$

where G_{Ic} is the fracture toughness in mode I (Tables 1.3–1.4; see discussion in Section 4.4.5). The transition thickness can be approximated as $t_t = 0.6\,\text{mm}, 0.8\,\text{mm}$, for E-glass–epoxy and carbon–epoxy composites, respectively (see discussion in Section 4.4.5). Finally, Λ_{22}^0 is given by

$$\Lambda_{22}^0 = 2\left(\frac{1}{E_2} - \frac{\nu_{12}^2 E_2^2}{E_1^3}\right) \tag{4.100}$$

Older empirical formulas were derived without consideration for fracture mechanics. One such formula for estimating the transverse tensile strength is given by [122, pp. 3–20, 21] and [164, pp. 51, 52], as

$$F_{2t} = F_{mt}C_v\left[1+\left(V_f-\sqrt{V_f}\right)\left(1-\frac{E_m}{E_T}\right)\right] \tag{4.101}$$

$$C_v = 1-\sqrt{\frac{4V_v}{\pi\left(1-V_f\right)}} \tag{4.102}$$

where V_f is the fiber volume fraction, and E_T is the transverse modulus of the fiber, which in the case of isotropic fibers (e.g., glass fibers) is $E_T = E_f$. The apparent tensile strength of the matrix F_{mt} has to be back-calculated from experimental data on F_{2t} using (4.101) and an empirical C_v is included to adjust for the presence of voids, with a void volume fraction V_v. Also, the empirical formula proposed in [165] can be further modified with the same empirical reduction coefficient C_v to account for voids, as follows

$$F_{2t} = F_{mt}C_v\left(1-V_f^{\frac{1}{3}}\right)\frac{E_2}{E_m} \tag{4.103}$$

where E_2 is the transverse modulus of the composite, and E_m is the modulus of the matrix. These equations usually predict values lower than the strength of the matrix and (4.103) yields lower values for most materials. The effect of voids is very detrimental to the transverse strength and this is reflected by the empirical formulas.

Experimental values of F_{2t} can be obtained by [118, ASTM D3039] but high dispersion of the data should be expected.

4.4.5 Mode I Fracture Toughness (*)[11]

When the lamina is part of a laminate, the lamina's transverse tensile strength appears to increase due to the constraining effect provided by the adjacent laminae. A parameter such as F_{2t} that changes its value as a function of such conditions cannot be a true material property; material properties are invariants. The reason for this dilemma is that transverse strength is a fracture problem. Therefore, the controlling material property is the fracture toughness[12] G_{Ic} in crack opening mode I (see discussion in Section 7.2.1).

Since the empirical equations available in the literature are not based on fracture mechanics, none of them are able to predict the transverse tensile strength accurately, except for particular cases. Instead values of fracture toughness G_{Ic} are required if transverse strength is the controlling mode of failure of the component (see Section 7.2.1).

[11] Sections marked with (*) can be omitted during the first reading but are recommended for further study and reference.

[12] In classical fracture mechanics, the term fracture toughness K_c refers to the critical value of stress intensity [30], but in this work the term fracture toughness refers to the critical energy release rate G_{Ic} because the term "stress intensity" is not commonly used in fracture mechanics analysis of composite materials.

A unidirectional (UD) lamina subject to σ_2 fails when a corner crack[13] propagates through the thickness, on the $1-3$ plane (Figure 4.20) [166–168]. If the value of F_{2t} is known from testing a UD lamina subjected to σ_2, then the fracture toughness in mode I can be *estimated* as follows

$$G_{Ic} = 1.12^2 \pi a_0 \Lambda_{22}^0 F_{2T}^2 \tag{4.104}$$

where Λ_{22}^0 is given in (4.100) and $2a_0$ is the size of a crack representative of the flaw sizes typical for the composite and the manufacturing process employed (Figure 7.8).

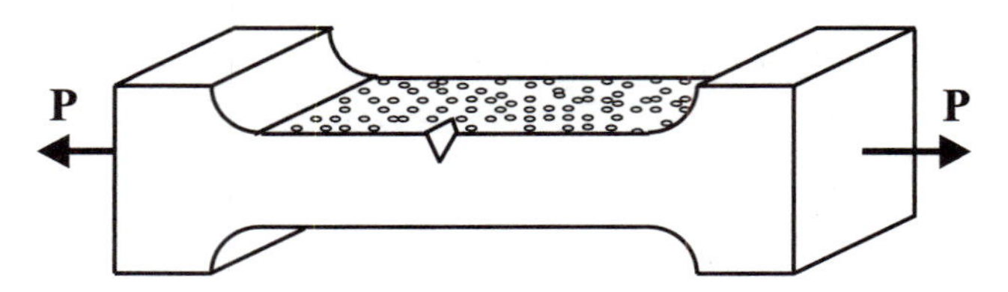

Figure 4.20: Corner crack in a UD lamina loaded in transverse tension.

The half-size of the representative crack can be estimated as $a_0 = t_t/4$, where the transition thickness t_t of a composite is either known or can be assumed (see discussion on page 239). Based on the experimental study in [169], as a first approximation one can assume $t_t = 0.6\,\text{mm}, 0.8\,\text{mm}$ for E-glass–epoxy and carbon–epoxy composites, respectively. Then, by inverting (4.104), the transverse strength of the unidirectional lamina can be predicted by (4.99).

Interlaminar fracture toughness can be measured using tests methods specifically designed for that purpose [118, ASTM D 5528], and [31, 170]. While interlaminar fracture occurs along the interface between two laminae, the crack that leads to transverse failure F_{2t} is intralaminar, i.e., it propagates first through the thickness of the lamina to occupy the entire thickness of the ply (see Figure 7.8) and then propagates along the fiber direction. Values of fracture toughness for intralaminar cracks have been reported to be higher than for interlaminar (delamination) cracks. Using values of interlaminar fracture toughness to estimate transverse tensile strength is likely to result in a conservative design.

Example 4.7 *Estimate the transverse tensile strength of IM7–8552 with properties given in Table 1.4.*

Solution to Example 4.7 *The transition thickness for a generic carbon–epoxy can be estimated to be $t_t = 0.8$ mm (see discussion on page 239). Then, the half-size of the representative crack (Figure 7.8) can be estimated to be $a = 0.8/4 = 0.2$ mm. Then, using (4.104)*

[13] In this configuration, a corner crack, representative of the initial defects in the material, requires the least energy to propagate. Other representative cracks, such as embedded penny shaped cracks, would require more energy to propagate in this case.

$$\Lambda_{22}^{0} = 2 * \left(\frac{1}{9,080} - \frac{0.32^2 \; 9,080^2}{171,400^3} \right) = \frac{1}{4,540}$$

$$F_{2t} = \sqrt{\frac{4,540 \; MPa \; (0.277 \; KJ/m^2)}{(1.12^2 \; 0.2 \; mm) \; \pi}} = 40 \, MPa$$

which compares reasonably well with the value $F_{2t} = 62.3 \, MPa$ given in Table 1.4, considering that the transition thickness was not determined using IM7–8552 composite, and that tensile strength values are very sensitive to details of sample processing, specimen preparation, and test protocol.

4.4.6 Transverse Compressive Strength

An empirical formula for estimating the transverse compressive strength is given by [122, pp. 3–20, 21] and [164, pp. 51, 52]

$$F_{2c} = F_{mc} C_v \left[1 + \left(V_f - \sqrt{V_f} \right) \left(1 - \frac{E_m}{E_T} \right) \right] \tag{4.105}$$

where the apparent compressive strength of the matrix F_{mc} has to be back-calculated from experimental data on F_{2c} using (4.105). As in (4.101), C_v is included to adjust, with yet another empirical factor, for the presence of voids, and E_T is the transverse modulus of the fiber, which in the case of isotropic fibers (e.g., glass fibers) is $E_T = E_f$.

Since empirical formulas are not accurate, experimental values are often required and obtained using, for example, [118, ASTM D695, D3410]. For a more detailed discussion of this topic, please see Section 4.4.7.

4.4.7 Mohr-Coulomb (*)[14]

Polymer matrix composites fail when a transverse compressive stress $-\sigma_2 = F_{2c}$ is applied, but the mode of failure is *shear* on a plane at an angle with respect to the load direction (2-direction in Figure 4.21). When testing for F_{2c}, it is useful to measure the angle of the fracture plane α_0 as well.

Brittle materials with internal friction such as concrete exhibit shear strength S_T that increases linearly with the amount of compressive stress $-\sigma_2$ applied (shear fracture line in Figure 4.22), as described by the Mohr-Coulomb (M-C) failure criterion

$$S_T = F_4 - \eta_T \sigma_n \tag{4.106}$$

$$\eta = tan(\phi)$$

[14]Sections marked with (*) can be omitted during the first reading but are recommended for further study and reference.

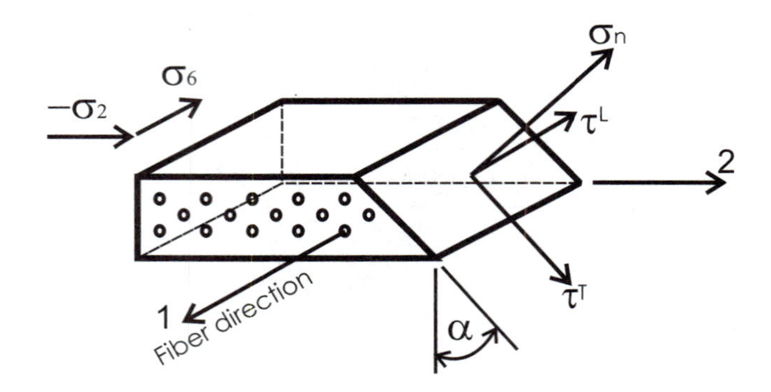

Figure 4.21: Fracture of a unidirectional lamina under compression and shear.

depicted by the *shear fracture line* in Figure 4.22. Here, F_4 is the out-of-plane (2-3, intralaminar) shear strength in the absence of normal stress, and ϕ is the angle of internal friction, both of which are material properties. Note that σ_n is negative in compression, so $S_T \geq F_4$. The parameter $\eta_T = tan(\phi)$ is called *coefficient of influence* because it is a weight factor for the influence of compression on the measured shear strength.

A typical fracture of a unidirectional lamina subject to transverse compression and in-plane shear is depicted in Figure 4.21, and as such it can be modeled with the Mohr-Coulomb criterion [167]. When the material has no internal friction ($\phi = 0$), the M-C failure criterion reduces to the Tresca failure criterion and the fracture plane in the shear-compression quadrant is at an angle $\alpha_0 = 45°$. Materials with internal friction exhibit fracture planes at angles larger than 45°. For the case of transverse compression of a unidirectional carbon-epoxy with no shear applied ($\sigma_{12} = \sigma_{23} = 0$), the lamina fails in shear along a fracture plane that contains the 1-axis and the normal to the fracture plane is at an angle $\alpha_0 \approx 53°$ with respect to the load direction [171]. The addition of in-plane shear σ_6 does not change the fact that the fracture plane remains parallel to the 1–direction because the high strength of the fibers precludes shear failure in any fracture plane that does not contain the fiber direction. In other words, the fracture plane cannot cut across fibers.

Figure 4.22 illustrates the situation in the 2-3 plane under pure transverse compression $\sigma_2 < 0, \sigma_3 = \sigma_4 = \sigma_6 = 0$. As the applied compressive stress σ_2 increases, the Mohr circle grows in size until it touches the line representing (4.106). Then, the material fails on a fracture plane at an angle α_0. Note that the rotation of the fracture plane takes place around the 1-axis. Also note that the lamina fails when the tangent to the Mohr circle intersects the ordinate axis at a value F_4.

The fracture plane is subjected to both compression σ and shear τ (Figures 4.21, 4.22), which are a function of the applied stress and the orientation of the fracture plane α, as follows

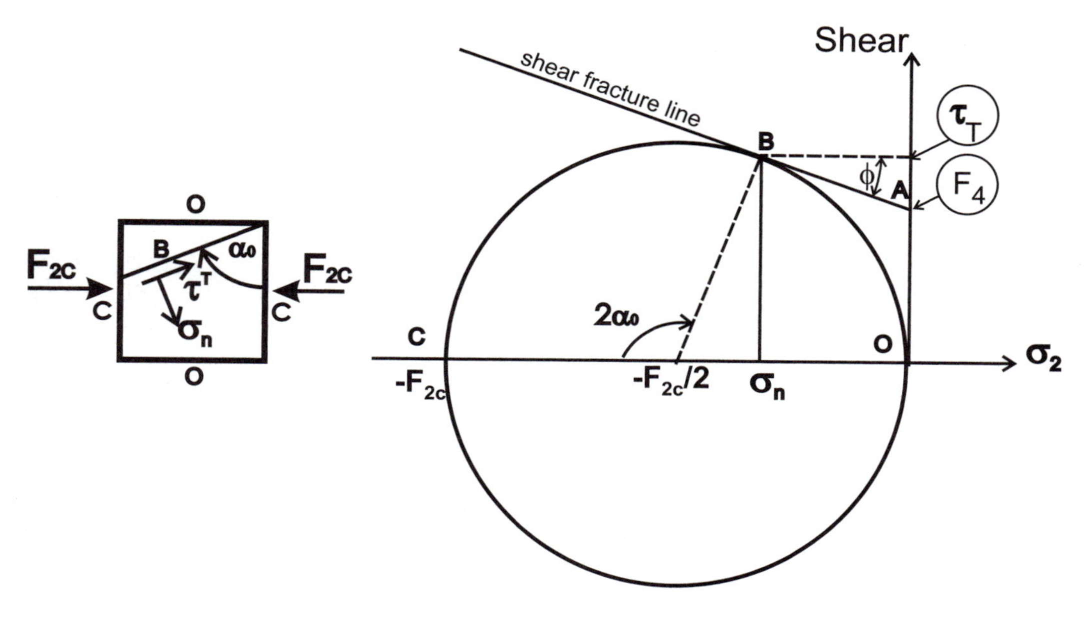

Figure 4.22: Mohr circle in the 2–3 plane under transverse compression.

$$
\begin{aligned}
\sigma_n &= \sigma_2 \cos^2 \alpha \leq 0 \\
\tau_T &= -\sigma_2 \sin \alpha \cos \alpha \\
\tau_L &= \sigma_6 \cos \alpha
\end{aligned}
\tag{4.107}
$$

where τ_L, τ_T, indicate longitudinal and transverse shear (Figure 4.21) on the fracture plane. The influence coefficient η_T can be easily measured by performing a test under transverse compression and measuring the angle of the fracture plane α_0. From Figure 4.22, $\phi = 2\alpha_0 - \pi/2$ and taking into account the definition $\eta = tan(\phi)$ in (4.106), it is clear that

$$
\eta_T = -cot\ 2\alpha_0 \tag{4.108}
$$

On the same test, one would also measure the transverse compressive strength F_{2c}, which allows for an estimate to be made for the transverse shear strength F_4. Using (4.106), (4.107), and (4.108), and noting that at failure $-\sigma_2 = F_{2t} > 0$, $\tau_T = S_T$, and $\alpha = \alpha_0$ yields

$$
F_4 = F_{2c} \cos \alpha_0 \left(sin\alpha_0 + \cos \alpha_0 \cot 2\alpha_0 \right) \tag{4.109}
$$

In the 1-2 plane, under transverse compression σ_2 and in-plane shear σ_6, the shear strength also increases linearly with the transverse compression but the fracture plane cannot rotate around the 3-axis because the fibers force the fracture plane to remain parallel to the fiber direction. Still, the Mohr-Coulomb criterion applies, similarly to (4.106), as follows

$$S_L = F_6 - \eta_L \sigma_n \tag{4.110}$$

where S_L is the shear strength on the fracture plane due to in-plane shear and η_L is the associated influence coefficient. In principle, the influence coefficient could be measured by testing for shear strength under various values of in-plane compression. Since such data is seldom available, the following estimate has been proposed [171]

$$\eta_L = -\frac{F_6 \cos 2\alpha_0}{F_{2c} \cos^2 \alpha_0} \tag{4.111}$$

In the absence of in-plane shear ($\sigma_2 \leq 0, \sigma_6 = 0$), the unidirectional lamina fails when τ_T reaches the *increased* shear strength $F_4 - \eta_T \sigma_n$. In the absence of transverse compression ($\sigma_2 = 0, \sigma_6 \neq 0$), a unidirectional lamina fails when τ_L reaches the in-plane shear strength F_6. When both transverse compression and in-plane shear are present, a failure criterion for the transverse-compression/shear quadrants (i.e., $\sigma_2 \leq 0, \sigma_6 \neq 0$) is constructed as a quadratic interaction between the two effects [167, Eq. (4)]. The no-fail region corresponds to values of $\sigma_2, \sigma_6, \alpha$, that yield $g \leq 0$, as follows

$$g(\sigma_2, \sigma_6, \alpha) = \left(\frac{\tau_T}{F_4 - \eta_T \sigma_n}\right)^2 + \left(\frac{\tau_L}{F_6 - \eta_L \sigma_n}\right)^2 - 1 \leq 0 \tag{4.112}$$

For a given state of stress ($\sigma_2 < 0, \sigma_6$), the lamina fractures with an angle that maximizes the failure criterion (4.112). As (negative) transverse compression increases, (4.112) predicts a linear increase of shear strength dominated by in-plane shear and thus controlled by (4.110), with the angle of the fracture plane remaining at $\alpha_0 = 0$ (Figure 4.23). Further increase of transverse compression causes the angle to quickly transition to much higher values; thus, the behavior is controlled by transverse compression (4.106) with shear strength quickly dropping.

Since (4.112) is a *stress-based* failure criterion, application to strength analysis of laminates requires to take into account the constraining effect of adjacent laminae. This is done by substituting the in-situ strength F_6^{is} (7.42) into (4.112), but for a unidirectional laminate F_6 remains unchanged (Section 7.2.1).

The failure index is computed as $I_F = g + 1$ and the strength ratio as $R = 1/I_F$ (see Section 7.1.1).

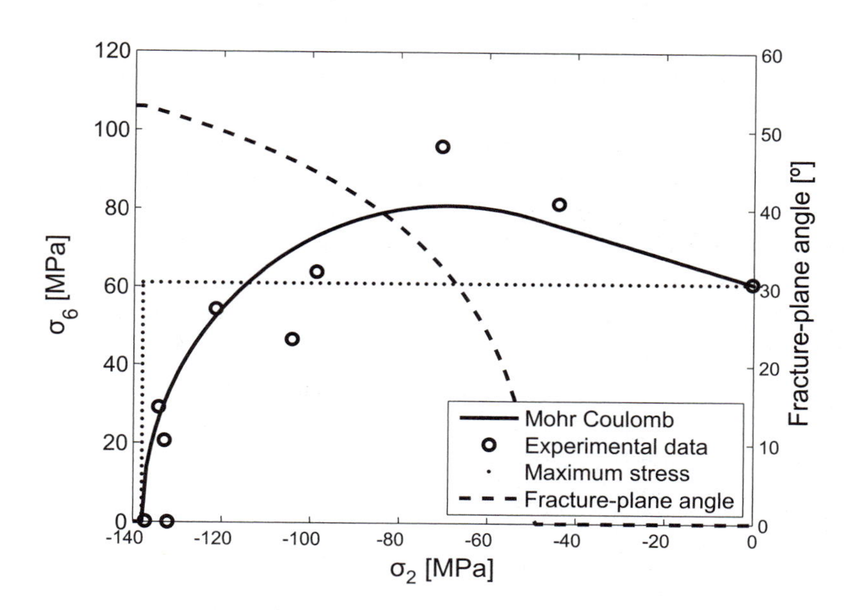

Figure 4.23: Comparison between the M-C criterion and experimental data for a UD E-glass–epoxy LY556 (Table 1.3) lamina subjected to transverse compression and in-plane shear.

Example 4.8 *Calculate the value of the damage activation function g (see (4.112)), failure index I_F, and strength ratio R, for a unidirectional lamina subjected to $\sigma_2 = -40\ MPa$, $\sigma_6 = 70\ MPa$. The material is glass–epoxy LY556 with properties:*

$E_1 = 53.48\ GPa$ $\quad F_{2T} = 35\ MPa$
$E_2 = 17.70\ GPa$ $\quad F_{2C} = 138\ MPa$ @ $\alpha_0 = 53°$
$G_{12} = 5.83\ GPa$ $\quad F_6 = 61\ MPa$
$\nu_{12} = 0.278$

Solution to Example 4.8 *Using (4.109), get*

$$F_4 = 138 \cos 53 \left(\sin 53 + \frac{\cos 53}{\tan 106} \right) = 52\ MPa$$

with (4.108), get

$$\eta_T = -\cot 106 = 0.287$$

and with (4.111) get

$$\eta_L = -\frac{61 \cos 106}{138 \cos^2 53} = 0.374$$

Then, finding the angle that maximizes (4.112) yields $\alpha = 0^o$, $g = -0.1161$. Then, $I_F = 1 + g = 0.884$ and $R = 1/I_F = 1.131$. Since $R > 1$, the lamina does not fail.

4.4.8 In-plane Shear Strength

Clear distinction between the various shear stress components, illustrated in Figure 4.24, must be made. The first subscript of a shear stress indicates the direction of the normal to the plane on which the stress acts. The second subscript indicates the direction of the stress. In Figure 4.24a, the normal to the plane considered is along the 1-axis. Both τ_{12} and τ_{13} would have to shear-off the fibers to produce failure, which is very unlikely to occur. In Figure 4.24b, the normal to the plane considered is along the 3-axis. Both τ_{31} and τ_{32} produce splitting of the matrix without shearing-off any fibers. In Figure 4.24c, both τ_{21} and τ_{23} produce splitting of the matrix without shearing-off the fibers.

Since the stresses τ_{21} is always equal to τ_{12} to satisfy equilibrium, both stresses are called σ_6. The corresponding shear strength, called *in-plane shear strength* is F_6. The composite fails when the in-plane stress τ_{21} reaches its ultimate value $F_6 = \tau_{21u}$. At this point matrix cracks appear along the fiber direction.

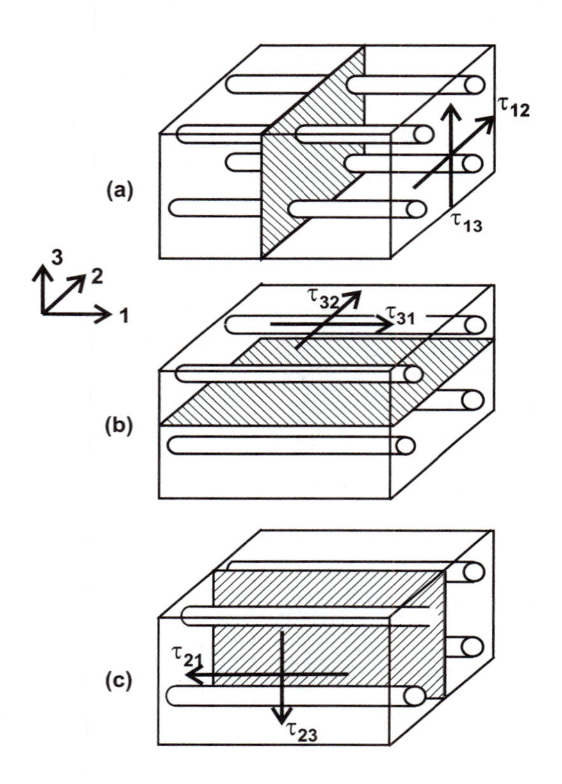

Figure 4.24: Shear components and their relationship to shear failure modes.

To illustrate further this situation, consider a tensile test of a composite as shown in Figure 4.25, with the fibers oriented at 45° with respect to the loading direction. It is well known that at 45° the shear stress is half of σ_x. Along the planes AB and CD, the material fails when $| \sigma_6 | > \tau_{21u}$. Along AD and BC, the material would fail at a higher load when $\mid \sigma_6 \mid > \tau_{12u}$.[15]

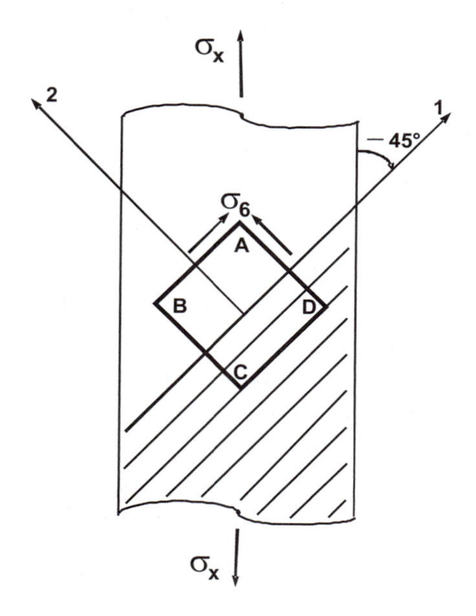

Figure 4.25: Tensile test of [−45°] lamina illustrating the most likely plane of shear failure.

In-plane shear failure of a unidirectional lamina occurs when a transverse crack propagates along the fiber direction, splitting the lamina. Therefore, shear strength is a fracture mechanics problem. In Section 4.4.9, the following equation is derived to predict the in-plane shear strength F_6 of a unidirectional lamina

$$F_6 = \sqrt{\frac{G_{IIc}}{\pi(t_t/4)\Lambda_{44}^0}} \tag{4.113}$$

where G_{IIc} is the fracture toughness in mode II (Tables 1.3–1.4; see discussion in Section 4.4.5). The transition thickness can be approximated as $t_t = 0.6\,\text{mm}, 0.8\,\text{mm}$, for E-glass–epoxy and carbon–epoxy composites, respectively (see discussion in Section 4.4.9). Finally, Λ_{44}^0 is given by

$$\Lambda_{44}^0 = \frac{1}{G_{12}} \tag{4.114}$$

The same comments made for transverse tensile strength (Section 4.4.4) apply in this case. Older empirical formulas [122, 164, 165] do not yield accurate predictions because they do not account for the propagation of a crack. They are useful only

[15] $\mid\mid$ denotes absolute value, which is used because the shear strength is independent of the sign of the shear stress in the material coordinate system. Interaction between $\sigma_6 = \sigma_x/2$ and normal stresses $\sigma_1 = \sigma_2 = \sigma_x/2$, which are also present, will be addressed in Chapter 7.

if the apparent matrix strength F_{ms} is back-calculated using the same formula and experimental data for F_6. Then, the formula can be useful to estimate changes in shear strength due to changes in fiber volume fraction, but only for the same system used to back-calculate F_{ms}. For example, one such formula is given by [122, pp. 3–20, 21] and [164, pp. 51, 52]

$$F_6 = F_{ms} C_v \left[1 + \left(V_f - \sqrt{V_f} \right) \left(1 - \frac{G_m}{G_A} \right) \right] \tag{4.115}$$

where the apparent shear strength of the matrix F_{ms} has to be back-calculated from experimental data on F_6 using (4.115). As in (4.101), C_v is included to adjust, with yet another empirical factor, for the presence of voids and G_A is the axial shear modulus of the fiber, which in the case of isotropic fibers (e.g., glass fibers) is $G_A = G_{12}$.

4.4.9 Mode II Fracture Toughness (*)[16]

When the lamina is part of a laminate, the lamina's in-plane shear strength appears to increase due to the constraining effect provided by the adjacent laminae. A parameter such as F_6 that changes its value as a function of such conditions cannot be a true material property; material properties are invariants. Since shear strength is a fracture problem, the controlling material property is the fracture toughness in crack opening mode II (see discussion in Section 7.2.1).

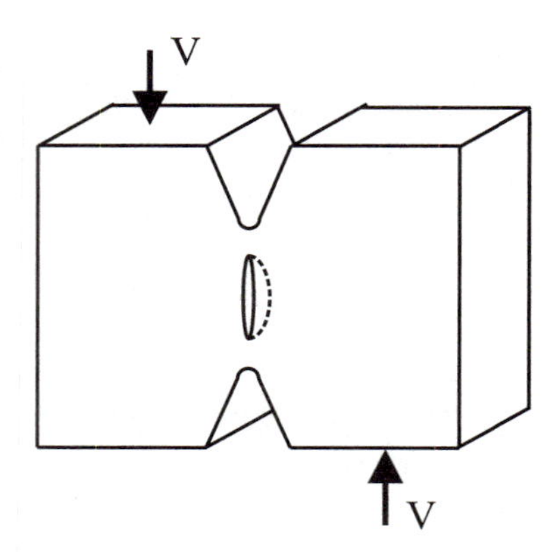

Figure 4.26: Surface crack in a UD lamina loaded in shear.

Values of fracture toughness[17] G_{IIc}, are therefore required if in-plane shear strength is the controlling mode of failure of the component. A unidirectional (UD)

[16] Sections marked with (*) can be omitted during the first reading but are recommended for further study and reference.

[17] In classical fracture mechanics, the term fracture toughness K_c refers to the critical value of stress intensity [30], but in this work the term fracture toughness refers to the critical energy release rate G_{IIc} because the term "stress intensity" is not commonly used in fracture mechanics analysis of composite materials.

lamina subject to σ_6 fails when a surface crack (Figure 4.26) propagates in the $1-3$ plane [166–168]. If the value of F_6 is known from testing a UD lamina subjected to σ_6, then the fracture toughness in mode II can be *estimated* as follows

$$G_{IIc} = \pi a_0 \Lambda_{44}^0 F_6^2 \tag{4.116}$$

where Λ_{44}^0 is given by (4.114) and $2a_0$ is the size of crack (Figure 7.8) representative of the flaw sizes typical for the composite and the manufacturing process employed.

If the transition thickness t_t of a composite is known (see discussion on page 7.2.1), one can estimate the half-size of the representative crack as $a_0 = t_t/4$. Based on the experimental study in [169], as a first approximation one can assume $t_t = 0.6\,\text{mm}, 0.8\,\text{mm}$, for E-glass–epoxy and carbon–epoxy composites, respectively. Conversely, one can measure the fracture toughness and use it to estimate the shear strength of the unidirectional lamina F_6 in terms of the measured fracture toughness G_{IIc} by inverting (4.116), which results in (4.113).

Experimental values of F_6 can be obtained by [118, ASTM D3518, D5379, D4255]. Experimental values of interlaminar fracture toughness G_{IIc} can be obtained by the procedure described in [31, 170]. For a discussion of the difference between interlaminar and intralaminar fracture, please refer to Section 4.4.5.

Example 4.9 *Estimate the shear strength of IM7–8551 with properties given in Table 1.4.*

Solution to Example 4.9 *The transition thickness for a generic carbon–epoxy can be estimated to be $t_t = 0.8\ mm$ (see discussion on page 239). Then, the half-size of the representative crack (Figure 7.8) can be estimated to be $a_0 = 0.8/4 = 0.2\ mm$. Then, using (4.116)*

$$F_6 = \sqrt{\frac{5,600\ MPa\ (0.788\ KJ/m^2)}{(0.2\ mm)\ \pi}} = 83.8\ MPa$$

which compares reasonably with the value $F_6 = 71\ MPa$ given in Table 1.4, considering that the transition thickness was not determined for the IM7–8551 composite processed in the same way as the material used measure the material properties in Table 1.4, and that some uncertainty is introduced by the type of experimental method and sample used to measure the shear strength reported in Table 1.4.

4.4.10 Intralaminar Shear Strength

Following the discussion in Section 4.4.8, the shear strengths F_4 and F_5 are discussed in this section. As it is clear from Figure 4.24, the shear stresses τ_{23} and τ_{32} cause splitting of the matrix without shearing-off any fibers. Since, by equilibrium, the shear stresses τ_{23} and τ_{32} are numerically equal, both are called σ_4. The corresponding shear strength is F_4, which can be predicted by (4.109).

The intralaminar shear strength F_4 is a matrix dominated property because the shear acts on a plane parallel to the fiber direction. In this case the fibers do not resist shear. On the contrary, the cross sections of the fibers can be viewed as circular inclusions creating stress concentrations in the matrix, thus debilitating the composite.

It can be seen in Figure 4.24 that the shear stress τ_{31} tends to split the matrix along the fiber direction. This is the usual mode of failure under stress σ_5. On the other hand, the fibers would have to be sheared-off by stress τ_{13} to produce failure, which is unlikely. Since, by equilibrium, the shear stresses τ_{13} and τ_{31} are numerically equal, both are called σ_5. The corresponding shear strength is F_5, which is numerically equal to the ultimate value of τ_{31u}. The shear stress σ_5 applies shear along the fiber direction. Therefore, the intralaminar shear strength F_5 is affected by the fiber–matrix bond strength.

When the whole thickness of the composite is unidirectional and homogeneous, there should be no difference between F_5 and F_6. This is true for a single lamina of unidirectional prepreg or for a unidirectional pultruded material. However, when transverse shear (σ_4 or σ_5) is applied to a laminate with distinct interfaces between laminae (such as prepreg lay-up), the resin rich interfaces may fail first. In this case, the intralaminar shear strength values F_4 and F_5 may be lower than the in-plane shear strength F_6.

Experimental values can be obtained by [118, ASTM D2344, D4475, D3914, D3846, D5379].

Exercises

Exercise 4.1 *Draw a representative volume element (RVE) for an epoxy matrix filled with cylindrical steel rods arranged in a rectangular array. Compute the volume fraction V_f as a function of the fiber diameter d_f and the spacing between the centers of the fibers a_x and a_y in the two directions (center to center).*

Exercise 4.2 *Write a definition for* Representative Volume Element. *Draw a set of fibers in an hexagonal array: six fibers at the vertices of an hexagon plus one in the center, with the fibers touching each other. Draw the smallest RVE for this arrangement. Compute the fiber volume fraction.*

Exercise 4.3 *Consider the following material called carbon–epoxy with a fiber volume fraction of 70% and $E_f = 379$ GPa, $\nu_f = 0.22$, $E_m = 3.3$ GPa, and $\nu_m = 0.35$. Compute E_1, E_2, ν_{12}, and G_{12} using both the rule of mixtures (ROM) and more accurate formulas recommended in this chapter. Consider the results and comment on which values are lower or upper bounds to the real value of each elastic property and what seems to be the best model in each case.*

Exercise 4.4 *The amount of fibers in a composite sample can be determined by a burn-out test. The burn-out eliminates all the resin and only the fibers remain. A composite sample plus its container weights 50.182 grams before burn-out and 49.448 grams after burn-out. The container weighs 47.650 grams. Compute the fiber weight fraction W_F and matrix weight fraction W_M.*

Exercise 4.5 *Compute E_1, E_2, G_{12}, and ν_{12} given $E_f = 230$ GPa, $E_m = E_f/50$, $G_f = E_f/2.5$, $G_m = E_m/2.6$, $\nu_f = 0.25$, $\nu_m = 0.3$, and $V_f = 40\%$. The fibers have a circular cross-section. Assume there are no voids.*

Exercise 4.6 *Compute all the elastic properties* ($E_1, E_2, G_{12}, G_{23}, \nu_{12}$) *for a unidirectional lamina for the following material combinations:*

(a) E-glass–polyester(isophthalic)

(b) S-glass–epoxy(9310)

(c) Carbon(T300)–vinyl ester

with $V_f = 0.55$*. If actual data of Poisson's ratio for either fiber or matrix is not available, take* $\nu_f = 0.22$*, and/or* $\nu_m = 0.38$*. Compare the results using the mechanics of materials (ROM) formulas (except for* G_{23}*) and other formulas recommended in this chapter.*

Exercise 4.7 *Estimate* F_{1t} *and* F_{2t} *of carbon–epoxy IM7–8552 with* $V_f = 0.5$ *and negligible void content. Assume* $\nu_f = 0.2, \nu_m = 0.38, t_k = 0.8\,mm$*.*

Exercise 4.8 *Estimate the tensile strength of Kevlar–epoxy composite with* $V_f = 0.5$ *at two temperatures, 23° C and 149° C, using the mechanics of materials approach. Use data from Tables 2.2 and 2.9, and Section 2.2. Assume that* E_f *and* F_{ft} *decrease linearly with temperature, up to 75% reduction at 171° C. The void content is negligible.*

Exercise 4.9 *With reference to Example 4.6 (p. 120), select a different matrix in order to achieve a compressive strength of at least* $1,100\ MPa$*. Assume that the standard deviation of fiber misalignment remains constant at the production value of 1.41 degrees. The fiber type and fiber volume fraction remain unchanged.*

Exercise 4.10 *The following data has been obtained experimentally for a composite based on an unidirectional carbon–epoxy prepreg (MR50 carbon fiber at 63% by volume in LTM25 epoxy). Determine if the restrictions on elastic constants are satisfied.*

$$E_1 = 156.403\ GPa, E_2 = 7.786\ GPa$$
$$\nu_{12} = 0.352, \nu_{21} = 0.016$$
$$G_{12} = 3.762\ GPa$$

Exercise 4.11 *Select a matrix, fiber, and fiber volume fraction to obtain a material with* $E_1 \geqslant 30\ GPa$*,* $E_1/E_2 \leqslant 3.5$*.*

Exercise 4.12 *Give an approximate value for the compressive strength of an Al–Boron composite with* $V_f = 0.4$*. Aluminum 2024 has* $E = 71 GPa$*, Poisson* $= 0.334$*,* $G = 26.6\ GPa$*, yield strength* $= 76\ MPa$*, fatigue limit* $= 90\ MPa$*, elongation* $= 22\%$*. Boron fibers have* $E = 400\ GPa$*, ultimate strength* $= 3.4\ GPa$*, Poisson* $= 0.25$*. Note that Boron fibers tend to be perfectly straight and that aluminum has a linear shear stress–strain plot before it yields.*

Exercise 4.13 *Compute approximate values of* E_1*,* E_2*,* G_{12}*,* ν_{12} *for the composite of exercise 4.12. Compare the results obtained using the mechanics of materials approach with the results obtained using more accurate formulas which are programmed in the accompanying software. Comment the results.*

Exercise 4.14 *Compute the tensile strength for the composite of exercise 4.12 assuming that all the fibers have the same tensile strength.*

Exercise 4.15 *Calculate the transverse tensile strength of AS4–3501-6 using the properties of Table 1.4. Assume* $t_t = 0.8\,mm$*.*

Exercise 4.16 *Calculate the in-plane shear strength of IM7–8552 using the properties of Table 1.4. Assume* $t_t = 0.8\,mm$.

Exercise 4.17 *Using the Mohr-Coulomb failure criterion, calculate the strength ratio* R *for a unidirectional lamina of glass–epoxy LY556 subjected to:*

1. $\sigma_2 = -60\,MPa,\ \sigma_6 = 60.0\,MPa$
2. $\sigma_2 = -60\,MPa,\ \sigma_6 = 80.255\,MPa$
3. $\sigma_2 = -60\,MPa,\ \sigma_6 = 90.0\,MPa$

Exercise 4.18 *Calculate the intralaminar shear strength* F_4 *of IM7–8552 using the properties of Table 1.4. Assume* $\alpha_0 = 53°$.

Chapter 5

Ply Mechanics

The main objective of this chapter is to present the constitutive equations of a lamina (also called *ply* or *lamina*) arbitrarily oriented with respect to a reference coordinate system. This is a necessary step before studying the mechanics of the laminate in the next chapter. First, a review of stress and strain is presented. Then, the assumption of plane stress is introduced to obtain a reduced version of the constitutive equations for a lamina. A review of the coordinate transformations precedes a section on off-axis constitutive equations for a lamina. Finally, the three dimensional constitutive equations are developed.

5.1 Coordinate Systems

There are two coordinate systems that are used in composites design. The material coordinate system (denoted by axes 1, 2, 3), is a cartesian coordinate system (Figure 5.1) that has the 1-axis aligned with the fiber direction. The 2-axis is on the surface of the composite shell and it is perpendicular to the 1-axis. The 3-axis is perpendicular to the surface of the composite shell and to the 1- and 2-axis. Each lamina has its own material coordinate system aligned with the fiber direction.

The laminate coordinate system (denoted by x, y, z) is common to all the laminae in the laminate. The orientation of the laminate system is chosen for convenience during the structural analysis. Therefore, it may be aligned with the boundary of the part being analyzed, with the direction of the major load, etc.

5.2 Stress and Strain

The concepts of stress and strain are reviewed in this section. The contracted notation for stress and strain is introduced. This notation is widely used for the analysis of composite materials and will be employed in the remainder of this book.

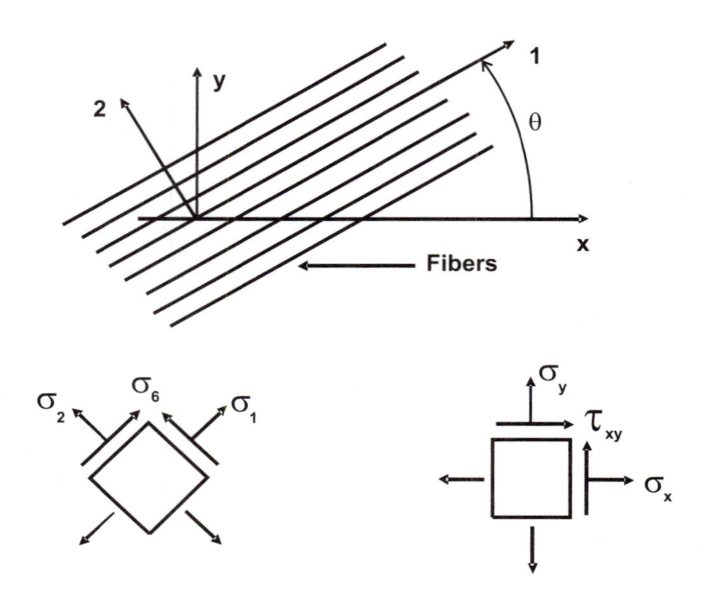

Figure 5.1: Laminate and material coordinate systems.

5.2.1 Stress

Consider a material body depicted in Figure 5.2 subject to a set of forces P_i, which are in equilibrium. The two portions limited by the plane shown are held together by internal forces. If the body is cut and the top portion is removed, a system of forces F needs to be added for each area ΔA (Figure 5.3) to restore the equilibrium. Each of the forces F acts over a portion of the area ΔA and the resultant of all the forces F equilibrates the external loads (P_1 and P_2 in Figure 5.3). The direction of the force F does not necessarily coincide with the normal to the plane because a shear force F_s may be needed in addition to a normal force F_n to restore equilibrium.

For a fixed set of external forces, the magnitude of the force depends on the area ΔA of the cut and the orientation of the cut. To obtain a quantity that is independent of the magnitude of the area ΔA, a stress vector is defined at each point P as the quotient of the force over the area ΔA when the area tends to zero

$$\vec{t} = \lim_{\Delta A \to 0} \frac{\vec{F}}{\Delta A} \tag{5.1}$$

where $\vec{t}$ indicates a vector with three components (e.g., $\vec{t} = (t_x, t_y, t_z)$). The stress vector is aligned with the force F. Unfortunately, the stress vector is not a quantity uniquely defined by the coordinates of each material point (point P) because there is one stress vector for every point and every different orientation of the cutting plane. The stress tensor is introduced to have a quantity that completely characterizes the state of stress at each point, valid for any cutting plane. For this, consider a rectangular piece of material, cut around the point of interest (Figure 5.4), and compute the stress vector on every face. Then, compute the normal and shear

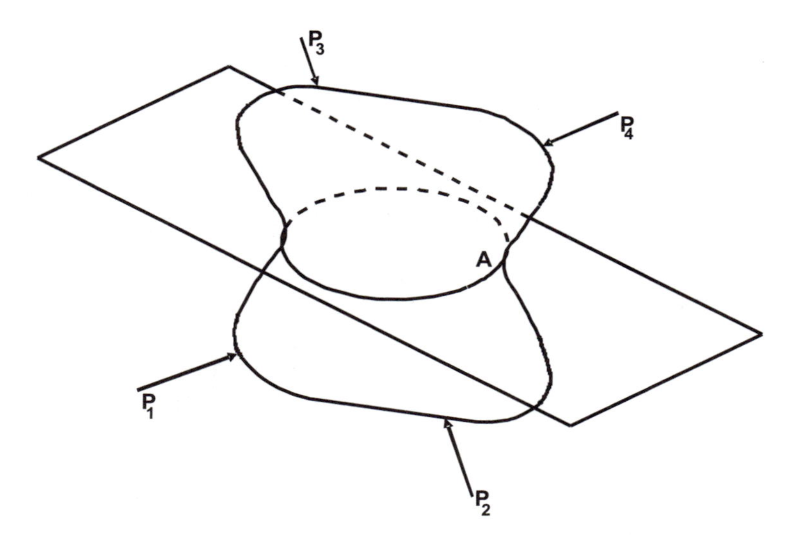

Figure 5.2: A body in equilibrium is cut with an arbitrary plane.

components of each stress vector, and order them in matrix form as

$$\boldsymbol{\sigma} = \begin{bmatrix} \sigma_{11} & \sigma_{12} & \sigma_{13} \\ \sigma_{21} & \sigma_{22} & \sigma_{23} \\ \sigma_{31} & \sigma_{32} & \sigma_{33} \end{bmatrix} \tag{5.2}$$

For each stress tensor component, the first subscript indicates the direction of the normal to the face on which the stress vector was computed and the second subscript indicates the direction of the component of the stress vector. Note that each stress vector is decomposed into a normal and two shear components.

Cauchy's law gives the projections of the stress vector (Figure 5.5) on the coordinate axis (t_1, t_2, t_3), as function of the stress tensor and the normal $\vec{n} = (n_1, n_2, n_3)$ to the surface (Figure 5.5), as

$$t_i = \sigma_{ij} n_j \; (i, j = 1..3) \tag{5.3}$$

where the dummy indexes i, j, can take the values of $1, 2, 3$.

In component form

$$\begin{aligned} t_1 &= \sigma_{11} n_1 + \sigma_{12} n_2 + \sigma_{13} n_3 \\ t_2 &= \sigma_{21} n_1 + \sigma_{22} n_2 + \sigma_{23} n_3 \\ t_3 &= \sigma_{31} n_1 + \sigma_{32} n_2 + \sigma_{33} n_3 \end{aligned} \tag{5.4}$$

The stress tensor is symmetric, which is indicated as

$$\sigma_{ij} = \sigma_{ji} \tag{5.5}$$

or in component form

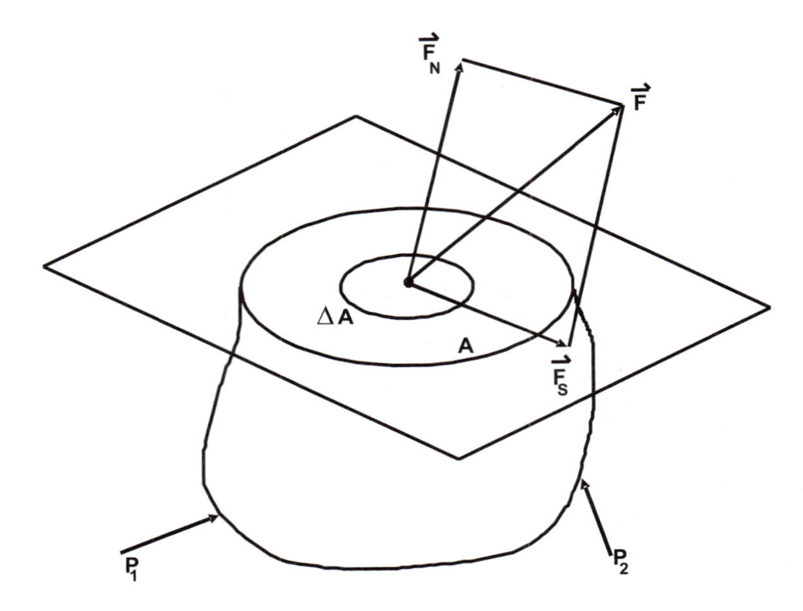

Figure 5.3: Forces required to restore equilibrium.

$$
\begin{aligned}
\sigma_{13} &= \sigma_{31} \\
\sigma_{12} &= \sigma_{21} \\
\sigma_{23} &= \sigma_{32}
\end{aligned}
\tag{5.6}
$$

Therefore, only six components are independent and (5.2) can be rewritten as

$$
\boldsymbol{\sigma} = \begin{bmatrix} \sigma_{11} & \sigma_{12} & \sigma_{13} \\ \sigma_{12} & \sigma_{22} & \sigma_{23} \\ \sigma_{13} & \sigma_{23} & \sigma_{33} \end{bmatrix} \tag{5.7}
$$

Then, the six independent components can be written in contracted notation as follows

Tensorial	**Contracted**
σ_{11}	σ_1
σ_{22}	σ_2
σ_{33}	σ_3
σ_{23}	σ_4
σ_{31}	σ_5
σ_{12}	σ_6

The following two rules are used to transform from tensorial notation to contracted notation as follows

1. repeated subscripts are written only once (e.g., 11 becomes 1).

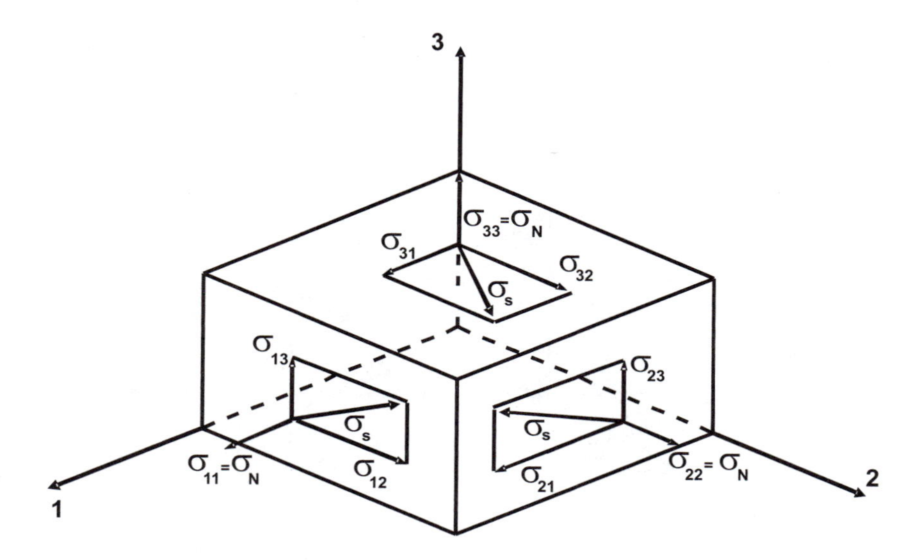

Figure 5.4: Definition of the nine components of the stress tensor.

2. if the two subscripts, say i and j, are different, then the contracted subscript is $9 - i - j$ (e.g., 12 becomes 6).

5.2.2 Strain

From elementary strength of materials, in one dimension (Figure 5.6), the definition of strain is

$$\epsilon_x = \lim_{\Delta L \to 0} \frac{\Delta L}{L} = \frac{du}{dx} \tag{5.8}$$

In three dimensions, the displacement of a point (Figure 5.7) is described by three components $\vec{u} = (u, v, w)$. The components of strain are [145]

$$\begin{aligned}
\epsilon_{xx} &= \frac{\partial u}{\partial x} \\
\epsilon_{yy} &= \frac{\partial v}{\partial y} \\
\epsilon_{zz} &= \frac{\partial w}{\partial z} \\
\gamma_{xy} &= \frac{\partial u}{\partial y} + \frac{\partial v}{\partial x} \\
\gamma_{xz} &= \frac{\partial u}{\partial z} + \frac{\partial w}{\partial x} \\
\gamma_{yz} &= \frac{\partial v}{\partial z} + \frac{\partial w}{\partial y}
\end{aligned} \tag{5.9}$$

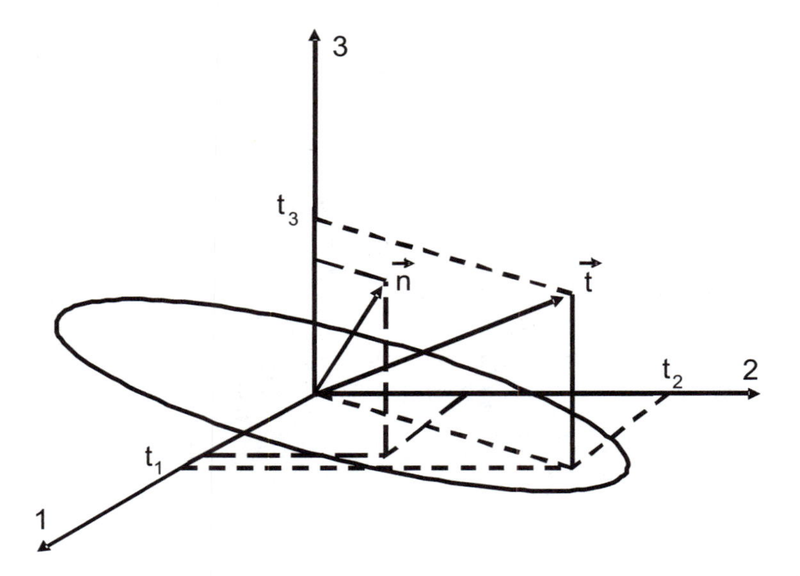

Figure 5.5: Decomposition of the stress vector into three components in a cartesian coordinate system.

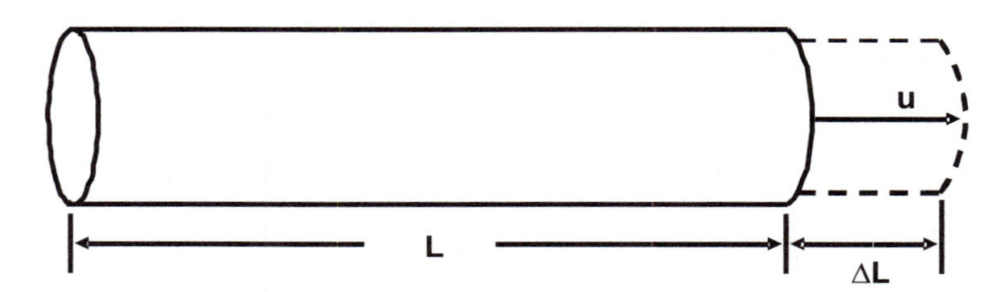

Figure 5.6: Definition of strain in the context of one-dimensional analysis.

The strain tensor is symmetric

$$\epsilon = \begin{bmatrix} \epsilon_{xx} & \gamma_{xy}/2 & \gamma_{xz}/2 \\ \gamma_{xy}/2 & \epsilon_{yy} & \gamma_{yz}/2 \\ \gamma_{xz}/2 & \gamma_{yz}/2 & \epsilon_{zz} \end{bmatrix} \tag{5.10}$$

Therefore, it can be written also in contracted notation. First, it is convenient to write it in terms of material coordinates

$$\epsilon = \begin{bmatrix} \epsilon_{11} & \gamma_{12}/2 & \gamma_{13}/2 \\ \gamma_{12}/2 & \epsilon_{22} & \gamma_{23}/2 \\ \gamma_{13}/2 & \gamma_{23}/2 & \epsilon_{33} \end{bmatrix} \tag{5.11}$$

Then, the six independent components can be written in contracted notation as follows

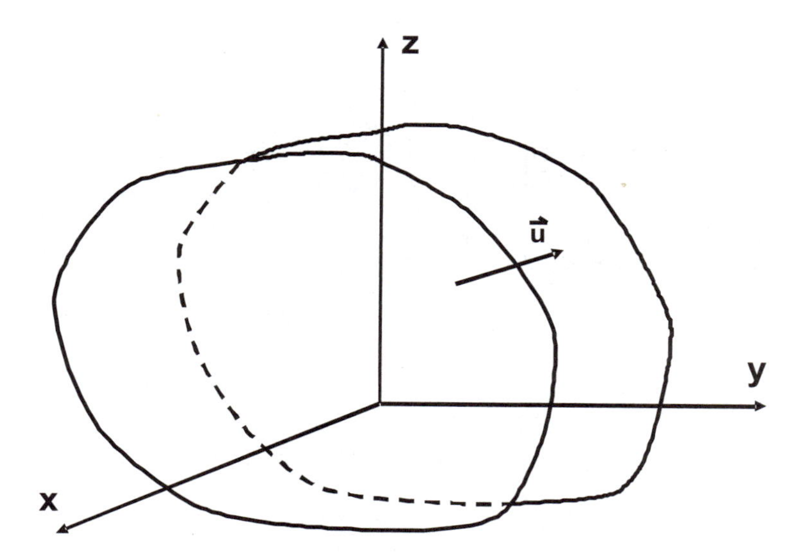

Figure 5.7: Definition of displacement vector in space.

Tensorial	**Contracted**
ϵ_{11}	ϵ_1
ϵ_{22}	ϵ_2
ϵ_{33}	ϵ_3
γ_{23}	γ_4
γ_{13}	γ_5
γ_{12}	γ_6

5.3 Stress–Strain Equations

Composite materials are used in the form of plates and shells, which have two dimensions (length and width) much larger that the third dimension (thickness). Even composite beams are thin-walled sections constructed with a collection of thin plates (e.g., I- or box-shaped cross-sections, Figure 10.1). When the thickness of a plate is small compared to the other dimensions, it is reasonable and customary to assume that the transverse stress is zero ($\sigma_3 = 0$).

The stress–strain equations for a lamina in a state of plane stress ($\sigma_3 = 0$) can be easily derived by performing the following thought experiment. First, apply a tensile stress σ_1 along the fiber direction (1-direction) as in Figure 5.8(a), with $\sigma_2 = \sigma_6 = \sigma_4 = \sigma_5 = 0$ and compute the strain produced

$$\epsilon_1 = \frac{\sigma_1}{E_1}$$

Then, apply a stress σ_2, as in Figure 5.8(b) with $\sigma_1 = \sigma_4 = \sigma_5 = \sigma_6 = 0$ and compute the strain in the fiber direction using the definition of Poisson's ratio given in Section 4.2.3 ($\nu_{ij} = -\epsilon_j/\epsilon_i$)

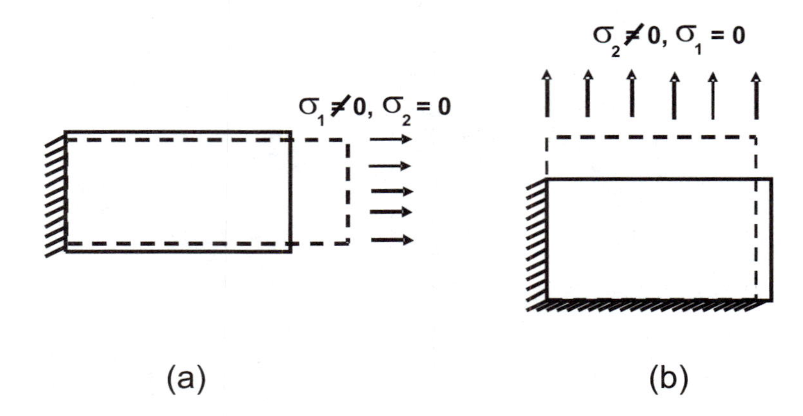

Figure 5.8: Determination of compliance coefficients in terms of engineering properties.

$$\epsilon_1 = -\nu_{21}\epsilon_2 = -\nu_{21}\frac{\sigma_2}{E_2}$$

The total strain in the fiber direction is the sum of the above two components

$$\epsilon_1 = \frac{1}{E_1}\sigma_1 - \frac{\nu_{21}}{E_2}\sigma_2 \tag{5.12}$$

Repeating the procedure along the transverse direction

$$\epsilon_2 = -\frac{\nu_{12}}{E_1}\sigma_1 + \frac{1}{E_2}\sigma_2 \tag{5.13}$$

The shear terms are obtained directly from the shear version of Hooke's law

$$\begin{aligned} \sigma_6 &= G_{12}\gamma_6 \\ \sigma_4 &= G_{23}\gamma_4 \\ \sigma_5 &= G_{13}\gamma_5 = G_{12}\gamma_5 \end{aligned} \tag{5.14}$$

Note that $G_{13} = G_{12}$ because, from a material point of view, the directions 2 and 3 are indistinguishable (see Figure 4.24). This type of material is called *transversely isotropic* (Section 4.1.6). Therefore, the compliance equations are

$$\begin{Bmatrix} \epsilon_1 \\ \epsilon_2 \\ \gamma_6 \end{Bmatrix} = \begin{bmatrix} 1/E_1 & -\nu_{21}/E_2 & 0 \\ -\nu_{12}/E_1 & 1/E_2 & 0 \\ 0 & 0 & 1/G_{12} \end{bmatrix} \begin{Bmatrix} \sigma_1 \\ \sigma_2 \\ \sigma_6 \end{Bmatrix} \tag{5.15}$$

and

$$\begin{Bmatrix} \gamma_4 \\ \gamma_5 \end{Bmatrix} = \begin{bmatrix} 1/G_{23} & 0 \\ 0 & 1/G_{13} \end{bmatrix} \begin{Bmatrix} \sigma_4 \\ \sigma_5 \end{Bmatrix} \tag{5.16}$$

Since the compliance matrix (5.15) must be symmetric, $\nu_{21}/E_2 = \nu_{12}/E_1$ (see also (4.47)). The stresses σ_4 and σ_5 are neglected if the plate is very thin or as an approximation during preliminary design. When $\sigma_3 = \sigma_4 = \sigma_5 = 0$, only four properties ($E_1$,$E_2$, G_{12}, and ν_{12}) are needed. When $\sigma_3 = 0$ but σ_4 and σ_5 are not zero, then the additional property G_{23} is needed (note that $G_{13} = G_{12}$). The compliance equations (5.15) and (5.16) can be written in compact form as

$$\begin{aligned} \{\epsilon\} &= [S]\,\{\sigma\} \\ \{\gamma\} &= [S^*]\,\{\tau\} \end{aligned} \tag{5.17}$$

where $\{\epsilon\} = \{\epsilon_1, \epsilon_2, \gamma_6\}^T$ are the in-plane strains, $\{\sigma\} = \{\sigma_1, \sigma_2, \sigma_6\}^T$ are the in-plane stresses, $\{\tau\} = \{\sigma_4, \sigma_5\}^T$ are the intralaminar stresses, $[S]$ is the 3×3 compliance matrix, and $[S^*]$ is the 2×2 intralaminar compliance matrix. The components of the $[S]$ and $[S^*]$ matrices are

$$\begin{aligned} S_{11} &= 1/E_1 \\ S_{12} &= S_{21} = -\nu_{12}/E_1 \\ S_{22} &= 1/E_2 \\ S_{66} &= 1/G_{12} \\ S^*_{44} &= 1/G_{23} \\ S^*_{55} &= 1/G_{13} \end{aligned} \tag{5.18}$$

The inversion of the 3×3 $[S]$ matrix leads to

$$\begin{Bmatrix} \sigma_1 \\ \sigma_2 \\ \sigma_6 \end{Bmatrix} = \begin{bmatrix} E_1/\Delta & \nu_{12}E_2/\Delta & 0 \\ \nu_{12}E_2/\Delta & E_2/\Delta & 0 \\ 0 & 0 & G_{12} \end{bmatrix} \begin{Bmatrix} \epsilon_1 \\ \epsilon_2 \\ \gamma_6 \end{Bmatrix} \tag{5.19}$$

where

$$\Delta = 1 - \nu_{12}\nu_{21} = 1 - \nu_{12}^2 E_2/E_1 \tag{5.20}$$

complemented with

$$\begin{Bmatrix} \sigma_4 \\ \sigma_5 \end{Bmatrix} = \begin{bmatrix} G_{23} & 0 \\ 0 & G_{13} \end{bmatrix} \begin{Bmatrix} \gamma_4 \\ \gamma_5 \end{Bmatrix} \tag{5.21}$$

In compact form, (5.19) and (5.21) can be written as

$$\begin{aligned} \{\sigma\} &= [Q]\,\{\epsilon\} \\ \{\tau\} &= [Q^*]\,\{\gamma\} \end{aligned} \tag{5.22}$$

where $[Q]$ is the reduced stiffness matrix and $[Q^*]$ is the 2×2 intralaminar stiffness matrix. The components of the 3×3 $[Q]$ matrix and the 2×2 $[Q^*]$ matrix are

$$
\begin{aligned}
Q_{11} &= E_1/\Delta \\
Q_{12} &= Q_{21} = \nu_{12} E_2/\Delta \\
Q_{22} &= E_2/\Delta \\
Q_{66} &= G_{12} \\
Q^*_{44} &= G_{23} \\
Q^*_{55} &= G_{13} \\
\Delta &= 1 - \nu_{12}\nu_{21}
\end{aligned}
\tag{5.23}
$$

where $G_{13} = G_{12}$ if the material is transversely isotropic.

For a narrow strip with the fibers along the length and loaded in tension, the stress–strain relationship is simply Hooke's law: $\sigma_1 = E_1\epsilon_1$. However, the stress–strain relationship for a plate similarly loaded is $\sigma_{11} = Q_{11}\epsilon_1$ with $Q_{11} > E_1$ because $\Delta < 1$. The additional stiffness represents the Poisson's effect. When a stress σ_1 is applied to a narrow strip (one-dimensional case), the only stiffness is E_1. When σ_1 is applied to the edge of a plate, the plate has to stretch in the direction of the load (1-direction in this case) and also it has to shrink in the perpendicular direction because of Poisson's effect. The resistance to the shrinkage is felt along the direction of the load and numerically represented by the magnification factor $1/\Delta$.

Example 5.1 *Compute the reduced stiffness matrix [Q] and the intralaminar shear coefficients using the following properties:* $E_1 = 19.981\ GPa$, $\nu_{12} = 0.274$, $E_2 = 11.389\ GPa$, $G_{12} = G_{13} = 3.789\ GPa$ *[172]. Lacking experimental data, assume the out-of-plane shear modulus* $G_{23} \cong G_m = 0.385\ GPa$.

Solution to Example 5.1 *Using (5.23)*

$$
\begin{aligned}
\Delta &= 1 - 0.274(0.274)\left(\frac{11.389}{19.981}\right) = 0.957 \\
Q_{11} &= \frac{19.981}{0.957} = 20.874\ GPa \\
Q_{22} &= \frac{11.389}{0.957} = 11.898\ GPa
\end{aligned}
$$

Note that $Q_{11} > E_1$ *and* $Q_{22} > E_2$*. This represents the stiffening observed in a plate with respect to the one-dimensional case because of the Poisson's effect. The remaining terms are*

$$
\begin{aligned}
Q_{12} = 0.274(11.898) &= 3.260\ GPa \\
Q_{66} &= 3.789\ GPa \\
Q^*_{44} &= 0.385\ GPa \\
Q^*_{55} &= 3.789\ GPa
\end{aligned}
$$

Example 5.2 *Verify that inversion of the reduced stiffness matrix [Q] gives the correct values for the coefficients of the reduced compliance matrix [S]. Use the numerical values from Example 5.1, p. 152.*

Solution to Example 5.2 *From Example 5.1, p. 152,*

$$[Q] = \begin{bmatrix} 20.874 & 3.260 & 0 \\ 3.260 & 11.898 & 0 \\ 0 & 0 & 3.789 \end{bmatrix} \; GPa$$

The inversion is simplified by the fact that the [Q] matrix has some coefficients equal to zero. In this case, it is possible to invert the 2×2 submatrix first. For a 2×2 matrix, the determinant D is computed as

$$D = Q_{11}Q_{22} - Q_{12}^2 = 237.731$$

and the coefficients of the inverse are

$$S_{11} = \frac{Q_{22}}{D} = 0.050 \; GPa^{-1}$$
$$S_{22} = \frac{Q_{11}}{D} = 0.0878 \; GPa^{-1}$$
$$S_{12} = \frac{-Q_{12}}{D} = -0.0137 \; GPa^{-1}$$

The inverse of the remaining term is simply

$$S_{66} = \frac{1}{Q_{66}} = 0.2639 \; GPa^{-1}$$

Now, using (5.18)

$$S_{11} = \frac{1}{19.981} = 0.0500 \; GPa^{-1}$$
$$S_{22} = \frac{1}{11.389} = 0.0878 \; GPa^{-1}$$
$$S_{12} = \frac{-0.274}{19.981} = -0.0137 \; GPa^{-1}$$
$$S_{66} = \frac{1}{3.789} = 0.2639 \; GPa^{-1}$$

which are the same results obtained before.

5.4 Off-axis Stiffness

Practical composite structures have more than one lamina because the properties in the transverse direction of a lamina are relatively low when compared to the longitudinal properties. Therefore, several laminae are stacked in different orientations so that reinforcements (fibers) are placed along all the directions of loading.

A laminate is a set of laminae with various fiber orientations (Figure 1.2) that are bonded together to form a plate or shell. Formulas that describe the mechanical properties of the laminate will be derived in Chapter 6. Before developing laminate properties, it is necessary to learn how to transform stresses, strains, compliances, and stiffness from the material coordinate system $(1, 2, 3)$ to a laminate coordinate system (x, y, z).

5.4.1 Coordinate Transformations

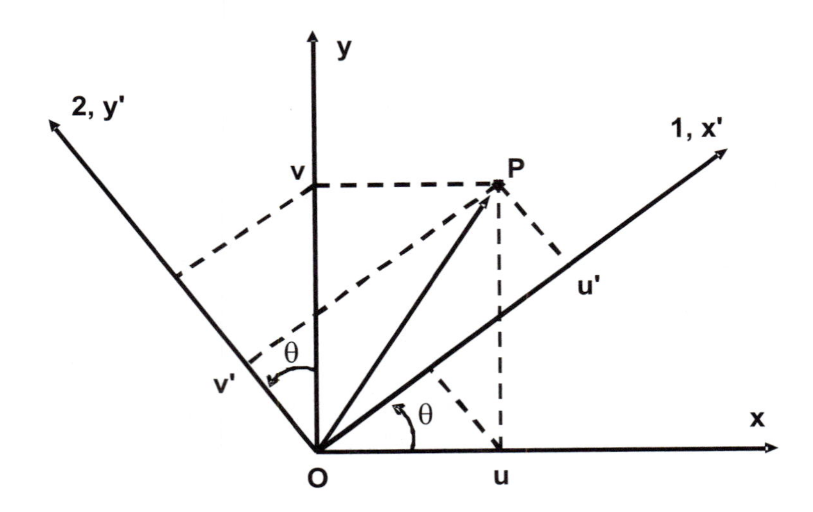

Figure 5.9: Coordinate transformations.

With reference to Figure 5.9, $\overrightarrow{OP}$ is a displacement vector with components (u,v) in laminate coordinates and components (u',v') in material coordinates, which are related by

$$\begin{aligned} u' &= u\cos\theta + v\sin\theta \\ v' &= -u\sin\theta + v\cos\theta \end{aligned} \tag{5.24}$$

or

$$\left\{ \begin{array}{c} u' \\ v' \end{array} \right\} = \left[\begin{array}{cc} m & n \\ -n & m \end{array} \right] \left\{ \begin{array}{c} u \\ v \end{array} \right\} \tag{5.25}$$

with $m = \cos(\theta)$ and $n = \sin(\theta)$. The angle θ is measured counterclockwise from the x-axis to the 1-axis. The inverse relationship is

$$\left\{ \begin{array}{c} u \\ v \end{array} \right\} = \left[\begin{array}{cc} m & -n \\ n & m \end{array} \right] \left\{ \begin{array}{c} u' \\ v' \end{array} \right\} \tag{5.26}$$

Similarly, the position of a point with coordinates x, y, can be described in the material coordinate system as

$$\left\{ \begin{array}{c} x' \\ y' \end{array} \right\} = \left[\begin{array}{cc} m & n \\ -n & m \end{array} \right] \left\{ \begin{array}{c} x \\ y \end{array} \right\} \tag{5.27}$$

and

$$\left\{ \begin{array}{c} x \\ y \end{array} \right\} = \left[\begin{array}{cc} m & -n \\ n & m \end{array} \right] \left\{ \begin{array}{c} x' \\ y' \end{array} \right\} \tag{5.28}$$

5.4.2 Stress and Strain Transformations

While the in-plane strains in laminate coordinates (x, y, Figure 5.9) are given by (5.9), in material coordinates (1,2, Figure 5.9) they are given by

$$\begin{aligned}
\epsilon_1 &= \frac{\partial u'}{\partial x'} \\
\epsilon_2 &= \frac{\partial v'}{\partial y'} \\
\gamma_6 = \gamma_{12} &= \frac{\partial u'}{\partial y'} + \frac{\partial v'}{\partial x'} \\
\gamma_5 = \gamma_{13} &= \frac{\partial u'}{\partial z'} + \frac{\partial w'}{\partial x'} \\
\gamma_4 = \gamma_{23} &= \frac{\partial v'}{\partial z'} + \frac{\partial w'}{\partial y'}
\end{aligned} \tag{5.29}$$

The strain in laminate coordinates is obtained using the chain rule

$$\epsilon_x = \frac{\partial u}{\partial x} = \frac{\partial u}{\partial x'}\frac{\partial x'}{\partial x} + \frac{\partial u}{\partial y'}\frac{\partial y'}{\partial x} \tag{5.30}$$

From (5.26)

$$\begin{aligned}
\frac{\partial u}{\partial x'} &= \frac{\partial u'}{\partial x'}m - \frac{\partial v'}{\partial x'}n \\
\frac{\partial u}{\partial y'} &= \frac{\partial u'}{\partial y'}m - \frac{\partial v'}{\partial y'}n
\end{aligned}$$

and using (5.27)

$$\begin{aligned}
\frac{\partial x'}{\partial x} &= m \\
\frac{\partial y'}{\partial x} &= -n
\end{aligned}$$

Then, it is possible to write

$$\epsilon_x = \frac{\partial u}{\partial x} = \frac{\partial u'}{\partial x'}m^2 - \frac{\partial v'}{\partial x'}mn - \frac{\partial u'}{\partial y'}mn + \frac{\partial v'}{\partial y'}n^2$$

or

$$\epsilon_x = m^2\epsilon_1 - 2mn\frac{\gamma_6}{2} + n^2\epsilon_2 \tag{5.31}$$

Expressions for ϵ_y and γ_{xy} are obtained in a similar way. Finally, the relationship between laminate and lamina strains is

$$\left\{ \begin{array}{c} \epsilon_x \\ \epsilon_y \\ \frac{1}{2}\gamma_{xy} \end{array} \right\} = \left[\begin{array}{ccc} m^2 & n^2 & -2mn \\ n^2 & m^2 & 2mn \\ mn & -mn & m^2-n^2 \end{array} \right] \left\{ \begin{array}{c} \epsilon_1 \\ \epsilon_2 \\ \frac{1}{2}\gamma_6 \end{array} \right\} = [T]^{-1} \left\{ \begin{array}{c} \epsilon_1 \\ \epsilon_2 \\ \frac{1}{2}\gamma_6 \end{array} \right\} \tag{5.32}$$

where $m = \cos(\theta)$, $n = \sin(\theta)$, and

$$[T]^{-1} = \left[\begin{array}{ccc} m^2 & n^2 & -2mn \\ n^2 & m^2 & 2mn \\ mn & -mn & m^2-n^2 \end{array} \right] \tag{5.33}$$

with the angle θ measured from the laminate to the material coordinate system as shown in Figure 5.9. Inverting (5.32), the relationship between lamina and laminate strains is

$$\left\{ \begin{array}{c} \epsilon_1 \\ \epsilon_2 \\ \frac{1}{2}\gamma_6 \end{array} \right\} = \left[\begin{array}{ccc} m^2 & n^2 & 2mn \\ n^2 & m^2 & -2mn \\ -mn & mn & m^2-n^2 \end{array} \right] \left\{ \begin{array}{c} \epsilon_x \\ \epsilon_y \\ \frac{1}{2}\gamma_{xy} \end{array} \right\} = [T(\theta)] \left\{ \begin{array}{c} \epsilon_x \\ \epsilon_y \\ \frac{1}{2}\gamma_{xy} \end{array} \right\} \tag{5.34}$$

with the transformation matrix written as

$$[T(\theta)] = \left[\begin{array}{ccc} m^2 & n^2 & 2mn \\ n^2 & m^2 & -2mn \\ -mn & mn & m^2-n^2 \end{array} \right] \tag{5.35}$$

Stresses are transformed in a similar way. The transformation from lamina to laminate axes is done using $[T]^{-1} = [T(-\theta)]$ as follows

$$\left\{ \begin{array}{c} \sigma_x \\ \sigma_y \\ \sigma_{xy} \end{array} \right\} = [T(-\theta)] \left\{ \begin{array}{c} \sigma_1 \\ \sigma_2 \\ \sigma_6 \end{array} \right\} \tag{5.36}$$

and $[T(\theta)]$ is used to rotate from laminate to lamina axes as follows

$$\left\{ \begin{array}{c} \sigma_1 \\ \sigma_2 \\ \sigma_6 \end{array} \right\} = [T(\theta)] \left\{ \begin{array}{c} \sigma_x \\ \sigma_y \\ \sigma_{xy} \end{array} \right\} \tag{5.37}$$

Note the use of $\frac{1}{2}$ in front of the shear strains but not in front of the shear stresses. This is because the shear strain γ_{xy} is not a tensor component, but $\epsilon_{xy} = \frac{1}{2}\gamma_{xy}$ is. Since only tensor components can be rotated with the transformation matrix $[T]$, the tensorial shear strain $\epsilon_{xy} = \frac{1}{2}\gamma_{xy}$ are used rather than the engineering shear strain γ_{xy}, but only for coordinate transformations. In the rest of the book, the engineering shear strain are used because the convenience of writing the shear version of Hooke's law as $\tau = G\gamma$. However, note that the symbol τ is not used because a shear stress like σ_6 can be clearly distinguished from a normal stress like σ_1.

Note that the angle θ in Figure 5.9 is measured counterclockwise from the x-axis to the 1-axis. The transformation matrix $[T]$ represents a transformation from axes x, y to axes $1, 2$. The matrix $[T]^{-1} = [T(-\theta)]$ represents a transformation in the opposite direction, from $1, 2$ axes to x, y axes.

The intralaminar shear terms transform according to

$$\left\{ \begin{array}{c} \sigma_4 \\ \sigma_5 \end{array} \right\} = \left[\begin{array}{cc} m & -n \\ n & m \end{array} \right] \left\{ \begin{array}{c} \sigma_{yz} \\ \sigma_{xz} \end{array} \right\} \tag{5.38}$$

and

$$\left\{ \begin{array}{c} \gamma_4 \\ \gamma_5 \end{array} \right\} = \left[\begin{array}{cc} m & -n \\ n & m \end{array} \right] \left\{ \begin{array}{c} \gamma_{yz} \\ \gamma_{xz} \end{array} \right\} \tag{5.39}$$

conversely

$$\left\{ \begin{array}{c} \sigma_{yz} \\ \sigma_{xz} \end{array} \right\} = \left[\begin{array}{cc} m & n \\ -n & m \end{array} \right] \left\{ \begin{array}{c} \sigma_4 \\ \sigma_5 \end{array} \right\} \tag{5.40}$$

and

$$\left\{ \begin{array}{c} \gamma_{yz} \\ \gamma_{xz} \end{array} \right\} = \left[\begin{array}{cc} m & n \\ -n & m \end{array} \right] \left\{ \begin{array}{c} \gamma_4 \\ \gamma_5 \end{array} \right\} \tag{5.41}$$

Example 5.3 *Transform the in-plane stresses $\sigma_1 = 100$, $\sigma_2 = 10$, $\sigma_6 = -5$, to a laminate coordinate system with the x-axis oriented at 55° counterclockwise (ccw) with respect to the lamina axis corresponding to the fiber direction (1-axis).*

Solution to Example 5.3 *The angle to be used in (5.33) and (5.35) is $\theta = -55°$ (angles θ are measured counterclockwise from the x-axis, Figure 5.9). Therefore*

$$[T] = \left[\begin{array}{ccc} 0.329 & 0.671 & -0.940 \\ 0.671 & 0.329 & 0.940 \\ 0.470 & -0.470 & -0.342 \end{array} \right]$$

The inverse can be obtained using (5.33)

$$[T]^{-1} = \left[\begin{array}{ccc} 0.329 & 0.671 & 0.940 \\ 0.671 & 0.329 & -0.940 \\ -0.470 & 0.470 & -0.342 \end{array} \right]$$

Finally

$$\left\{\begin{array}{c}\sigma_x\\ \sigma_y\\ \sigma_{xy}\end{array}\right\} = [T]^{-1}\left\{\begin{array}{c}100\\ 10\\ -5\end{array}\right\} = \left\{\begin{array}{c}34.911\\ 75.089\\ -40.576\end{array}\right\}$$

Example 5.4 *Transform the strains* $\epsilon_1 = 3.635(10^{-3})$, $\epsilon_2 = 7.411(10^{-3})$, $\gamma_4 = \gamma_5 = \gamma_6 = 0$, *to a laminate coordinate system with the x-axis oriented at 55° counterclockwise (ccw) with respect to the lamina axis corresponding to the fiber direction (1-axis).*

Solution to Example 5.4 *First, note that the shear strain must be divided by 2 before it can be transformed. Then, according to (5.32) and using* $[T]^{-1}$ *from Example 5.3, p. 157,*

$$\left\{\begin{array}{c}\epsilon_x\\ \epsilon_y\\ \frac{1}{2}\gamma_{xy}\end{array}\right\} = \left[\begin{array}{ccc}0.329 & 0.671 & 0.940\\ 0.671 & 0.329 & -0.940\\ -0.470 & 0.470 & -0.342\end{array}\right]\left\{\begin{array}{c}3.635\\ 7.411\\ \frac{1}{2}\cdot 0\end{array}\right\}10^{-3} = \left\{\begin{array}{c}6.169\\ 4.877\\ 1.774\end{array}\right\}10^{-3}$$

Then, $\epsilon_x = 6.169(10^{-3})$, $\epsilon_y = 4.877(10^{-3})$ *and* $\gamma_{xy} = 2(1.774)10^{-3} = 3.547(10^{-3})$.

Example 5.5 *Check the safety factor for the plate of Example 1.1, p. 6, if the applied shear can have either sign.*

Solution to Example 5.5 *In Example 1.1, p. 6, a unidirectional lamina oriented at* 26.57° *was selected to carry the loads* $\sigma_x = 400\ MPa$, $\sigma_y = 100\ MPa$, $\sigma_{xy} = 200\ MPa$. *Using principal stress design, it was found that a lamina at 26.57° would be optimum. If there is uncertainty in the sign of the shear stress, it was found that the principal stress* $\sigma_I = 500\ MPa$ *may end up applied along a direction 53.3° off the fiber direction when* $\sigma_{xy} = -200\ MPa$. *To evaluate this situation, the stresses in the material coordinates are computed using (5.37), as*

$$m = \cos(26.57) = 0.894$$
$$n = \sin(26.57) = 0.447$$

$$\left\{\begin{array}{c}\sigma_1\\ \sigma_2\\ \sigma_6\end{array}\right\} = \left[\begin{array}{ccc}.799 & .2 & .799\\ .2 & .799 & -.799\\ -.4 & .4 & .599\end{array}\right]\left\{\begin{array}{c}400\\ 100\\ -200\end{array}\right\} = \left\{\begin{array}{c}180\\ 320\\ -240\end{array}\right\} MPa$$

The corresponding strength values are found from Table 1.3: tension along the fibers $F_{1t} = 1020\ MPa$, *tension across fibres* $F_{2t} = 40\ MPa$, *shear strength* $F_6 = 60\ MPa$. *Using the maximum stress criterion (Section 7.1.2), each stress is compared to its corresponding strength, finding the following*

$$
\begin{aligned}
1020 &> 180 \\
40 &< 320 \\
60 &< 240
\end{aligned}
$$

The safety factor is

$$R = \frac{40}{320} = 0.125$$

It can be seen that a sign reversal in the shear load has at least two important effects: the safety factor was reduced from 2.04 *to* $0.125 < 1$ *which is unacceptable, and the critical property is now transverse tensile strength. This is undesirable because the transverse tensile strength* F_{2t} *is a matrix dominated property. Because of the number and variety of defects in the matrix, the variability in experimental data of* F_{2t} *is very large compared to the variability of* F_{1t}*. This calls for extra precautions and a larger safety factor.*

5.4.3 Stiffness and Compliance Transformations

The stress–strain equations (5.15–5.16), and (5.19–5.21) are limited to the case of having the stresses and strains oriented along material coordinates. To simplify the analysis as much as possible, it is convenient to relate stresses and strains in laminate coordinates (Figure 5.1) directly. This can be done by using the following equation

$$
\begin{Bmatrix} \sigma_x \\ \sigma_y \\ \sigma_{xy} \end{Bmatrix} = \begin{bmatrix} \overline{Q}_{11} & \overline{Q}_{12} & \overline{Q}_{16} \\ \overline{Q}_{12} & \overline{Q}_{22} & \overline{Q}_{26} \\ \overline{Q}_{16} & \overline{Q}_{26} & \overline{Q}_{66} \end{bmatrix} \begin{Bmatrix} \epsilon_x \\ \epsilon_y \\ \gamma_{xy} \end{Bmatrix} \tag{5.42}
$$

where the $\overline{Q}_{ij}$ are the components of the *transformed* reduced stiffness matrix. The $[\overline{Q}]$ matrix can be obtained from the $[Q]$ matrix as follows. The stress–strain relationship in material coordinates is given by (5.22–5.23) in terms of the reduced stiffness $[Q]$, which can be rewritten to accommodate $\frac{1}{2}\gamma_6$ as

$$
\begin{Bmatrix} \sigma_1 \\ \sigma_2 \\ \sigma_6 \end{Bmatrix} = \begin{bmatrix} Q_{11} & Q_{12} & 0 \\ Q_{12} & Q_{22} & 0 \\ 0 & 0 & 2Q_{66} \end{bmatrix} \begin{Bmatrix} \epsilon_1 \\ \epsilon_2 \\ \frac{1}{2}\gamma_6 \end{Bmatrix} = [Q]\,[R] \begin{Bmatrix} \epsilon_1 \\ \epsilon_2 \\ \frac{1}{2}\gamma_6 \end{Bmatrix} \tag{5.43}
$$

with

$$
[R] = \begin{bmatrix} 1 & 0 & 0 \\ 0 & 1 & 0 \\ 0 & 0 & 2 \end{bmatrix} \tag{5.44}
$$

Substitute (5.34) and 5.37) to get

$$
[T] \begin{Bmatrix} \sigma_x \\ \sigma_y \\ \sigma_{xy} \end{Bmatrix} = [Q][R][T] \begin{Bmatrix} \epsilon_x \\ \epsilon_y \\ \frac{1}{2}\gamma_{xy} \end{Bmatrix}
$$

Pre-multiply by $[T]^{-1}$ to get

$$\begin{Bmatrix} \sigma_x \\ \sigma_y \\ \sigma_{xy} \end{Bmatrix} = [T]^{-1}[Q][R][T] \begin{Bmatrix} \epsilon_x \\ \epsilon_y \\ \frac{1}{2}\gamma_{xy} \end{Bmatrix}$$

$$= [T]^{-1}[Q][R]\,[T]\,[R]^{-1} \begin{Bmatrix} \epsilon_x \\ \epsilon_y \\ \gamma_{xy} \end{Bmatrix} \tag{5.45}$$

or

$$\begin{Bmatrix} \sigma_x \\ \sigma_y \\ \sigma_{xy} \end{Bmatrix} = [\overline{Q}] \begin{Bmatrix} \epsilon_x \\ \epsilon_y \\ \gamma_{xy} \end{Bmatrix} \tag{5.46}$$

which is the compact form of (5.42). A similar formula can be found for the intralaminar components

$$\begin{Bmatrix} \sigma_{yz} \\ \sigma_{xz} \end{Bmatrix} = \begin{bmatrix} \overline{Q}^*_{44} & \overline{Q}^*_{45} \\ \overline{Q}^*_{45} & \overline{Q}^*_{55} \end{bmatrix} \begin{Bmatrix} \gamma_{yz} \\ \gamma_{xz} \end{Bmatrix} \tag{5.47}$$

Noting that $[R][T][R]^{-1} = [T]^{-T}$, the transformed reduced stiffness is computed as

$$[\overline{Q}] = [T]^{-1}[Q][T]^{-T} = [T(-\theta)][Q][T(-\theta)]^T \tag{5.48}$$

where $[T]^{-T}$ denotes the transpose of the inverse of $[T]$. Finally, the components of $[\overline{Q}]$ and $[\overline{Q}^*]$ in the laminate coordinate system are given explicitly in terms of the components of $[Q]$ and $[Q^*]$ in the material coordinate system as

$$\begin{aligned}
\overline{Q}_{11} &= Q_{11}\cos^4\theta + 2(Q_{12} + 2Q_{66})\sin^2\theta\cos^2\theta + Q_{22}\sin^4\theta \\
\overline{Q}_{12} &= (Q_{11} + Q_{22} - 4Q_{66})\sin^2\theta\cos^2\theta + Q_{12}(\sin^4\theta + \cos^4\theta) \\
\overline{Q}_{22} &= Q_{11}\sin^4\theta + 2(Q_{12} + 2Q_{66})\sin^2\theta\cos^2\theta + Q_{22}\cos^4\theta \\
\overline{Q}_{16} &= (Q_{11} - Q_{12} - 2Q_{66})\sin\theta\cos^3\theta + (Q_{12} - Q_{22} + 2Q_{66})\sin^3\theta\cos\theta \\
\overline{Q}_{26} &= (Q_{11} - Q_{12} - 2Q_{66})\sin^3\theta\cos\theta + (Q_{12} - Q_{22} + 2Q_{66})\sin\theta\cos^3\theta \\
\overline{Q}_{66} &= (Q_{11} + Q_{22} - 2Q_{12} - 2Q_{66})\sin^2\theta\cos^2\theta + Q_{66}(\sin^4\theta + \cos^4\theta) \\
\overline{Q}^*_{44} &= Q^*_{44}\cos^2\theta + Q^*_{55}\sin^2\theta \\
\overline{Q}^*_{55} &= Q^*_{44}\sin^2\theta + Q^*_{55}\cos^2\theta \\
\overline{Q}^*_{45} &= (Q^*_{55} - Q^*_{44})\sin\theta\cos\theta
\end{aligned} \tag{5.49}$$

Similarly, the constitutive equations in terms of compliance (5.17) can be transformed to laminate coordinates

$$\left\{\begin{array}{c}\epsilon_x\\ \epsilon_y\\ \gamma_{xy}\end{array}\right\} = [\overline{S}]\left\{\begin{array}{c}\sigma_x\\ \sigma_y\\ \sigma_{xy}\end{array}\right\} \tag{5.50}$$

and

$$\left\{\begin{array}{c}\gamma_{yz}\\ \gamma_{xz}\end{array}\right\} = \left[\begin{array}{cc}\overline{S}^*_{44} & \overline{S}^*_{45}\\ \overline{S}^*_{45} & \overline{S}^*_{55}\end{array}\right]\left\{\begin{array}{c}\sigma_{yz}\\ \sigma_{xz}\end{array}\right\} \tag{5.51}$$

with components given by

$$\begin{aligned}
\overline{S}_{11} &= S_{11}\cos^4\theta + (2S_{12}+S_{66})\sin^2\theta\cos^2\theta + S_{22}\sin^4\theta\\
\overline{S}_{12} &= (S_{11}+S_{22}-S_{66})\sin^2\theta\cos^2\theta + S_{12}(\sin^4\theta+\cos^4\theta)\\
\overline{S}_{22} &= S_{11}\sin^4\theta + (2S_{12}+S_{66})\sin^2\theta\cos^2\theta + S_{22}\cos^4\theta\\
\overline{S}_{16} &= (2S_{11}-2S_{12}-S_{66})\sin\theta\cos^3\theta - (2S_{22}-2S_{12}-S_{66})\sin^3\theta\cos\theta\\
\overline{S}_{26} &= (2S_{11}-2S_{12}-S_{66})\sin^3\theta\cos\theta - (2S_{22}-2S_{12}-S_{66})\sin\theta\cos^3\theta\\
\overline{S}_{66} &= 2(2S_{11}+2S_{22}-4S_{12}-S_{66})\sin^2\theta\cos^2\theta + S_{66}(\sin^4\theta+\cos^4\theta)\\
\overline{S}^*_{44} &= S^*_{44}\cos^2\theta + S^*_{55}\sin^2\theta\\
\overline{S}^*_{55} &= S^*_{44}\sin^2\theta + S^*_{55}\cos^2\theta\\
\overline{S}^*_{45} &= (S^*_{55}-S^*_{44})\sin\theta\cos\theta
\end{aligned} \tag{5.52}$$

Note that the coefficients $\overline{Q}_{16} \neq 0$ and $\overline{Q}_{26} \neq 0$ after the transformation. Also, the coefficients $\overline{S}_{16} \neq 0$ and $\overline{S}_{26} \neq 0$ after the transformation. If the laminate axes do not coincide with the lamina axes, when shear stress is applied, not only shear strain but also normal strains are produced.

Example 5.6 *Compute the stresses in laminate coordinates corresponding to the strains in laminate coordinates* $\epsilon_x = 6.169(10^{-3})$, $\epsilon_y = 4.877(10^{-3})$, $\gamma_{xy} = 3.548(10^{-3})$, $\gamma_{xz} = 0.7(10^{-3})$, $\gamma_{yz} = 1.5(10^{-3})$. *Use the material properties of Example 5.1, p. 152, to compute the transformed reduced stiffness matrix* $[\overline{Q}]$ *of a lamina with the fibers oriented at* $\theta = -55°$.

Solution to Example 5.6 *From Example 5.1, p. 152,*

$$[Q] = \left[\begin{array}{ccc}20.874 & 3.260 & 0\\ 3.260 & 11.898 & 0\\ 0 & 0 & 3.789\end{array}\right] GPa$$

For $\theta = -55°, m^2 = 0.329, n^2 = 0.671$. *Then using (5.49)*

$$\overline{Q}_{11} = 12.40\ GPa$$
$$\overline{Q}_{12} = 5.71\ GPa$$
$$\overline{Q}_{22} = 15.50\ GPa$$
$$\overline{Q}_{16} = -1.22\ GPa$$
$$\overline{Q}_{26} = -3.00\ GPa$$
$$\overline{Q}_{66} = 6.24\ GPa$$

and using (5.46)

$$\left\{ \begin{array}{c} \sigma_x \\ \sigma_y \\ \sigma_{xy} \end{array} \right\} = [\overline{Q}] \left\{ \begin{array}{c} 6.169 \\ 4.877 \\ 3.548 \end{array} \right\} 10^{-3} = \left\{ \begin{array}{c} 0.1 \\ 0.1 \\ 0 \end{array} \right\} GPa$$

Finally, the intralaminar shear coefficients for the orthotropic lamina are, using data from Example 5.1, p. 152,

$$Q^*_{44} = G_{23} = 0.385\ GPa$$
$$Q^*_{55} = G_{13} = G_{12} = 3.789\ GPa$$

Using (5.49) for $\theta = -55°$

$$\overline{Q}^*_{44} = 0.385 \cos^2(-55) + 3.789 \sin^2(-55) = 2.669\ GPa$$
$$\overline{Q}^*_{55} = 0.385 \sin^2(-55) + 3.789 \cos^2(-55) = 1.505\ GPa$$
$$\overline{Q}^*_{45} = (3.789 - 0.385) \sin(-55) \cos(-55) = -1.599\ GPa$$

and using (5.47)

$$\left\{ \begin{array}{c} \sigma_{yz} \\ \sigma_{xz} \end{array} \right\} = \left[\overline{Q}^*\right] \left\{ \begin{array}{c} 1.5 \\ 0.7 \end{array} \right\} 10^{-3} = \left\{ \begin{array}{c} 2.8842 \\ -1.345 \end{array} \right\} 10^{-3}\ GPa$$

5.4.4 Specially Orthotropic Lamina

Any lamina is orthotropic in its own material coordinate system $(1, 2, 3)$. A lamina is called *specially orthotropic* when it is also orthotropic in the laminate coordinate system (x, y, z), which happens only for orientations $\theta = 0°$ and $\theta = 90°$, or for laminae reinforced with balanced fabrics (see Example 5.7). An orthotropic lamina that is oriented at an angle not a multiple of 90° from the laminate coordinate system is called *generally orthotropic*. Therefore $\overline{Q}_{16}$, $\overline{Q}_{26}$, $\overline{S}_{16}$, and $\overline{S}_{26}$ are different from zero for a generally orthotropic lamina.

Example 5.7 *Compute the $[\overline{Q}]$ matrix of a lamina reinforced with a ±45 woven fabric. The fabric weight is $w = 600\ g/m^2$ of which 300 g are at +45° and 300 g are at −45°. The matrix is epoxy HTP-1072 (Table 2.9, assume $\nu_m = 0.38$). The fiber is Kevlar 49™ (Table 2.2) and the fiber volume fraction is 50%.*

Solution to Example 5.7 *First assume that the +45 and the −45 fibers are separated in two laminae. Compute the [Q] matrix of the unidirectional material. Using (4.24), (4.31), (4.37), and (4.33), results in* $E_1 = 67.2\ GPa$, $E_2 = 12.1\ GPa$, $G_{12} = 3.447\ GPa$, *and* $\nu_{12} = 0.365$. *Then, using (5.23) results in*

$$[Q] = \begin{bmatrix} 68.8 & 4.54 & 0 \\ 4.54 & 12.4 & 0 \\ 0 & 0 & 3.45 \end{bmatrix} GPa$$

Then compute the $[\overline{Q}]$ *matrices using (5.49)*

$$[\overline{Q}]_{45} = \begin{bmatrix} 26.0 & 19.1 & 14.1 \\ 19.1 & 26.0 & 14.1 \\ 14.1 & 14.1 & 18.1 \end{bmatrix} GPa; \ [\overline{Q}]_{-45} = \begin{bmatrix} 26.0 & 19.1 & -14.1 \\ 19.1 & 26.0 & -14.1 \\ -14.1 & -14.1 & 18.1 \end{bmatrix} GPa$$

Finally, average them to get

$$[\overline{Q}]_{fabric} = \frac{300}{600}[\overline{Q}]_{45} + \frac{300}{600}[\overline{Q}]_{-45} = \begin{bmatrix} 26.0 & 19.1 & 0 \\ 19.1 & 26.0 & 0 \\ 0 & 0 & 18.1 \end{bmatrix} GPa$$

Note that the effect of using a balanced fabric is to cancel $\overline{Q}_{16}$ *and* $\overline{Q}_{26}$. *Therefore, balanced fabrics produce a specially orthotropic lamina.*

Example 5.8 *Compute the* $[\overline{Q}]$ *matrix of one lamina of stitched fabric XM2408 (Table 2.8). The fiber is E-glass (Table 2.1), the fiber volume fraction is 50%, and the matrix is epoxy HTP-1072 (Table 2.9, assume* $v_m = 0.38$*).*

Solution to Example 5.8 *The fabric XM2408 has 400* g/m^2 *at 45°, 400* g/m^2 *at -45° and 225* g/m^2 *of chopped strand mat. Even though these are three distinct laminae, which are not woven, it is usual to treat them as a unit with specially orthotropic properties. The contribution of each fiber orientation to the [Q] matrix is proportional to its weight.*

First, compute the properties of the unidirectional lamina using (4.24), (4.31), (4.37), and (4.33), results in

$$E_1 = 37.864\ GPa$$
$$E_2 = 11.224\ GPa$$
$$G_{12} = 3.317\ GPa$$
$$\nu_{12} = 0.3$$

Then, compute the properties of the chopped strand mat lamina using (9.67)

$$E = 21.214\ GPa$$
$$G = 7.539\ GPa$$
$$\nu = 0.407$$

Now, compute the $[Q]$ matrices, for the unidirectional and for the continuous strand mat (CSM) material, using (5.23)

$$[Q]_{UNI} = \begin{bmatrix} 38.9 & 3.46 & 0 \\ 3.46 & 11.5 & 0 \\ 0 & 0 & 3.32 \end{bmatrix} GPa; [Q]_{CSM} = \begin{bmatrix} 25.4 & 10.3 & 0 \\ 10.3 & 25.4 & 0 \\ 0 & 0 & 7.54 \end{bmatrix} GPa$$

Next, compute the $[\overline{Q}]$ matrix for each lamina using (5.49)

$$[\overline{Q}]_{45} = \begin{bmatrix} 17.7 & 11.0 & 6.84 \\ 11.0 & 17.7 & 6.84 \\ 6.84 & 6.84 & 10.9 \end{bmatrix} GPa;\ [\overline{Q}]_{-45} = \begin{bmatrix} 17.7 & 11.0 & -6.84 \\ 11.0 & 17.7 & -6.84 \\ -6.84 & -6.84 & 10.9 \end{bmatrix} GPa$$

Note that rotations do not affect the CSM lamina, so

$$[\overline{Q}]_{CSM} = [Q]_{CSM}$$

Finally, the $[\overline{Q}]$ matrix for XM2408 with 50% of epoxy is

$$[\overline{Q}]_{XM2408} = \frac{400}{1025}[\overline{Q}]_{45} + \frac{400}{1025}[\overline{Q}]_{-45} + \frac{225}{1025}[\overline{Q}]_{CSM}$$

$$[\overline{Q}]_{XM2408} = \begin{bmatrix} 19.39 & 10.85 & 0 \\ 10.85 & 19.39 & 0 \\ 0 & 0 & 10.16 \end{bmatrix} GPa$$

Since the fabric ±45 fibers are balanced, the resulting composite lamina is specially orthotropic ($\overline{Q}_{16} = \overline{Q}_{26} = 0$).

Exercises

Exercise 5.1 *Explain contracted notation for stresses and strains.*

Exercise 5.2 *Complete the table below with the contracted notation symbols for the component of stress and strain shown in the table. Use engineering notation for strains.*

Tensor notation (given)	σ_{11}	σ_{12}	σ_{13}	ϵ_{21}	σ_{22}	σ_{23}	ϵ_{31}	ϵ_{32}	ϵ_{33}
Contracted notation									

Exercise 5.3 *Define the plane stress assumption used in the context of laminated composite materials?*

Exercise 5.4 *Compute the reduced stiffness $[Q]$ and compliance $[S]$ matrices, and the intralaminar shear stiffness $[Q^*]$ and compliance $[S^*]$ matrices for a lamina with $E_1 = 35\ GPa$, $E_2 = 3.5\ GPa$, $\nu_{12} = 0.3$, $G_{12} = 1.75\ GPa$, $G_{23} = 0.35\ GPa$.*

Exercise 5.5 *A lamina of composite with fiber orientation $\theta = 60°$ is subject to $\sigma_x = 10\,MPa, \sigma_y = 5\,MPa, \sigma_{xy} = 2.5\,MPa$. Using stress transformation, compute the safety factor. Use the material data for carbon–epoxy (AS4–3501-6) in Table 1.4. Consider all possible modes of failure.*

Exercise 5.6 *Compute the transformation matrix for a lamina oriented at 45° with respect to the x-axis. Use it to find the transformed reduced stiffness matrix $[\overline{Q}]$ in laminate coordinates, corresponding to the reduced stiffness matrix $[Q]$ in material coordinates computed in Exercise 5.4.*

Exercise 5.7 *Plot the variation of* $\overline{Q}_{11}$ *as a function of the angle* θ *between* 0 *and* $90°$*, using the material properties of Exercise 4.10. When does* $\overline{Q}_{11}$ *becomes numerically equal to* $\overline{Q}_{22}$*?*

Exercise 5.8 *A tensile test specimen of the material in Exercise 5.4 is* $10\,mm$ *wide,* $2\,mm$ *thick, and* $150\,mm$ *long and it subjected to an axial force of* $200\,N$*. The fibers are oriented at* $\theta = 10°$ *with respect to the loading axis.*

(a) *Compute the in-plane and shear strains in the material coordinate system.*

(b) *Compute the in-plane and shear strains in the laminate coordinate system.*

(c) *Is the cross-section under a uniaxial state of strain in the laminate coordinates? Explain the origin of each nonzero strain.*

Exercise 5.9 *Compute the transformed reduced stiffness for the material of Exercise 5.4 at* $\theta = 45°$ *and* $\theta = 90°$*. Comment on the results.*

Exercise 5.10 *Consider a plane defined in parametric form as*

$$F(x, y, z) = \frac{x}{a} + \frac{y}{b} + \frac{z}{c} - 1 = 0$$

where a, b, c *are the intercepts on the* x, y, z *axes, and take* $a = 3, b = 4, c = 5$*. The stress tensor at that point, in contracted notation, is* $\sigma_1 = 100, \sigma_2 = 10, \sigma_3 = -20, \sigma_4 = 1, \sigma_5 = 2, \sigma_6 = -5$*. Compute the stress vector at* $x = y = z = 60/47$ *using Cauchy's law.*

Chapter 6

Macromechanics

Micromechanics formulas that link the properties of the composite material to the properties of the constituents (fiber and matrix) were developed in Chapter 4. The variations of lamina properties with orientation were explored in Chapter 5. The relationships between the structural properties of the final laminate and those of the laminae and their orientations are developed in this chapter. While using a laminated composite allows the designer to optimize the material/structural system, it complicates the analysis. Therefore, a simple relationship between the forces and moments applied to a laminate and the strains and curvatures induced is required for the design of laminated plates and shells, which is the objective of this chapter.

Practical composite structures are built with laminates having several laminae with various orientations, as illustrated in Figure 1.2. The lamina orientations are chosen to provide adequate stiffness and strength in the direction of the applied loads, taking into account that the composite material is much stronger and stiffer in the fiber direction than in any other direction.

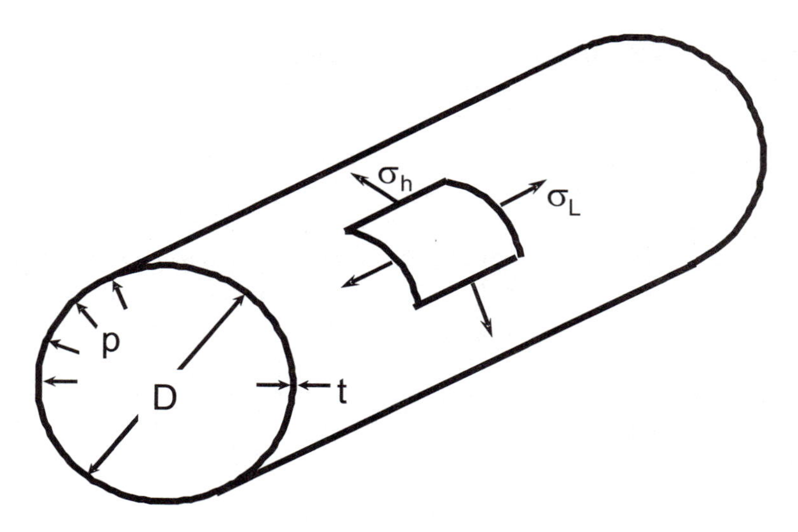

Figure 6.1: Cylindrical pressure vessel.

To illustrate the need for a laminated structure to carry loads, consider a cylindrical pressure vessel of diameter D subject to internal pressure p (Figure 6.1). The stresses in the wall are

$$\sigma_l = \frac{pD}{4t} \; ; \sigma_h = \frac{pD}{2t}$$

Since the hoop stress σ_h is twice the longitudinal stress σ_l, a metal design would call for a thickness based on σ_h, namely

$$t = \frac{pD}{2\sigma_y}$$

where σ_y is the yield strength of the metal. Such design operates with $\sigma_h = \sigma_y$ and $\sigma_l = \sigma_y/2$, which means the material strength is 100% underutilized in the longitudinal direction. On the other hand, a composite pressure vessel can be optimized to carry both stresses σ_l and σ_h at the material allowable strength, simply by placing twice as much hoop-oriented fibers as longitudinal fibers.

6.1 Plate Stiffness and Compliance

Most applications of composites involve thin laminated plates or shells, under the action of bending moments and stretching loads. Even composite beams are thin-walled sections composed of plate elements Figure 10.1. Therefore, the basic building block of a composite structure is a plate element. The constitutive equations for such an element are presented in this section.

6.1.1 Assumptions

The following hypotheses are used in the derivation of the plate stiffness and compliance equations:

1. A line originally straight and perpendicular to the middle surface remains straight after the plate is deformed (line A–D in Figure 6.2). This assumption is based on experimental observation and implies that the shear strains γ_{xz} and γ_{yz} are constant through the thickness. This assumption is accurate for thin laminates and is a good approximation in most cases except when the laminate is thick and the laminae have very different shear stiffness [173].

2. The length of the line A–D in Figure 6.2 is constant. This implies that the normal strain $\epsilon_{zz} \simeq 0$. This assumption is also based on experimental observation. It is a very good assumption that is satisfied in most cases, except for very thick laminates.

Consider the coordinate system illustrated in Figures 6.2 and 6.3. The middle surface is located halfway through the thickness of the plate, as shown in Figure

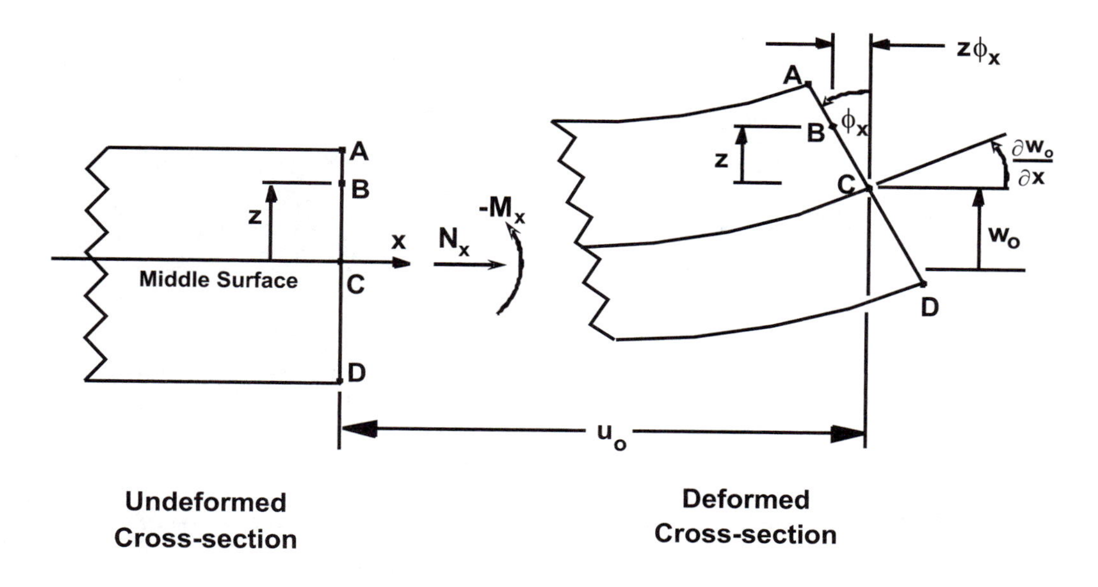

Figure 6.2: Geometry of deformation in the x-z plane.

6.2. Only the middle surface of the plate is shown in Figure 6.3. Using assumption (1), the displacements at every point through the thickness are[1]

$$
\begin{aligned}
u(x,y,z) &= u_0(x,y) - z\phi_x(x,y) \\
v(x,y,z) &= v_0(x,y) - z\phi_y(x,y)
\end{aligned}
\tag{6.1}
$$

where $u(x,y,z)$ is the displacement along the x-direction at each point (x,y,z) and $v(x,y,z)$ is the displacement along the y-direction (see Figure 6.3). Using assumption (2), and noting that practical values of ϕ_x and ϕ_y are very small, the transverse deflection is the same for every point through the thickness of the plate

$$
w(x,y,z) = w_0(x,y) \tag{6.2}
$$

Note that the independent variables in (6.1) and (6.2) are $u_0(x,y)$, $v_0(x,y)$, $w_0(x,y)$, $\phi_x(x,y)$, and $\phi_y(x,y)$. The fact that all of these are independent of the thickness coordinate z is the main characteristic of plate theory in contrast to a fully three-dimensional elasticity problem, which would have variables u, v, w that depend on all three coordinates x, y, z. The functions u_0, v_0, w_0 represent the displacements of every point (x,y) of the middle surface of the plate. The functions ϕ_x and ϕ_y are the rotations (positive counterclockwise) of the normal to the middle surface (line A-D) in Figure 6.2 at each point x, y.

[1]Notation: $f(x,y,z)$ represents a function that takes different values depending on the values of x,y,z. For example, the displacement along the x-direction of a point located at coordinates x,y,z is given by $u(x,y,z)$.

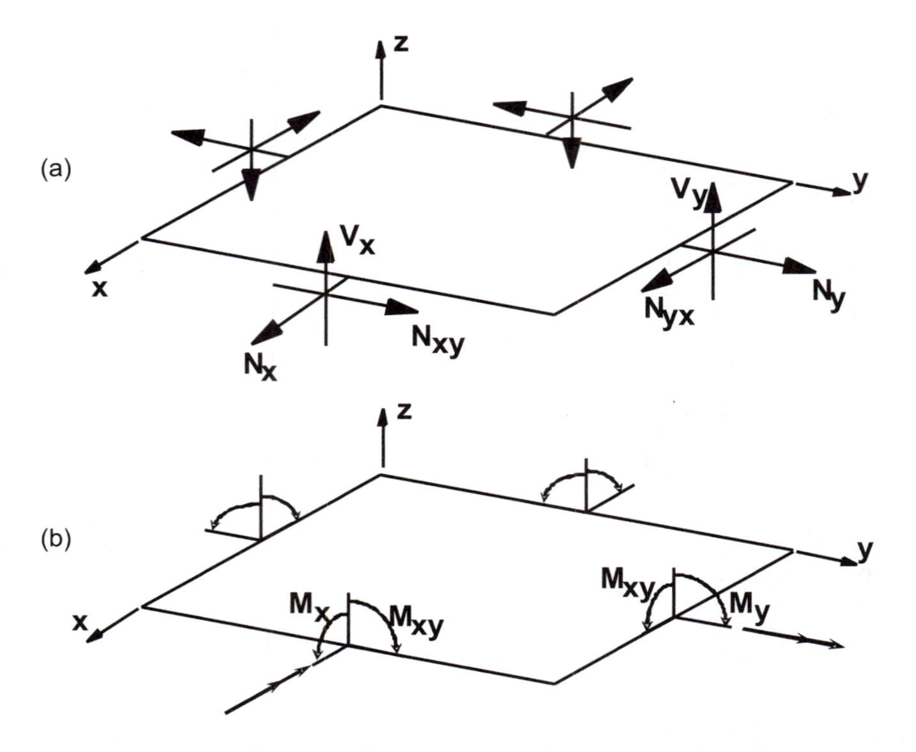

Figure 6.3: Force and moment resultants on a flat plate.

6.1.2 Strains

The strains at any point in the plate are a function of the displacements as given by (5.9), namely

$$\begin{aligned}
\epsilon_x(x,y,z) &= \frac{\partial u}{\partial x} \\
\epsilon_y(x,y,z) &= \frac{\partial v}{\partial y} \\
\gamma_{xy}(x,y,z) &= \frac{\partial u}{\partial y} + \frac{\partial v}{\partial x} \\
\gamma_{yz}(x,y,z) &= \frac{\partial v}{\partial z} + \frac{\partial w}{\partial y} \\
\gamma_{xz}(x,y,z) &= \frac{\partial u}{\partial z} + \frac{\partial w}{\partial x}
\end{aligned} \tag{6.3}$$

Then, using (6.1) and (6.2), the strains can be written as

$$
\begin{aligned}
\epsilon_x(x,y,z) &= \frac{\partial u_0}{\partial x} - z\frac{\partial \phi_x}{\partial x} \\
\epsilon_y(x,y,z) &= \frac{\partial v_0}{\partial y} - z\frac{\partial \phi_y}{\partial y} \\
\gamma_{xy}(x,y,z) &= \frac{\partial u_0}{\partial y} + \frac{\partial v_0}{\partial x} - z\left(\frac{\partial \phi_x}{\partial y} + \frac{\partial \phi_y}{\partial x}\right) \\
\gamma_{yz}(x,y) &= -\phi_y + \frac{\partial w_0}{\partial y} \\
\gamma_{xz}(x,y) &= -\phi_x + \frac{\partial w_0}{\partial x}
\end{aligned} \tag{6.4}
$$

Note that γ_{yz} and γ_{xz} are independent of z (or constant through the thickness) because of assumption (1). The first three of (6.4) can be written as

$$
\left\{ \begin{array}{c} \epsilon_x \\ \epsilon_y \\ \gamma_{xy} \end{array} \right\} = \left\{ \begin{array}{c} \epsilon_x^0 \\ \epsilon_y^0 \\ \gamma_{xy}^0 \end{array} \right\} + z \left\{ \begin{array}{c} \kappa_x \\ \kappa_y \\ \kappa_{xy} \end{array} \right\} \tag{6.5}
$$

where, ϵ_x^0, ϵ_y^0,γ_{xy}^0 are called the *midsurface strains.* They represent the stretching and shear of the plate, and are defined as

$$
\begin{aligned}
\epsilon_x^0(x,y) &= \frac{\partial u_0}{\partial x} \\
\epsilon_y^0(x,y) &= \frac{\partial v_0}{\partial y} \\
\gamma_{xy}^0(x,y) &= \frac{\partial u_0}{\partial y} + \frac{\partial v_0}{\partial x}
\end{aligned} \tag{6.6}
$$

The curvatures of the plate due to bending κ_x and κ_y, and due to twisting κ_{xy}, are given by

$$
\begin{aligned}
\kappa_x(x,y) &= -\frac{\partial \phi_x}{\partial x} \\
\kappa_y(x,y) &= -\frac{\partial \phi_y}{\partial y} \\
\kappa_{xy}(x,y) &= -\left(\frac{\partial \phi_x}{\partial y} + \frac{\partial \phi_y}{\partial x}\right)
\end{aligned} \tag{6.7}
$$

and the transverse[2] shear strains are given in the last two lines of (6.4) or

$$
\begin{aligned}
\gamma_{yz}(x,y,z) &= -\phi_y + \frac{\partial w_0}{\partial y} \\
\gamma_{xz}(x,y,z) &= -\phi_x + \frac{\partial w_0}{\partial x}
\end{aligned} \tag{6.8}
$$

[2]These are laminate strains that act both inside and between laminae.

The main reason for separating the strains into in-plane strains (ϵ_x^0, ϵ_y^0, γ_{xy}^0) and curvatures (κ_x, κ_y, κ_{xy}) is convenience. For most plates, if only in-plane forces are applied, only (ϵ_x^0, ϵ_y^0, γ_{xy}^0) are induced and it is not necessary to consider the curvatures. This is true for the pressure vessel example considered earlier and many more applications. If only moments or transverse loads are applied to the plate, only the curvatures need to be computed (see Section 6.3.2).

For a single-lamina plate, the material is the same through the thickness, and (6.5) leads to the typical distribution of stress shown in Figure 6.4. The linear distribution of stress in Figure 6.4 is caused by the curvatures and the constant term is caused by the in-plane strains. In Figure 6.4, the distribution of stress is linear through the thickness because the strains, given by (6.5), are linear through the thickness and the constitutive equations (stiffness) are constant through the thickness for a single-lamina plate. While the strains remain linear for a laminated plate, the stresses become piece-wise linear (see Figure 6.5) because the stiffness coefficients are different in each lamina.

If the plate thickness is much smaller than the dimensions of the plate, the shear deformations γ_{xz} and γ_{yz} can be neglected. Setting $\gamma_{xz} = \gamma_{yz} = 0$ in (6.8), the rotations of the normal line A–D in Figure 6.2 are equal to the slopes of the middle surface, or

$$\begin{aligned} \phi_x &= \frac{\partial w_0}{\partial x} \\ \phi_y &= \frac{\partial w_0}{\partial y} \end{aligned} \tag{6.9}$$

Substituting back into (6.7), the curvatures can be written in terms of w_0 only as

$$\begin{aligned} \kappa_x(x,y) &= -\frac{\partial^2 w_0}{\partial x^2} \\ \kappa_y(x,y) &= -\frac{\partial^2 w_0}{\partial y^2} \\ \kappa_{xy}(x,y) &= -2\frac{\partial^2 w_0}{\partial x \partial y} \end{aligned} \tag{6.10}$$

Equations 6.10 along with (6.6) form the basis for Classical Plate Theory (CPT) while (6.6–6.8) are the basis for First Order Shear Deformation Theory (FSDT). Classical Plate Theory is often used because most existing solutions for isotropic plates are based on CPT. Although transverse shear deformations (γ_{xz} and γ_{yz}) are neglected in CPT, it still gives accurate results for isotropic plates because isotropic materials are very stiff in shear ($G \approx E/2.5$), yielding shear deformations that, if not zero, are very small. However, composites have low shear modulus ($G < E/10$), thus requiring us to account for transverse shear deformations.

For preliminary design, and if the laminate is thin, CPT usually provides a reasonable approximation. However, most existing solutions for isotropic plates are not accurate for laminated plates, except in special cases (see Example 6.3). Perhaps the main motivation for considering FSDT is that all commercial finite element programs are formulated with the more accurate FSDT. In the remainder of this chapter and in Chapter 11, FSDT is developed first, and then reduced to CPT. In this way, the reader will be able to use the most appropriate theory with a minimum of additional effort.

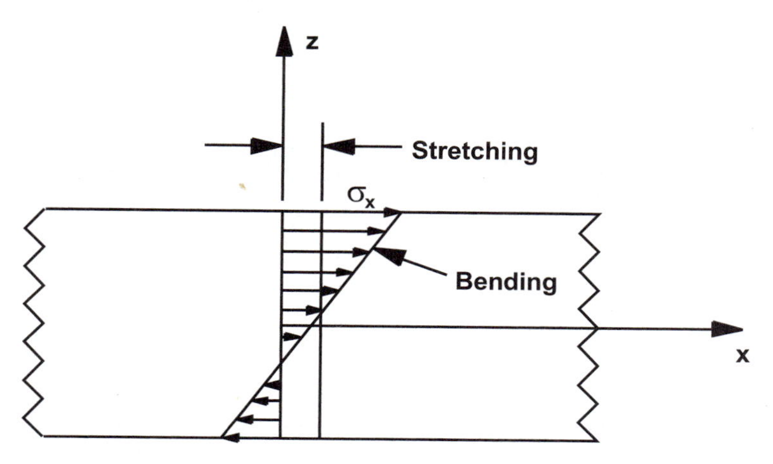

Figure 6.4: Distributions of stress due to bending and extension in an isotropic plate.

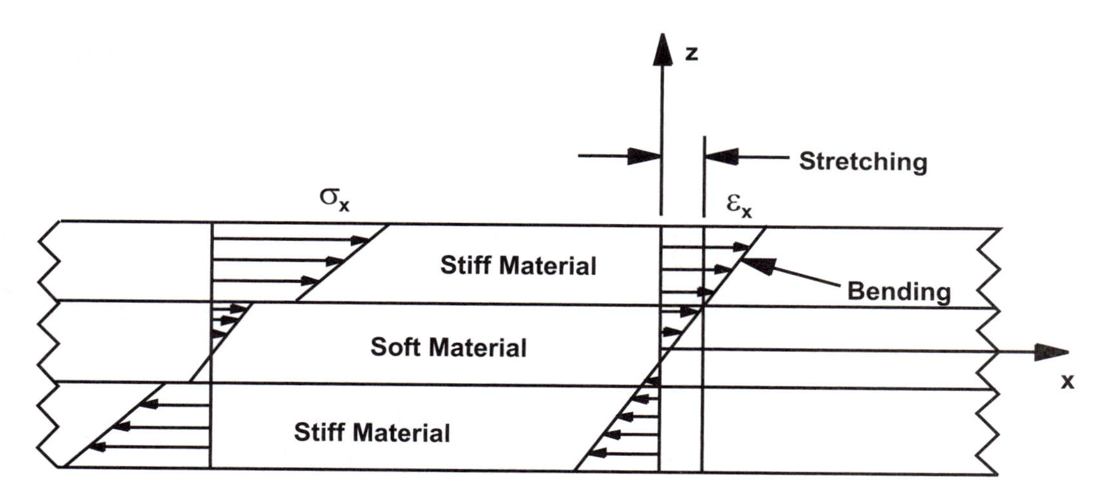

Figure 6.5: Distribution of strain and stress due to bending and extension in a laminated plate.

6.1.3 Stress Resultants

So far, the strains at every point x, y, z, in the plate were replaced in terms of the midplane strains $\epsilon_x^0, \epsilon_y^0, \gamma_{xy}^0$, and the curvatures $\kappa_x, \kappa_y, \kappa_{xy}$ using (6.5). The underlaying motivation is to write all the functions (displacements, stresses, and strains) in terms of only two variables (x and y) to simplify the analysis. Next, it is convenient to replace the stress components (σ_x, σ_y, σ_z, σ_{xy}, σ_{yz}, and σ_{xz}) at every point x, y, z, of the plate in terms of functions of only two variables x, y. To do this, the stresses are integrated over the thickness t of the plate to obtain the resultant forces and moments on a laminate (Figure 6.3) as

$$\left\{\begin{array}{c} N_x \\ N_y \\ N_{xy} \end{array}\right\} = \int_{-t/2}^{t/2} \left\{\begin{array}{c} \sigma_x \\ \sigma_y \\ \sigma_{xy} \end{array}\right\} dz$$

$$\left\{\begin{array}{c} V_y \\ V_x \end{array}\right\} = \int_{-t/2}^{t/2} \left\{\begin{array}{c} \sigma_{yz} \\ \sigma_{xz} \end{array}\right\} dz$$

$$\left\{\begin{array}{c} M_x \\ M_y \\ M_{xy} \end{array}\right\} = \int_{-t/2}^{t/2} \left\{\begin{array}{c} \sigma_x \\ \sigma_y \\ \sigma_{xy} \end{array}\right\} zdz \tag{6.11}$$

where N_x, N_y, and N_{xy} are the tensile and shear forces per unit length along the boundary of the plate element shown in Figure 6.3 with units [N/m], V_x and V_y are the shear forces per unit length of the plate with units [N/m], and M_x, M_y, and M_{xy} are the moments per unit length with units [N]. The moments per unit length M_x and M_y are positive when they produce a concave deformation looking from the negative z-axis (Figure 6.3). Since $\sigma_{yx} = \sigma_{xy}$, only M_{xy} is used in this book, with the orientation given in Figure 6.3.

The integrations in (6.11) span over several laminae. Therefore, the integrals can be divided into summations of integrals over each lamina

$$\left\{\begin{array}{c} N_x \\ N_y \\ N_{xy} \end{array}\right\} = \sum_{k=1}^{N} \int_{z_{k-1}}^{z_k} \left\{\begin{array}{c} \sigma_x \\ \sigma_y \\ \sigma_{xy} \end{array}\right\}^k dz$$

$$\left\{\begin{array}{c} V_y \\ V_x \end{array}\right\} = \sum_{k=1}^{N} \int_{z_{k-1}}^{z_k} \left\{\begin{array}{c} \sigma_{yz} \\ \sigma_{xz} \end{array}\right\}^k dz$$

$$\left\{\begin{array}{c} M_x \\ M_y \\ M_{xy} \end{array}\right\} = \sum_{k=1}^{N} \int_{z_{k-1}}^{z_k} \left\{\begin{array}{c} \sigma_x \\ \sigma_y \\ \sigma_{xy} \end{array}\right\}^k zdz \tag{6.12}$$

where k is the lamina number counting from the bottom up, N is the number of laminae in the laminate, and z_k is the coordinate of the top surface of the $k-th$ lamina (Figure 6.6).

6.1.4 Plate Stiffness and Compliance

Since the thickness of the laminate is small compared to the in-plane dimensions of the plate, every lamina is in a state of plane stress. Therefore, the stress–strain

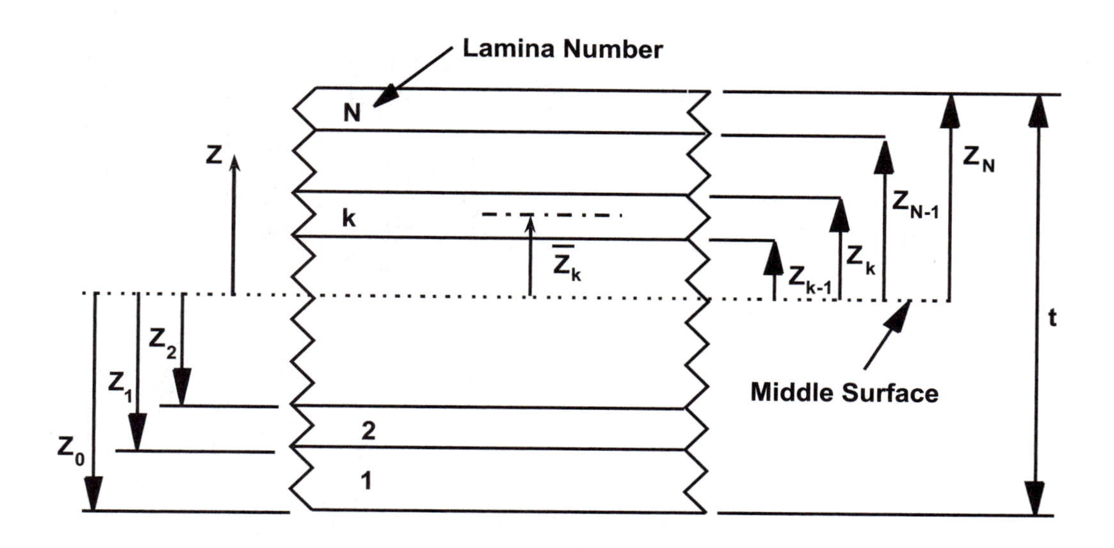

Figure 6.6: Geometry of a laminate with N laminae.

relations in material coordinates (5.19) and (5.21) are

$$\begin{Bmatrix} \sigma_1 \\ \sigma_2 \\ \sigma_6 \end{Bmatrix}^k = \begin{bmatrix} Q_{11} & Q_{12} & 0 \\ Q_{12} & Q_{22} & 0 \\ 0 & 0 & Q_{66} \end{bmatrix}^k \begin{Bmatrix} \epsilon_1 \\ \epsilon_2 \\ \gamma_6 \end{Bmatrix}^k$$
$$\begin{Bmatrix} \sigma_4 \\ \sigma_5 \end{Bmatrix}^k = \begin{bmatrix} Q^*_{44} & 0 \\ 0 & Q^*_{55} \end{bmatrix}^k \begin{Bmatrix} \gamma_4 \\ \gamma_5 \end{Bmatrix}^k \tag{6.13}$$

where the superscript k indicates the lamina number (Figure 6.6). The stress–strain relations in laminate coordinates are given by (5.42) and (5.47), namely

$$\begin{Bmatrix} \sigma_x \\ \sigma_y \\ \sigma_{xy} \end{Bmatrix}^k = \begin{bmatrix} \overline{Q}_{11} & \overline{Q}_{12} & \overline{Q}_{16} \\ \overline{Q}_{12} & \overline{Q}_{22} & \overline{Q}_{26} \\ \overline{Q}_{16} & \overline{Q}_{26} & \overline{Q}_{66} \end{bmatrix}^k \begin{Bmatrix} \epsilon_x \\ \epsilon_y \\ \gamma_{xy} \end{Bmatrix}^k \tag{6.14}$$
$$\begin{Bmatrix} \sigma_{yz} \\ \sigma_{xz} \end{Bmatrix}^k = \begin{bmatrix} \overline{Q}^*_{44} & \overline{Q}^*_{45} \\ \overline{Q}^*_{45} & \overline{Q}^*_{55} \end{bmatrix}^k \begin{Bmatrix} \gamma_{yz} \\ \gamma_{xz} \end{Bmatrix}^k$$

Replacing (6.5) into (6.14), and the result into (6.12), the following plate stiffness equations are obtained

$$
\left\{\begin{array}{c} N_x \\ N_y \\ N_{xy} \\ M_x \\ M_y \\ M_{xy} \end{array}\right\} = \left[\begin{array}{cccccc} A_{11} & A_{12} & A_{16} & B_{11} & B_{12} & B_{16} \\ A_{12} & A_{22} & A_{26} & B_{12} & B_{22} & B_{26} \\ A_{16} & A_{26} & A_{66} & B_{16} & B_{26} & B_{66} \\ B_{11} & B_{12} & B_{16} & D_{11} & D_{12} & D_{16} \\ B_{12} & B_{22} & B_{26} & D_{12} & D_{22} & D_{26} \\ B_{16} & B_{26} & B_{66} & D_{16} & D_{26} & D_{66} \end{array}\right] \left\{\begin{array}{c} \epsilon_x^0 \\ \epsilon_y^0 \\ \gamma_{xy}^0 \\ \kappa_x \\ \kappa_y \\ \kappa_{xy} \end{array}\right\} \tag{6.15}
$$

$$
\left\{\begin{array}{c} V_y \\ V_x \end{array}\right\} = \left[\begin{array}{cc} H_{44} & H_{45} \\ H_{45} & H_{55} \end{array}\right] \left\{\begin{array}{c} \gamma_{yz} \\ \gamma_{xz} \end{array}\right\}
$$

with

$$
\begin{aligned}
A_{ij} &= \sum_{k=1}^{N} \left(\overline{Q}_{ij}\right)_k (z_k - z_{k-1}) = \sum_{k=1}^{N} \left(\overline{Q}_{ij}\right)_k t_k; i,j = 1,2,6 \\
B_{ij} &= \frac{1}{2}\sum_{k=1}^{N} \left(\overline{Q}_{ij}\right)_k (z_k^2 - z_{k-1}^2) = \sum_{k=1}^{N} \left(\overline{Q}_{ij}\right)_k t_k \bar{z}_k; i,j = 1,2,6 \\
D_{ij} &= \frac{1}{3}\sum_{k=1}^{N} \left(\overline{Q}_{ij}\right)_k (z_k^3 - z_{k-1}^3) = \sum_{k=1}^{N} \left(\overline{Q}_{ij}\right)_k \left(t_k \bar{z}_k^2 + \frac{t_k^3}{12}\right); i,j = 1,2,6 \\
H_{ij} &= \frac{5}{4}\sum_{k=1}^{N} \left(\overline{Q}_{ij}^*\right)_k \left[t_k - \frac{4}{t^2}\left(t_k \bar{z}_k^2 + \frac{t_k^3}{12}\right)\right]; i,j = 4,5
\end{aligned} \tag{6.16}
$$

where $\bar{z}_k$ is the coordinate of the middle surface of the k-th lamina (Figure 6.6), A_{ij} are the extensional stiffness, B_{ij} are the bending-extension coupling stiffness, D_{ij} are the bending stiffness, H_{ij} are the transverse shear stiffness,[3] and t is the total thickness of the laminate.

The 6×6 stiffness matrix in (6.15) is composed of three submatrices, $[A]$, $[B]$, $[D]$, each 3×3 in size. All three are symmetric matrices, that is

$$A_{ij} = A_{ji} \tag{6.17}$$

$$B_{ij} = B_{ji} \tag{6.18}$$

$$D_{ij} = D_{ji} \tag{6.19}$$

The coefficients A_{ij}, B_{ij}, D_{ij} (6.16) are function of the thickness, orientation, stacking sequence, and material properties of the laminae. Each matrix has a particular role in the analysis of the laminate, as follows:

- The $[A]$ matrix is called *in-plane stiffness matrix* because it directly relates in-plane strains (ϵ_x^0, ϵ_y^0, γ_{xy}^0) to in-plane forces (N_x, N_y, N_{xy}).

[3] These coefficients incorporate a correction to account for a parabolic distribution of shear stress that vanish on the surface of the plate, as explained in [174].

- The $[D]$ matrix is called *bending stiffness matrix* because it relates curvatures (κ_x, κ_y, κ_{xy}) to bending moments (M_x, M_y, M_{xy}).
- The $[B]$ matrix relates in-plane strains to bending moments and curvatures to in-plane forces. This coupling effect does not exist for homogeneous plates (e.g. metallic plates). Therefore, $[B]$ is called *bending-extension coupling matrix.*
- The $[H]$ matrix relates transverse shear strains γ_{yz}, γ_{xz}, to transverse shear forces V_y, V_x. It is then called *transverse shear stiffness matrix.* This matrix is used only within the context of first order shear deformation theory (FSDT). The $[H]$ matrix is not used with classical plate theory (CPT) because both γ_{yz} and γ_{xz} are assumed to be zero in CPT (see discussion in Section 6.1.2)

Equations 6.15 are called *stiffness equations* because of the analogy with Hooke's law $\sigma = E\ \epsilon$, where E is the modulus or stiffness of the material. The compliance equations of the plate are obtained inverting the matrices in (6.15) to obtain

$$\left\{\begin{array}{c} \epsilon_x^0 \\ \epsilon_y^0 \\ \gamma_{xy}^0 \\ \kappa_x \\ \kappa_y \\ \kappa_{xy} \end{array}\right\} = \left[\begin{array}{cccccc} \alpha_{11} & \alpha_{12} & \alpha_{16} & \beta_{11} & \beta_{12} & \beta_{16} \\ \alpha_{12} & \alpha_{22} & \alpha_{26} & \beta_{12} & \beta_{22} & \beta_{26} \\ \alpha_{16} & \alpha_{26} & \alpha_{66} & \beta_{16} & \beta_{26} & \beta_{66} \\ \beta_{11} & \beta_{12} & \beta_{16} & \delta_{11} & \delta_{12} & \delta_{16} \\ \beta_{12} & \beta_{22} & \beta_{26} & \delta_{12} & \delta_{22} & \delta_{26} \\ \beta_{16} & \beta_{26} & \beta_{66} & \delta_{16} & \delta_{26} & \delta_{66} \end{array}\right] \left\{\begin{array}{c} N_x \\ N_y \\ N_{xy} \\ M_x \\ M_y \\ M_{xy} \end{array}\right\} \tag{6.20}$$
$$\left\{\begin{array}{c} \gamma_{yz} \\ \gamma_{xz} \end{array}\right\} = \left[\begin{array}{cc} h_{44} & h_{45} \\ h_{45} & h_{55} \end{array}\right] \left\{\begin{array}{c} V_y \\ V_x \end{array}\right\}$$

If the laminate is symmetric with respect to the middle surface all the bending-extension coupling coefficients B_{ij} are zero and all the coefficients β_{ij} are also zero. In this case, (6.20) can be separated into

$$\left\{\begin{array}{c} \epsilon_x^0 \\ \epsilon_y^0 \\ \gamma_{xy}^0 \end{array}\right\} = \left[\begin{array}{ccc} \alpha_{11} & \alpha_{12} & \alpha_{16} \\ \alpha_{12} & \alpha_{22} & \alpha_{26} \\ \alpha_{16} & \alpha_{26} & \alpha_{66} \end{array}\right] \left\{\begin{array}{c} N_x \\ N_y \\ N_{xy} \end{array}\right\} \tag{6.21}$$

and

$$\left\{\begin{array}{c} \kappa_x \\ \kappa_y \\ \kappa_{xy} \end{array}\right\} = \left[\begin{array}{ccc} \delta_{11} & \delta_{12} & \delta_{16} \\ \delta_{12} & \delta_{22} & \delta_{26} \\ \delta_{16} & \delta_{26} & \delta_{66} \end{array}\right] \left\{\begin{array}{c} M_x \\ M_y \\ M_{xy} \end{array}\right\} \tag{6.22}$$

According to (6.21) and (6.22), when a set of in-plane forces is applied (Figure 6.3a), only midsurface strains are induced. When only a set of bending moments is applied (Figure 6.3b), only curvatures are induced. This is the typical behavior of metallic plates.

If the laminate is not symmetric, the bending-extension coupling matrix $[B]$ and compliance $[\beta]$ are not zero. According to (6.20), application of any force or moment may result in a number of strains and curvatures being induced. Several cases are illustrated in Figure 6.7: (a) [0/90] result in bending-extension coupling, (b) thermal expansion of a [0/90] laminate results in a saddle shape, (c) two laminae at $\pm\theta$ result in torsion-extension coupling, and (d) all laminae at an angle θ result in shear-extension coupling (see also Section 6.3).

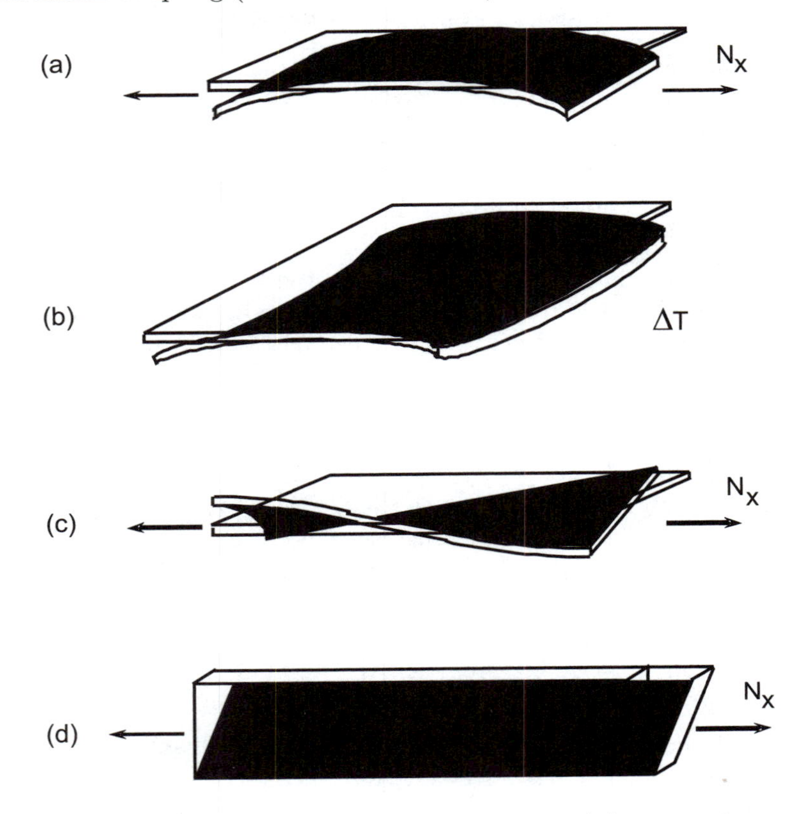

Figure 6.7: Coupling effects: (a) bending-extension, (b) thermal expansion of a [0/90] laminate, (c) torsion-extension, and (d) shear-extension.

Example 6.1 *Compute the coefficients in the plate stiffness equations (6.9) for a two-lamina laminate with $\theta_1 = 55°$, $\theta_2 = -55°$, $t_1 = t_2 = 0.635$ mm, with material properties given in Example 5.1, p. 152.*

Solution to Example 6.1 *From Example 5.1, p. 152, the reduced stiffness matrix of one lamina is*

$$[Q] = \begin{bmatrix} 20.874 & 3.260 & 0 \\ 3.260 & 11.898 & 0 \\ 0 & 0 & 3.789 \end{bmatrix} \; GPa$$

Following the procedure in Example 5.6, p. 161, the transformed reduced stiffness matrices in the two laminae are, at $\theta = 55°$

$$[\overline{Q}]^{(1)} = \begin{bmatrix} 12.40 & 5.71 & 1.22 \\ 5.71 & 15.50 & 3.00 \\ 1.22 & 3.00 & 6.24 \end{bmatrix} \; GPa$$

and at $\theta = 55°$

$$[\overline{Q}]^{(2)} = \begin{bmatrix} 12.40 & 5.71 & -1.22 \\ 5.71 & 15.50 & -3.00 \\ -1.22 & -3.00 & 6.24 \end{bmatrix} \; GPa$$

where the superscript indicates the lamina number. Then, using (6.16)

$$[A] = \begin{bmatrix} 15.8 & 7.25 & \varepsilon \\ 7.25 & 19.6 & \varepsilon \\ \varepsilon & \varepsilon & 7.92 \end{bmatrix} \; GPa\ mm$$

$$[B] = \begin{bmatrix} \varepsilon & \varepsilon & -0.491 \\ \varepsilon & \varepsilon & -1.21 \\ -0.491 & -1.21 & \varepsilon \end{bmatrix} \; GPa\ mm^2$$

$$[D] = \begin{bmatrix} 2.12 & 0.975 & \varepsilon \\ 0.975 & 2.64 & \varepsilon \\ \varepsilon & \varepsilon & 1.06 \end{bmatrix} \; GPa\ mm^3$$

Because of rounding errors during the computations, ε *may show as a very small number when compared to the other numbers in the same matrix, when it should be identically zero. Using data from Example 5.1, p. 152, the intralaminar shear coefficients for the orthotropic lamina are*

$$Q^*_{44} = G_{23} = 0.385 \; GPa$$
$$Q^*_{55} = G_{13} = 3.789 \; GPa$$

Using (5.49) for $\theta = 55°$

$$\overline{Q}^{*(1)}_{44} = 2.669$$
$$\overline{Q}^{*(1)}_{45} = 1.599$$
$$\overline{Q}^{*(1)}_{55} = 1.505$$

and for $\theta = -55°$

$$\overline{Q}^{*(2)}_{44} = 2.669$$
$$\overline{Q}^{*(2)}_{45} = -1.599$$
$$\overline{Q}^{*(2)}_{55} = 1.505$$

Finally, the transverse shear coefficients are obtained using (6.16)

$$H_{44} = 2.82$$
$$H_{55} = 1.59$$
$$H_{45} = 0$$

6.2 Computation of Stresses

Once the $[A], [B], [D]$, and $[H]$ matrices have been formulated, strains and curvatures at the middle surface can be computed by solving (6.20) for a given set of stress resultants. Once the midsurface strains ϵ_x^0, ϵ_y^0, γ_{xy}^0, the curvatures κ_x, κ_y, κ_{xy}, and the transverse shear strains γ_{xy}, γ_{xz}, are known, the strains ϵ_x, ϵ_y, γ_{xy}, can be computed at any point through the thickness of the plate using (6.5), which is repeated here for convenience

$$\left\{ \begin{array}{c} \epsilon_x \\ \epsilon_y \\ \gamma_{xy} \end{array} \right\} = \left\{ \begin{array}{c} \epsilon_x^0 \\ \epsilon_y^0 \\ \gamma_{xy}^0 \end{array} \right\} + z \left\{ \begin{array}{c} \kappa_x \\ \kappa_y \\ \kappa_{xy} \end{array} \right\} \tag{6.23}$$

From (6.23), the strains (ϵ_x, ϵ_y, γ_{xy}) are a linear function of the z-coordinate, while γ_{yz} and γ_{xz} are obtained directly from (6.20) and they are constant through the thickness. The stresses can be computed on each lamina using the constitutive equations of that particular lamina, (5.42) and (5.47), which are repeated here for convenience

$$\begin{aligned} \left\{ \begin{array}{c} \sigma_x \\ \sigma_y \\ \sigma_{xy} \end{array} \right\}^k &= \left[\begin{array}{ccc} \overline{Q}_{11} & \overline{Q}_{12} & \overline{Q}_{16} \\ \overline{Q}_{12} & \overline{Q}_{22} & \overline{Q}_{26} \\ \overline{Q}_{16} & \overline{Q}_{26} & \overline{Q}_{66} \end{array} \right]^k \left\{ \begin{array}{c} \epsilon_x \\ \epsilon_y \\ \gamma_{xy} \end{array} \right\} \\ \left\{ \begin{array}{c} \sigma_{yz} \\ \sigma_{xz} \end{array} \right\}^k &= \left[\begin{array}{cc} \overline{Q}^*_{44} & \overline{Q}^*_{45} \\ \overline{Q}^*_{45} & \overline{Q}^*_{55} \end{array} \right]^k \left\{ \begin{array}{c} \gamma_{yz} \\ \gamma_{xz} \end{array} \right\} \end{aligned} \tag{6.24}$$

where the superscript k indicates that these equations apply to the k-th lamina. Note that the strain components given in (6.23) are continuous through the thickness while the stresses are discontinuous at each interface between laminae due to the change of material properties from one lamina to another (see Figure 6.5). The transverse shear stresses σ_{yz} and σ_{xz} can be computed more accurately using theory of elasticity [175], but the complexity of such computations is usually not necessary in preliminary design.

Example 6.2 *Compute the stresses at the bottom surface of a single-lamina plate subjected to* $M_x = 1\ KNm/m$, $N_x = N_y = N_{xy} = V_x = V_y = M_y = M_{xy} = 0$. *The plate thickness is* $t = 0.635\ mm$ *and the material properties are given in Example 5.1, p. 152. The fiber direction coincides with the global* x*-axis.*

Solution to Example 6.2 *The reduced stiffness matrix* $[Q]$ *of Example 5.1, p. 152, can be used because the axes of orthotropy of the material coincide with the laminate coordinates*

$$[Q] = \left[\begin{array}{ccc} 20.874 & 3.260 & 0 \\ 3.260 & 11.898 & 0 \\ 0 & 0 & 3.789 \end{array} \right] GPa$$

There is no need to compute the $[H]$ matrix because the shear forces are zero. Also, the $[A]$ matrix is not needed because the in-plane forces are zero. The $[B]$ matrix is zero because a single-lamina plate is, of course, symmetric. Then, using (6.16)

$$[D] = \begin{bmatrix} 0.445 & 0.0696 & 0 \\ 0.0696 & 0.254 & 0 \\ 0 & 0 & 0.0808 \end{bmatrix} \; GPa\; mm^3$$

In this particular example, because $D_{16} = D_{26} = 0$, the bending compliance $[\delta]$ can be computed as

$$\begin{aligned} Det &= 0.445(0.254) - 0.0696^2 = 0.108 \\ \delta_{11} &= \frac{D_{22}}{Det} = 2.348\; (GPa\; mm^3)^{-1} \\ \delta_{22} &= \frac{D_{11}}{Det} = 4.113\; (GPa\; mm^3)^{-1} \\ \delta_{12} &= \frac{-D_{12}}{Det} = -0.643\; (GPa\; mm^3)^{-1} \\ \delta_{66} &= \frac{1}{D_{66}} = 12.376\; (GPa\; mm^3)^{-1} \end{aligned}$$

Using (6.20), the curvatures are

$$\begin{Bmatrix} \kappa_x \\ \kappa_y \\ \kappa_{xy} \end{Bmatrix} = \begin{bmatrix} 2.348 & -0.643 & 0 \\ -0.643 & 4.113 & 0 \\ 0 & 0 & 12.376 \end{bmatrix} \begin{Bmatrix} 1 \\ 0 \\ 0 \end{Bmatrix} = \begin{Bmatrix} 2.348 \\ -0.643 \\ 0 \end{Bmatrix}$$

Using (6.23), the strains at the bottom lamina ($z = -t/2$) are

$$\begin{Bmatrix} \epsilon_x \\ \epsilon_y \\ \gamma_{xy} \end{Bmatrix} = \frac{-0.635}{2} \begin{Bmatrix} 2.348 \\ -0.643 \\ 0 \end{Bmatrix} = \begin{Bmatrix} -0.745 \\ 0.204 \\ 0 \end{Bmatrix}$$

where the midplane strains are zero because there are no in-plane forces and the laminate is symmetric. Using (6.24), the stresses at the bottom surface are

$$\begin{Bmatrix} \sigma_x \\ \sigma_y \\ \sigma_{xy} \end{Bmatrix} = \begin{bmatrix} 20.874 & 3.260 & 0 \\ 3.260 & 11.898 & 0 \\ 0 & 0 & 3.789 \end{bmatrix} \begin{Bmatrix} -0.745 \\ 0.204 \\ 0 \end{Bmatrix} = \begin{Bmatrix} -14.886 \\ tol \\ 0 \end{Bmatrix} \; GPa$$

where, because of rounding, tol is a very small quantity, instead of zero.

6.3 Common Laminate Types

The constitutive equations of the most general type of laminate are given by (6.15) with the coefficients computed according to (6.16). There are 21 different coefficients in (6.15), 18 in the $[ABD]$ matrix and 3 in the $[H]$ matrix. The first six equations are coupled through the $[ABD]$ matrix, but they are uncoupled from the two transverse shear equations. If all 18 coefficients in the $[ABD]$ matrix are different from zero, the laminate response is fully coupled, meaning that the application of just one load (say N_x) makes all six strains different from zero. A desire to eliminate to some extent some of the coupling terms has motivated the use of particular types of laminates, which are described in this section.

Many laminate configurations exist for which some of the terms in the $[ABD]$ matrix vanish. The importance of analyzing each particular laminate with a particular set of equations has decreased with the availability of computer analysis tools like the ones presented here. Using a computer program, any laminate can be analyzed with the same effort. However, symmetric laminates (see 6.3.2) deserve special consideration because the physical behavior of these is more desirable. Also, quasi-isotropic laminates (see 6.3.5) are also special because a large number of design equations, available for isotropic plates, can be used to approximate the solution of these. The most important laminate configurations are described in this section.

6.3.1 Laminate Description

The notation used to describe laminates has its roots in the description used to specify the lay-up sequence for hand lay-up using prepreg. When using prepreg, all laminae have the same thickness and only the angles need to be specified. In the hand lay-up process the laminate is built starting at the tool surface, or bottom of the laminate, adding laminae on top. Therefore, the laminae are numbered starting at the bottom and the angles are given from bottom up. For example, a two-lamina laminate may be $[30/-30]$, a three-lamina one $[-45/45/0]$, etc.

If the laminate is symmetric, like $[30/0/0/30]$, an abbreviated notation is used where only half of the stacking sequence is given and a subscript (S) is added to specify symmetric. The last example becomes $[30/0]_S$. Since most laminates used are symmetric, there is the possibility of confusion when no subscript is used. To eliminate the possibility of confusion, a subscript (T) may be used to indicate that the sequence given is the total (e.g., $[45/-45/45/-45]_T$) or $[\pm 45/\pm 45]_T$. A subscript may be used to indicate a repeating pattern. In a such a way the last example becomes $[(\pm 45)_2]_T$.

If the thicknesses of the laminae are different, they are specified for each lamina. For example $[\theta_{t_1}/-\theta_{t_2}]$. If the different thicknesses are multiples of a single thickness t, the notation simplifies to $[\theta/-\theta_2]$ which indicates one lamina of thickness t at an angle θ and two laminae of the same thickness t (or one lamina twice as thick) at an angle $-\theta$. Angle-ply combinations like $[\theta/-\theta]$ can be denoted as $[\pm\theta]$. If all laminae have the same thickness, the laminate is called *regular.*

6.3.2 Symmetric Laminates

A laminate is symmetric if laminae of the same material, thickness, and orientation are symmetrically located with respect to the middle surface of the laminate. For example $[30/0/0/30]$ is symmetric but not balanced (see Section 6.3.4), while $[30/-30/-30/30]$ is symmetric and balanced.

Symmetric laminates experience no bending-extension coupling $B_{ij} = 0$. This is very important during fabrication because curing and subsequent cooling of the composite induces thermal forces N_x and N_y. If the laminate is not symmetric, these forces will induce warping (κ_x, κ_y) of the final part; see (6.20) with $B_{ij} \neq 0$.

A $[0/90]_T$ laminate is not symmetric but both laminae are specially orthotropic. Since specially orthotropic laminae have $\overline{Q}_{16} = \overline{Q}_{26} = 0$, the laminate has $A_{16} = A_{26} = B_{16} = B_{26} = B_{66} = D_{16} = D_{26} = 0$ but the remaining B_{ij} terms are not zero. Using (6.20), it can be shown that the laminate bends as shown in Figure 6.7a when loaded with only an in-plane force N_x. That is, a curvature $\kappa_x = \beta_{11} N_x$ is induced in addition to stretching $\epsilon_x^0 = \alpha_{11} N_x$.

In Figure 6.7b, the $[0/90]_T$ laminate is subject to in-plane forces N_x and N_y caused by thermal contraction during cool-down after curing. It can be shown that β_{11} and β_{22} (as well as B_{11} and B_{22}) have opposite signs, resulting in the saddle-shape deformation shown. Since residual thermal forces due to curing are unavoidable, only a symmetric laminate will remain flat after cool-down.

A $[\pm\theta]_T$ laminate is not symmetric but balanced (Section 6.3.4). It can be shown that for this laminate the only zero terms are

$$
\begin{aligned}
A_{16} &= A_{26} = \alpha_{16} = \alpha_{26} = 0 \\
D_{16} &= D_{26} = \delta_{16} = \delta_{26} = 0 \\
B_{11} &= B_{22} = B_{12} = \beta_{11} = \beta_{22} = \beta_{12} = 0
\end{aligned}
\tag{6.25}
$$

Therefore, an in-plane load N_x makes the laminate to twist ($\kappa_{xy} = \beta_{16} N_x$) as shown in Figure 6.7c.

Finally, a single lamina with fibers oriented at an angle θ ($\theta \neq 0°$ and $\theta = 90°$), is symmetric but not balanced. Simple application of (6.16) results in nonzero values for A_{16} and A_{26}, which in turn give α_{16} and α_{26} different from zero. Such material will shear ($\gamma_{xy}^{\circ} = A_{16} N_x$) when subjected to in-plane load N_x, as shown in Figure 6.7c.

From (6.16), the bending-extension coupling coefficients are

$$
B_{ij} = \sum_{k=1}^{N} \left(\overline{Q}_{ij}\right)_k t_k \bar{z}_k \ ; \ i,j = 1,2,6
$$

With reference to Figure 6.6, note that for a symmetric laminate there is a negative $\bar{z}_k$ for each positive $\bar{z}_k$ (due to symmetry) corresponding to the same material properties and same thickness. Therefore, pairs of terms in the equation above corresponding to symmetrically located laminae cancel each other, resulting in $B_{ij} = 0$.

The fact that $[B]$ matrix is zero uncouples all bending terms from the extension terms in (6.15). This means that a symmetric laminate subjected to bending moments will not stretch or shear. Also, a symmetric laminate will not bend or twist if subject to in-plane forces. This is the typical behavior from isotropic plates, which of course are symmetric. Symmetric laminates are used whenever possible because the analysis is simplified and they do not behave in extraneous ways. For example, unsymmetrical laminates bend and warp as a result of thermal expansion. In a symmetric laminate, the symmetric laminae expand the same amount, and the laminate will expand but not warp. Also, the analysis of symmetric laminates is simpler. The first three of (6.15) can be solved independently of the last three. If the compliance (6.20) are desired, the 3×3 $[A]$ matrix can be inverted to get the $[\alpha]$ matrix (see (6.21)) and the 3×3 $[D]$ matrix can be inverted to get the $[\delta]$ matrix (see (6.22)) without having to invert the whole 6×6 $[ABD]$ matrix.

6.3.3 Antisymmetric Laminate

An antisymmetric laminate has pairs of laminae of opposite orientation but same material and thickness symmetrically located with respect to the middle surface of the laminate. For example $[30/-30/30/-30]$ is an antisymmetric angle-ply and $[0/90/0/90]$ is an antisymmetric cross-ply. Antisymmetric laminates have $A_{16} = A_{26} = D_{16} = D_{26} = 0$, but they are not particularly useful nor they are easier to analyze than general laminates because the bending extension coefficients B_{16} and B_{26} are not zero for these laminates.

6.3.4 Balanced Laminate

In a balanced laminate, for every lamina at $+\theta$ there is another at $-\theta$ with the same thickness and material; and for each 0°-lamina there is a complementary 90°-lamina, also of the same thickness and material. By virtue of (5.49),

$$\begin{aligned} \overline{Q}_{16}(\theta) &= -\overline{Q}_{16}(-\theta) \\ \overline{Q}_{26}(\theta) &= -\overline{Q}_{26}(-\theta) \end{aligned} \tag{6.26}$$

a balanced laminate always has $A_{16} = A_{26} = 0$. A symmetric balanced laminate has also $B_{ij} = 0$ but $D_{16} \neq 0$ and $D_{26} \neq 0$. An antisymmetric laminate has $A_{16} = A_{26} = D_{16} = D_{26} = 0$ but $B_{16} \neq 0$ and $B_{26} \neq 0$. Balanced laminates having only 0-, 45-, and 90-deg. laminae have identical laminate moduli in two directions ($E_x = E_y$, see Section 6.4).

6.3.5 Quasi-isotropic Laminates

Quasi-isotropic laminates are constructed in an attempt to create a composite laminate that behaves like an isotropic plate. The in-plane behavior of quasi-isotropic laminates is similar to that of isotropic plates but the bending behavior of quasi-isotropic laminates is quite different than the bending behavior of isotropic plates. The use of quasi-isotropic laminates usually leads to an inefficient, conservative design. In a quasi-isotropic laminate, each lamina has an orientation given by

$$\theta_k = \frac{k\pi}{N} + \theta_o \tag{6.27}$$

where k indicates the lamina number, N is the number of laminae (at least 3), and θ_o is an arbitrary initial angle. When writing the stacking sequence, any angle larger than 90 degrees is replaced by its complement (for $\theta_k > 90 \Rightarrow \theta_k = \theta_k - 180$). For example, [60/120/180] is rewritten as [60/ − 60/0]. The laminae can be ordered in any order, like [60/0/ − 60], and the laminate is still quasi-isotropic.

Quasi-isotropic laminates are in general not symmetric but they can be made symmetric by doubling the number of laminae in a mirror (symmetric) fashion. For example, the [60/ − 60/0] can be made into a [60/ − 60/0/0/ − 60/60], which is still quasi-isotropic. The advantage of symmetric quasi-isotropic laminates is that $[B] = 0$.

Quasi-isotropic laminates do not behave exactly like isotropic plates. The in-plane stiffness matrix $[A]$ and the bending stiffness matrix $[D]$ of isotropic plates can be written in terms of the thickness (t) of the plate and only two material properties, E and ν as

$$[A] = \frac{Et}{1-\nu^2}\begin{bmatrix} 1 & \nu & 0 \\ \nu & 1 & 0 \\ 0 & 0 & \frac{1-\nu}{2} \end{bmatrix}$$
$$[D] = \frac{Et^3}{12\,(1-\nu^2)}\begin{bmatrix} 1 & \nu & 0 \\ \nu & 1 & 0 \\ 0 & 0 & \frac{1-\nu}{2} \end{bmatrix} \tag{6.28}$$

Note that for isotropic plates, $A_{11} = A_{22} = Et/(1-\nu^2)$ and $A_{16} = A_{26} = 0$. Also, $D_{11} = D_{22} = Et^3/12(1-\nu^2)$ and $D_{16} = D_{26} = 0$. On the other hand, a quasi-isotropic laminate has

$$[A] = \begin{bmatrix} A_{11} & A_{12} & 0 \\ A_{12} & A_{11} & 0 \\ 0 & 0 & A_{66} \end{bmatrix}$$
$$[D] = \begin{bmatrix} D_{11} & D_{12} & D_{16} \\ D_{12} & D_{22} & D_{26} \\ D_{16} & D_{26} & D_{66} \end{bmatrix} \tag{6.29}$$

Like isotropic plates, quasi-isotropic laminates have $A_{11} = A_{22}$, but the later have $D_{11} \neq D_{22}$, $D_{16} \neq 0$, and $D_{26} \neq 0$, which makes quasi-isotropic laminates to be quite different from isotropic plates. Therefore, formulas for bending, buckling, and vibrations of isotropic plates can be used for quasi-isotropic laminates only as an approximation. However, a laminate can be designed trying to approach the characteristics of isotropic plates, with $D_{11} \approx D_{22}$ and D_{16} and D_{26} as small as possible. This can be done by using a symmetric quasi-isotropic laminate with balanced 0/90 and $\pm\theta$ laminae, and a large number of laminae. In that case, the formulas for isotropic plates provide a reasonable approximation, as illustrated in Example 6.3.

Example 6.3 *Compute the maximum deflection of a rectangular plate of length $a = 1.6$ m (along x), width $b = 0.8$ m (along y), and thickness $t = 0.016$ m, with all sides simply supported and loaded with a distributed load $q = 0.02$ MPa. The laminate is made of E-glass–epoxy with properties given in Table 1.3. Use a symmetric quasi-isotropic configuration.*

Solution to Example 6.3 *First, try three different quasi-isotropic laminates to observe how D_{11} approaches D_{22} and how D_{16} and D_{26} become smaller as the number of laminae increases. The three laminates have the same total thickness.*

For a $[0/90/\pm 45]_S$ with the thickness of each lamina $t_k = 0.002$ m, the bending stiffness matrix is

$$[D] = \begin{bmatrix} 11.1 & 1.12 & 0.267 \\ & 7.89 & 0.267 \\ sym. & & 2.21 \end{bmatrix} 10^{-3}\ MPa\ m^3$$

For a $[0/90/\pm 45]_{2s}$ with $t_k = 0.001$ m

$$[D] = \begin{bmatrix} 9.66 & 1.61 & 0.200 \\ & 8.33 & 0.200 \\ sym. & & 2.70 \end{bmatrix} 10^{-3}\ MPa\ m^3$$

where the subscript 2s means that the lay-up in the bracket is repeated twice, then made symmetric, to end up with a 16-lamina laminate.

For a $[0/90/\pm 45]_{4s}$ with $t_k = 0.0005$ m

$$[D] = \begin{bmatrix} 9.05 & 1.86 & 0.117 \\ & 8.45 & 0.117 \\ sym. & & 2.95 \end{bmatrix} 10^{-3}\ MPa\ m^3$$

Clearly, the difference between D_{11} and D_{22} reduces as the number of laminae increases. The same is true about the magnitude of the coefficients D_{16} and D_{26}.

To use formulas for isotropic plates, it is necessary to find equivalent values of ν and E. Comparing Eqs. 6.28 to 6.29, the following formulas can be derived (see also Section 6.4)

$$\nu = \frac{D_{12}}{D_{22}}$$

$$E = \frac{12(1-\nu^2)}{t^3}\left(\frac{D_{11}+D_{22}}{2}\right)$$

where the average of D_{11} and D_{22} has been used because $D_{11} \neq D_{22}$.

For the $[0/90/\pm 45]_{4s}$, the equivalent properties are computed as

$$\nu = \frac{1.86}{8.45} = 0.220$$

$$E = \frac{12(1-0.220^2)}{(0.016)^3}\left(\frac{9.05+8.45}{2}\right) = 24,394\ MPa$$

Using [176, Table 26], for isotropic plates with $\nu = 0.3$*, the maximum deflection is approximated by*

$$w_{\max} = -\frac{\alpha q b^4}{E t^3} = -\frac{0.1110(0.02)(0.8)^4}{24,394(0.016)^3} = 9.10\ mm$$

with $\alpha = 0.1110$ *for* $a/b = 2$ *from [176]. However, computation of stresses cannot be done with the formulas in [176] because those formulas are for homogeneous isotropic materials. Computation of stresses in laminated plates is presented in Section 6.2.*

To further illustrate the difference between a quasi-isotropic laminate and an isotropic plate, note that the equivalent shear stiffness could be computed in two ways, either as

$$G = \frac{E}{2(1+\nu)} = 9,997.5\ MPa$$

or, assuming $D_{16} = D_{26} = 0$*, as*

$$G = \frac{12 D_{66}}{t^3} = 8,642.5\ MPa$$

The difference between the computed values provides additional indication that treating a quasi-isotropic laminate as an isotropic plate is only an approximation.

The approximate equation for the maximum deflection from [176] is for a fixed value of the Poisson's ratio, $\nu = 0.3$*. A similar equation is presented in [177] as*

$$w_{\max} = \delta_1 \frac{q b^4}{D}$$
$$D = \frac{E t^3}{12(1-\nu^2)}$$

where $\delta_1 = 0.01013$ *for* $a/b = 2$*, from table 3.1 in [177]. Using* $\nu = 0.220$*, the resulting value of deflection is* $w_{\max} = 9.48\ mm$*. The exact solution for a simply supported plate with* $D_{16} = D_{26} = 0$ *can be obtained by the Navier method [175]*

$$w_{\max} = \sum_{n=1}^{\infty} \sum_{m=1}^{\infty} W_{mn} \sin(m\pi/2) \sin(n\pi/2)$$

where $s = b/a$ *and*

$$W_{mn} = \frac{16 q b^4}{\pi^6 m n \left[D_{11} m^4 s^4 + 2(D_{12} + 2D_{66}) m^2 n^2 s^2 + D_{22} n^4 \right]}$$

For this example, the computed deflection is $w_{\max} = 9.62\ mm$*. It can be seen that, for this example, approximating the plate as an isotropic material led to an error of only 5.7%.*

Example 6.4 *Compute the stress at the bottom surface of the plate in Example 6.3, p. 186.*

Solution to Example 6.4 *The deflected shape of a plate simply supported on all edges and subject to uniform load, can be approximated as*

$$w(x,y) = w_{\max} \sin\left(\frac{\pi x}{a}\right) \sin\left(\frac{\pi y}{b}\right)$$

using the coordinate system in Figure 6.3. The maximum deflection $w_{\max}$ was obtained in Example 6.3, p. 186. Using (6.10), the curvatures of the plate are

$$\kappa_x = -\frac{\partial^2 w}{\partial x^2} = \left(\frac{\pi}{a}\right)^2 w(x,y)$$

$$\kappa_y = -\frac{\partial^2 w}{\partial y^2} = \left(\frac{\pi}{b}\right)^2 w(x,y)$$

$$\kappa_{xy} = -2\frac{\partial^2 w}{\partial x \partial y} = -\frac{2\pi^2}{ab} w_{\max} \cos\left(\frac{\pi x}{a}\right) \cos\left(\frac{\pi y}{b}\right)$$

For this case, the maximum values of curvature are at the center of the plate ($x = a/2$, $y = b/2$)

$$\left\{ \begin{array}{l} \kappa_x \\ \kappa_y \\ \kappa_{xy} \end{array} \right\} = \left\{ \begin{array}{l} (\pi/a)^2 \\ (\pi/b)^2 \\ 0 \end{array} \right\} w_{\max} = \left\{ \begin{array}{l} 0.035 \\ 0.140 \\ 0 \end{array} \right\} m^{-1}$$

Using (6.23), the strains at the bottom lamina ($z = -t/2$) are

$$\left\{ \begin{array}{l} \epsilon_x \\ \epsilon_y \\ \gamma_{xy} \end{array} \right\} = -\frac{t}{2} w_{\max} \left\{ \begin{array}{l} (\pi/a)^2 \\ (\pi/b)^2 \\ 0 \end{array} \right\} = -\left\{ \begin{array}{l} 280 \\ 1120 \\ 0 \end{array} \right\} 10^{-6}$$

Using (6.24) and the $[\overline{Q}]$ matrix of the 0-degree lamina at the bottom of the laminate

$$[\overline{Q}] = \left[\begin{array}{lll} 45437 & 2302 & 0 \\ & 12116 & 0 \\ sym. & & 5500 \end{array} \right] MPa$$

the stresses at the bottom surface are

$$\left\{ \begin{array}{l} \sigma_x \\ \sigma_y \\ \sigma_{xy} \end{array} \right\} = -\left\{ \begin{array}{l} 15.300 \\ 14.214 \\ 0 \end{array} \right\} MPa$$

It is possible to check these results using the accompanying software (CADEC). First compute the bending moments using the $[D]$ matrix of the $[0/90/\pm 45]_{4S}$ in Example 6.3 (p. 186) into (6.15)

$$M_x = 577(10^{-6})\ MN$$

$$M_y = 1,248(10^{-6})\ MN$$

$$M_{xy} = 20(10^{-6})\ MN$$

Next, introduce the values of M_x, M_y, and M_{xy} into CADEC to obtain the stresses at all points through the thickness. Note that $M_{xy} \neq 0$ because the laminate has D_{16} and D_{26} different from zero, but the value of M_{xy} is very small because an effort was made in

Example 6.3 (p. 186) to make D_{16} and D_{26} small. A better approximation can be obtained using a large number of terms in the series expansion of $w(x, y)$ in the Navier method [175]

$$\sigma_x = z \sum \sum W_{mn} \left(\overline{Q}_{11}^{(k)} \alpha^2 + \overline{Q}_{12}^{(k)} \beta^2 \right) \sin \alpha x \sin \beta y$$
$$\sigma_y = z \sum \sum W_{mn} \left(\overline{Q}_{12}^{(k)} \alpha^2 + \overline{Q}_{22}^{(k)} \beta^2 \right) \sin \alpha x \sin \beta y$$
$$\sigma_{xy} = -2z \sum \sum W_{mn} \overline{Q}_{66}^{(k)} \alpha \beta \cos \alpha x \cos \beta y$$

where $\alpha = m\pi/a$, $\beta = n\pi/b$ and W_{mn} is given in Example 6.3, p. 186. As the number of terms used increases, the exact solution at $z = -t/2$ approaches the values $\sigma_x = 12.96\ MPa$ and $\sigma_y = 9.47\ MPa$. In this example, the approximate solution gives a conservative estimate of the stress.

6.3.6 Cross-ply Laminate

A cross-ply laminate has only laminae oriented at 0 and 90 degrees. If the construction is symmetric, it is called a *symmetric cross-ply laminate.* The main advantage of a cross-ply laminate is simplicity of construction, plywood being the typical example. Since all laminae in the laminate have $\overline{Q}_{16}$ and $\overline{Q}_{26}$ equal to zero, the 16 and 26 entries in $[A]$ and $[D]$ are also zero, as well as the H_{45} term. Therefore, cross-ply laminates do not have coupling between shear and extension in the $[A]$ matrix, nor do they have coupling between twisting and bending in the $[D]$ matrix. This is not really a great advantage from analysis or performance point of view, unless the laminate is also symmetric.

If only transverse loads and bending moments are applied, the bending of a thin plate depends only on the $[D]$ matrix. A symmetric cross-ply laminate has the same zero entries in the $[A]$ and $[D]$ matrices as homogeneous (single lamina) plates ($D_{16} = D_{26} = 0$), but the values of the nonzero coefficients are different. Therefore, transverse deflections of a symmetric cross-ply laminate can be found treating the laminate as an orthotropic, homogeneous plate. Solutions for orthotropic plates are given in the literature [177] in terms of plate rigidities D_x, D_y, and H, which are related to the $[D]$ matrix as follows

$$\begin{aligned} D_x &= D_{11} \\ D_y &= D_{22} \\ H &= D_{12} + 2D_{66} \end{aligned} \tag{6.30}$$

For a homogeneous plate, using (6.16)

$$
\begin{aligned}
D_{11} &= \frac{t^3}{12} Q_{11} = \frac{t^3 E_1}{12\Delta} \\
D_{22} &= \frac{t^3}{12} Q_{22} = \frac{t^3 E_2}{12\Delta} \\
D_{12} &= \frac{t^3}{12} Q_{12} = \frac{t^3 E_2 \nu_{12}}{12\Delta} \\
D_{66} &= \frac{t^3}{12} Q_{66} = \frac{t^3 G_{12}}{12} \\
\Delta &= 1 - \nu_{12}\nu_{21}
\end{aligned}
\tag{6.31}
$$

where t is the thickness of the plate. If the plate is not homogeneous but $[B] = 0$ and $D_{16} = D_{26} = 0$, (6.30) still applies. These conditions are satisfied by symmetric cross-ply laminates and by symmetric specially orthotropic laminates (Section 6.3.8). Therefore, existing solutions for orthotropic plates can be used to analyze these types of laminates. However, the stresses must be computed by the procedure described in Section 6.2.

6.3.7 Angle-ply Laminate

Angle-ply laminates are built with pairs of laminae of the same material and thickness, oriented at $\pm\theta$. For example $[30/-30/-30/30]$ is a symmetric angle-ply laminate and $[30/-30/30/-30]$ is an antisymmetric angle-ply laminate. Angle-ply laminates are also balanced (see Section 6.3.4). A symmetric angle-ply has $A_{16} = A_{26} = 0$ and $B_{ij} = 0$ but $D_{16} \neq 0$ and $D_{26} \neq 0$. An antisymmetric angle-ply has $A_{16} = A_{26} = D_{16} = D_{26} = 0$ but $B_{16} \neq 0$ and $B_{26} \neq 0$.

6.3.8 Specially Orthotropic Laminate

Each fiber reinforced composite lamina is orthotropic in material coordinates, resulting in $Q_{16} = Q_{26} = 0$. If the lamina is oriented at 0 or 90 degrees, it is called *specially orthotropic* (Section 5.4.4) because the $[\overline{Q}]$ matrix has the same zero entries as the $[Q]$ matrix; that is $\overline{Q}_{16} = \overline{Q}_{26} = 0$. A lamina reinforced with woven or stitched bidirectional fabric is also specially orthotropic if the amount of fabric in both directions $(\pm\theta)$ is the same. In this case, the fabric is called *balanced.* The contribution of the $+\theta$ fibers to $\overline{Q}_{16}$ and $\overline{Q}_{26}$ cancel out the contribution of the $-\theta$ fibers because $\overline{Q}_{16}(-\theta) = -\overline{Q}_{16}(\theta)$ (see (5.49) and Example 5.7).

A specially orthotropic laminate is constructed with specially orthotropic laminae. It can be symmetric (e.g., $[0/90]_S$) or not (e.g., $[0/90]$). Specially orthotropic laminates are not limited to cross-ply types, as they may include balanced $\pm\theta$ fabrics as well. The advantage of a specially orthotropic laminate is that it has

$$
\begin{aligned}
A_{16} &= A_{26} = \alpha_{16} = \alpha_{26} = 0 \\
B_{16} &= B_{26} = \beta_{16} = \beta_{26} = 0 \\
D_{16} &= D_{26} = \delta_{16} = \delta_{26} = 0
\end{aligned} \tag{6.32}
$$

Balanced laminates in general do not satisfy (6.32). If the off-axis laminae in a balanced laminate are symmetric, then the first two equations in (6.32) are satisfied but the third is not. However, δ_{16} and δ_{26} become smaller in comparison to δ_{11} and δ_{22} when the number of laminae increases. A negligible value of δ_{16} is advantageous for the analysis of torsion of thin walled beams (Section 10.2).

Example 6.5 *Compute the [A], [B], and [D] matrices of a two-lamina laminate reinforced with one lamina of Kevlar^{TM}woven fabric (described in Example 5.7, p. 162) and one lamina of E-glass fabric XM2408 (described in Example 5.8, p. 163).*

Solution to Example 6.5 *The woven fabric lamina is specially orthotropic because the fabric is balanced. From Example 5.7, p. 162,*

$$
[\overline{Q}]^{(1)} = \begin{bmatrix} 26.0 & 19.1 & 0 \\ 19.1 & 26.0 & 0 \\ 0 & 0 & 18.1 \end{bmatrix} \; GPa
$$

where the superscript indicates the lamina number. The lamina with XM2408 fabric is also specially orthotropic. From Example 5.8, p. 163,

$$
[\overline{Q}]^{(2)} = \begin{bmatrix} 18.9 & 10.85 & 0 \\ 10.85 & 19.39 & 0 \\ 0 & 0 & 10.16 \end{bmatrix} \; GPa
$$

The thickness of the two laminae are computed as a function of the weight of the fabric by using (4.13)

$$
\begin{aligned}
t_1 &= \frac{w}{1000\rho_f V_f} = \frac{600}{1000(1.45)0.5} = 0.833 \; mm \\
t_2 &= \frac{1025}{1000(2.5)0.5} = 0.82 \; mm
\end{aligned}
$$

The density of Kevlar^{TM}and E-glass are obtained from Table 1.3. The total thickness is $t = t_1 + t_2 = 1.653 \; mm$. The middle surface of the laminate is located at $t/2 = 0.8265 \; mm$ from the bottom of the laminate. Taking into account that the woven fabric lamina (number 1) is at the bottom of the laminate, the coordinates of the interfaces are $z_0 = -t/2 = -0.8265 \; mm$, $z_1 = -t/2 + t_1 = 0.0065 \; mm$, and $z_2 = t/2 = 0.8265 \; mm$. The coordinate of the middle surface of both laminae (see Figure 6.6) are

$$
\begin{aligned}
\bar{z}_1 &= (z_0 + z_1)/2 = (-0.8265 + 0.0065)/2 = -0.41 \; mm \\
\bar{z}_2 &= (z_1 + z_2)/2 = (0.0065 + 0.8265)/2 = 0.4165 \; mm
\end{aligned}
$$

Finally, using (6.16)

$$[A] = \begin{bmatrix} 37.15 & 24.80 & 0.00 \\ & 37.56 & 0.00 \\ sym. & & 23.41 \end{bmatrix} \; GPa \; mm$$

$$[B] = \begin{bmatrix} -2.43 & -2.82 & 0.00 \\ & -2.26 & 0.00 \\ sym. & & -2.71 \end{bmatrix} \; GPa \; mm^2$$

$$[D] = \begin{bmatrix} 8.45 & 5.64 & 0.00 \\ & 8.54 & 0.00 \\ sym. & & 5.32 \end{bmatrix} \; GPa \; mm^3$$

Note that the laminate is not symmetric, but the reader can easily verify that (6.32) are satisfied.

6.4 Laminate Moduli

A set of equivalent laminate moduli E_x, E_y, G_{xy}, ν_{xy}, can be defined for balanced symmetric laminates. These moduli represent the stiffness of a fictitious, equivalent, orthotropic plate that behaves like the actual laminate under in-plane loads. For a balanced symmetric laminate under in-plane loads, $\alpha_{16} = \alpha_{26} = 0$, and (6.20) reduces to

$$\begin{Bmatrix} \epsilon_x^0 \\ \epsilon_y^0 \\ \gamma_{xy}^0 \end{Bmatrix} = \begin{bmatrix} \alpha_{11} & \alpha_{12} & 0 \\ \alpha_{12} & \alpha_{22} & 0 \\ 0 & 0 & \alpha_{66} \end{bmatrix} \begin{Bmatrix} N_x \\ N_y \\ N_{xy} \end{Bmatrix} \tag{6.33}$$

If there are no bending loads, all the curvatures are zero, and the strains are constant through the thickness (6.5), that is $\epsilon_x = \epsilon_x^0$, $\epsilon_y = \epsilon_y^0$, $\gamma_{xy} = \gamma_{xy}^0$.

For a homogeneous orthotropic plate under in-plane load, the stresses are constant through the thickness: $\sigma_x = N_x/t$, $\sigma_y = N_y/t$, $\sigma_{xy} = N_{xy}/t$. Therefore, the stress–strain relation, (6.33) can be written for an orthotropic plate as

$$\begin{Bmatrix} \epsilon_x \\ \epsilon_y \\ \gamma_{xy} \end{Bmatrix} = \begin{bmatrix} 1/E_x & -\nu_{xy}/E_x & 0 \\ -\nu_{xy}/E_x & 1/E_y & 0 \\ 0 & 0 & 1/G_{xy} \end{bmatrix} \begin{Bmatrix} N_x/t \\ N_y/t \\ N_{xy}/t \end{Bmatrix} \tag{6.34}$$

provided that the axes of orthotropy coincide with the global coordinates (x, y, z). Comparing (6.33 and (6.34), the equivalent moduli of the laminate are

$$
\begin{aligned}
E_x &= \frac{1}{t\alpha_{11}} = \frac{A_{11}A_{22} - A_{12}^2}{tA_{22}} \\
E_y &= \frac{1}{t\alpha_{22}} = \frac{A_{11}A_{22} - A_{12}^2}{tA_{11}} \\
G_{xy} &= \frac{1}{t\alpha_{66}} = \frac{A_{66}}{t} \\
\nu_{xy} &= -\frac{\alpha_{12}}{\alpha_{11}} = \frac{A_{12}}{A_{22}}
\end{aligned}
\tag{6.35}
$$

The laminate moduli in (6.35) are valid for in-plane loads only and they should not be used to predict bending response (see Example 6.6). A set of laminate bending moduli can be derived for bending of balanced symmetric laminates by neglecting the terms D_{16} and D_{26}. However, it must be noted that these terms vanish only if all laminae are specially orthotropic (see Section 6.3.8). The coefficients D_{16} and D_{26} also vanish for antisymmetric laminates (see Section 6.3.3) but in that case the $[B]$ matrix is not zero and the behavior of the laminate is quite complex. For a symmetric, specially orthotropic laminate (see Section 6.3.8), the laminate bending moduli (also called *flexural muduli*) are (see Example 6.3)

$$
\begin{aligned}
E_x^b &= \frac{12}{t^3\delta_{11}} = \frac{12(D_{11}D_{22} - D_{12}^2)}{t^3D_{22}} \\
E_y^b &= \frac{12}{t^3\delta_{22}} = \frac{12(D_{11}D_{22} - D_{12}^2)}{t^3D_{11}} \\
G_{xy}^b &= \frac{12}{t^3\delta_{66}} = \frac{12D_{66}}{t^3} \\
\nu_{xy}^b &= -\frac{\delta_{12}}{\delta_{11}} = \frac{D_{12}}{D_{22}}
\end{aligned}
\tag{6.36}
$$

where the superscript b is used to indicate that these values are only used in bending. If the laminate does not have $D_{16} = D_{26} = 0$, the laminate bending moduli will only approximate the true response. A dimensionless measure of how close a laminate is to an equivalent orthotropic material is given by how small the coefficients A_{16}, A_{26}, D_{16}, and D_{26} are.

For in-plane loads

$$
r_N = \sqrt{\left(\frac{A_{16}}{A_{11}}\right)^2 + \left(\frac{A_{26}}{A_{22}}\right)^2} \tag{6.37}
$$

For bending loads

$$
r_M = \sqrt{\left(\frac{D_{16}}{D_{11}}\right)^2 + \left(\frac{D_{26}}{D_{22}}\right)^2} \tag{6.38}
$$

A dimensionless measure of the lack of symmetry of a laminate is provided by the coefficients in the B matrix

$$r_B = \frac{3}{(A_{11} + A_{22} + A_{66})t}\sqrt{\sum_i \sum_j (B_{ij})^2} \tag{6.39}$$

The dimensionless ratios r_N, r_M, and r_B can be used to assess the quality of the approximation obtained using apparent moduli given by (6.35) and (6.36). The closer these ratios are to zero, the more accurate the corresponding laminate moduli are in representing the true laminate response.

6.5 Design Using Carpet Plots

The analysis of deformations of laminated composites can be done accurately using (6.15) once the material, lamina orientations, laminate total thickness, etc., have been chosen. However, the *analysis* methodology presented so far does not indicate how to *design* the laminate; that is, it does not provide a simple procedure to estimate the required thickness, lamina orientations, etc. Ideally, the designer would like to follow a procedure, which starting with the loads and boundary conditions leads to the complete preliminary design of the laminate. A laminate design consists of: material, number of laminae, lamina thickness and orientations, and laminate stacking sequence. The material specification includes resin and fiber types, fiber volume fraction, and fiber architecture in various laminae (unidirectional, fabric, continuous strand mat, etc.). In the case of beams, the design also includes the geometry and dimensions of the cross-section (box, I-beam, etc.).

Most engineers have considerable training and experience in the design of simple structural components, such as bars, beams, and shafts, using isotropic materials (see [178], [14], [179]). Laminate moduli, defined in Section 6.4, can be used to take advantage of this knowledge for the design of composite structures. Using laminate moduli, a laminate is thought of as a homogeneous plate with apparent properties (E_x, E_y, etc.). However, the laminate moduli can be computed only after selecting the laminate configuration. To simplify the design process, plots of apparent moduli for various laminate configurations can be produced beforehand. These are called *carpet plots* and they can be produced using the equations in Section 6.4 or directly from experimental data.

While producing carpet plots, a decision is made regarding the type of fiber architecture and orientations to be used. For example, it is customary to limit the type of laminates to a combination of 0-degree, 90-degree, and ±45-degree laminae. The 0-degree and 90-degree laminae are included to give stiffness and strength in the x- and y-directions and the ±45-degree laminae to provide shear stiffness and strength. This family of laminates is described by

$$[0_\alpha/90_\beta/(\pm 45)_\gamma]_S \tag{6.40}$$

where α, β, and γ are the ratio of thickness of each lamina type to the total laminate thickness t

$$\begin{aligned}\alpha &= t_0/t \\ \beta &= t_{90}/t \\ \gamma &= t_{\pm 45}/t \\ \alpha + \beta + \gamma &= 1\end{aligned} \tag{6.41}$$

with t_0, t_{90}, and $t_{\pm 45}$ being the total thickness of 0, 90, and ± 45 laminae, and the subscript S indicates that the laminate is symmetric. Other combinations of lamina types can be included, such as fabrics, continuous strand mats, etc. Carpet plots are constructed for a specific material type and fiber volume fraction. E-glass fibers and isophthalic-polyester matrix with $V_f = 0.5$ have been used to construct the carpet plots in Figures 6.8–6.9 using (6.35) for in-plane loading. The material properties for an individual lamina are given in Table 1.3. For laminates loaded primarily in bending, carpet plots can be produced using (6.36), as shown in Figures 6.10–6.12 (see Section 6.5.2).

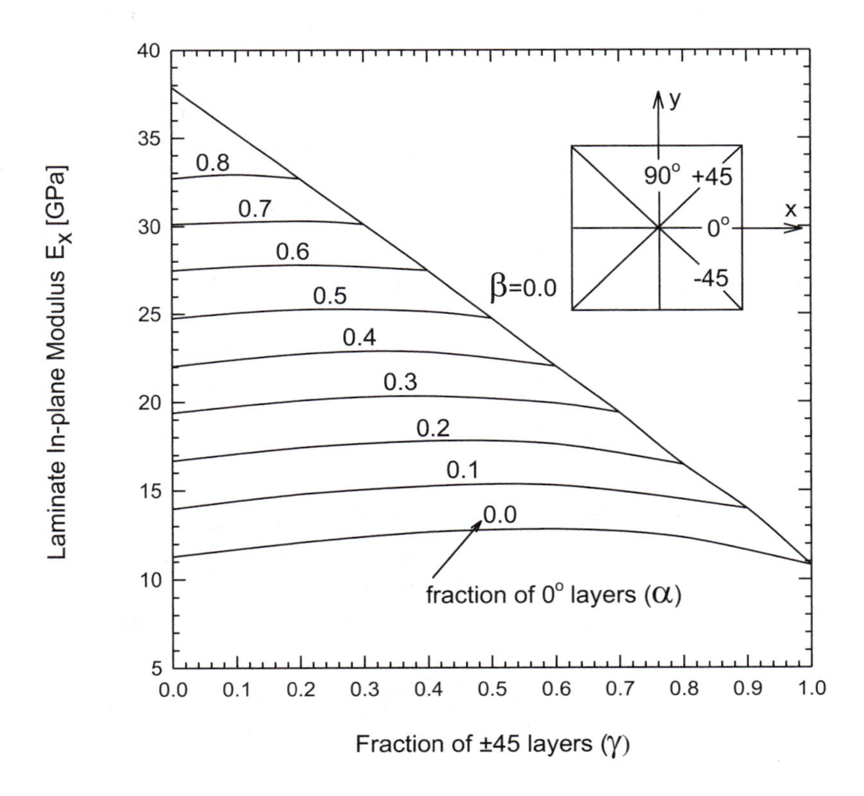

Figure 6.8: Carpet plot for laminate in-plane modulus E_x under in-plane load N_x.

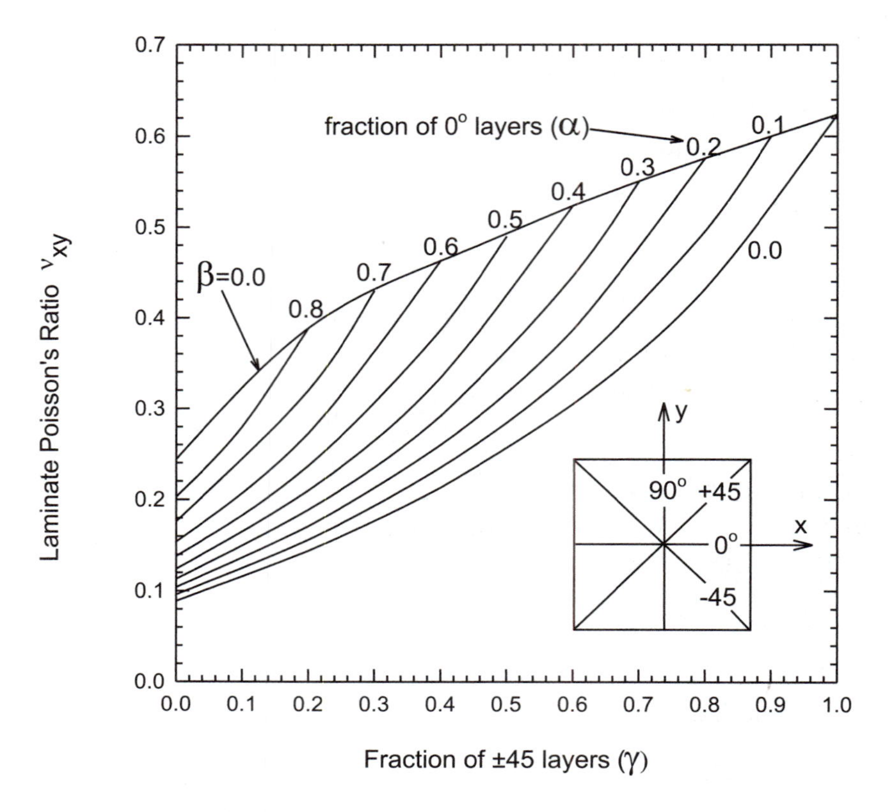

Figure 6.9: Carpet plot for laminate in-plane Poisson's ratio ν_{xy}.

While the carpet plots seem to constrain the design choices significantly, it must be noted that they are used only for preliminary design. It is always possible to refine the laminate design at a later stage. A redesigned laminate, different from those included in the carpet plots, can be analyzed using the general analysis tools described in this book (e.g., (6.15)).

6.5.1 Stiffness Controlled Design

Carpet plots of apparent in-plane moduli E_x, G_{xy}, and ν_{xy} are shown in Figure 6.8–6.9. Note in Figure 6.8 that the apparent modulus of elasticity E_x grows with the ratio α, as expected, because more fibers are orientated along the x-axis. The apparent modulus E_x is dominated by the amount of 0-degree laminae. For a given value of α, E_x is almost unchanged by γ. Since $\alpha + \beta + \gamma = 1$ all the curves end at the line for $\beta = 0$, which represent laminates with no transverse laminae.

The values of E_y can be read from the plot of E_x by interchanging the values of α and β. The plot of G_{xy} depends only on the amount of ± 45 laminae. For $\gamma = 0$, G_{xy} coincides with G_{12} of the unidirectional material.

The carpet plots of E_x and G_{xy} provide a glimpse of the design process. Lami-

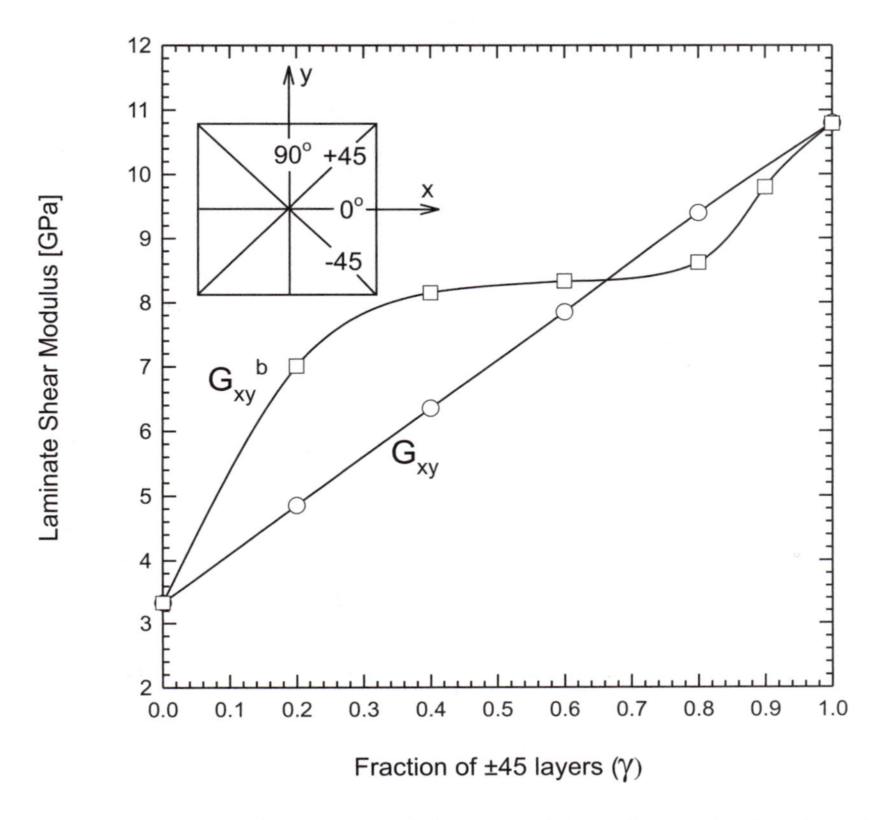

Figure 6.10: Carpet plot for laminate shear modulus G_{xy} under in-plane load N_{xy} and laminate shear modulus G^b_{xy} under bending load M_{xy}.

nates will be selected by interchanging α and γ to satisfy the requirements. Higher values of α (or β) to achieve higher moduli E_x (or E_y) and higher values of γ to get higher shear modulus G_{xy}. Since $\alpha + \beta + \gamma = 1$, a compromise must be reached by the designer. Once a laminate configuration is selected (by selecting values of α and γ), the total laminate thickness will be determined using standard strength of materials formulas and the apparent moduli read from the carpet plots.

Example 6.6 *Design a box section to be used as a cantilever beam of length $L = 2$ m subject to a tip load $P = 1000$ N. Limit the tip deflection δ to 1/300 of the span L of the beam.*

Solution to Example 6.6 *This is a stiffness controlled design because the requirement is $\delta = L/300$. The tip deflection can be computed using Figure 10.2*

$$\delta = \delta_b + \delta_s = \frac{PL^3}{3EI} + \frac{PL}{GA}$$

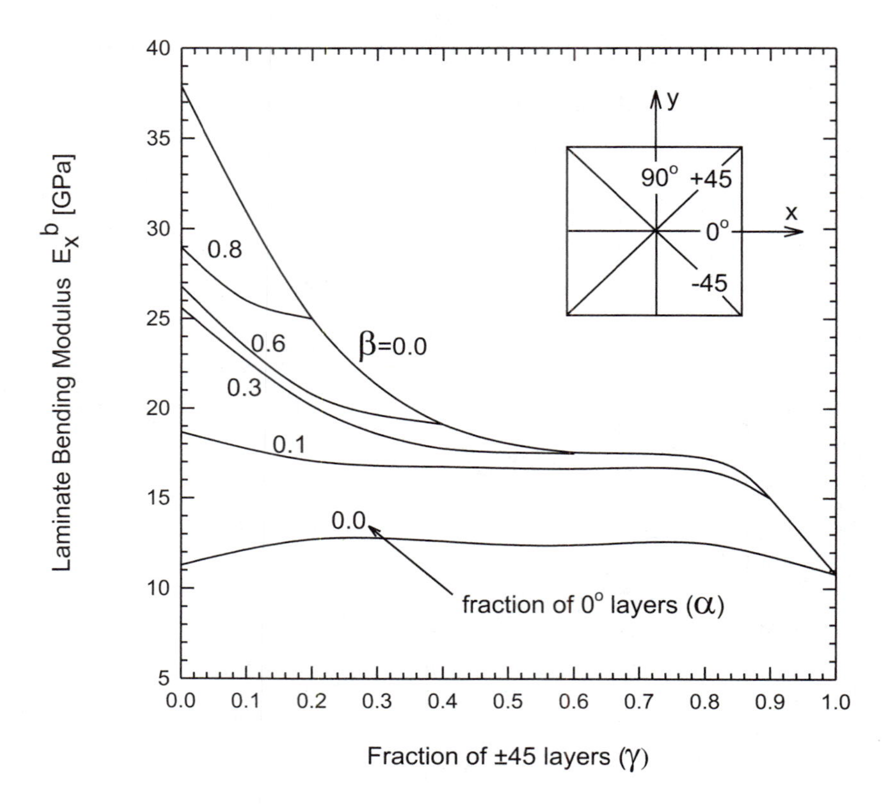

Figure 6.11: Carpet plot for laminate bending modulus E_x^b (loading: M_x).

where δ_b *and* δ_s *are the bending and shear deflections respectively. The design is controlled by the bending stiffness* EI *and the shear stiffness* GA.

Bending produces axial stress σ_x *in the beam. For a double symmetric section loaded through the centroid, the strain distribution is as shown in Figure 6.13. If all the cross-section is of the same material, the stress is also linear through the thickness and given by*

$$\sigma_x = \frac{M\ z}{I}$$

The major stress (σ_x) *is almost uniform through the thickness of the flanges (Figure 6.13) and it is uniform through the thickness in the web. Therefore, the material is subject mainly to in-plane stress* σ_x, *not bending stress (see also Section 6.5.2). In this case the apparent moduli to be used is* E_x, *from Figure 6.8.*

The shear load $V = P$ *produces in-plane shear stress in the webs, the apparent moduli to be used is the in-plane shear modulus* G_{xy} *from Figure 6.10. Note that the subscripts* xy *are related to the laminate coordinates and for the case of the web, they do not coincide with the beam coordinates.*

To simplify the design and the manufacturing of the beam, the same material is selected for

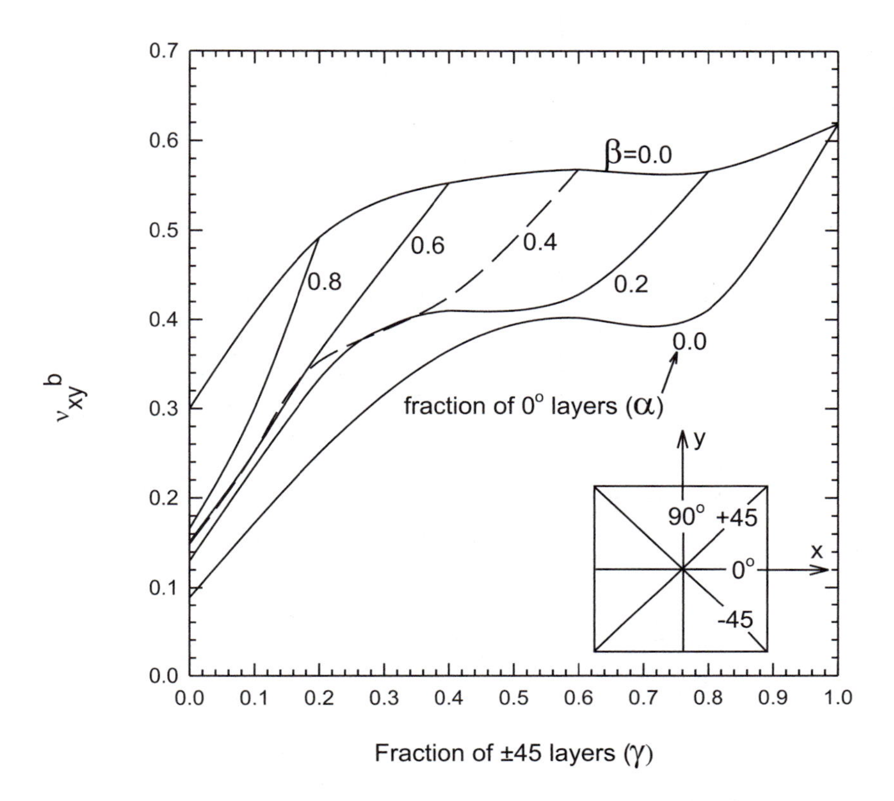

Figure 6.12: Carpet plot for laminate in-plane Poisson's ratio ν_{xy}^b in bending (loading: M_x).

flanges and webs. This is not an optimum choice because the flanges carry axial stress σ_x and the webs carry mostly shear stress σ_{xy}. Therefore, a higher percentage of ±45 laminae in the webs could be beneficial.

Assuming that bending deflections are more important than shear deflections, a laminate with $\alpha = 0.8$ and $\gamma = 0.2$ is selected. A value $\alpha = 0.8$ is chosen to maximize the axial stiffness E_x. However, a small amount of ±45-degree laminae is included to provide integrity to the box section, resist shear, and handle any secondary stresses that may not be accounted for in the design.

Using E-glass–polyester with $V_f = 0.5$, the following values are obtained from Figure 6.8 and 6.10

$$E_x = 32.8 \; GPa$$
$$G_{xy} = 4.8 \; GPa$$

Then, restricting $\delta_b < L/300$, the required moment of inertia is

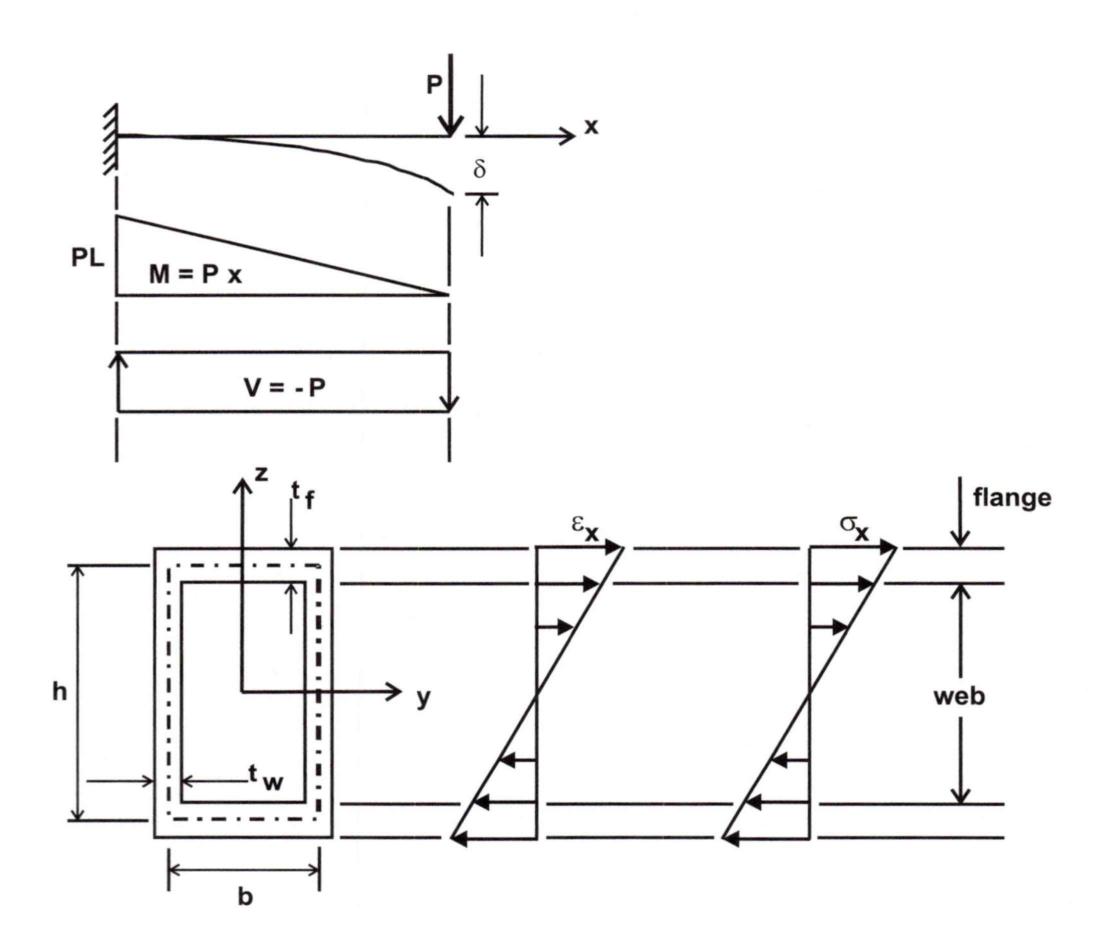

Figure 6.13: Design of a doubly-symmetric box section.

$$I = \frac{P\ L^3}{3E_x(L/300)} = \frac{1000(2)^3}{(3)32.8(10^9)(2/300)} = 12.2(10^{-6})\ m^4$$

The moment of inertia with respect to the neutral axis of a doubly-symmetric box section as shown in Figure 6.13 is approximately

$$I = 2\left[\frac{t_w h^3}{12} + \frac{b t_f^3}{12} + (b t_f)\left(\frac{h}{2}\right)^2\right]$$

Assuming $t_f = t_w = t$ *and neglecting the second term, which is small in comparison to the other two terms, the moment of inertia is approximated as*

$$I \approx t\left[\frac{h^3}{6} + \frac{bh^2}{2}\right]$$

Assuming ratios $h/b = 2$ *and* $b/t = 20$ *based on buckling and lateral stability considerations*

$$I \cong \frac{h^4}{96}$$

Then, using the required value of I, the depth results in

$$h = [96\ (12.2)\ 10^{-6}]^{\frac{1}{4}} = 185(10^{-3})m = 185\ mm$$

Conservatively, take

$$h = 188\ mm$$
$$b = 94\ mm$$
$$t = 4.7\ mm$$

Next, check for shear deformation assuming all shear deformation takes place in the webs. The area of the web is

$$A_w = 2ht = 1767\ mm^2 = 1.767(10^{-3})\ m^2$$
$$I = h^4/96 = 13.01(10^{-6})\ m^4$$

and the shear deflection

$$\delta_s = \frac{PL}{G_{xy}A_w} = \frac{1000(2)}{4.8\ (10^9)(1.767)(10^{-3})} = 235.8(10^{-6})\ m$$

The actual bending deflection is

$$\delta_b = \frac{PL^3}{3E_xI} = \frac{1000(2)^3}{(3)\ 32.8(10^9)\ 13.01(10^{-6})} = 6.248(10^{-3})\ m$$

and the total deflection is

$$\delta = \delta_b + \delta_s = 6.484(10^{-3})\ m$$

which is smaller than the constraint value $L/300 = 2/300 = 6.667(10^{-3})$ m. It remains to check the tensile strength of the top flange, compressive strength of the bottom flange, shear strength of the webs, and buckling of the walls of the box section (see Example 10.1).

6.5.2 Design for Bending

A different set of laminate moduli are used when the state of stress through the thickness of the laminate is mainly due to bending. In this case, the distribution of strain through the thickness is linear (Figure 6.5) and the outer laminae contribute more to the laminate stiffness than the inner laminae. Also, in bending, the outer laminae are more stressed than the inner laminae (Figure 6.5). For this reason, the apparent bending moduli (6.36) are different from the apparent in-plane moduli (6.35). The difference can be illustrated by considering two laminates, a $[0/90]_S$ and a $[90/0]_S$ with the same material and lamina thicknesses. In the x-direction, the 0-degree laminae are much stiffer than the 90-degree laminae. If the laminate is used for a beam with rectangular cross-section (shown at the left of Figure 6.14,) the bending stiffness of the $[0/90]_S$ laminate is clearly higher than the $[90/0]_S$ laminate.

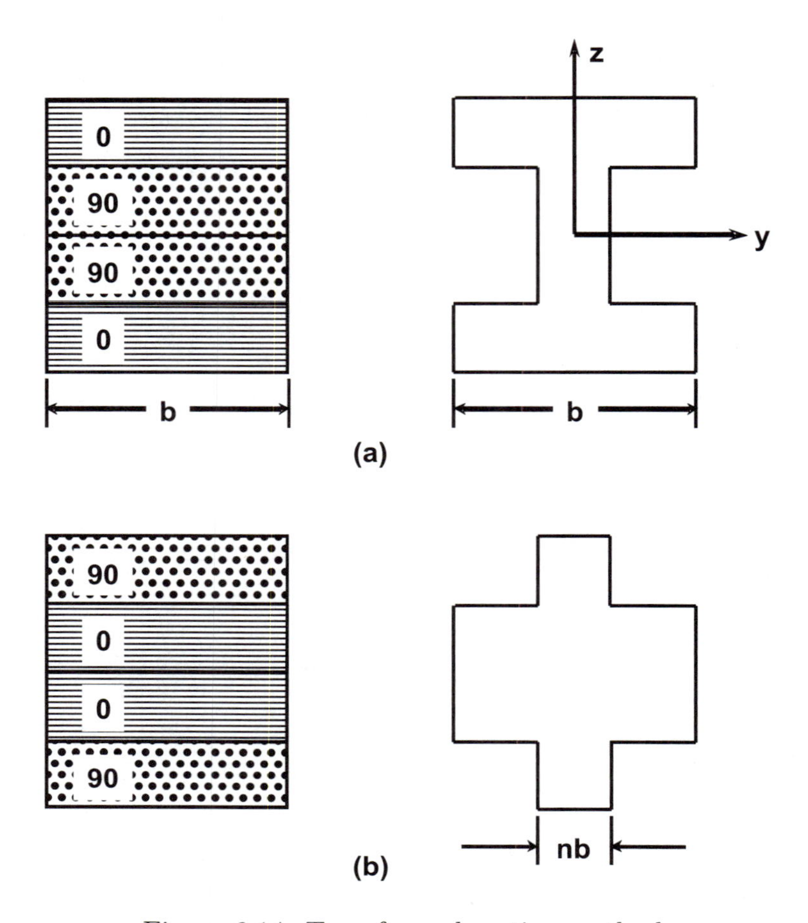

Figure 6.14: Transformed section method.

The transformed section method has been used in Figure 6.14 to illustrate the difference between the two laminates.[4] In the transformed section method [178, p. 304], the width of each lamina is adjusted by the modular ratio η_i

$$\eta_i = \frac{E_i}{E} \tag{6.42}$$

where E_i is the modulus of lamina i along the length of the beam (x-axis) and E is the modulus of one lamina taken as a reference (Figure 6.14 is constructed using a 0-degree lamina as reference). Then, the width of the lamina i in the transformed section is

$$b_i = \eta_i b \tag{6.43}$$

where b is the true width of the beam. The modulus of the transformed section

[4]The transformed section method can be used only when the laminae are specially orthotropic (Section 6.3.8). More general methods for analysis of composite beams are discussed in Chapter 10.

beam is uniform and equal to the modulus of the reference lamina. The bending stiffness of the original beam can be computed now as the product of the E times the moment of the inertia of the transformed section I_t. Clearly, the bending stiffness $E\ I_t$ of the $[0/90]_S$ is higher than the $[90/0]_S$ because the moment of inertia of the transformed section I_t is higher for the first case.

The transformed section method can be used also to compute the stress σ_x in each lamina of the beam using the classical flexure formula corrected with the appropriate modular ratio

$$\sigma_x^i = \frac{M\ z}{I_t}\eta_i \tag{6.44}$$

where i indicates the lamina number and z is the coordinate, with respect to the centroid of the transformed section, of the point where the stress is computed.

The laminate bending moduli (E_x^b, ν_{xy}^b, G_{xy}^b) defined in (6.36) are used to construct the carpet plots in Figures 6.10–6.12. Values of E_y^b can be approximated by reading the plot of E_x^b with α exchanged by β. These values can be used for the design of laminates in bending, stiffened panels (Section 11.3), etc.

Using the laminate bending moduli given in Figures 6.10–6.12, the coefficients of the bending stiffness matrix of the laminate can be estimated without conducting all the laminate computations described in Chapter 6, but simply as

$$\begin{aligned}
D_{11} &= \frac{t^3 E_x^b}{12\Delta} \\
D_{22} &= \frac{t^3 E_y^b}{12\Delta} \\
D_{12} &= \frac{t^3 E_x^b \nu_{yx}^b}{12\Delta} = \frac{t^3 E_y^b \nu_{xy}^b}{12\Delta} \\
D_{66} &= \frac{t^3 G_{xy}^b}{12} \\
\Delta &= 1 - \nu_{xy}^b \nu_{yx}^b \\
\nu_{yx} &= \nu_{xy}^b \frac{E_y^b}{E_x^b}
\end{aligned} \tag{6.45}$$

where t is the thickness of the laminate. Equations 6.45 are also necessary when experimental values of laminate bending moduli are available. These values are sometimes given by material suppliers for a limited set of laminate configurations, such as unidirectional, quasi-isotropic, random, etc.

The values of laminate in-plane E_x, E_y, G_{xy}, and ν_{xy} are independent of the stacking sequence; that is, of the order in which the various laminae are placed through the thickness, as long as the laminate remains symmetric. On the other hand, the values of the laminate bending moduli E_x^b, E_y^b, G_{xy}^b, and ν_{xy}^b depend on the stacking sequence. Therefore, actual values for the coefficients of the bending stiffness matrix $[D]$ should be computed using (6.16) once the precise stacking sequence has been chosen.

When carpet plots of laminate bending moduli are not available, the carpet plots of the laminate in-plane moduli can be used to provide a rough approximation. The error introduced is minimum for quasi-isotropic configurations (e.g., $\alpha = \beta = 0.25$, $\gamma = 0.5$, $\theta = 45°$) and it grows to a maximum for a $[0/90]_S$ laminate with $\alpha = \beta = 0.5$. The difference between laminate in-plane and laminate bending moduli is reduced by using a large number of laminae and alternating the lamina orientations as much as possible.

6.6 Hygrothermal Stresses (*)[5]

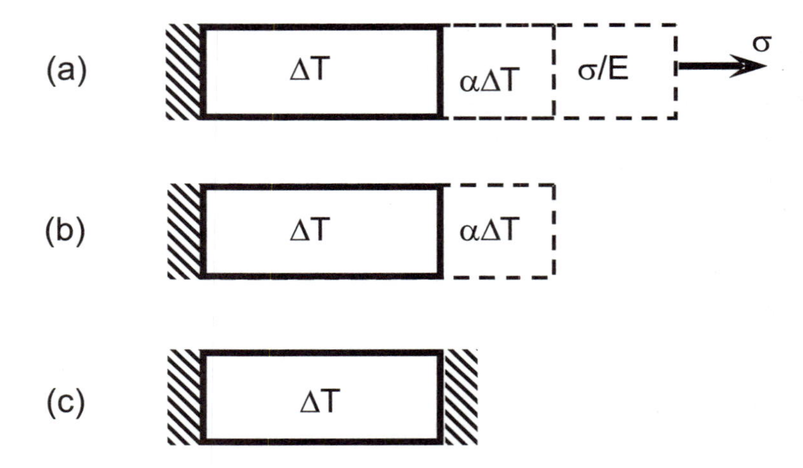

Figure 6.15: One-dimensional thermal expansion.

As explained in Section 4.3, the total strain in a composite material is the summation of mechanical strain and hygrothermal strain (thermal expansion and swelling strain due to moisture). When both mechanical and hygrothermal effects are present, the computation of stresses is more involved. The problem will be illustrated considering a one-dimensional bar (Figure 6.15) subject to mechanical and thermal load. Moisture induced strains and stresses can be incorporated following the same procedure. The applied stress σ produces a strain σ/E and the change in temperature induces additional strain $\alpha\Delta T$ (Figure 6.15a). The total strain is

$$\epsilon = \frac{\sigma}{E} + \alpha\Delta T \tag{6.46}$$

Then, the stress–strain relation for the one-dimensional case is

$$\sigma = E(\epsilon - \alpha\Delta T) \tag{6.47}$$

where E and α are the modulus of elasticity and coefficient of thermal expansion, respectively. If the bar in Figure 6.15b is free to expand, the stress is zero and the

[5]Sections marked with (*) can be omitted during the first reading but are recommended for further study and reference.

thermal strain is equal to $\epsilon = \alpha\Delta T$. Equation 6.47 models correctly the situation. If the bar is clamped at both ends (Figure 6.15c), the strain is zero and the stress induced by ΔT is correctly predicted by (6.47) as $\sigma = -E\alpha\Delta T$. In the latter case, the supports are equivalent to mechanical forces that force the strain to be zero. Note that the free bar (case b) can be made to expand an amount $\alpha\Delta T$ either by increasing the temperature by ΔT or by applying a stress $\sigma = E\alpha\Delta T$. This equivalence between temperature and stress will be used later in this section to obtain the thermal expansion of laminates that may have laminae with different coefficients of thermal expansion.

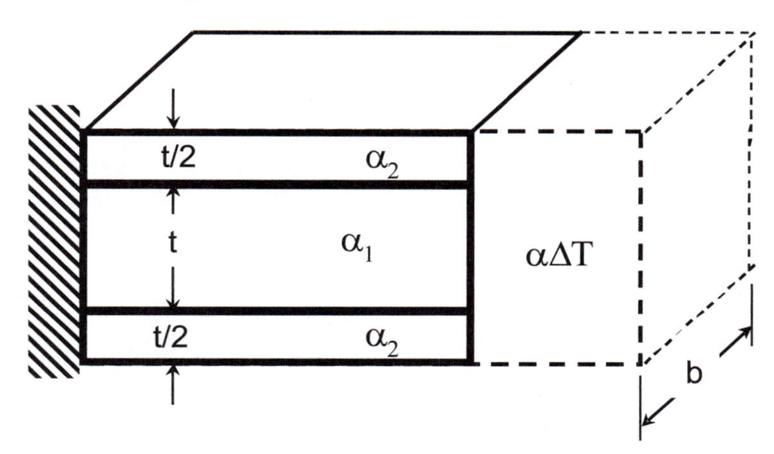

Figure 6.16: Partially restrained thermal expansion because of CTE mismatch.

Next, consider the case of a rectangular bar made of two materials perfectly bonded as illustrated in Figure 6.16. Both materials have the same moduli E, and same thickness t, but different coefficients of thermal expansion, $\alpha_1 > \alpha_2$. Material #1 has been divided in two laminae to obtain a symmetric laminate. If the bar is clamped at both ends, the force developed at the supports is

$$bN^T = b\int_{-t}^{t} \sigma dz = b\int_{-t}^{t} E(\epsilon - \alpha\Delta T)dz \tag{6.48}$$

where b is the width of the bar. Since the bar is clamped at both ends, $\epsilon = 0$, and

$$bN^T = -b\sum_{k=1}^{2} E^{(k)}\alpha^{(k)}\Delta T\; t^{(k)} \tag{6.49}$$

where k indicates the lamina number. In this example both materials have the same total thickness ($t^{(1)} = t^{(2)} = t$).

If the force bN^T were to be applied to the laminate (not clamped) instead of the change of temperature ΔT, it would produce the same strain as the change of temperature ΔT produces on the free expanding laminate. Therefore, the free thermal strain, which is the same in the three layers in Figure 6.16, can be computed as

$$\epsilon^T = \frac{-N^T}{E^{(1)}t^{(1)} + E^{(2)}t^{(2)}} \tag{6.50}$$

Note that ϵ^T is different from the free thermal strain of either of the two laminae. In fact

$$\alpha_1 \Delta T > \epsilon^T \quad \text{and} \quad \alpha_2 \Delta T < \epsilon^T \tag{6.51}$$

for $\alpha_1 > \alpha_2$. For this particular example $\epsilon^T = \frac{1}{2}(\alpha_1 + \alpha_2)\Delta T$ because $E^{(1)} = E^{(2)}$ and $t^{(1)} = t^{(2)}$. Since one lamina tends to expand more than the other, internal stresses are generated. These can be computed for each lamina using (6.47) as

$$\sigma^{(k)} = E^{(k)}(\epsilon^T - \alpha^{(k)}\Delta T) \tag{6.52}$$

where k is the lamina number. Since $\alpha_1 > \alpha_2$, the center layer is in compression and the outer layers are in tension. Since the layers are bonded together, the center layer forces the outer layers to expand more than what the would thermally expand by themselves. Likewise, the outer layers force the center layer to expand less than what it would thermally expand by itself.

If mechanical loads are applied simultaneously with the change in temperature, they produce mechanical strains ϵ^M in addition to the thermal strains ϵ^T. Therefore, the stress in each lamina is

$$\sigma^{(k)} = E^{(k)}(\epsilon^M + \epsilon^T - \alpha^{(k)}\Delta T) \tag{6.53}$$

For a laminated composite, using (5.15) to compute the strains due to mechanical loads $(\sigma_1, \sigma_2, \sigma_6)$, and using the coefficients of thermal and moisture expansion described in Section 4.3, the total strains for a lamina free to expand are

$$\left\{ \begin{array}{c} \epsilon_1 \\ \epsilon_2 \\ \gamma_6 \end{array} \right\} = \left[\begin{array}{ccc} S_{11} & S_{12} & 0 \\ S_{12} & S_{22} & 0 \\ 0 & 0 & S_{66} \end{array} \right] \left\{ \begin{array}{c} \sigma_1 \\ \sigma_2 \\ \sigma_6 \end{array} \right\} + \left\{ \begin{array}{c} \alpha_1 \\ \alpha_2 \\ 0 \end{array} \right\} \Delta T + \left\{ \begin{array}{c} \beta_1 \\ \beta_2 \\ 0 \end{array} \right\} \Delta m \tag{6.54}$$

Equation 6.54 for a laminate is equivalent to (6.46) for a bar. Rotation of the strains in (6.54) to laminate coordinates is done using the transformation matrix $[T]^{-1}$ (5.32)

$$\left\{ \begin{array}{c} \epsilon_x \\ \epsilon_y \\ \gamma_{xy} \end{array} \right\} = \left[\begin{array}{ccc} \overline{S}_{11} & \overline{S}_{12} & \overline{S}_{16} \\ \overline{S}_{12} & \overline{S}_{22} & \overline{S}_{26} \\ \overline{S}_{16} & \overline{S}_{26} & \overline{S}_{66} \end{array} \right] \left\{ \begin{array}{c} \sigma_x \\ \sigma_y \\ \sigma_{xy} \end{array} \right\} + \left\{ \begin{array}{c} \alpha_x \\ \alpha_y \\ \alpha_{xy} \end{array} \right\} \Delta T + \left\{ \begin{array}{c} \beta_x \\ \beta_y \\ \beta_{xy} \end{array} \right\} \Delta m \tag{6.55}$$

where the coefficients $\overline{S}_{ij}$ are given in (5.52). Since $\{\alpha\}\Delta T$ and $\{\beta\}\Delta m$ in (6.54) are strains, the $\{\alpha\}$ and $\{\beta\}$ coefficients transform like strains

$$\left\{\begin{array}{c}\alpha_x\\ \alpha_y\\ \frac{1}{2}\alpha_{xy}\end{array}\right\} = [T]^{-1}\left\{\begin{array}{c}\alpha_1\\ \alpha_2\\ 0\end{array}\right\};\left\{\begin{array}{c}\beta_x\\ \beta_y\\ \frac{1}{2}\beta_{xy}\end{array}\right\} = [T]^{-1}\left\{\begin{array}{c}\beta_1\\ \beta_2\\ 0\end{array}\right\} \tag{6.56}$$

where $[T]^{-1}$ is given by (5.33).

Thermal and moisture variations do not induce shear in material coordinates (see (6.54)) but they do induce shear in laminate coordinates (see (6.55)) if the lamina is not specially orthotropic (see Section 5.4.4 and 6.3.8).

The stresses on a lamina constrained from expansion either by the supports or by adjacent laminae can be computed inverting (6.55)

$$\left\{\begin{array}{c}\sigma_x\\ \sigma_y\\ \sigma_{xy}\end{array}\right\} = \left[\begin{array}{ccc}\overline{Q}_{11} & \overline{Q}_{12} & \overline{Q}_{16}\\ \overline{Q}_{12} & \overline{Q}_{22} & \overline{Q}_{26}\\ \overline{Q}_{16} & \overline{Q}_{26} & \overline{Q}_{66}\end{array}\right]\left(\left\{\begin{array}{c}\epsilon_x\\ \epsilon_y\\ \gamma_{xy}\end{array}\right\} - \left\{\begin{array}{c}\alpha_x\\ \alpha_y\\ \alpha_{xy}\end{array}\right\}\Delta T - \left\{\begin{array}{c}\beta_x\\ \beta_y\\ \beta_{xy}\end{array}\right\}\Delta m\right) \tag{6.57}$$

with the coefficients $\overline{Q}_{ij}$ given by (5.49). Equation 6.57 is equivalent to (6.47) for a bar.

For a laminate, the strains in (6.57) vary linearly through the thickness of the laminate according to (6.5). Therefore, a set of constitutive equations analogous to (6.15) is obtained inserting (6.5) into (6.57) and the result into (6.11) to obtain

$$\left\{\begin{array}{c}N_x\\ N_y\\ N_{xy}\\ M_x\\ M_y\\ M_{xy}\end{array}\right\} = \left[\begin{array}{cc}[A] & [B]\\ [B] & [D]\end{array}\right]\left\{\begin{array}{c}\epsilon_x^\circ\\ \epsilon_y^\circ\\ \gamma_{xy}^\circ\\ \kappa_x\\ \kappa_y\\ \kappa_{xy}\end{array}\right\} - \left\{\begin{array}{c}N_x^T\\ N_y^T\\ N_{xy}^T\\ M_x^T\\ M_y^T\\ M_{xy}^T\end{array}\right\} \tag{6.58}$$

where $[A]$, $[B]$, and $[D]$ are the 3×3 matrices of (6.15), and the hygrothermal forces and moments per unit length are

$$\begin{aligned}\left\{\begin{array}{c}N_x^T\\ N_y^T\\ N_{xy}^T\end{array}\right\} &= \Delta T\sum_{k=1}^{N}[\overline{Q}]^k\left\{\begin{array}{c}\alpha_x\\ \alpha_y\\ \alpha_{xy}\end{array}\right\}^k t_k + \Delta m\sum_{k=1}^{N}[\overline{Q}]^k\left\{\begin{array}{c}\beta_x\\ \beta_y\\ \beta_{xy}\end{array}\right\}^k t_k\\ \left\{\begin{array}{c}M_x^T\\ M_y^T\\ M_{xy}^T\end{array}\right\} &= \Delta T\sum_{k=1}^{N}[\overline{Q}]^k\left\{\begin{array}{c}\alpha_x\\ \alpha_y\\ \alpha_{xy}\end{array}\right\}^k t_k\ \bar{z}_k + \Delta m\sum_{k=1}^{N}[\overline{Q}]^k\left\{\begin{array}{c}\beta_x\\ \beta_y\\ \beta_{xy}\end{array}\right\}^k t_k\ \bar{z}_k\end{aligned} \tag{6.59}$$

where t_k and $\bar{z}_k$ are the thickness and middle surface coordinate of the k-th lamina respectively (Figure 6.6) and ΔT and Δm are the change in temperature and moisture content, respectively.

The hygrothermal loads are then moved to the left hand side of (6.58), or added to the mechanical loads in (6.20), to find the midsurface strains and curvatures of the laminate. With these values, the strains at any location through the thickness are computed using (6.5). Finally, the stresses at any location can be obtained with (6.57) and failure is evaluated as explained in Section 7.1.5.

Note that the hygrothermal stresses must be computed with (6.57); not with (6.24). The constitutive equations for the hygrothermal case are different from those for the mechanical loading case, as illustrated by (6.47). If only hygrothermal loads are used in (6.20), without any mechanical loads, the resulting strains are the hygrothermal strains.

A laminate free to expand will expand/contract at its midsurface with thermal strains given in terms of the (apparent) laminate thermal expansion coefficients $\alpha_x^\circ, \alpha_y^\circ, \alpha_{xy}^\circ$ as follows

$$\{\epsilon^\circ\} = \left\{ \begin{array}{c} \alpha_x^\circ \\ \alpha_y^\circ \\ \alpha_{xy}^\circ \end{array} \right\} \Delta T \tag{6.60}$$

and, if it the laminate is not symmetric, it will bend/twist with thermal curvatures

$$\{\kappa\} = \left\{ \begin{array}{c} \alpha_x^* \\ \alpha_y^* \\ \alpha_{xy}^* \end{array} \right\} \Delta T \tag{6.61}$$

Substituting (6.60–6.61) in (6.58), with the thermal loads calculated with (6.59), and no mechanical loads ($\{N\} = \{M\} = 0$), we obtain an formula for the (apparent) laminate thermal expansion coefficients

$$\left\{ \begin{array}{c} \alpha_x^\circ \\ \alpha_y^\circ \\ \alpha_{xy}^\circ \\ \alpha_x^* \\ \alpha_y^* \\ \alpha_{xy}^* \end{array} \right\} = \frac{1}{\Delta T} \left[\begin{array}{cc} [A] & [B] \\ [B] & [D] \end{array} \right]^{-1} \left\{ \begin{array}{c} N_x^T \\ N_y^T \\ N_{xy}^T \\ M_x^T \\ M_y^T \\ M_{xy}^T \end{array} \right\} \tag{6.62}$$

Example 6.7 *Compute the thermal strains and stresses of the laminate in Example 6.1 (p. 178) when it cools down to room temperature (21°C) from its processing temperature (121°C). Assume $\alpha_1 = 8.42\ 10^{-6}/°C$ and $\alpha_2 = 29.8\ 10^{-6}/°C$.*

Solution to Example 6.7 *The coefficients of thermal expansion (CTE) in laminate coordinates are obtained by using (6.56)*

	55°	*-55°*
α_x $[10^{-6}/°C]$	*22.77*	*22.77*
α_y $[10^{-6}/°C]$	*15.45*	*15.45*
α_{xy} $[10^{-6}/°C]$	*-20.1*	*20.1*

Then, thermal loads are obtained by using (6.59)

$$\left\{ \begin{array}{c} N_x^T \\ N_y^T \\ N_{xy}^T \end{array} \right\} = \left\{ \begin{array}{l} -44.0 \\ -39.2 \\ 0 \end{array} \right\} 10^{-3} \, ; \, \left\{ \begin{array}{c} M_x^T \\ M_y^T \\ M_{xy}^T \end{array} \right\} = \left\{ \begin{array}{l} 0 \\ 0 \\ 2.07 \end{array} \right\} \times 10^{-3}$$

Using the $[A]$, $[B]$, and $[D]$ matrices of Example 6.1 (p. 178), the thermal strains and curvatures at the middles surface are

$$\left\{ \begin{array}{c} \epsilon_x^0 \\ \epsilon_y^0 \\ \gamma_{xy}^0 \end{array} \right\} = \left\{ \begin{array}{l} -2.27 \\ -1.44 \\ 0 \end{array} \right\} 10^{-3} \, ; \, \left\{ \begin{array}{c} \kappa_x \\ \kappa_y \\ \kappa_{xy} \end{array} \right\} = \left\{ \begin{array}{l} 0 \\ 0 \\ 4.63 \end{array} \right\} \times 10^{-3}$$

The strains through the thickness are obtained using (6.23)

Ply	*Face*	ϵ_x	ϵ_y	γ_{xy}
2	*top*	*-2.27*	*-1.44*	*-2.94*
2	*bottom*	*-2.27*	*-1.44*	*0*
1	*top*	*-2.27*	*-1.44*	*0*
1	*bottom*	*-2.27*	*-1.44*	*2.94*

$\times 10^{-3}$

Finally, the thermal stresses are obtained using (6.57)

Ply	*Face*	σ_x	σ_y	σ_{xy}
2	*top*	*1.79*	*4.41*	*-6.11*
2	*bottom*	*-1.79*	*-4.41*	*12.2*
1	*top*	*-1.79*	*-4.41*	*-12.2*
1	*bottom*	*1.79*	*4.41*	*6.11*

$\times 10^{-3}$ *MPa*

Example 6.8 *Design a cylindrical pressure vessel to hold compressed natural gas (CNG) to fuel automobiles. The diameter of the vessel is ø= 330 mm. To prevent CNG from leaking through the composite wall, a flexible membrane of high density poly ethylene (HDPE), is placed inside as a liner. The liner adds no strength to the vessel but is sufficiently impervious to CNG. The working pressure is $p = 20.67$ MPa with a variability given by a COV $C_p = 0.2$. Even though the vessel will be loaded in cycles, when the vehicle is refueled, the pressure is applied for most of the time, thus it is considered to be a permanent (dead) load for the life of the vehicle. The material chosen is E-glass–epoxy. The properties of E-glass fibers are: $E_f = 72.345$ GPa, and $\nu_f = 0.22$. The properties of the epoxy matrix are: $E_m = 2.495$ GPa, and $\nu_m = 0.38$. Because of the long duration of the load, creep rupture is taken into account by code of practice requiring a resistance (reduction) factor $\phi = 1/3.5$. A reliability of 99% is required.*

Solution to Example 6.8 *Pressure vessels are thin or moderately thin shells subjected to internal pressure. They may be spherical but usually they are cylindrical with spherical, flat, or geodesic shaped heads. The stresses induced by internal pressure are mostly membrane stresses, with little bending stresses. The cylindrical portion is subjected to membrane stress resultants in the longitudinal x- and hoop y-directions*

$$N_x = \frac{p\,\o}{4} = \frac{20.67 \times 10^6 (0.33)}{4} = 1.705 \times 10^6 \ N/m = 1.705 \ kN/mm$$
$$N_y = \frac{p\,\o}{2} = \frac{20.67 \times 10^6 (0.33)}{2} = 3.410 \times 10^6 \ N/m = 3.410 \ kN/mm$$
$$N_{xy} = 0$$
$$M_x = M_y = M_{xy} = 0$$

Assuming the heads are hemispherical, the stress resultant on any direction on a sphere under internal pressure is equal to the longitudinal one on the cylindrical portion. Cylindrical pressure vessels are among the most highly optimized composite structures because composite materials are the only materials that can be tailored to yield twice the strength in the hoop direction than in the longitudinal direction. This is easily accomplished in a variety of ways, including adding more hoop-oriented fibers than longitudinally oriented fibers. If a single winding angle is used, the optimum angle is ± 55. Therefore, for this design a $[(\pm 55°)_n]_S$ laminate is chosen, with $4n$ laminae, each $t_k = 1.27$ mm thick. Once the required laminate thickness is calculated, the number of laminae needed is computed. Then, it is up to the manufacturer to decide how to achieve the total thickness during actual winding, perhaps using a smaller number of thicker laminae.

The analysis of the cylindrical portion can be easily performed because there are no bending moments applied. In this case, a small portion of the cylindrical shell can be analyzed as a plate under membrane loads. Using the equations for plates and taking $N_x = N_L$ and $N_y = N_H$, all the stresses can be obtained as explained in Section 6.2.

The analysis of the spherical portion is also similar. Since only membrane forces are present, the analysis of a small portion of the spherical head can be done as if it were a plate, with $N_x = N_y = N_s$.

Most pressure vessels are manufactured by filament winding, which always leads to pairs of laminae of opposite angles. The lamina thickness depends on how the winding machine is set up, but for design purposes it is not essential to know the actual lamina thickness, but only the total thickness at a given angle. When only membrane forces are present, the individual lamina thickness and the stacking sequence have little influence in the analysis. Therefore, the total thickness could be achieved during processing as any combination of number of laminae and individual lamina thickness. Furthermore, the filament winding process does not really produce distinct laminae but rather an overlapped and intermingled set of filament paths are created to cover the surface of the mandrel. However, the analogy to a laminated structure is valid for design purposes and it is used because it correctly accounts for the fiber orientations.

The fiber volume fraction depends on the winding process. The value selected for design is based on past experience or measured from prototype samples. Since the contribution of the matrix to the strength of the composite is small compared to that of the fibers, it is not very important to have the exact fiber volume fraction. If a high value of fiber volume is used, the elastic moduli will be high (see Chapter 4) and the resulting thickness will be small, but the amount of fibers on the final design will be about the same as if a low volume fraction was assumed. This is only applicable for membrane structures like this pressure vessel. In other words, if there are no bending loads, the thickness (which is influenced by the fiber volume) is not important. This is the reason for the use of netting analysis [180] for the design of pressure vessels. However, netting analysis is not recommended because it neglects the contribution of the fibers to the transverse and shear strengths.

Therefore, a volume fraction is assumed during the preliminary design of the structure. Then, after a prototype is built, the fiber volume can be measured along with the elastic moduli and the composite tensile strength. With these experimental values, detailed analysis or preliminary design of other similarly built structures can be done with confidence. Assuming the fiber volume fraction to be 55%, and using (4.24)

$$E_1 = 72.345(0.55) + 2.495(0.45) = 40.912\ GPa$$

Using Halpin-Tsai (4.31), and taking $\zeta = 2$*, the transverse modulus is predicted as*

$$\eta = \frac{(72.345/2.495) - 1}{(72.345/2.495) + 2} = 0.903$$
$$E_2 = 2.495 \left[\frac{1 + 2(0.903)0.55}{1 - 0.903(0.55)} \right] = 9.880\ GPa$$

Using (4.42), the shear moduli of fiber and matrix are computed as

$$G_f = \frac{72.345}{2\,(1 + 0.22)} = 29.650\ GPa$$
$$G_m = \frac{2.495}{2\,(1 + 0.38)} = 0.904\ GPa$$

which used in (4.37) and (4.33) results in a composite shear modulus $G_{12} = 2.844\ GPa$ *and Poisson's ratio* $\nu_{12} = 0.292$.

Using (4.82) and taking the average tensile strength of the fibers to be $1,378\ MPa$*, the tensile strength of the composite is predicted as*

$$F_{1t} = 1,378 \left[0.55 + \frac{2.495}{72.345} (1 - 0.55) \right] = 779\ MPa$$

but this value needs to be reduced by the resistance factor, so effectively the strength reduces to

$$\phi\ F_{1t} = (1/3.5)\ 779 = 222.571\ MPa$$

The stiffness matrix of a lamina $[Q]$ *is*

$$[Q] = \begin{bmatrix} 41.8 & 2.95 & 0 \\ 2.95 & 10.1 & 0 \\ 0 & 0 & 2.84 \end{bmatrix} \quad GPa$$

With $n = 1$*, the extensional-stiffness matrix* $[A]$ *of a symmetric,* $[(\pm 55)_n]_S$ *laminate is*

$$[A] = \begin{bmatrix} 65.4 & 53.8 & 0 \\ 53.8 & 120.0 & 0 \\ 0 & 0 & 53.2 \end{bmatrix} MN/m$$

Since the laminate is symmetric, the middle surface strains can be easily computed from

$$\begin{Bmatrix} \epsilon_x^0 \\ \epsilon_y^0 \\ \gamma_{xy}^0 \end{Bmatrix} = \begin{bmatrix} 0.024 & -0.011 & 0 \\ -0.011 & 0.013 & 0 \\ 0 & 0 & 0.019 \end{bmatrix} \begin{Bmatrix} 1.705 \\ 3.410 \\ 0 \end{Bmatrix} = \begin{Bmatrix} 0.004429 \\ 0.026343 \\ 0.0 \end{Bmatrix}$$

Once the middle surface strains are known, the stresses at the top and bottom of each lamina are computed. First, the strains at the points of interest are found. For example, compute the stresses at the bottom surface lamina number 1, which is also the bottom surface of the laminate. Since the curvatures are all zero, the strains are uniform through the thickness

$$\begin{Bmatrix} \epsilon_x \\ \epsilon_y \\ \gamma_{xy} \end{Bmatrix} = \begin{Bmatrix} \epsilon_x^0 \\ \epsilon_y^0 \\ \gamma_{xy}^0 \end{Bmatrix} + z \begin{Bmatrix} 0 \\ 0 \\ 0 \end{Bmatrix} = \begin{Bmatrix} 0.004429 \\ 0.026343 \\ 0.0 \end{Bmatrix}$$

Now, the stresses vary through the thickness because the stiffness varies from lamina to lamina. For lamina number 1, located at the bottom of the laminate

$$\begin{Bmatrix} \sigma_x \\ \sigma_y \\ \sigma_{xy} \end{Bmatrix}^{(1)} = \begin{bmatrix} 12.9 & 10.6 & 4.66 \\ 10.6 & 23.7 & 10.2 \\ 4.66 & 10.2 & 10.5 \end{bmatrix}^{(1)} \begin{Bmatrix} \epsilon_x \\ \epsilon_y \\ \gamma_{xy} \end{Bmatrix} = \begin{Bmatrix} 337.143 \\ 671.429 \\ 291.429 \end{Bmatrix} MPa$$

$$\begin{Bmatrix} \sigma_{yz} \\ \sigma_{xz} \end{Bmatrix}^{(1)} = \begin{bmatrix} \overline{Q}_{44} & \overline{Q}_{45} \\ \overline{Q}_{45} & \overline{Q}_{55} \end{bmatrix}^{(k)} \begin{Bmatrix} 0 \\ 0 \end{Bmatrix} = \begin{Bmatrix} 0 \\ 0 \end{Bmatrix} MPa$$

Transforming the stresses to material coordinates (at $55°$*) yields*

$$\begin{Bmatrix} \sigma_1 \\ \sigma_2 \\ \sigma_6 \end{Bmatrix}^{(1)} = \begin{Bmatrix} 834 \\ 174 \\ 291 \end{Bmatrix} MPa$$

Once the stresses and strains are known, one of the failure criteria described in Chapter 7 is used to determine if the structure fails or if it is over designed. But first, the load factor should be considered. For the required reliability $R_e = 0.99$*, using Table 1.1 yields a standard variable* $z = -2.326$*. With the load having a COV of 20%, the load factor is (1.17)*

$$\alpha = 1 - z\, C_p = 1 + 2.326(0.3) = 1.465$$

Now, using the maximum stress criterion along the fiber direction

$$R_1 = \frac{\phi F_{1t}}{\alpha\, \sigma_1} = \frac{(1/3.5)779\ MPa}{(1.465)834\ MPa} = 0.182$$

Assuming that this is the controling mode of failure, the thickness has to be increased by 1/0.182 to get $R_1 = 1$*. Therefore, the design calls for a total thickness of*

$$t_T = 4(1.27)/0.182 = 27.88\ mm$$

Other modes of failure should be checked as well, as described in Chapter 7. Note that no COV for the material strength was used in this example because the code of practice dictates *the value of resistance factor to be used, in this case $\phi = 1/3.5$. Old codes of practice may refer to this as a safety factor, but if only the variability and reduction of strength (such as creep rupture in this example) were addressed by the code, it is likely that the factor is a resistance factor. Therefore, the variability of the load must be taken into consideration to obtain the desired reliability.*

Exercises

Exercise 6.1 *Consider a homogeneous, isotropic beam.*

1. *Make a sketch (not to scale) of the deformed shape of a beam subject to a positive bending moment M_x and positive axial load N_x. Make sure your drawing shows the thickness of the beam; i.e., do not idealize the beam by a line.*
2. *On the same drawing, sketch the distribution of strain (not to scale).*
3. *On a similar drawing, sketch the distribution of stress (not to scale).*

Exercise 6.2 *Consider a symmetric laminated beam with three laminae of equal thickness. The top and bottom laminae are made out of steel. The center one of aluminum.*

1. *Make a sketch (not to scale) of the deformed shape of a beam subject to a positive bending moment M_x. Make sure your drawing shows the thickness of the beam; i.e., do not idealize the beam by a line.*
2. *On the same drawing, sketch the distribution of strain (not to scale).*
3. *On a similar drawing, sketch the distribution of stress (not to scale).*

Exercise 6.3 *Compute the extensional [A], coupling [B], and bending stiffness [D] matrices of a bimetallic (copper and aluminum) strip with each lamina 2 mm thick. For aluminum $E = 71\ GPa$, $\alpha = 23(10^{-6})/^\circ C$. For copper $E = 119\ GPa$, $\alpha = 17(10^{-6})/^\circ C$. Assume the Poisson's ratio is 0.3 for both materials.*

Exercise 6.4 *Given the following displacements and rotation functions*

$$
\begin{aligned}
u_o &= A\cos(\alpha x)\sin(\beta y) \\
v_o &= B\sin(\alpha x)\cos(\beta y) \\
w &= C\sin(\alpha x)\sin(\beta y) \\
\phi_x &= Q\cos(\alpha x)\sin(\beta y) \\
\phi_y &= R\sin(\alpha x)\cos(\beta y)
\end{aligned}
$$

where A, B, C, Q, and R are constants.

(a) *Derive the middle-surface strains and curvatures.*

(b) *Derive the strain functions as a function of x, y, and z.*

(c) *For $A = B = C = 0.001$, $Q = R = 2$, $\alpha = \beta = \pi$, $x = y = 1/2$, $t = 0.002$, evaluate and plot the strain as a function of the thickness coordinate z.*

Exercise 6.5 *Compute the stresses in laminate coordinates at top and bottom of each lamina in Example 6.1 (p. 178) when the laminate is subjected to $\epsilon_x^0 = 765(10^{-6})$ mm/mm, $\epsilon_y^0 = -280(10^{-6})$ mm/mm, $\kappa_{xy} = 34.2(10^{-6})1/mm$ and all remaining strains and curvatures equal to zero.*

Exercise 6.6 *Given the following stresses*

$$\begin{aligned}
\sigma_x &= A^k z \sin(\alpha x) \sin(\beta y) \\
\sigma_y &= B^k z \sin(\alpha x) \sin(\beta y) \\
\sigma_{xy} &= 0 \\
\sigma_{yz} &= Q^k \sin(\alpha x) \cos(\beta y) \\
\sigma_{xz} &= R^k \cos(\alpha x) \sin(\beta y)
\end{aligned}$$

where $k = 1 \ldots N$, N *is the number of laminae, and* A^k, B^k, Q^k, *and* R^k *are constants, derive the seven stress resultants integrating over the thickness of a laminated plate with* $N = 3$, $t_1 = t_2 = t_3$.

Exercise 6.7 *Find* σ_1, σ_2, *and* σ_6 *at* $z = 0$, *on both laminae of a laminate* $[\pm 30]_T$ *subject to* $N_x = 1\ N/m$ *(all remaining forces and moments equal to zero), with* $t_1 = t_2 = 1.27\ mm$, $E_1 = 137.8\ GPa$, $E_2 = 9.6\ GPa$, $G_{12} = 5.2\ GPa$, $\nu_{12} = 0.3$.

Exercise 6.8 *Compute the* $[A]$, $[B]$, $[D]$, *and* $[H]$ *matrices for the following laminates. All laminae are* 1.0 *mm thick. Comment on the coupling of the constitutive equations for each case.*

(a) one lamina of aluminum ($E = 71\ GPa$, $\nu = 0.3$*)*

(b) one lamina $[0]$*; carbon–epoxy (AS4–3501-6).*

(c) $[30]$*; carbon–epoxy*

(d) $[0/90]_T$*; carbon–epoxy*

(e) $[0/90]_S$*; carbon–epoxy*

(f) $[\pm 30]_S$*; carbon–epoxy*

Exercise 6.9 *Compute the* $[A]$, $[B]$, $[D]$, *and* $[H]$ *matrices for the following laminates. In all cases, the total laminate thickness is* 4.0 *mm. Comment on the coupling of the constitutive equations for each case. Comment on the relative magnitude of the coefficients as the number of laminae increases.*

(a) $[\pm 45/0/90/\pm 30]$*; carbon–epoxy (AS4–3501-6)*

(b) $[\pm 45]_2 = [\pm 45/\pm 45]$*; carbon–epoxy*

(c) $[\pm 45]_4$*; carbon–epoxy*

(d) $[\pm 45]_8$*; carbon–epoxy*

(e) $[0/90]_2$*; carbon–epoxy*

(f) $[0/90]_4$*; carbon–epoxy*

(g) $[0/90]_8$*; carbon–epoxy*

Exercise 6.10 *Consider a* $[0/90]_S$ *laminate made out of Kevlar 49–epoxy (see Table 1.3), each lamina* 1.0 *mm thick. Compute the stresses in material coordinates when* $N_x = 1\ N/m$.

Exercise 6.11 *Consider a* $[+45/-45]_S$ *laminate with the material properties of Exercise 6.10. Compute the middle-surface strains* ϵ_x^0, ϵ_y^0, γ_{xy}^0, *and the curvatures* κ_x, κ_y, κ_{xy} *for the following load cases*

(a) $N_x = 1000\ N/m$; all remaining components are equal to zero

(b) $M_{xy} = 1\ Nm/m$; all remaining components are equal to zero

(c) Comment on the coupling effects observed.

Exercise 6.12 *Compute the strains ϵ_x, ϵ_y, and γ_{xy} at the interface between the 45° and −45° laminae above the middle surface by using the results of case (b) of Exercise 6.11 and (6.23).*

Exercise 6.13 *Compute the stresses σ_x and σ_1 in the 45° and in the −45° laminae right next to the interface based on the strains computed in Exercise 6.12 using (6.24). Discuss the results.*

Exercise 6.14 *Write those terms that are equal to zero in the constitutive equations (A, B, and D) for the following special laminates: (a) Symmetric, (b) Balanced antisymmetric, (c) Cross-ply, and (d) Specially orthotropic.*

Exercise 6.15 *Define and explain the following: (a) Regular laminate, (b) Balanced laminate, and (c) Specially orthotropic laminate.*

Exercise 6.16 *Indicate whether the following are continuous or discontinuous functions through the thickness. Also, indicate why some values are continuous or discontinuous and state which locations through the thickness this occurs.*

(a) Strains ϵ_x, ϵ_y, and γ_{xy}

(b) Stresses σ_x, σ_y, and σ_{xy}

Exercise 6.17 *Demonstrate that an angle-ply laminate has $A_{16} = A_{26} = 0$ using an example laminate of your choice. In which case $D_{16} = 0$? In which case $D_{16} \neq 0$?*

Exercise 6.18 *Given $\epsilon_x^0 = \epsilon_y^0 = \gamma_{xy}^0 = 0$ and $\kappa_x = 1.0\ m^{-1}$, $\kappa_y = 0.5\ m^{-1}$, $\kappa_{xy} = 0.25\ m^{-1}$, compute the strains (in laminate coordinates) at $z = -1.27\ mm$ in the laminate of Exercise 6.7.*

Exercise 6.19 *Compute the stresses (laminate coordinates) at $z = -1.27\ mm$ using the results of Exercise 6.18. Note that $\theta = 30°$.*

Exercise 6.20 *Show that a balanced laminate, not necessarily symmetric, has $A_{16} = A_{26} = 0$. In which case $D_{16} = 0$?*

Exercise 6.21 *Compute the [Q] matrix of a lamina reinforced with a balanced bidirectional fabric selected from Table 2.8. Use $V_f = 0.37$, E-glass fibers, and isophthalic polyester resin.*

Exercise 6.22 *Compute [A], [B], and [D] matrices of the lamina in Exercise 6.21. Also, compute the (in-plane) laminate moduli E_x, E_y, G_{xy}, and ν_{xy}.*

Exercise 6.23 *Compute the curvature of the bimetallic in Exercise 6.3 when it is subjected to a change of temperature $\Delta T = 10\ °C$.*

Exercise 6.24 *Plot the strain and stress distribution (ϵ_x and σ_y) through the thickness of the bimetallic analyzed in Exercise 6.23.*

Exercise 6.25 *For each of the following laminates, specify the laminate type. Note that a laminate may have more than one characteristic, e.g., balanced and symmetric.*

(a) $[30/-45/45/-30]$

(b) $[30/0/-45/45/90/-30]$

(c) $[30/0/-30/-30/90/30]$

(d) $[\pm 30/\pm 45]$

(e) $[0/90/\pm 45]_T$

Exercise 6.26 *For each of the laminates described in Table 2.8, specify if they are balanced and/or specially orthotropic. Explain why all these laminates have* $A_{16} = A_{26} = B_{16} = B_{26} = D_{16} = D_{26} = 0$.

Exercise 6.27 *Consider laminates reinforced with one lamina of a fabric from Table 2.8. Describe each of the eleven laminates using the laminate description code in Section 6.3.1 (see also Table 1.5). For each fabric type, specify if the laminate is balanced and/or specially orthotropic.*

Exercise 6.28 *Compute the laminate moduli* (E_x, E_y, G_{xy}, *and* ν_{xy}) *and the laminate bending moduli* (E_x, E_y, G_{xy}, *and* ν_{xy}^b) *for the laminate in Exercise 6.10.*

Exercise 6.29 *Compute the laminate moduli for a* $[0_2/90/\pm 45]_S$ *laminate and compare your results with the carpet plots (Figure 6.8–6.10). The material is E-glass–polyester with properties given in Table 1.3. Use an individual lamina thickness* t_k *of your choice. Note that* E_y *can be read from the* E_x *plot by interchanging the values of* α *and* β.

Chapter 7

Strength

Failure of a structural element occurs when it cannot perform its intended function. Material fracture is the obvious type of failure but not the only one. Excessive deflection is also a type of failure if the performance or the aesthetic value of the structure are compromised. Even if the part does not collapse, partially damaged materials may be considered failed, for example if cracking affects the aesthetic value of the structural element. Deflections are easier to predict than damage and fracture. Prediction of deflections are addressed in Chapters 10 to 13 and in examples throughout this book. The objective of this chapter is to describe strength design; that is, to develop tools to predict damage and fracture of composites and to use those tools to select materials, configurations, and dimensions that allow the structure to safely perform its function.

Damage and fracture of composite materials may occur in a variety of failure modes including:

- Fiber breaking: occurs mainly under tensile loads. It may initiate at stress levels well below the ultimate strength of the material.
- Matrix crazing: indicates the appearance of microscopic cracks in the polymer matrix. It may be caused by mechanical loading, curing-induced residual stress, thermal stresses, moisture ingress, or aging.
- Matrix cracking: is similar to matrix crazing but the cracks are larger, with dimensions of the order of magnitude of the fiber diameter or more.
- Fiber debonding: occurs when the fiber–matrix bond fails.
- Delamination: or separation between the laminae in a laminate.
- Many others.

It is difficult to incorporate these many modes of failure into the design. A simpler alternative is to use empirical, lamina failure criteria, similar to the failure criteria used in metal design, but modified for composites. While many criteria have been proposed [25, 181]the most useful lamina failure criteria are:

1. Maximum stress criterion.
2. Maximum strain criterion.
3. Interacting failure criterion.
4. Truncated-maximum-strain criterion.
5. Damage mechanics

Lamina failure criteria are equations with parameters adjusted to fit experimental data of failure of single-lamina composites. The criteria are then used for design situations for which experimental data is not available including the design of laminates. Furthermore, failure criteria are complemented with laminate failure analysis techniques to predict first ply failure (FPF) from single-lamina data. Analysis beyond FPF cannot be handled accurately without a good understanding of the progressive degradation of the laminate under the particular load redistribution and constraining provided by the laminate as a whole. For this purpose, prediction of laminate strength (Section 7.3) is done either with approximate design methods (Section 7.3.2) or with damage mechanics methods (Chapter 8). The ply discount methods (7.3.1), requiring empirical damage factors, are being replaced by more general formulations such as the one described in Chapter 8.

Lamina failure criteria are adjusted using experimental data obtained on simple tests of unidirectional composites (Figure 7.1). As discussed in Section 4.4, the strength of a unidirectional lamina is described by the following strength values:

- Tensile strength in the fiber direction F_{1t} (see Section 4.4.1).
- Compressive strength in the fiber direction F_{1c} (see Section 4.4.2).
- Tensile strength in the transverse direction F_{2t}; not a true material property (see Section 4.4.4).
- Compressive strength in the transverse direction F_{2c} (see Section 4.4.6).
- Angle of the fracture plane α_0 in transverse compression (see Section 4.4.6).
- In-plane shear strength F_6; not a true material property (see Section 4.4.8 and 7.2.1).
- Intralaminar shear strength F_4 (see Section 4.4.10).
- Intralaminar shear strength F_5 (see Section 4.4.10).
- Mode I fracture toughness[1] G_{Ic} (see Figure 7.2 and Section 1.5).

[1]In classical fracture mechanics, the term *fracture toughness* K_c refers to the critical value of stress intensity [30], but in this work the term fracture toughness refers to the critical energy release rate G_c because stress intensity is not used for fracture mechanics analysis of composite materials.

- Mode II fracture toughness G_{IIc} (see Figure 7.2 and Section 1.5).
- Mode III fracture toughness G_{IIIc} (see Figure 7.2 and Section 1.5).

If experimental data is not available, some of the material strength values listed above can be predicted using micromechanics (Section 4.4).

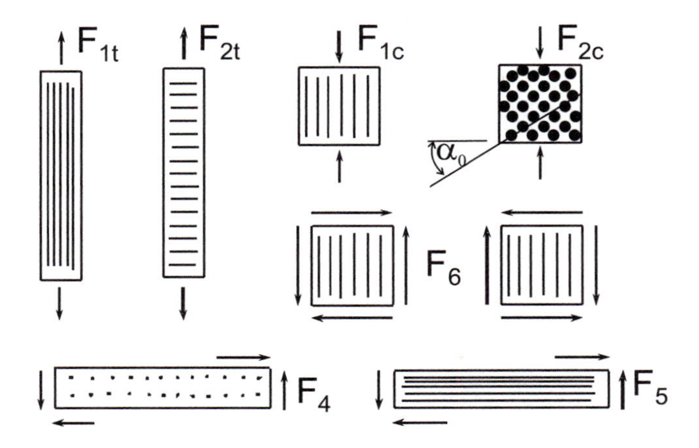

Figure 7.1: Simple test configurations used to determine strength values.

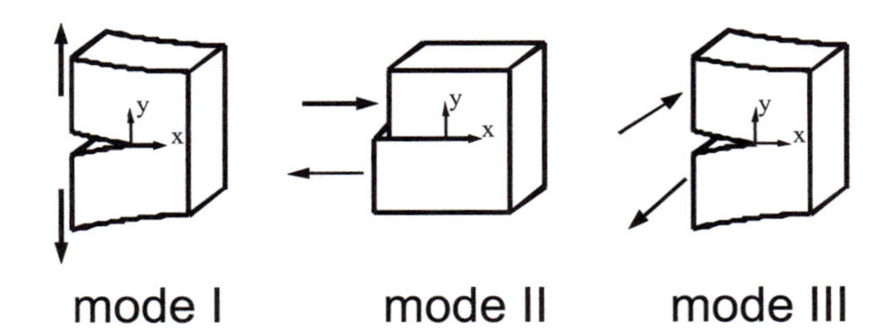

Figure 7.2: Modes of fracture. I: opening, II: in-plane shear, III: out-of-plane shear.

In metal design there are two limits on the stress that the material can carry. The first limit is the yield strength, which marks the end of the elastic behavior of the material. The second limit is the ultimate strength that indicates fracture of the material. Both situations produce irreversible changes in the material. In the case of yield, permanent deformations are induced. Metal structures may be considered permanently damaged if the yield strength is exceeded because after yield the structure does not recover its original shape. However, the structure may not catastrophically collapse until the ultimate strength is reached.

In composites reinforced with long fibers, the yield and ultimate strength usually coincide; the material usually having elastic behavior up to fracture except when subjected to shear (Figure 4.11). However, internal damage usually takes place below the ultimate stress. Thus, damage initiation may be taken as the first limit, equivalent to the yield of metals. Even though permanent deformations are not necessarily induced, internal damage may induce or accelerate undesirable secondary effects. For example, matrix cracking facilitates moisture ingress that in turn reduces the life of the material. Therefore, the designer may want to limit the stress values below those that trigger internal damage. Increasingly accurate design methodologies to predict initial damage and ultimate strength of laminates are introduced progressively in this chapter, culminating with a more realistic and accurate model in Chapter 8.

7.1 Lamina Failure Criteria

In this section, empirical equations are used to fit experimental data for failure of a unidirectional lamina. The samples used for testing may have a multilamina construction to obtain the desired thickness but all laminae have the same fiber architecture and same orientation. It is usual to perform these tests on unidirectional composites. However, with the advent of 2D and 3D textile fiber systems, it is necessary to incorporate data for these as well.

Eleven parameters, most of them material properties, can be identified immediately from simple experiments (Figure 7.1): F_{1t},F_{1c},F_{2t},F_{2c}, α_0,F_4, F_5,F_6,G_{Ic},G_{IIc}, and G_{IIIc}. If a lamina is subjected to a simple state of stress, then it is easy to predict failure just comparing the value of stress to that of strength. But most structures are subject to complex states of stress, with several stress components acting simultaneously. The objective of this section is to present a methodology for predicting failure of an isolated lamina under a complex state of stress using strength information from simple tests.

The applicability of *lamina failure criteria* to the analysis of *laminates* is limited to prediction of FPF. Once damage starts, lamina failure criteria do not provide useful information for predicting damage evolution, stress redistribution to other laminae, and ultimate failure.

7.1.1 Strength Ratio

In order to use any failure criteria more efficiently, the strength ratio R is defined as the strength over the applied stress

$$R = \frac{\sigma_{ultimate}}{\sigma_{applied}} \tag{7.1}$$

The strength ratio is similar to a safety factor. If $R > 1$ the stress level is below the strength of the material. If $R < 1$, the stress value is higher than the strength and failure is predicted. Since linear elastic behavior is assumed in this section, multiplying all the applied loads by a factor R increases or reduces the stresses proportionally, so that the new value of strength ratio becomes $R = 1$. The failure load can be computed simply by analyzing the structure with an arbitrary reference load, then multiplying the reference load by the value of R. To include reliability in the design, see Section 1.4 and examples throughout this book (e.g., Example 6.8).

7.1.2 Maximum Stress Criterion

This criterion predicts failure of a lamina when at least one of the stresses in material coordinates $(\sigma_1, \sigma_2, \sigma_6, \sigma_4, \sigma_5)$ exceeds the corresponding experimental value of strength. The criterion states that failure occurs if any of the following is true

$$\begin{array}{lll} \sigma_1 > F_{1t} & \text{if} & \sigma_1 > 0 \\ abs(\sigma_1) > F_{1c} & \text{if} & \sigma_1 < 0 \\ \sigma_2 > F_{2t} & \text{if} & \sigma_2 > 0 \\ abs(\sigma_2) > F_{2c} & \text{if} & \sigma_2 < 0 \\ abs(\sigma_4) > F_4; & & abs(\sigma_5) > F_5; \quad abs(\sigma_6) > F_6. \end{array} \tag{7.2}$$

Note that compressive strength values are taken as positive numbers. From a phenomenological point of view, the quantities $F_{1t}, F_{1c}, F_{2t}, F_{2c}, F_4, F_5, F_6$ are ultimate values. However, in design (see Section 1.4.3), the stress values on the left-hand side must be increased by load factors $\alpha > 1$ (1.17) and the strength on the right-hand side must be reduced by resistance factors $\phi < 1$ (1.16), or replaced by allowable values that have the resistance factors built in.

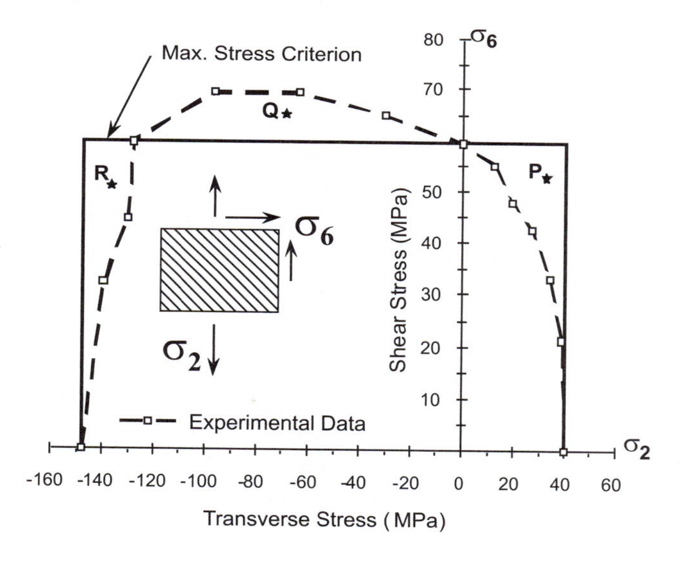

Figure 7.3: The maximum stress criterion in stress space $\sigma_2 - \sigma_6$.

A graphical representation of the maximum stress criterion can be obtained when only two stress components are different from zero. The special case of $\sigma_1 = \sigma_4 = \sigma_5 = 0$ is illustrated in Figure 7.3. Failure is not anticipated as long as the stresses σ_2, σ_6 define a point (called *design state*) inside the rectangle. Failure is predicted when the stress values σ_2, σ_6 reach any point on the rectangle (failure envelope). The rectangle in Figure 7.3 limits the design space.

If the ultimate strength values are multiplied by ϕ/α, then the design space becomes the safe region of operation. Considering all six components of stress, the design space is a hypercube with sides $F_{1t}, F_{1c}, F_{2t}, F_{2c}, F_4, F_5, F_6$, of which Figure

7.3 is a projection on the σ_2, σ_6 plane. In terms of strength ratio, the maximum stress criterion reads

$$\begin{array}{lll} R_1 = F_{1t}/\sigma_1 \ ; \text{ if } \sigma_1 > 0 \text{ or} & R_1 = -F_{1c}/\sigma_1 \ ; \text{ if } \sigma_1 < 0 & \\ R_2 = F_{2t}/\sigma_2 \ ; \text{ if } \sigma_2 > 0 \text{ or} & R_2 = -F_{2c}/\sigma_2 \ ; \text{ if } \sigma_2 < 0 & \\ R_4 = F_4/abs(\sigma_4); & R_5 = F_5/abs(\sigma_5); & R_6 = F_6/abs(\sigma_6). \end{array} \tag{7.3}$$

Then, the strength ratio for the lamina i is the minimum of all the R_i computed with (7.3). Storing strength values in two column arrays, $F_T = [F_{1t}; F_{2t}; F_4; F_5; F_6]$ for tensile strengths, and $F_C = [F_{1c}; F_{2c}; F_4; F_5; F_6]$ for compressive strengths, with the stress in lamina coordinates also stored in a column array $\boldsymbol{\sigma} = [\sigma_1; \sigma_2; \sigma_4; \sigma_5; \sigma_6]$, (7.3) reduces to

$$R_i = 1/\max(\langle \boldsymbol{\sigma} \rangle ./F_T + \langle -\boldsymbol{\sigma} \rangle ./F_C) \tag{7.4}$$

where $x./y$ denotes[2] division element by element of the arrays x and y, both of which must have the same size and structure, and the McAuley operator $\langle x \rangle$ denotes the positive part of the argument x, which can be computed as follows

$$\langle x \rangle = \frac{1}{2} \left(x + |x| \right)) \tag{7.5}$$

where $|x|$ denotes the absolute value of the argument x.

One advantage of the maximum strength criterion is that it gives information about the mode of failure. The minimum R_i corresponds to a particular mode of failure that can be identified. Then, the laminate can be changed to reduce the likelihood of failure in that mode. For example, the fiber angles can be changed to reinforce the laminate against the identified mode of failure, or more material may be added in the desired orientation, or a change in material system may be necessary. Also, by looking at the relative values of the various R_i, it is possible to detect overdesign, which is indicated by a large strength ratio R_i.

The maximum stress criterion is not conservative for stress states that are not dominated by one component of stress. Consider point P in Figure 7.3, both σ_2 and σ_6 are positive. Their values are close but still lower than the corresponding strength values F_{2t} and F_6. The maximum stress criterion predicts no failure. But experimental data from biaxial strength tests [182] show that there are interaction effects that produce failure when two or more stress components are close to their limits. When σ_2 becomes negative (compression), the maximum stress criterion may be too conservative (see point Q) or it may over predict the strength of the structure (see point R) as it does in this example for both tensile (P) and compressive (R) transverse load. Interacting failure criteria (Section 7.1.4) have been developed specifically to address this shortcoming.

[2]In MATLAB® notation. Note that some of the σ-values could be zero; thus (7.4) cannot be written as $\min(F/\sigma)$.

Example 7.1 *Use the maximum stress failure criterion to compute the strength ratio for a lamina subject to* $\sigma_1 = 200$ *MPa*, $\sigma_6 = 100$ *MPa*, $\sigma_2 = -50$ *MPa*, $\sigma_4 = \sigma_5 = 0$. *The material is carbon–epoxy with the following properties:*

$$E_1 = 156,400\ MPa,\ E_2 = 7,786\ MPa$$
$$G_{12} = 3,762\ MPa,\ \nu_{12} = 0.352\ MPa$$
$$F_{1t} = 1,826\ MPa,\ F_{1c} = 1,134\ MPa$$
$$F_{2t} = 19\ MPa,\ F_{2c} = 131\ MPa$$
$$F_6 = 75\ MPa,\ t_k = 2.54\ mm$$

Solution to Example 7.1 *First, compute the values of strength ratio corresponding to each orientation. Since the stress is tensile along the fiber direction,* R_1 *is*

$$R_1 = \frac{F_{1t}}{\sigma_1} = \frac{1826\ MPa}{200\ MPa} = 9.13$$

In the transverse direction, the stress is in compression, so

$$R_2 = -\frac{F_{2c}}{\sigma_2} = -\frac{131\ MPa}{-50\ MPa} = 2.62$$

Since there is only one shear stress different from zero

$$R_6 = \frac{F_6}{\sigma_6} = \frac{75\ MPa}{100\ MPa} = 0.75$$

and R_4 *and* R_5 *are not used. Finally, the minimum strength ratio for the lamina is* $R = R_6 = 0.75$ *and the mode of failure is in-plane shear. The lamina will fail when the applied load is only 0.75 of the reference value but the laminate is overdesigned for the applied longitudinal load because it could carry 9.13 times the applied longitudinal load.*

7.1.3 Maximum Strain Criterion

The maximum strain criterion [183] is still the most popular failure criterion in industry today. Using the concept of strength ratio (see 7.1.1), the maximum strain criterion reads

$$\begin{array}{lll} R_1 = \epsilon_{1t}/\epsilon_1\ ;\ \text{if } \epsilon_1 > 0 \text{ or} & R_1 = -\epsilon_{1c}/\epsilon_1\ ;\ \text{if } \epsilon_1 < 0 & \\ R_2 = \epsilon_{2t}/\epsilon_2\ ;\ \text{if } \epsilon_2 > 0 \text{ or} & R_2 = -\epsilon_{2c}/\epsilon_2\ ;\ \text{if } \epsilon_2 < 0 & \\ R_4 = \gamma_{4u}/abs(\epsilon_4) & R_5 = \gamma_{5u}/abs(\epsilon_5) & R_6 = \gamma_{6u}/abs(\epsilon_6). \end{array} \tag{7.6}$$

where $\epsilon_{1t}, \epsilon_{1c}, \epsilon_{2t}, \epsilon_{2c}, \gamma_{4u}, \gamma_{5u}, \gamma_{6u}$ are strains to failure. Since, elongations to failure are usually reported in percent values in the literature, they would have to be divided by 100 to be used in (7.6). Note that compressive strains to failure are positive numbers and all R values are positive.

Storing the strain-to-failure values in two arrays, $\epsilon_T = [\epsilon_{1t}; \epsilon_{2t}; \gamma_{4u}; \gamma_{4u}; \gamma_{6u}]$ for tensile values, and $\epsilon_C = [\epsilon_{1c}; \epsilon_{2c}; \gamma_{4u}; \gamma_{4u}; \gamma_{6u}]$ for compressive values, with the strains in lamina coordinates also stored in an array $\epsilon = [\epsilon_1; \epsilon_2; \gamma_4; \gamma_5; \gamma_6]$, (7.6) reduces to

$$R_i = 1/\max(\langle\epsilon\rangle./\epsilon_T + \langle-\epsilon\rangle./\epsilon_C) \tag{7.7}$$

where $x./y$ denotes[3] division element by element of the arrays x and y, both of which must have the same size and structure, and $\langle x\rangle$ denotes the positive part of the argument x, which can be computed by (7.5).

If the material is linear elastic up to failure, the strains to failure are directly related to the ultimate strength values by

$$\begin{array}{lll} \epsilon_{1t} = F_{1t}/E_1 & \epsilon_{1c} = F_{1c}/E_1 & \\ \epsilon_{2t} = F_{2t}/E_2 & \epsilon_{2c} = F_{2c}/E_2 & \\ \gamma_{4u} = F_4/G_{23} & \gamma_{5u} = F_5/G_{13} & \gamma_{6u} = F_6/G_{12} \end{array} \tag{7.8}$$

In this case both criteria, maximum stress and maximum strain give close but not identical predictions of failure (Figure 7.4). However, actual values of elongation to failure are usually larger than predicted by (7.8) except in the fiber direction because of the nonlinear behavior shown in Figure 4.11 (see also Exercise 7.12).

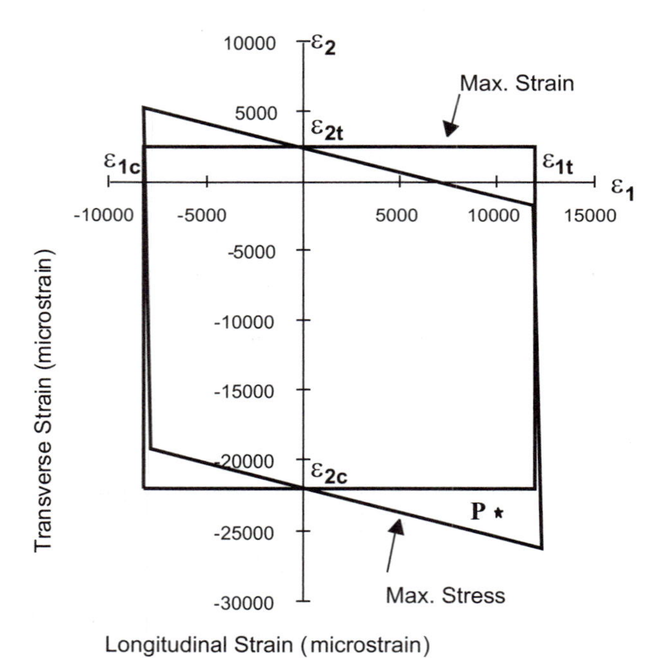

Figure 7.4: Failure envelopes using the maximum strain and maximum stress criteria in strain space $\epsilon_1 - \epsilon_2$ for carbon–epoxy.

For the special case $\gamma_4 = \gamma_5 = \gamma_6 = 0$, Figure 7.4 represents the failure envelope in the space ϵ_1, ϵ_2. The maximum strain criterion is a rectangle defined by the elongations to failure ϵ_{1t}, ϵ_{2t}, ϵ_{1c}, ϵ_{2c}. Figure 7.4 is drawn using the data given in Example 7.1, for carbon–epoxy with 63% fiber volume.

[3]In MATLAB notation.

To illustrate some differences between maximum stress and maximum strain criteria the maximum stress criterion is represented in the $\epsilon_1 - \epsilon_2$ diagram (Figure 7.4) assuming that the material behavior is linear up to failure (7.8). Note that a material point with a stress state given by point P in Figure 7.4 is predicted to fail by the maximum stress criterion but considered not to fail by the maximum strain criterion. To construct the failure envelope, note that the stress in the longitudinal direction is (5.19)

$$\sigma_1 = \frac{1}{\Delta}\left[E_1\epsilon_1 + \nu_{21}E_1\epsilon_2\right] \qquad (7.9)$$
$$\Delta = 1 - \nu_{12}\nu_{21}$$

If the material is linear up to failure, the maximum stress criterion predicts failure when

$$\sigma_1 = \epsilon_{1t}E_1 \qquad (7.10)$$

Combining (7.9) and (7.10)

$$\epsilon_{1t}\Delta = \epsilon_1 + \nu_{21}\epsilon_2 \qquad (7.11)$$

Then, the failure envelope for $\sigma_1 > 0$ is defined by

$$\epsilon_2 = \frac{\epsilon_{1t}\Delta}{\nu_{21}} - \frac{\epsilon_1}{\nu_{21}} \qquad (7.12)$$

This is a line of slope $(-1/\nu_{21})$ that crosses the ϵ_1-axis $(\epsilon_2 = 0)$ at $(1 - \nu_{12}\nu_{21})\,\epsilon_{1t}$, as shown in Figure 7.4. Similarly to (7.9), the stress in the transverse direction is

$$\sigma_2 = \frac{1}{\Delta}\left[\nu_{12}E_2\epsilon_1 + E_2\epsilon_2\right] \qquad (7.13)$$

For the case of transverse compressive failure, the maximum stress criterion predicts failure when $\sigma_2 = -\epsilon_{2c}E_2$, with the transverse compressive strain to failure ϵ_{2c} being a positive number. Then, the bottom portion of the failure envelope is defined by

$$\epsilon_2 = -\epsilon_{2c}\Delta - \nu_{12}\epsilon_1 \qquad (7.14)$$

which is a line of slope $(-\nu_{12})$ with intercept at $\epsilon_1 = 0, \epsilon_2 = -\epsilon_{2c}\,(1 - \nu_{12}\nu_{21})$. The other lines are found in a similar way.

One reason for the use of the maximum strain criterion is the nonlinear behavior of the composite (see Figure 4.11). In this case, strains to failure are larger than the strains computed using Hooke's law and the strength ratio values (7.8). The difference is large for pure resins (see Chapter 3) but small for composites in longitudinal and transverse tension [182, p. 173].

7.1.4 Interacting Failure Criterion

Most strength data available is for unidirectional composites under uniaxial load, as follows:

1. Tensile strength in the fiber direction F_{1t} which is dominated by the strength of the fibers (section 4.4.1).

2. The compressive strength in the fiber direction F_{1c}, which is a fiber buckling failure (Section 4.4.2).

3. The tensile strength transverse to the fibers F_{2t}, which is matrix-dominated (Section 4.4.4).

4. The compressive strength in the transverse direction F_{2c}, which is qualitatively and quantitatively different from all of the above (Section 4.4.6).

5. The in-plane shear strength of the composite F_6, which is matrix-dominated (Section 4.4.8).

This data gives four points on the σ_1, σ_2, axes plus two points on the pure-shear line ($\sigma_{12} = -\sigma_{12} = \pm F_6$). For uniaxial states of stress, the applied stress can be compared directly with the corresponding strength to predict failure. For multiaxial states of stress, some assumption about the material behavior is needed in order to predict strength. Two distinct assumptions, or approaches, are discussed next.

The failure criteria presented in Sections 7.1.2 and 7.1.3 treat each mode of failure independently of the rest. That is, a multiaxial state of stress is treated as a set of uniaxial states that are independent of each other. If any of those uncoupled, uniaxial states of stress exceed their corresponding strength values, then that material is said to have failed. The fact that the material is subject to *combined* stresses is ignored. This may lead to unconservative (or overconservative) designs when two modes of failure interact to cause failure at a state of stress (or strain) lower (or higher) than predicted by each of the two modes acting separately. Typical cases are the interaction among:

- Transverse tensile $\sigma_2 > 0$ and shear stress σ_6. In this case the interaction is detrimental because both modes cause the same type of matrix cracks, thus negatively affecting each other. For a detailed discussion see Chapter 8.

- Transverse compressive $\sigma_2 < 0$ and shear stress σ_6. In this case, transverse compression usually enhances shear strength because matrix cracks caused by shear are have to overcome the friction imposed by the compression that closes them. This effect can be modeled empirically by Mohr-Coulomb equations as described in Section 4.4.6.

- Longitudinal compressive $\sigma_1 < 0$ and shear stress σ_6 (detrimental). In this case shear stress causes shear damage that reduces the shear modulus G_{12}, which in turn brings about a reduction of the longitudinal compressive strength. This effect can be seen clearly in (4.85)–(4.86).

- Other.

Quadratic failure criteria such as Tsai-Hill [184, Section 7.1.4] and Tsai-Wu [184, Section 7.1.5] were developed long time ago to address the interaction among modes, but they went too far by artificially imposing interaction among *all* the modes. For example, longitudinal tension is a fiber-dominated mode that is unaffected by the remaining modes and thus, it should be left alone. Quadratic failure criteria model the entire failure envelope as a smooth surface in stress space. A projection of that surface on the plane $\sigma_2-\sigma_6$ is shown in Figure 7.5 where it can be seen that the interaction between matrix-dominated modes, in this case shear and transverse tension, can be predicted satisfactorily. However, the interaction between fiber modes (i.e., longitudinal tension and compression) and the remaining modes is unsatisfactory.

The noninteracting failure criteria, such as those presented in Sections 7.1.2 and 7.1.3, produce multiple values of strength ratio R_i, one for each mode of failure. This is inconvenient during visualization of finite element analysis because it forces visualization of multiple contour plots, one for each mode. Quadratic failure criteria are popular in part because they produce a single value of R that summarizes the state of stress in relation to strength, and thus require a single contour plot. However, this convenience comes at a steep price because the quadratic criteria do not provide any information regarding the nature of the mode of failure that is dominant, thus depriving the designer of important guidance on how to improve the design. Non-interacting failure criteria, by providing separate values for strength ratio, explicitly indicate the mode that is likely to cause failure and the level of underutilization of the material with respect to the modes that are not critical. In fact, many designers that rely on quadratic criteria also display the results of noninteracting criteria in order to identify the failure modes, thus complicating the visualization process even further.

In summary, the major shortcomings of quadratic failure criteria are (a) the artificial interaction between fiber and matrix modes of failure and (b) the lack of identification of the critical modes of failure. Conversely, the major shortcoming of the noninteracting criteria is the lack of interaction among modes that do interact. In order to ameliorate the shortcomings mentioned so far, the *interacting* criterion described below attempts to separate matrix modes from fiber modes, as follows

- Fiber-dominated failure:

$$\langle \sigma_1^f / F_{1t} \rangle + \langle -\sigma_1^f / F_{1c} \rangle - 1 = 0 \tag{7.15}$$

  where $\langle x \rangle$ denotes the positive part of x (see (7.5)), and F_{1t}, F_{1c} are both positive values. Equation (7.15) is the maximum stress criterion (see Exercise 7.14). In the longitudinal direction, it does not recognize any interaction with any other mode.

- Matrix-dominated failure:

$$f_{22}(\sigma_2^f)^2 + f_{44}(\sigma_4^f)^2 + f_{55}(\sigma_5^f)^2 + f_{66}(\sigma_6^f)^2 + f_2\sigma_2^f - 1 = 0 \tag{7.16}$$

with

$$
\begin{aligned}
f_2 &= \frac{1}{F_{2t}} - \frac{1}{F_{2c}} \\
f_{22} = \frac{1}{F_{2t}F_{2c}} \; &; \; f_{66} = \frac{1}{(F_6)^2} \\
f_{44} = \frac{1}{(F_4)^2} \; &; \; f_{55} = \frac{1}{(F_5)^2}
\end{aligned} \tag{7.17}
$$

Note that in the definition of the coefficients f_i and f_{ij}, the compressive strength values are defined as positive values. As it can be seen, the interacting criterion (7.15–7.16) accounts for different behavior in tension and compression.

In (7.15–7.16), the superscript f is used to indicate that any state of stress $\sigma_i^f\,; i = 1, 2, 4, 5, 6$ is a state of stress that produces failure. All the failure states of stress obtained by combinations of those five stress components generate a closed surface, the failure envelope, that separates the no-fail region from the failure region. The failure envelope generated by (7.16) is an ellipse as shown in Figure 7.5. Good comparison is shown between the predicted failure stresses and experimental data from [182]. Note that both stresses σ_2 and σ_6 cause the same type of failure, namely matrix-dominated failure. Interaction between the in-plane stresses σ_1 and σ_2 is purposely avoided in (7.15–7.16).

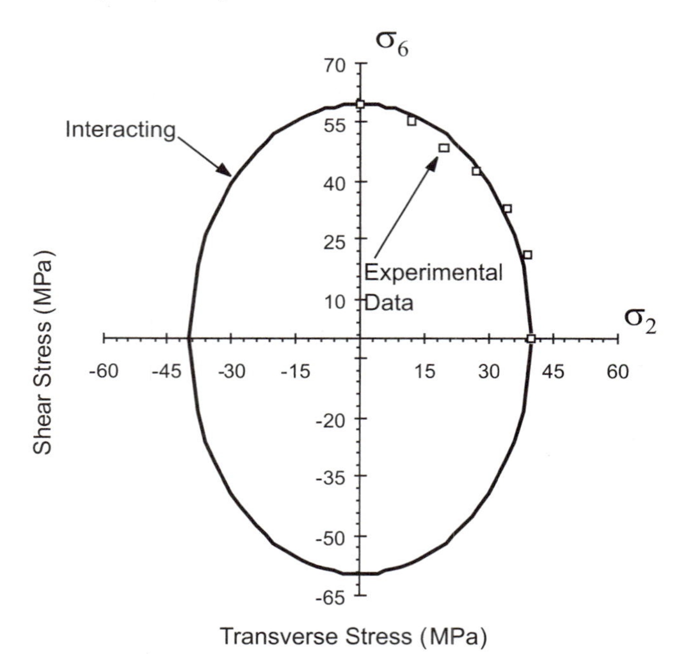

Figure 7.5: Interacting failure envelope in $\sigma_6 - \sigma_2$ space compared with experimental data in the first quadrant.

For a particular design situation, a set of stresses are computed, that will be either inside or outside of the failure envelope defined by (7.15–7.16) and illustrated by the ellipse in Figure 7.5. If the design point is inside, the failure envelope indicates that the material does not fail but there is no indication of how conservative the design is. If the point is outside, the material will fail and some changes are needed, but the magnitude of the necessary changes is not known. In order to have a useful failure criteria, (7.15–7.16) is rewritten using the strength ratio R. Replacing the stress components on the failure envelope σ_i^f by $R\sigma_i$, (7.15–7.16) become:

- Fiber-dominated failure with strength ratio R_1

$$R_1 = \langle F_{1t}/\sigma_1 \rangle + \langle -F_{1c}/\sigma_1 \rangle \tag{7.18}$$

  where $\langle x \rangle$ denotes the positive part of x (see (7.5)); F_{1t}, F_{1c} are both positive values, and $\sigma_1 \neq 0$.

- Matrix-dominated failure with strength ratio R_2

$$\left(f_{22}\sigma_2^2 + f_{66}\sigma_6^2 + f_{44}\sigma_4^2 + f_{55}\sigma_5^2\right) R_2^2 + \left(f_2\sigma_2\right) R_2 - 1 = 0 \tag{7.19}$$

  which can be written in the form

$$aR_2^2 + bR_2 - 1 = 0 \tag{7.20}$$

  and solved for the strength ratio, which is the one with smallest positive value of the two values of R calculated with

$$R_2 = \frac{-b \pm \sqrt{b^2 + 4a}}{2a} \tag{7.21}$$

This gives the designer the choice to evaluate the two major modes of failure separately. In this way, fiber orientations can be added to the laminate to assure that matrix failure does not lead to the collapse of the laminate. However, the interacting criterion (7.18–7.19) is still deficient in at least two ways. First, completely unrelated modes, such as tensile and compressive modes, are still mixed up in the prediction of failure. Second, each failure mode is assumed to have a gradually increasing influence on the neighboring modes, proportional to the values of the stress components and inversely proportional to the strength values. This is unlikely to be the case, given the drastic differences among the physics behind the various failure modes. For a more in-depth discussion, please refer to Chapter 8.

Example 7.2 *Use the interacting criterion to compute the burst pressure of a cylindrical pressure vessel with a diameter d = 25 cm, and a wall thickness t = 2.5 mm. Use the properties of carbon–epoxy in Example 7.1, p. 223. To simplify the computations, assume the cylindrical portion of the vessel is constructed with a single fiber orientation, filament wound in the hoop direction. Note however that this is not an efficient design.*

Solution to Example 7.2 *To take advantage of the concept of strength ratio, define a reference pressure $p_{ref} = 1.0\ MPa$. The ratio $d/t = 100$ indicates that this is a thin pressure vessel, for which the following membrane equations can be used.*

In the hoop direction, the stress is

$$\sigma_1 = \sigma_h = \frac{pd}{2t} = \frac{1(0.25)}{2(0.0025)} = 50\ MPa$$

Along the length of the cylinder, the stress is

$$\sigma_2 = \sigma_l = \frac{pd}{4t} = 25\ MPa$$

Since the vessel is hoop wound, $\sigma_1 = \sigma_h$ and $\sigma_2 = \sigma_l$. With only internal pressure, there is no shear stresses. All the normal stresses are positive. Next, compute the coefficients in (7.17)

$$f_2 = \frac{1}{19} - \frac{1}{131} = \frac{1}{22.2}$$
$$f_{22} = \frac{1}{19(131)} = \frac{1}{2,489}$$
$$f_{66} = \frac{1}{75^2}$$

The interacting criterion (7.18–7.19) predicts the minimum value $R_2 = 0.76$, corresponding to failure in the transverse direction, while in the fiber direction $R_1 = 36.52$. Then, the predicted burst pressure is

$$p_{burst} = R_2\, p_{ref} = 0.76(1) = 0.76\ MPa$$

The maximum stress criterion (7.18) predicts

$$R_1 = 1826/50 = 36.52$$
$$R_2 = 19/25 = 0.76$$

Since there is no shear, there is no interaction. Therefore, the interacting criterion and the maximum stress criterion predict the same values. Note that a large value of R_1 indicates that hoop fiber orientation is not ideal for a pressure vessel. In fact, the optimum fiber orientation for a cylindrical pressure vessel is 55°.

Example 7.3 *Predict the maximum negative pressure that can be applied if the pressure vessel in Example 7.2 (p. 229) is now used to hold vacuum.*

Solution to Example 7.3 *Although this thin vessel is likely to fail by buckling (see 11.2 and [2, Chapter 4]), the analysis below is limited to the evaluation of the strength of the material only. As in Example 7.2 (p. 229), taking $p_i = p_{ref} = 1\ MPa$, we have $\sigma_1 =$*

$-50\ MPa, \sigma_2 = -25\ MPa$.

Since there is no shear, there is no interaction and thus the maximum stress criterion can be used to obtain $R_1 = 1134/50 = 22.68$ *and* $R_2 = 131/25 = 5.24$. *Finally, the burst pressure is* $p_b = Rp_{ref} = 5.24\ MPa$.

The increase of burst pressure with respect to Example 7.2 can be explained as follows. The failure in Example 7.2 was dominated by the tensile strength $F_{2t} = 19\ MPa$, *as it was indicated by the maximum stress criterion. By changing the state of stress to a compressive stress, a much higher compressive strength* $F_{2c} = 131\ MPa$ *allows a higher burst pressure of the vessel. The interacting criterion and the maximum stress criterion predict identical values for Examples 7.2 and 7.3 because there is no shear, so there is no interaction.*

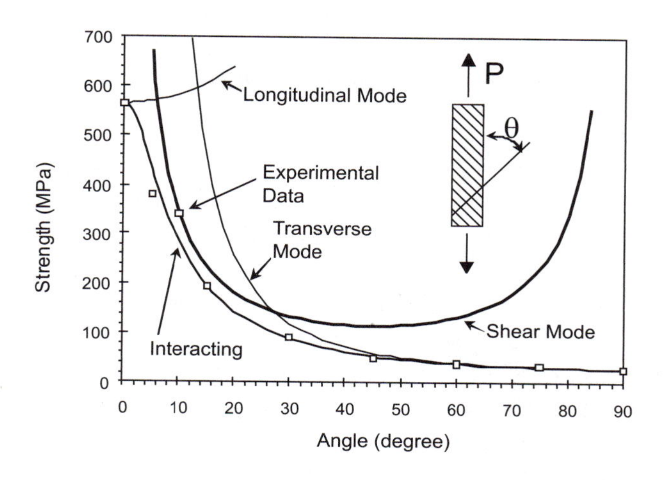

Figure 7.6: The off-axis test.

Example 7.4 ***The Off-Axis Test***
The off-axis test (Figure 7.6) consists of applying a tensile load on a rectangular sample with fibers oriented at an angle θ *with respect to the loading direction [185]. The analysis of the off-axis test is presented below as an illustration of the the interacting criterion and to gain some insight into its applicability.*

Solution to Example 7.4 *Away from the grips, the normal, transverse and shear stresses (5.37) are*

$$\sigma_1 = \sigma_x \cos^2\theta$$
$$\sigma_2 = \sigma_x \sin^2\theta$$
$$\sigma_6 = -\sigma_x \sin\theta\cos\theta$$

The longitudinal tensile strength $F_{1t} = 565\ MPa$ *is obtained in a test with* $\theta = 0°$. *The transverse tensile strength* $F_{2t} = 30\ MPa$ *is obtained from a test with* $\theta = 90°$. *This data, as* *well as data for intermediate values of θ are shown with symbols in Figure 7.6 for carbon–epoxy with $V_f = 0.5$ [182].*

Inspection of Figure 7.6 reveals that the failure is dominated by three isolated modes in three regions. For $\theta < 6°$, the longitudinal strength controls. For $\theta > 28°$, the transverse strength controls. For $6° < \theta < 28°$, shear strength controls. In the region $10° < \theta < 16°$, the shear strength clearly controls the failure while both longitudinal and transverse stresses are under 50% of their strength values. This suggests that the shear strength could be obtained from the off-axis test in this range of angles. In fact, the experimental value for $\theta = 10°$ was used to determine $F_6 = 340 \sin(10) \cos(10) = 58$ MPa, where the minus sign has been omitted because the shear strength in material coordinates F_6 is independent of the sign of the applied shear stress σ_6.

Using experimental values for longitudinal and transverse strength along with the computed shear strength into the interacting criterion, and assuming $F_{1c} = F_{1t}$ and $F_{2c} = F_{2t}$ in (7.17), an excellent prediction is obtained for all the off-axis data as shown in Figure 7.6.

7.1.5 Hygrothermal Failure (*)[4]

In order to account for hygrothermal expansion, it is necessary to introduce temperature and moisture into the lamina failure criteria so that they account for the additional stresses and strains induced by temperature and moisture. In this section, it is assumed that the designer has no control over the environment, and that the temperature change ΔT and the moisture concentration change Δm are given values, out of the designer's control. Therefore, the strength ratio R is defined so that it is a safety factor for the mechanical loads. Then, the strength ratio can be used to find the maximum mechanical loads that the lamina can resist under constant environmental changes in temperature ΔT and moisture Δm. For example, taking the first equation of the maximum stress criterion (7.3), it can be written as

$$R_1 \sigma_1 + \sigma_1^{HT} = F_{1t} \tag{7.22}$$

or

$$R_1 = \frac{F_{1t} - \sigma_1^{HT}}{\sigma_1} \tag{7.23}$$

where σ_1^{HT} is the hygrothermal stress (Section 6.6). Note that the effect of hygrothermal loading is equivalent to a reduction of the tensile strength. Therefore, the maximum stress criterion becomes

$$\begin{array}{lcl}
R_1 = (F_{1t} - \sigma_1^{HT})/\sigma_1 \text{ if } \sigma_1 > 0 & \text{or} & R_1 = (-F_{1c} - \sigma_1^{HT})/\sigma_1 \text{ if } \sigma_1 < 0 \\
R_2 = (F_{2t} - \sigma_2^{HT})/\sigma_2 \text{ if } \sigma_2 > 0 & \text{or} & R_2 = (-F_{2c} - \sigma_2^{HT})/\sigma_2 \text{ if } \sigma_2 < 0 \\
R_6 = (F_6 - \sigma_6^{HT})/\sigma_6 \text{ if } \sigma_6 > 0 & \text{or} & R_6 = (F_6 + \sigma_6^{HT})/\sigma_6 \text{ if } \sigma_6 < 0 \\
R_4 = (F_4 - \sigma_4^{HT})/\sigma_4 \text{ if } \sigma_4 > 0 & \text{or} & R_4 = (F_4 + \sigma_4^{HT})/\sigma_4 \text{ if } \sigma_4 < 0 \\
R_5 = (F_5 - \sigma_5^{HT})/\sigma_5 \text{ if } \sigma_5 > 0 & \text{or} & R_5 = (F_5 + \sigma_5^{HT})/\sigma_5 \text{ if } \sigma_5 < 0.
\end{array} \tag{7.24}$$

[4]Sections marked with (*) can be omitted during the first reading in an introductory course but are recommended for further study and reference.

where the compressive strength values F_{1c} and F_{2c} are taken as positive numbers.

All the remaining failure criteria can be modified similarly. For example, the interacting criterion (7.18–7.19) can be written as (see Exercise 7.14)

- Fiber-dominated:

$$R_1 = \langle \frac{F_{1t} - \sigma_1^{HT}}{\sigma_1} \rangle + \langle - \frac{F_{1c} + \sigma_1^{HT}}{\sigma_1} \rangle \tag{7.25}$$

- Matrix-dominated:

$$f_i \left(\sigma_i R + \sigma_i^{HT} \right) + f_{ij} \left(\sigma_i R + \sigma_i^{HT} \right) \left(\sigma_j R + \sigma_j^{HT} \right) - 1 = 0 \tag{7.26}$$

  where Einstein's summation convention[5] is used with $i, j = 2, 4, 5, 6$ and the f_{ij} coefficients are given by (7.17).

After expanding and collecting like terms, (7.26) can be written as

$$aR_2^2 + 2bR_2 + c = 0 \tag{7.27}$$

with

$$\begin{aligned} a &= f_{22}\sigma_2^2 + f_{44}\sigma_4^2 + f_{55}\sigma_5^2 + f_{66}\sigma_6^2 \\ b &= f_{22}\sigma_2\sigma_2^{HT} + f_{44}\sigma_4\sigma_4^{HT} + f_{55}\sigma_5\sigma_5^{HT} + f_{66}\sigma_6\sigma_6^{HT} + \frac{1}{2} f_2\sigma_2 \\ c &= f_{22}(\sigma_2^{HT})^2 + f_{44}(\sigma_4^{HT})^2 + f_{55}(\sigma_5^{HT})^2 + f_{66}(\sigma_6^{HT})^2 + f_2\sigma_2^{HT} - 1 \end{aligned} \tag{7.28}$$

and

$$R_2 = \frac{1}{a} \left(-b + \sqrt{b^2 - ac} \right) \tag{7.29}$$

A negative radical in (7.29) indicates that the hygrothermal loads alone are sufficient to cause failure of the laminate. Finally, the strength ratio is the minimum value of R_1 and R_2.

7.2 Laminate First Ply Failure

First ply failure (FPF) load is the load at which the first lamina failure occurs in a laminate. Since the transverse strength F_{2t} and shear strength F_6 of the polymer matrix composites (PMC) are usually much lower than the tensile strength F_{1t}, FPF is usually due to transverse and/or shear failure modes, which are normally caused by matrix cracking.

To find the first ply (or lamina) that fails and the corresponding first ply failure (FPF) load, one can use any of the *strength failure criteria* described in Section 7.1. First compute the stress in all laminae before the first lamina fails using the

[5]Repeated subscripts in a term imply a summation over the range of the subscript, i.e., $f_i\sigma_i = \sum_{i=2,4,5,6} f_i\sigma_i$.

procedure described in Section 6.2. Since the stress distribution is piece-wise linear through the thickness of a laminate (see Figure 6.5), the maximum value of a component of stress is found either on the top or bottom of each lamina. Next, a strength failure criterion is used at those points to compute the strength ratios $R^{(k)}$ for all n laminae, with $k = 1...n$, as per Section 7.1.1. The minimum value $R = min(R^{(k)})$ of all those computed corresponds to the most compromised lamina. If $R^{(k)} < 1$, that lamina fails.

In design, provided that the strength values have been reduced by resistance factors ϕ and the loads increased by load factors α (Section 1.4.3), the minimum value of R of all those computed is the safety factor of the laminate. If R is less than one, redesign must be done. Also, if R is too large, it indicates that the laminate is overdesigned and savings can be realized by redesign.

Unlike the unidirectional (UD) lamina considered in Section 7.1, the laminae in a laminate are constrained by adjacent laminae; i.e., a cracked lamina is *held together* by adjacent laminae making it difficult for cracks to propagate. More properly, the constraining effect retards the propagation of inherent microdefects into a fully developed first crack and further retards the appearance of multiple cracks necessary to cause major loss of stiffness and strength in the lamina. For an unconstrained UD lamina, the initiation of matrix cracks happens at a lower stress (strain) than when the same lamina is embedded in a laminate. Furthermore, in a UD lamina, crack initiation coincides with fracture of the lamina because there are no constraining laminae to prevent the propagation of the first matrix crack to the point that the UD lamina fractures completely. Therefore, in a UD lamina, F_{2t} and F_6 coincide with the formation of the first crack.

The initiation and propagation of cracks is controlled by two fracture toughness values, which are represented by material properties G_{Ic} and G_{IIc}. Since strength failure criteria do not use fracture toughness but rather use strength values F_{2c} and F_6 of a UD lamina, they are incapable of predicting the constraining effect of adjacent laminae. In order to continue using strength failure criteria, one must *adjust* the strength values to take into account the constraining effect, thus calculating so called *in-situ strength* values F_{2c}^k and F_6^k for each lamina k [167]. In-situ strengths are not material properties because their values depend on the thickness of the lamina and the stiffness of the laminate in which the lamina is embedded. Calculation of in-situ strengths is presented in Section 7.2.1.

Calculation the FPF load is only the beginning of the analysis because more often one needs to determine the laminate strength (see Section 7.3). In the *ply discount method* (Section 7.3.1), it is assumed that once a lamina fails in transverse/shear mode, the lamina looses all of its stiffness except in the fiber direction. Such sudden discontinuity in stiffness causes numerical convergence problems in algorithms designed to try to follow the progressive failure of the laminate, one lamina at a time. In reality transverse/shear failure does not happen suddenly but gradually. Transverse/shear damage in the form of matrix cracks, parallel to the fiber direction, appear at a certain level of stress or strain, both of which are a function of the applied load (Figure 7.7). The crack density increases with load and both

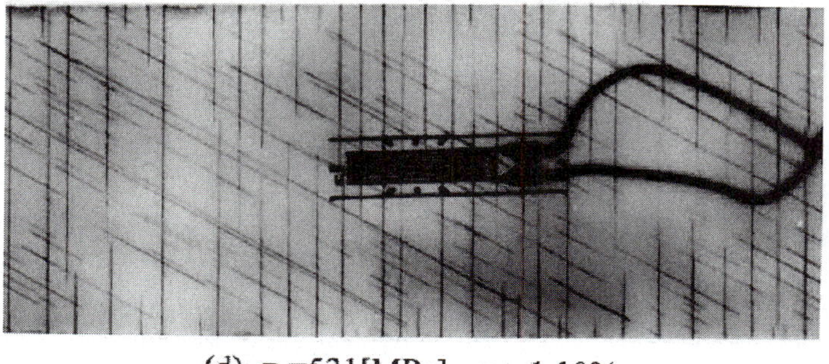

Figure 7.7: Matrix cracks in the 30° and 90° laminae; $[0/30_2/90]_S$ laminate loaded along the 0° direction. (From T. Yokozeki et al., *Consecutive Matrix Cracking in Contiguous Plies of Composite Laminates,* Int. J. Solids Structures, vol. 42, p. 2795, Elsevier 2005.)

the transverse and shear stiffness of the lamina decrease until they virtually vanish. During this gradual process, stress is redistributed to the adjacent laminae. An analysis methodology based on this, a more realistic description of material behavior, is presented in Chapter 8.

Matrix cracking is more important and pronounced for material systems with brittle matrix and compliant fibers such as glass–polyester (i.e., when $\epsilon_{fu} > \epsilon_{mu}$). Multidirectional laminate composites using carbon fibers and though matrices such as epoxy have the strains in every direction controlled by the stiff and relatively brittle carbon fibers ($\epsilon_{fu} < \epsilon_{mu}$). If, based on experience, matrix cracking can be neglected or ruled out, the strength of multidirectional carbon–fiber laminates can be predicted by a simpler approach (Section 7.3.2) based on the fiber strain to failure (ϵ_{fu}).

Example 7.5 *Using the maximum stress criterion, compute the first ply failure load of a* $[0/90]_S$ *laminate subject to* $N_x \neq 0$, $N_y = N_{xy} = 0$. *Each lamina is* 2.54 *mm thick. Use the following material properties*

$$E_1 = 54\ GPa\ ;\ E_2 = 18\ GPa$$
$$G_{12} = 9\ GPa\ ;\ \nu_{12} = 0.25$$
$$F_{1t} = 1{,}034\ MPa\ ;\ F_{1c} = 1{,}034\ MPa$$
$$F_{2t} = 31\ MPa\ ;\ F_{2c} = 138\ MPa$$
$$F_6 = 41\ MPa$$

Solution to Example 7.5 *For glass–epoxy, the transition thickness can be assumed to be* 0.6 *mm, so the in-situ values (7.42) are*

$$te = min(2.54, 0.6) = 0.6\ mm \quad \textit{Note: the laminae are "thick"}$$
$$F_{2t}^{is} = 1.12\sqrt{2}\ 31 = 49.1\ MPa$$
$$F_6^{is} = \sqrt{2}\ 58 = 82.0\ MPa$$

Since the laminate is a symmetric cross-ply and only in-plane loads are present. Therefore, only the [A] matrix is necessary. Using the macro-mechanics analysis presented in Chapters 5 and 6 and (5.23)

$$[Q] = \begin{bmatrix} 55.1 & 4.6 & 0 \\ 4.6 & 18.4 & 0 \\ 0 & 0 & 9 \end{bmatrix} GPa$$

and using (6.16)

$$[A] = \begin{bmatrix} 374 & 46.7 & 0 \\ 46.7 & 374 & 0 \\ 0 & 0 & 91.44 \end{bmatrix} GPa\ mm$$

Note that the laminate has the same amount of material oriented in the 0-degree and 90-degree directions, so $A_{11} = A_{22}$. Only two of the middle-surface strains are different from zero, which can be computed with (6.21). Using a reference load $N_x = 1000\ N/mm$ we have

$$\epsilon_x^0 = 2.72\ (10^{-3})$$
$$\epsilon_y^0 = -034\ (10^{-3})$$

These are also the only strains throughout the laminate since the curvatures are zero. Since there are only two nonzero stresses (6.23–6.24), they can be easily tabulated along with the values of the strength ratios (7.3).

Lamina	σ_1[MPa]	σ_2[MPa]	R_1	R_2
OUTER	148.4	6.2	6.9	7.8
MIDDLE	-6.2	48.4	165.4	1.013

The top and bottom laminae are 0-degree plies, which experience the same stress. The two middle laminae are the 90-degree plies. The minimum value is $R = 1.013$. Therefore, the FPF load is

$$N_{FPF} = 1000\ (1.013) = 1013\ N/mm$$

At this load, the two middle laminae experience some kind of transverse failure. The stress in the middle lamina at failure is

$$\sigma_2 = 48.4\ (1.013) = 49.0 MPa \equiv F_{2t}^{is}$$

which is equal to the in-situ transverse strength of the material. After first ply failure, that stress must be transferred to the outer laminae. See [1] for MATLAB code for this example. See also Examples 7.6 and 7.8 (pp. 241, 245).

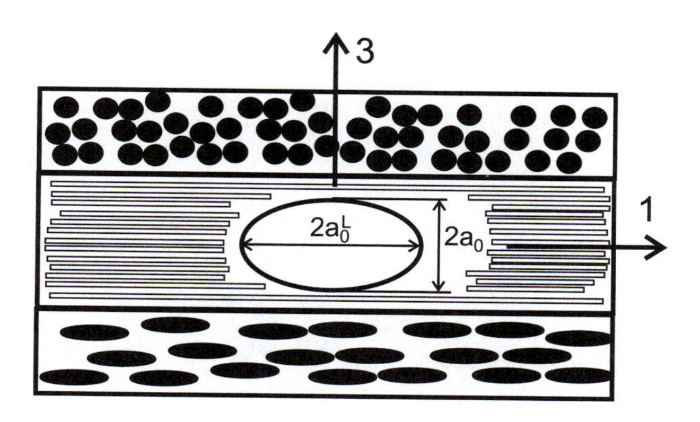

Figure 7.8: Representative crack geometry [168].

7.2.1 In-Situ Strength

Regardless of how much care is taken in production, all materials contain inherent defects, including voids, microcracks, sites with weak fiber–matrix interfaces, and so on. These defects act as nucleation sites where cracks can initiate and propagate. The fracture mechanics analysis of solid media assumes that a representative crack exist that represents the effects of various defects present in the material. For a fiber-reinforced lamina embedded in a laminate, one such representative crack is illustrated in Figure 7.8. A crack parallel to the fiber direction will grow in the longitudinal (L) or thickness (T) direction as soon as the energy release rate (ERR) exceeds the fracture toughness[6] G_{Ic}, G_{IIc}, of the material in one of those directions. The ERR for a laminate such as the one displayed in Figure 7.8 were calculated in [168] as a function of the transverse dimension $2a_0$.

In the transverse direction, the ERR are

$$
\begin{aligned}
G_I^T &= \frac{\pi a_0}{2}\Lambda_{22}^0 \sigma_2^2 \\
G_{II}^T &= \frac{\pi a_0}{2}\Lambda_{44}^0 \sigma_6^2
\end{aligned}
\tag{7.30}
$$

were

$$
\begin{aligned}
\Lambda_{22}^0 &= 2\left(\frac{1}{E_2} - \frac{\nu_{21}^2}{E_1}\right) \\
\Lambda_{44}^0 &= \frac{1}{G_{12}}
\end{aligned}
\tag{7.31}
$$

[6]In classical fracture mechanics, the term *fracture toughness* K_c refers to the critical value of stress intensity [30], but in this work the term fracture toughness refers to the critical energy release rate G_c because stress intensity is not used in the fracture mechanics analysis of composite materials.

Therefore, a crack propagates in the thickness direction x_3 when the ERRs reach the fracture toughnesses ($G_I = G_{Ic}, G_{II} = G_{IIc}$) and at that moment the values of stress are, by definition, equal to the in-situ strength values ($\sigma_2 = F_{2t}^{is}, \sigma_6 = F_6^{is}$). Therefore, (7.30) become

$$\begin{aligned} G_{Ic} &= \frac{\pi a_0}{2} \Lambda_{22}^0 \left(F_{2t}^{is}\right)^2 \\ G_{IIc} &= \frac{\pi a_0}{2} \Lambda_{44}^0 \left(F_6^{is}\right)^2 \end{aligned} \tag{7.32}$$

In the longitudinal direction, the ERR are

$$\begin{aligned} G_I^L &= \frac{\pi a_0}{4} \Lambda_{22}^0 \sigma_2^2 \\ G_{II}^L &= \frac{\pi a_0}{4} \Lambda_{44}^0 \sigma_6^2 \end{aligned} \tag{7.33}$$

Since the transverse ERR (7.30) are twice higher than the longitudinal ERR (7.33), a crack will grow first through the thickness until it is constrained by the adjacent laminae; then it has no choice but to grow along the length. Furthermore, (7.30) also indicates that the energy release rate is proportional to the crack length $2a_0$, so the crack will grow *unstably* in the transverse direction. If the lamina containing the crack is thin, the crack quickly occupies the entire thickness of the lamina $t = 2a_0$. Then, it must grow in the longitudinal direction when $G_I = G_{Ic}$, and/or $G_{II} = G_{IIc}$. In other words, a crack occupying the entire thickness of the lamina ($t = 2a_0$ in Figure 7.8) cannot propagate further in the thickness direction because it would have to penetrate the adjacent laminae, so it propagates in the longitudinal direction (7.33) when the ERR reaches the fracture toughness and at that moment the values of stress are, by definition, the in-situ strength values F_{2t}^{is}, F_6^{is}, as follows

$$\begin{aligned} G_{Ic} &= \frac{\pi t}{8} \Lambda_{22}^0 \left(F_{2t}^{is}\right)^2 \\ G_{IIc} &= \frac{\pi t}{8} \Lambda_{44}^0 \left(F_6^{is}\right)^2 \end{aligned} \tag{7.34}$$

were, since there is no evidence to the contrary, we have assumed that the fracture toughness in the transverse and longitudinal directions are identical. Therefore, for a *thin* lamina, assuming the crack occupies the entire thickness of the lamina $t = 2a_0$, the apparent *in-situ* strength [167] can be calculated as a function of the lamina thickness t and fracture toughness as

$$\begin{aligned} F_{2t}^{is-thin} &= \sqrt{\frac{8G_{Ic}}{\pi t \Lambda_{22}^0}} \\ F_6^{is-thin} &= \sqrt{\frac{8G_{IIc}}{\pi t \Lambda_{44}^0}} \end{aligned} \tag{7.35}$$

Note that the in-situ values are a function of the thickness of the lamina, and thus they are not material properties.

If the lamina containing the crack is *thick*, the crack (with size $2a_0$ in Figure 7.8) will grow unstably through thickness until it reaches the interface. So, the apparent *in-situ* strength can be calculated from (7.32) as

$$F_{2t}^{is-thick} = \sqrt{\frac{2G_{Ic}}{\pi a_0 \Lambda_{22}^0}}$$
$$F_{6}^{is-thick} = \sqrt{\frac{2G_{IIc}}{\pi a_0 \Lambda_{44}^0}} \tag{7.36}$$

Note that (7.35) yields the same values as (7.36) when $t = 4a_0$. This means that the transition from thin to thick occurs when $t_t = 4a_0$, thus defining the *transition thickness* t_t for the material. Plots of experimental data [168, Figures 5 and 6] representing the behavior of thin laminae (7.35) and thick laminae (7.36) intersect when the lamina thickness is equal to the *transition thickness*

$$t_t = 4a_0 \tag{7.37}$$

Experimental data from [169] yields $t_t = 0.6$ mm for glass–epoxy and $t_t = 0.8$ mm for carbon–epoxy composites. Laminae thinner than t_t are considered *thin* and those thicker than t_t are said to be *thick*. Further, if the transition thickness of a composite is known, one can estimate the half-size of the *representative* crack as $a_0 = t_t/4$.

In a thick laminate, a crack smaller than the thickness of the lamina is initially not constrained, so it can grow under conditions similar to those encountered in a unidirectional laminate. Testing a unidirectional laminate yields strength values F_{2t} and F_6. For a transverse tensile test, e.g., [118, ASTM D 3039], the most critical crack is a corner crack of dimension $2a_0$ that propagates when $G_I = G_{Ic}$ (Figure 4.20). For an in-plane shear test, e.g., [118, ASTM D 5379], the most critical crack is a surface crack of dimension $2a_0$ that propagates when $G_{II} = G_{IIc}$ (Figure 4.26). The ERR for these two cases are available from [166] as[7]

$$G_I = 1.12^2 \pi a_0 \Lambda_{22}^0 \sigma_2^2$$
$$G_{II} = \pi a_0 \Lambda_{44}^0 \sigma_6^2 \tag{7.38}$$

and the unidirectional laminate fails when

$$G_{Ic} = 1.12^2 \pi a_0 \Lambda_{22}^0 \left(F_{2t}\right)^2$$
$$G_{IIc} = \pi a_0 \Lambda_{44}^0 \left(F_6\right)^2 \tag{7.39}$$

[7]Note that (7.30,7.33) for a constrained crack are different from (7.38) for an unconstrained crack.

Equation (7.39) provides a way to estimate the fracture toughness from strength data for unidirectional laminates. However, transverse tensile strength data F_{2t} of unidirectional laminates is often unreliable due to the large scatter in the data and other factors associated with the brittle nature of the unconstrained laminate. If the intralaminar fracture toughness is known, or it can be estimated, then, strength values for a unidirectional lamina can be estimated by inverting (7.39) and assuming a_0 to be 1/4 of the transition thickness t_t, i.e.,

$$
\begin{aligned}
F_{2t} &= \sqrt{G_{Ic}/\left(1.12^2 \pi a_0 \Lambda^0_{22}\right)} \\
F_6 &= \sqrt{G_{IIc}/\left(\pi a_0 \Lambda^0_{44}\right)}
\end{aligned}
\tag{7.40}
$$

Comparing (7.32) with (7.39), explicit expressions for the in-situ strength values [167] for a *thick lamina* are obtained as

$$
\begin{aligned}
F_{2t}^{is-thick} &= 1.12\sqrt{2}F_{2t} \\
F_6^{is-thick} &= \sqrt{2}F_6
\end{aligned}
\tag{7.41}
$$

Substituting (7.39) into (7.35), comparing to (7.36), and defining the effective thickness $t_e = min(t, t_t)$, unique expressions for the in-situ values, valid for thick and thin lamina, are obtained as follows

$$
\begin{aligned}
t_e &= min(t, t_t) \\
F_{2t}^{is} &= 1.12\sqrt{\frac{2t_t}{t_e}}F_{2t} \\
F_6^{is} &= \sqrt{\frac{2t_t}{t_e}}F_6
\end{aligned}
\tag{7.42}
$$

Values of the transition thickness can be estimated from [169] as $t_t = 0.6$ mm for glass–epoxy and $t_t = 0.8$ mm for carbon–epoxy composites. However, adjusting strength values as a function of ply thickness via in-situ values such as (7.42)[8] do not provide any information about crack density (damage) evolution; they only provide the load at which the first crack appears. In a UD laminate, the first crack leads to collapse, but for most practical laminates, the first crack has barely any immediate effect on the mechanical behavior of the laminate. Design-critical issues such as loss of laminate stiffness, stress redistribution to the remaining laminae, and eventual collapse requires prediction of damage evolution (see Chapter 8).

The LSS has an effect on the first crack load (or strain) and on damage evolution, as exemplified in Figure 7.9. The damage mechanics calculations, labeled DDM, were performed with the model described in [187]. The in-situ FPF were calculated

[8]Equation (7.42) does not take into account shear nonlinearity and, consequently may overestimate the in-situ shear strength [186].

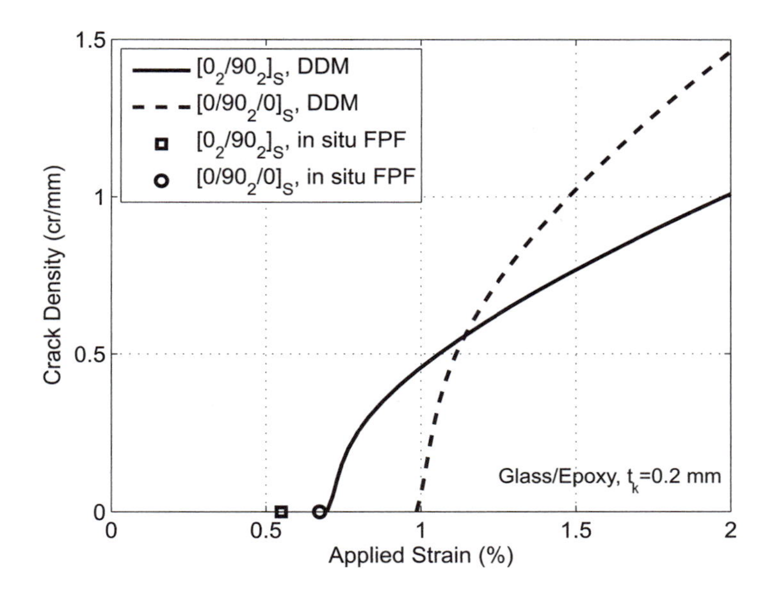

Figure 7.9: Crack initiation and evolution in two laminates with the same in-plane stiffness, i.e., $[0_2/90_2]_S$ and $[0/90_2/0]_S$.

with classical lamination theory (CLT) with strength values adjusted with (7.42) to reflect the thickness of each lamina. The material is glass–epoxy (see Example 8.1) and the lamina thickness is $t_k = 0.2$ mm. It can be seen that the same thickness of 90° material cracks at lower applied strain when it is at the at the midsurface than when it is constrained between a pair of 0° laminae.

While using (7.42), the **total** thickness t of the center lamina in $[0_2/90_4/0_2]_T$ must be used. Also, **twice** the thickness of the surface lamina in $[90_2/0_2]_S$ must be used. The values of crack initiation strain predicted by using (7.42) are shown with a square symbol in Figure 7.9.

Example 7.6 *Compute the FPF load $[N_x, 0, 0]$ and FPF strain ϵ_x^{FPF} for a $[0_3/90_4]_S$ made of glass–epoxy (Table 1.3) with lamina thickness $t_k = 0.1$ mm using in-situ values into the maximum strength criterion. Compare with discrete damage mechanics (Chapter 8).*

Solution to Example 7.6 *Since the laminate is symmetric, only the $[A]$ matrix is needed*

$$[A] = \begin{bmatrix} 36.96 & 3.22 & 0 \\ 3.22 & 43.62 & 0 \\ 0.0 & 0.0 & 7.7 \end{bmatrix} \; GPa$$

For glass–epoxy, from Table 1.3, $F_{2t} = 40$. The transition from thin to thick lamina for glass–epoxy is assumed to be $t_t = 0.6$ mm [169]. Since the laminate is symmetric, the in-situ strength of the center lamina is a calculated by using the full thickness of the center lamina as

$$te = min(t_t, t_k) = min(0.6, 8 \times 0.1) = 0.6\ mm \quad \text{Note: the lamina is "thick."}$$
$$F_{2t}^{is} = 1.12\sqrt{2}\ 40 = 63.36\ MPa$$

Applying $N = [1, 0, 0]$ *and using CADEC [1] results in a strength ratio* $R_{FPF} = 195$. *Applying* $N = [195, 0, 0]$ *results in* $\epsilon_x^{FPF} = 0.53\%$ *with the center lamina failing at* $\sigma_2 = 63.36\ MPa$.

Example 7.7 *Design a cylindrical pressure vessel with diameter* $d = 4\ m$, *internal pressure* $p = 3.0\ MPa$ *with 10% COV, for 95% reliability with respect to first ply failure. Use the maximum stress criterion and a coordinate system on the surface of the cylinder with the y-axis oriented in the hoop direction and the x-axis oriented along the length. Use E-glass–epoxy with properties from Table 1.3. Assume the strength properties have a 12% COV. Use ply thickness* $t_k = 1.27\ mm$.

Solution to Example 7.7 *For a 95% reliability, from Table (1.1), the standard variable is* $z = -1.645$. *Then, the load and resistance factors are calculated (1.16)–(1.17) as*

$$\alpha = 1 - zC_p = 1 + 1.645\ (10/100) = 1.1645$$
$$\phi = 1 + zC_F = 1 - 1.645\ (12/100) = 0.8026$$

The loads per unit length (stress resultants), in the longitudinal and hoop directions, that equilibrate the internal pressure are

$$N_x = \frac{\alpha\ p\ d}{4} = \frac{1.1645 \times 3.0 \times 4000}{4} = 3500\ N/mm$$
$$N_y = \frac{\alpha\ p\ d}{2} = \frac{1.1645 \times 3.0 \times 4000}{2} = 7000\ N/mm$$

Using an angle ply laminate and since the ratio $N_y/N_x = 2$ *it is easy to demonstrate by trial and error that the optimum angle is* 55°. *Selecting the LSS to be* $[55_n/-55_n]_S$, *it only remains to find* n. *Using CADEC [1] and the max. stress criterion, a laminate with* $n = 1$ *loaded with* $N = \{3500, 7000, 0\}$ *has and a preliminary* $R_{FPF} = 0.1$ *and mode of failure: transverse failure. So, the required number of plies can be selected to be* $n = 10$. *In this case,* $t_k = n \times 1.27 = 12.7\ mm$. *At this point it is necessary to calculate the in-situ values (7.42). With the transition thickness for glass–epoxy taken to be* $t_t = 0.6\ mm$, $t_k > t_t$, *the laminae are considered to be* thick. *Therefore,*

$$t_e = min(t_k, t_t) = min(12.7, 0.6) = 0.6\ mm$$
$$F_{2t}^{is} = 1.12\sqrt{2}F_{2t} = 1.584F_{2t} = 63.357\ MPa$$
$$F_6^{is} = \sqrt{2}F_6 = 1.414F_6 = 84.553\ MPa$$

To take into account the variation of strength,

$$\phi F_{2t}^{is} = 0.8026\ (63.357) = 50.68\ MPa$$
$$\phi F_{6}^{is} = 0.8026\ (84.553) = 67.64\ MPa$$

Using these values, $R_{FPF} = 1.26$ *and the number of laminae can be reduced to* $n = 10/1.26 = 7.9 \approx 8$*, resulting in* $t_k = 8(1.27) = 10.16\ mm$*. Since still* $t_k < t_t$ *the in-situ values do not change and the design is finished for* $n = 8$*, resulting in* $R_{FPF} = 1.32(8/10) = 1.05$*.*

Note that matrix cracking at FPF load may allow seepage of gas and liquid through the laminate and, unless a liner is used, this would dictate the limit pressure for the vessel. Also note that a more efficient laminate would be $[\pm(55)_n]$*.*

7.3 Laminate Strength

Computation of the first ply failure load was presented in Section 7.2. For most polymer matrix composites, FPF usually involves matrix degradation, such as matrix cracks. As a lamina degrades, its stiffness is reduced and its stress redistributed to the remaining laminae, which may then become overloaded and fail. Prediction of laminate strength requires tracking the degradation of most if not all the laminae until the laminate is no longer capable of carrying the applied load. Three strategies are presented in this book for the prediction of laminate strength. The first one, called *ply discount*, is the most obvious but not necessarily most accurate. Next, the *truncated maximum strain criterion*, which is popular for the design of carbon-fiber reinforced composites. Finally, *discrete damage mechanics* (DDM) is the more complex and accurate method, is presented in Chapter 8.

7.3.1 Ply Discount

Once transverse/shear failure has been detected in one lamina, the stiffness of that lamina needs to be degraded according to the severity of the damage to that lamina. In the *damage mechanics method* described in Chapter 8, the crack density λ and degraded stiffness $E(\lambda)$ are calculated at each load step. In the *ply discount method* discussed here, the stiffness is degraded entirely in all but the fiber direction. Such selectivity is justified because matrix cracks do not significantly affect the longitudinal stiffness E_1 as long as the fibers remain intact. However, the drastic change of stiffness called upon by ply discount may cause convergence problems in the numerical implementation of the method, specially in the context of finite element analysis. In an attempt to avoid drastic changes of stiffness, a scalar damage factor $0 \leq d < 1$ is sometimes advocated to reduce the stiffness as follows[9]

[9]The damage factor d introduced here relates to the degradation factor d_f used in the first edition of this book as follows: $d_f = (1 - d)$.

$$
\begin{aligned}
E_1 &= \widetilde{E}_1 \\
E_2 &= (1-d)\,\widetilde{E}_2 \\
G_{12} &= (1-d)\,\widetilde{G}_{12} \\
G_{23} &= (1-d)\,\widetilde{G}_{23} \\
\nu_{12} &= (1-d)\,\widetilde{\nu}_{12}
\end{aligned}
\tag{7.43}
$$

were *tilde* indicates original properties of the undamaged material. When $d = 1$, the method reduces to the original ply discount method. Note that no reliable method exists to predict the value of d. Although it is possible to select a value of d so that the predicted laminate strength matches an experimental value of laminate strength, such value is usually inadequate to predict the strength when the laminate stacking sequence (LSS) or other characteristics of the laminate change. This is because all components of stiffness are reduced equally. A better approach is described in Chapter 8.

Once a lamina has been discounted (or degraded), only fiber strength failure criteria are applied because the composite is assumed to have failed in all the matrix-dominated modes (transverse and shear); only the strength of the fibers remain. Therefore, the lamina strength criteria in Section (7.1) reduce to

- Maximum stress criterion, (7.4) reduces to

$$
R_1 = \langle F_{1t}/\sigma_1 \rangle + \langle -F_{1c}/\sigma_1 \rangle \tag{7.44}
$$

- Maximum strain criterion, (7.7) reduces to

$$
R_1 = \langle \epsilon_{1t}/\epsilon_1 \rangle + \langle -\epsilon_{1c}/\epsilon_1 \rangle \tag{7.45}
$$

- Interacting failure criterion, same as (7.18)

$$
R_1 = \langle F_{1t}/\sigma_1 \rangle + \langle -F_{1c}/\sigma_1 \rangle \tag{7.46}
$$

where $\langle x \rangle$ denotes the positive part of x (see (7.5)).

Once the lamina stiffness has been degraded, the state of stress in the laminate can be recomputed. Additional transverse/shear failures in other laminae are treated similarly by degrading the stiffness of the affected laminae and recomputing the stress in the laminate until the *last ply failure* (LPF) occurs, thus defining the strength of the laminate.

Should a lamina fail in a longitudinal mode of failure, either tensile or compressive, it is very likely that the strength of the entire laminate is compromised. This situation is called *fiber failure* and it is assumed to represent the (ultimate) strength of the laminate.

Fiber failure in a lamina will most likely correspond to the final fracture of the laminate or at least be very close to it, but for some laminates a matrix failure may determine the ultimate load of the laminate. For example, an angle-ply laminate under uniaxial load fails suddenly at the first ply failure (FPF) load because there are no fibers in the load direction to carry the load after the matrix fails. Such design should be avoided in practice by adding laminae oriented closely to the load direction. The designer should be able to avoid these undesirable situations by adding lamina orientations designed to constrain the matrix failures. Once proper design practices are followed, more reliable failure estimates can be obtained, as shown in Section 7.3.2.

Example 7.8 *Compute the LPF load of the laminate in Example 7.5, p. 235.*

Solution to Example 7.8 *At the FPF load of $N_x = 1013\ N/mm$ ($R = 1.013$ with reference to $N_x = 1000\ N/mm$), the two center laminae fail. Using a damage factor $d = 0.999$ in (7.43) and using (7.44) to recompute R, fiber failure (LPF) occurs with $R = 5.254$ when the outer laminae fail along the fiber direction. Therefore the LPF load is*

$$N_{LPF} = 1000\ (5.254) = 5254\ N/mm$$

and the stress at failure is

$$\sigma_1 = 196.78\ (5.254) = 1034\ MPa$$

which is equal to the tensile strength of the material. Note that the LPF load is larger than the FPF load because the outer laminae are capable of carrying all the load after FPF. Also note that "d" is not set exactly to one, to avoid numerical instability but not too large to avoid introducing artificial stiffness to the laminate. However, using a damage factor less than 0.99 is not recommended because the degraded laminate would maintain an artificial residual stiffness, equal to $1 - d$ times the original stiffness, which also adds an artificial strength. For example, using $d = 0.8$, the calculated LPF strength ratio would be $R = 5.603$, thus overpredicting the tensile strength of the laminate by 7%.

Example 7.9 *Modify a cross-ply laminate $[(0/90)_2]_S$ to be able to carry a pure shear load $N_{xy} = 100\ N/mm$ without collapsing as a result of a matrix-dominated failure. Each lamina is 0.1 mm thick and made of carbon–epoxy with properties*

$E_1 = 156.4\ GPa$	$E_2 = 7.786\ GPa$
$\nu_{12} = 0.352$	$\nu_{21} = 0.016$
$G_{12} = 3.762\ GPa$	
$F_{1t} = 1.826\ GPa$	$F_{1c} = 1.134\ GPa$
$F_{2t} = 0.019\ GPa$	$F_{2c} = 0.131\ GPa$
$F_6 = 0.075\ GPa$	
$G_{Ic} = 0.073\ KJ/m^2$	$G_{IIc} = 0.939\ KJ/m^2$

Solution to Example 7.9 *The in-situ shear strength (7.42) is calculated as*

$$t_e = min(0.1, 0.8) = 0.1$$

$$F_{2t}^{is} = 1.12\sqrt{\frac{2(0.8)}{0.1}}19 = 85.12\ MPa$$

$$F_6^{is} = \sqrt{\frac{2(0.8)}{0.1}}75 = 300.00\ MPa$$

Using the maximum strain criterion (Section 7.1.3), the [0/90]$_{2S}$ yields $R_{FPF} = 2.4$. The mode of failure is in-plane shear, with all the laminae failing simultaneously. After the shear failure, there are no laminae with angle fibers to take the load. This is evident from the result $R_{FPF} = \infty$. It is important to notice that when the laminate has no remaining strength after FPF, the in-situ correction of strength cannot be used because the laminae in the laminate behave as UD lamina without laminate constraints. A cross-ply laminate like this fails at the same shear stress that a UD lamina. In fact [118, ASTM D 5379] recommends using a cross-ply laminate to test for F_6 of a UD lamina. Since the in-situ values should not be used in this case, recalculating the FPF without using in-situ values yields $R_{FPF} = 0.6$.

To improve the laminate behavior, the same thickness and number of laminae are rearranged in a $[0/90/\pm 45]_S$ configuration. If in-situ values are not used, the maximum-strain criterion yields $R_{FPF} = 0.706$, which is a 17.7% improvement. But more important is the change of behavior after FPF. The FPF occurs as a result of transverse failure in all the $\pm 45°$ laminae simultaneously. After FPF, fiber failure is detected in the 45 laminae at $R_{LPF} = 2.268$. This represents an improvement in the ultimate laminate strength of 278%, which is accomplished by changing the mode of failure to a fiber-dominated one. When in-situ values are used, $R_{FPF} = 2.097$ and $R_{LPF} = 2.268$.

7.3.2 Truncated-Maximum-Strain Criterion (*)[10]

As discussed previously, there are two major events during loading of a laminate that can be used as indication of strength: matrix cracking and fiber failure. Fiber failure is often viewed as the collapse load of the laminate. Matrix cracking usually occurs at a lower load level, and it does not lead to collapse if the laminate is properly designed. If the transverse deformation of each lamina is constrained by another lamina having fibers perpendicular to the first one, the reduction in strength in one lamina due to matrix cracking is compensated by a load transfer into the fibers of the perpendicular lamina. Furthermore, if the fibers are more brittle than the matrix, it can be argued that the fibers prevent transverse failure because the fibers would have to fail first. Based on these observations, the truncated-maximum-strain criterion [188] is formulated assuming that:

- All laminae have a complementary lamina with perpendicular fiber orientation. The laminate does not need to be balanced, but matrix cracking in any lamina should be constrained by perpendicular fibers. This requirement effectively

[10]Sections marked with (*) can be omitted during the first reading but are recommended for further study and reference.

translates in the need to have laminae with 0, 90 and ±45 orientations with at least 10% of the laminate thickness in any orientation (10% rule).

- Laminae with various fiber orientations are interspersed as much as possible to avoid clusters of laminae with the same orientation where a matrix crack can grow easily.

- The strains in the fiber are those of the lamina (ϵ_1, ϵ_2). This assumption is satisfied in the fiber direction but it is only approximate in the transverse direction. The strains ϵ_1, ϵ_2 are computed for each lamina using standard methods.

The truncated-maximum-strain criterion is applied lamina by lamina but it is a *laminate failure criterion*; i.e., it cannot be used for a single unidirectional lamina. Fiber failure is detected by comparing the strain in the fiber ϵ_1 to the strain to failure of the unidirectional composite ϵ_{1t} in tension and ϵ_{1c} in compression

$$-\epsilon_{1c} < \epsilon_1 < \epsilon_{1t} \tag{7.47}$$

where ϵ_{1c} is a positive number. It must be noted that the criterion looks only at fiber failure, but uses effective fiber strain to failure ϵ_{1t} and ϵ_{1c} obtained from tests of unidirectional composites, not from fiber tests. Since each lamina is assumed to be accompanied by a perpendicular lamina, the fiber strain in the perpendicular lamina can be checked directly using the transverse strain in the current laminae as

$$-\epsilon_{1c} < \epsilon_2 < \epsilon_{1t} \tag{7.48}$$

This last condition makes the failure envelope in Figure 7.10 quite different from the maximum strain criterion shown in Figure 7.4. The truncated-maximum-strain criterion does not uses ϵ_2 to check matrix cracking but to check fiber failure in the complementary lamina, which is built into the laminate precisely to constrain the transverse deformation ϵ_2.

Up to this point, the failure criterion can be summarized as to just checking every lamina for fiber strain. This procedure, called *maximum-fiber-strain criterion,* is popular in the aerospace industry because of simplicity, and it is reported to be quite accurate. The truncated-maximum-strain criterion goes one step further by limiting the strains in the tension-compression quadrants to account for shear failure of the fibers (second and fourth quadrants in Figure 7.10). The shear strain in the fiber is computed as

$$\gamma = \frac{\epsilon_1 - \epsilon_2}{2} \tag{7.49}$$

which may differ from γ_6 of the lamina. This is equivalent to assuming that ϵ_1, ϵ_2 are principal strains, which is the same as neglecting γ_6. Under this condition, the 45° lines in Figure 7.10 are represented by

$$|\epsilon_1 - \epsilon_2| = (1 + \nu_{12}) \max(\epsilon_{1t}, \epsilon_{1c}) \tag{7.50}$$

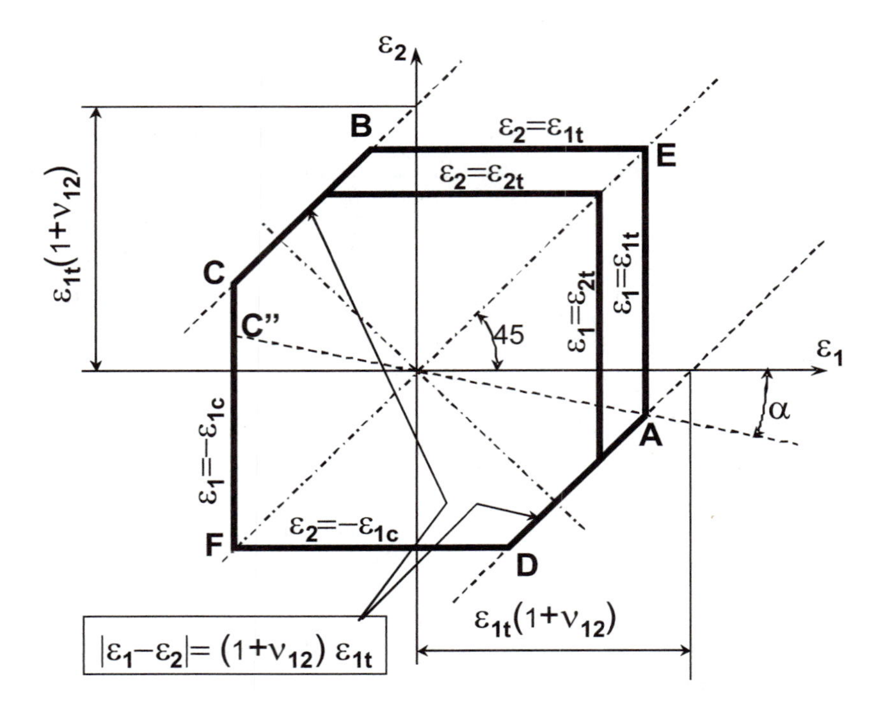

Figure 7.10: Truncated-maximum-strain Criterion.

Note that both fiber shear cutoffs, lines AD and CB are equidistant from the origin, as they should be since shear failure is independent of the sign of the shear stress. Either the maximum of tensile or compressive strain is used to avoid undercutting the true tensile or compressive strain of the unidirectional material, shown in Figure 7.10 as points A and C'', respectively. In Figure 7.10, it is assumed that $\epsilon_{1t} > \epsilon_{1c}$, which is the usual case for polymer matrix composites. The unidirectional tensile or compressive test is represented by the line AC'' which forms an angle α with the fiber strain axis

$$\alpha = \arctan \nu_{12} \tag{7.51}$$

where the Poisson's ratio ν_{12} of a unidirectional lamina must be used. If a fabric reinforced composite (Chapter 9) needs to be analyzed, it must be analytically decomposed into four, symmetrically arranged, unidirectional laminae with 1/4 the thickness each and the same fiber volume fraction of the original lamina in order to compute ν_{12}. The factor $(1 + \nu_{12})$ in (7.50) is necessary because the critical shear strain is defined as one half the diameter of the Mohr circle of strain for the unidirectional test, as in the Tresca criterion [14, p. 211]. Since the unidirectional tensile test, represented by point A, has two strain components, ϵ_{1t} and $\epsilon_2 = -\nu_{12}\epsilon_{1t}$, the critical shear strain of the fiber is then

$$\gamma_f^{CR} = \frac{\epsilon_{1t}(1+\nu_{12})}{2} \tag{7.52}$$

It must be noted that experimental shear strength data (of fiber or composite) is not used to construct the diagram. The only two strength data used in this criterion are the tensile and compressive strain to failure of the unidirectional material.

Additional modes of failure can be added by inserting additional cutoffs. For example, a microcracking cutoff was inserted in Figure 7.10 to limit the transverse strains using the transverse strain to failure ϵ_{2t} of the composite. This may be necessary when the matrix is brittle relative to the fiber ($\epsilon_{mu} < \epsilon_{fu}$, e.g. glass–polyester), when fatigue life is affected by microcracks, or when moisture ingress is accelerated by microcracks. In this case the failure criterion (7.47), (7.48), (7.50), becomes

$$\begin{aligned} -\epsilon_{1c} &< \epsilon_1 < \min(\epsilon_{1t}, \epsilon_{2t}) \\ -\epsilon_{1c} &< \epsilon_2 < \min(\epsilon_{1t}, \epsilon_{2t}) \\ |\epsilon_1 - \epsilon_2| &< (1+\nu_{12}) \max(\epsilon_{1t}, \epsilon_{1c}) \end{aligned} \tag{7.53}$$

In terms of the strength ratio (Section 7.1.1) the *truncated-maximum-strain criterion with matrix cutoff* can be summarized as

$$\begin{aligned} R_1 &= \min(\epsilon_{1t},\, \epsilon_{2t})/\epsilon_1 \text{ if } \epsilon_1 > 0 \\ R_1 &= -\epsilon_{1c}/\epsilon_1 \text{ if } \epsilon_1 < 0 \\ R_2 &= \min(\epsilon_{1t},\, \epsilon_{2t})/\epsilon_2 \text{ if } \epsilon_2 > 0 \\ R_2 &= -\epsilon_{1c}/\epsilon_2 \text{ if } \epsilon_2 < 0 \\ R_6 &= \frac{(1+\nu_{12}) \max(\epsilon_{1t}, \epsilon_{1c})}{|\epsilon_1 - \epsilon_2|} \end{aligned} \tag{7.54}$$

If transverse matrix cracking is not a concern, then ϵ_{2t} is simply not used in the computations. The strength ratio is the minimum of R_1, R_2, and R_6. A value $R_1 < 1$ means tensile failure of the fibers in the current lamina (or transverse matrix cracking in the complementary lamina if $\epsilon_{2t} < \epsilon_{1t}$). A value $R_2 < 1$ means tensile fiber failure in the complementary lamina (or transverse matrix cracking in the current lamina if $\epsilon_{2t} < \epsilon_{1t}$). Finally, a value $R_6 < 1$ means shear failure of the fibers in the current lamina.

Prediction of matrix failure is important to identify undesirable laminate designs; that is laminates with fracture controlled by matrix failure. Matrix cracking of the individual laminae of a PMC laminate does not usually lead to overall rupture of the laminate. If load is primarily carried by matrix, such as a $[0/90]_S$ loaded in shear, matrix cracking leads to collapse. But $[0/90]_S$ is an inefficient laminate for shear and ± 45 laminae should be added. In general, when the laminate is efficiently designed and it contains at least laminae with all three angles (0, 90, and ± 45)

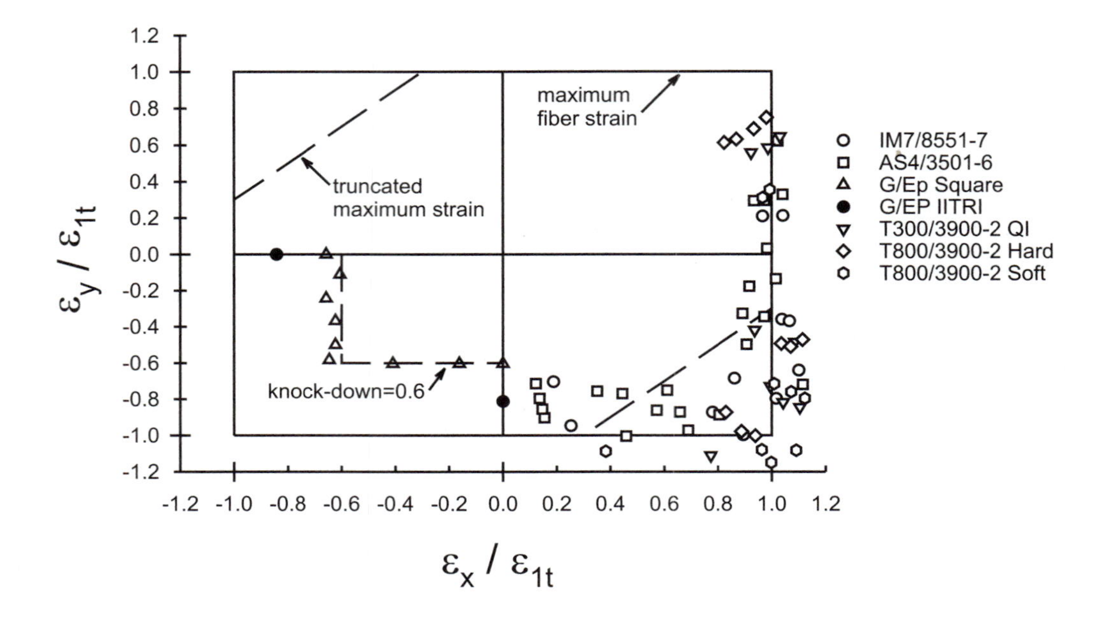

Figure 7.11: Comparison of experimental data with maximum fiber strain and truncated-maximum-strain criterion.

matrix cracking does not affect significantly the static strength of the laminate. A common design rule is to use at least 10% of each 0, 90, and ±45 in any laminate.

Note that the interaction among modes of failure are purposively avoided in the truncated-maximum-strain criterion [188]. This failure criterion is used in the aircraft industry for carbon–epoxy composites for which microcracking seems not to cause significant reductions in the static strength. Furthermore, some degree of approximation can be tolerated because in the design of subsonic aircraft, the operating strain level is further restricted to accommodate holes, impact damage, and repairs. Special care should be taken when the matrix becomes brittle at low temperatures or when curing residual stresses are high because in these cases the fibers may no longer be able to control microcracking. If matrix cracking becomes relevant, additional strain cutoffs are needed and the effect of matrix cracks on the strength and stiffness of the laminate must be assessed.

A comparison is presented on Figure 7.11 between experimental data and two failure criteria discussed in this section, namely the *maximum fiber strain* and the *truncated-maximum-strain criteria*. The experimental data for laminates has been normalized with the strain to failure ϵ_{1t} or ϵ_{1c} depending on the sign of the applied strain.

The data for $[90/\pm 45/0]_S$ IM7–8555-7 carbon–epoxy from [21] was normalized with $\epsilon_{1t} = 0.0130$ and $\epsilon_{1c} = 0.0114$. These values incorporate a knock-down factor of about 20% with respect to the strain to failure of unidirectional material ($\epsilon_{1t} = 0.016$). The data for AS4–3501-6 carbon–epoxy from [21], also $[90/\pm 45/0]_S$, is normalized with $\epsilon_{1t} = 0.0129$ and $\epsilon_{1c} = 0.00772$; which are also lower than typical values (Table 1.4). While the data is reported in [21] as an average stress (over the laminate thickness), it can be easily transformed into strain by using

$$\begin{Bmatrix} \epsilon_x \\ \epsilon_y \end{Bmatrix} = \begin{bmatrix} 1/E_x & -\nu_{xy}/E_x \\ -\nu_{xy}/E_x & 1/E_y \end{bmatrix} \begin{Bmatrix} \sigma_x \\ \sigma_y \end{Bmatrix} \tag{7.55}$$

where E_x, E_y, and ν_{xy} are the laminate moduli that can be computed from lamina properties using (6.35).

For E-glass–epoxy in a $[0/90/0/90/0]_S$ configuration, the biaxial strain to failure data from [189] was normalized with the compressive strain to failure of a $[0_{10}]$ unidirectional sample ($\epsilon_{1c} = 0.02$). A reduction in the laminate strength is also observed here for the cross-ply laminate; about 20% in uniaxial tests with the IITRI fixture [118, ASTM D 3410] and 40% when the biaxial test fixture was used. In the later case, the reduction is due to change into the mode of failure. While the unidirectional material fails by compression crushing of the material, the laminate experiences delamination, which hastens the failure process. Data is also presented for T800–3900-2 carbon–epoxy from [20] for three lay-ups: $[0/\pm 45/90]_S$ (quasi-isotropic; $\epsilon_{1t} = 0.0123$, $\epsilon_{1c} = 0.00846$), $[0_3/\pm 45/90]_S$ (hard; $\epsilon_{1t} = 0.0123$, $\epsilon_{1c} = 0.00801$), and $[0/(\pm 45)_2/90]_S$ (soft; $\epsilon_{1t} = 0.0123$, $\epsilon_{1c} = 0.00755$).

The maximum-fiber-strain criterion is represented in the Figure 7.4 by a square in a solid line. The truncated-maximum-strain criterion coincides with the former except in the tension-compression quadrants, as indicated by the dashed line. A knock-down factor equal to 0.6 is also shown in the compression-compression quadrant to model the glass–epoxy data, as shown in Figure 7.11.

Example 7.10 *Use the truncated-maximum-strain criterion to predict the strength of the* $[0/90/\pm 45]_S$ *laminate analyzed in Example 7.4, p. 231.*

Solution to Example 7.10 *The analysis of this laminate reveals that the maximum normal strains occur in the* ± 45 *laminae, with values*

$$\epsilon_1 = -\epsilon_2 = 0.00286$$

using (7.8) to estimate the values of strain to failure, results in

$$\epsilon_{1t} = \frac{1.826}{156.4} = 0.011675$$
$$\epsilon_{1c} = \frac{1.134}{156.4} = 0.007251$$

Then using (7.54) (without matrix cutoff), the strength ratios are

$$R_2 = \frac{0.007251}{0.00286} = 2.5$$
$$R_6 = \frac{(1 + 0.352)0.011675}{0.00286 + 0.00286} = 2.76$$

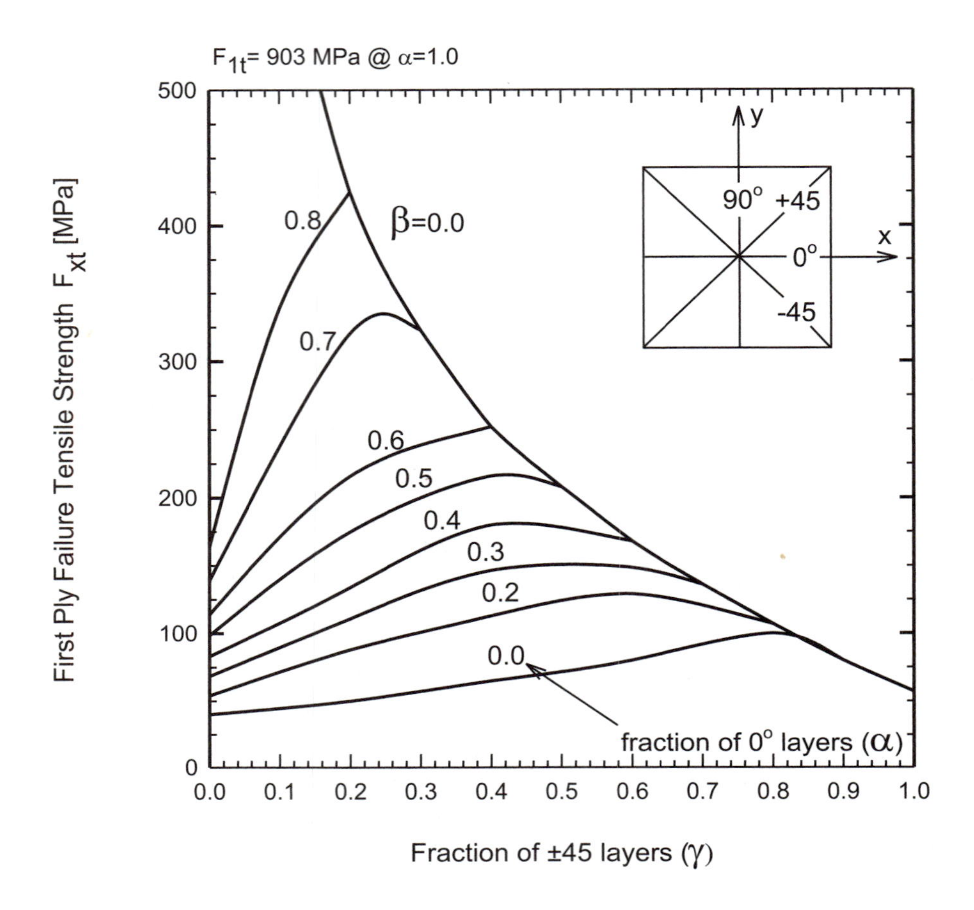

Figure 7.12: Carpet plot of FPF tensile strength F_{xt} (loading: N_x).

resulting in a strength ratio to failure $R = 2.5$. Note that this value is very close to the LPF value computed in Example 7.4 ($R_{LPF} = 2.33$). Also note that ϵ_{1c} was used to compute R_1 because one of the laminae in the ± 45 pair is in compression and $\epsilon_{1c} < \epsilon_{1t}$ for the material being used.

7.4 Strength Design Using Carpet Plots

The first ply failure (FPF) strength can be estimated with the procedure presented in Section 7.2. Last ply failure (LPF) strength can be estimated using the ply discount method (Section 7.3). Both FPF and LPF of a general laminate subjected to general load can be predicted accurately using the methodology discussed in Chapter 8. Although the required computations can be easily performed using available computer programs [1], a graphical representation of the strength values for a family of laminates is useful for preliminary design. Carpet plots were introduced in Section 6.5 to represent the laminate moduli of a family of laminates described by

$$[0_\alpha/90_\beta/(\pm 45)_\gamma]_S \tag{7.56}$$

where

$$\begin{aligned} \alpha &= t_0/t \\ \beta &= t_{90}/t \\ \gamma &= t_{\pm 45}/t \\ \alpha + \beta + \gamma &= 1 \end{aligned} \tag{7.57}$$

represent the ratio of the thickness of each fiber orientation to the total thickness t. In carpet plots, data is presented as a function of one variable, using the other two variables as parameters, and taking into account that $\alpha + \beta + \gamma = 1$.

The load per unit length N_x that causes failure on a laminate, while all other loads are zero, is used to construct Figure 7.12 for FPF and Figure 7.13 for LPF. The failure loads were predicted using discrete damage mechanics (DDM, Chapter 8) for E-glass–polyester (Table 1.3) with ply thickness $t_k = 0.125$ mm. To achieve the necessary values of α, β, γ, the LSS chosen are variants of $[45_r/-45_r/0_p/90_q/45_s/-45_s]_S$ with $2r + 2s + p + q = 20$ and $r \geq 0, s \geq 0, p \geq 0, q \geq 0$ chosen appropriately. Notice that this type of LSS is not optimal because clusters of laminae with the same angle are formed when any of the subscripts r, s, p, q takes a value larger than one. This is usually undesirable because it reduces the in-situ strength (see Section 7.2.1) and leads to higher free-edge stresses (see [2, Chapter 5]).

The FPF strength F_{xt} is obtained by dividing the applied N_x by the thickness of the laminate, i.e., $F_{xt} = N_x/t$. The value reported is an average of the actual stress values in the various laminae that form the laminate and it does not necessarily coincide with the stress at failure on the first ply that fails. The laminate strength represents the tensile strength along the x-direction of an equivalent, homogeneous, orthotropic plate that behaves similarly to the actual laminate. The laminate strength is useful in design because it allows for a quick estimate of the thickness required to carry a given load.

The *strain* at FPF can be easily calculated from the FPF strength because the material is linear elastic until FPF. After FPF, the material behavior is nonlinear and the strain to failure can be estimated only by tracking the progressive damage of the laminae as discussed in Chapter 8.

The in-plane strength of a balanced symmetric laminate is not independent of the stacking sequence because cracking on a lamina is affected by its thickness and position in the LSS. For example, laminae located on the surface and straddling the midplane have lower FPF that similar laminae located elsewhere in the LSS (see Figure 7.9). For additional insight into this issue, see Section 7.2.1 and perform simulations using the methodology presented in Chapter 8. However, as a first

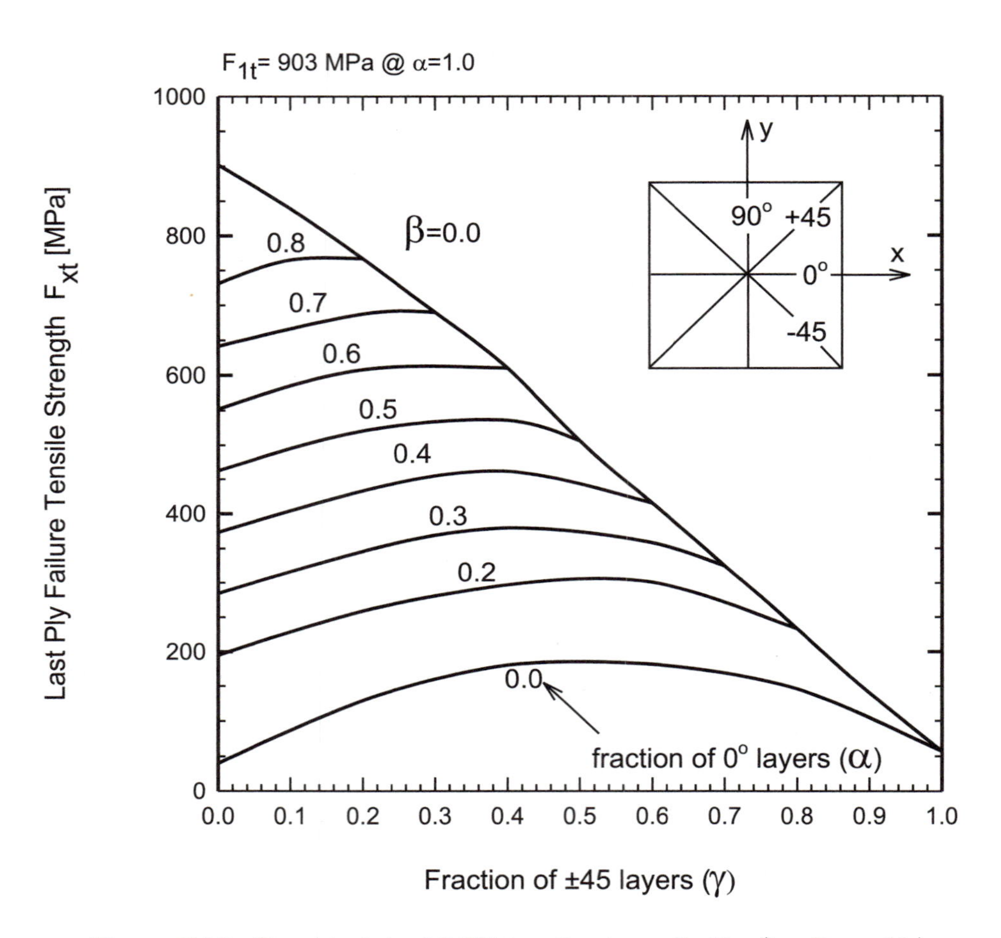

Figure 7.13: Carpet plot of LPF tensile strength F_{xt} (loading: N_x).

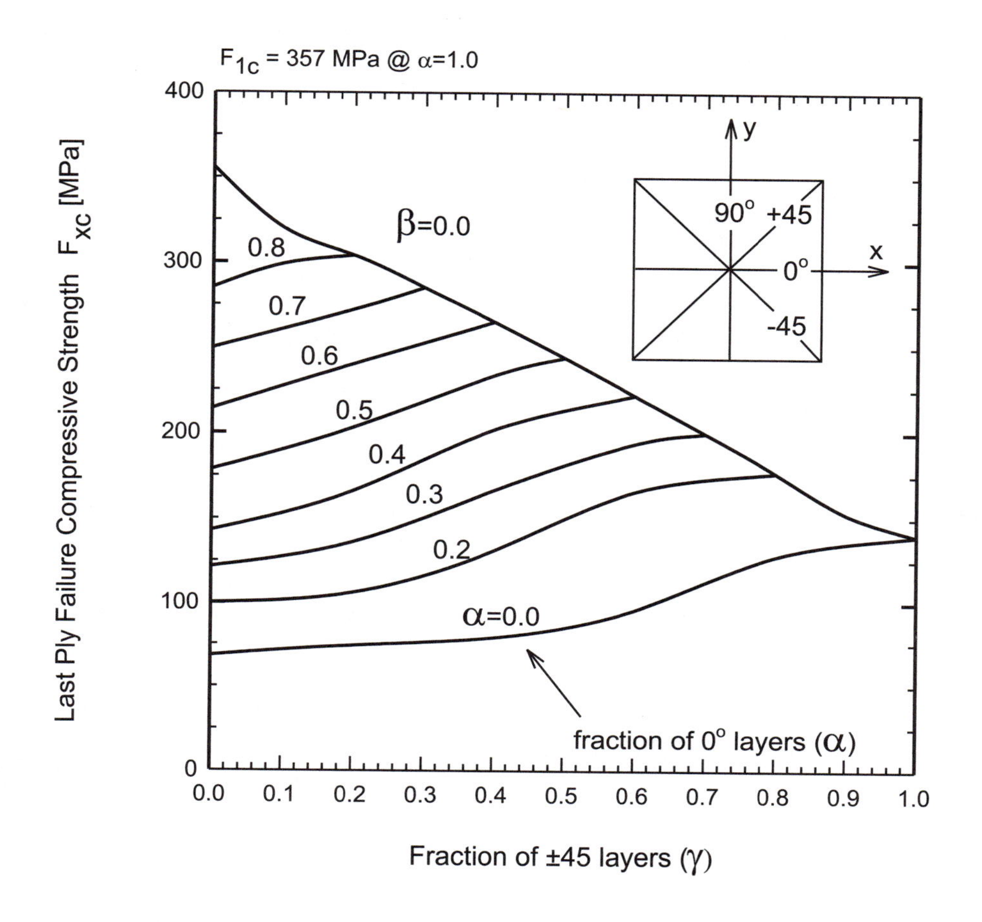

Figure 7.14: Carpet plot of LPF compressive strength F_{xc} (loading: $-N_x$).

approximation, values of transverse strength F_{yt} can be read from the carpet plots for F_{xt} by interchanging the values α and β.

Carpet plots for *compressive* strength F_{xc} are constructed by applying a compressive load $-N_x$ (Figure 7.14). Carpet plots for shear strength F_{xy} are constructed by applying a shear load N_{xy} (Figure 7.15). Additional carpet plots are available in the Web site [1].

Using carpet plots, the design proceeds as follows. Given the loads N_x, N_y, N_{xy}, the laminate must satisfy the following (see also (1.13)),

$$
\begin{aligned}
\phi F_x &> \alpha N_x/t \\
\phi F_y &> \alpha N_y/t \\
\phi F_{xy} &> \alpha N_{xy}/t
\end{aligned}
\tag{7.58}
$$

where α is the load factor (1.17), ϕ is the resistance factor (1.16), and F_x, F_y, and F_{xy} are the laminate strength values obtained from the carpet plots (Figures 7.12–7.15).

The laminate strengths F_x, F_y, and F_{xy} depend on the relative amounts of 0-degree, 90-degree, and $\pm\theta$ laminae, which are defined by the ratios α, β, and γ. The strengths F_x and F_y represent either the tensile or compressive strength of the laminate, that is F_{xt} or F_{xc} and F_{yt} or F_{yc}. At this point, the designer selects α, β, and γ taking into account the magnitude of the loads N_x, N_y, and N_{xy}. Larger values of α could be selected if N_x is the main load, larger values of β would be required if N_y is the main load, and larger values of γ would be used to resist shear load N_{xy}. Then, one of (7.58) is used to estimate the thickness and the other two are used to check the design.

Usually, the largest component of the load is used to find the thickness and the other two components are used to check the design. The design may be unsatisfactory either because some of (7.58) are not satisfied or because some of (7.58) indicate an overdesign. If the design is not satisfactory, different values of α, β, and γ, are selected trying to obtain similar values of thickness with all three of (7.58).

Sometimes the resistance factor ϕ is set to one because it has been included in the material properties, which are then given as allowable values. These are values lower than the actual strength of the material, which have been reduced to incorporate resistance factors into the various strength values. Allowable values are used when it is desired to use different reduction factors for various strength components. Load factors α are introduced in the process of arriving at the applied loads (N_x, M_x, etc.) The particular design process largely depends on the field of application (Civil, Aerospace, Automotive, etc.) but in any case (7.58) must be satisfied.

Example 7.11 *Design a cylindrical pressure vessel with diameter $d = 4\ m$, internal pressure $p = 3.0\ MPa$, with a COV $C_p = 0.10$ to achieve a reliability $R_e = 0.95$ with respect to the burst pressure. Use a coordinate system on the surface of the cylinder with the y-axis oriented in the hoop direction and the x-axis oriented along the length. Assume that the strength reported in Figures 7.12–7.15 has a variability represented by COV $C_F = 0.12$.*

Solution to Example 7.11 *For $R_e = 0.95$, from Table (1.1), the standard variable is $z = -1.645$. Then, the load and resistance factors are calculated (1.16)–(1.17) as*

$$\alpha = 1 - zC_p = 1 + 1.645\ (10/100) = 1.1645 \approx 1.2$$
$$\phi = 1 + zC_F = 1 - 1.645\ (12/100) = 0.8026 \approx 0.8$$

The safety factor can be calculated using either (1.12) or (1.19)

$$R = \frac{1 \pm \sqrt{1 - (1 - 1.645^2\ 0.12^2)(1 - 1.645^2\ 0.10^2)}}{1 - 1.645^2\ 0.12^2} = 1.3$$
$$R_{LRFD} = \frac{1.1645}{0.8026} = 1.45 \approx 1.5$$

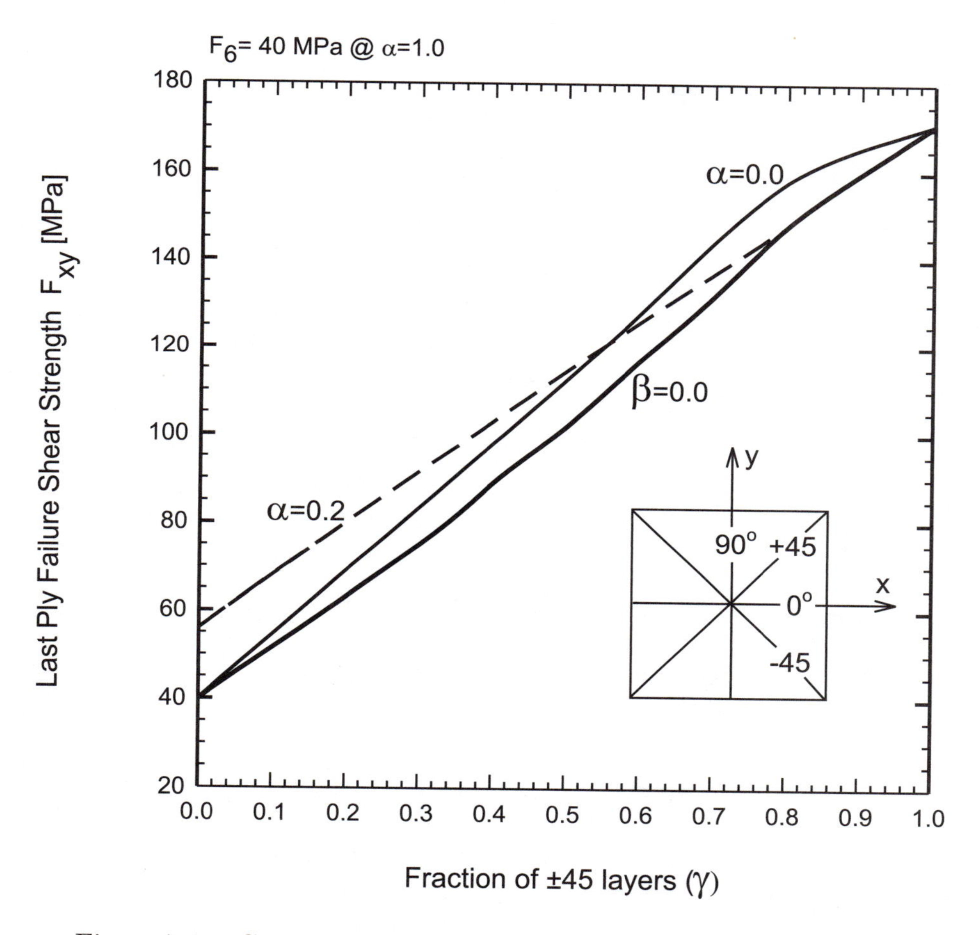

Figure 7.15: Carpet plot of LPF *shear* strength F_{xy} (loading: N_{xy}).

where it can be observed that separating the safety factor into load and resistance factors is conservative.

The loads per unit length, in the hoop and longitudinal directions, are easily computed in such a way as to equilibrate the internal pressure "p"

$$N_x = \frac{\alpha\ p\ d}{4} = \frac{1.2\ (3.0)\ 4.0}{4} = 3.6\ MN/m$$
$$N_y = \frac{\alpha\ p\ d}{2} = \frac{1.2\ (3.0)\ 4.0}{2} = 7.2\ MN/m$$

Therefore, the required thickness is the largest of

$$t = \frac{N_x}{\phi\ F_x}$$

or

$$t = \frac{N_y}{\phi\ F_y}$$

Since the design is based on burst pressure, the laminate strength values F_x and F_y to be used correspond to last ply failure (LPF). A value of $\gamma = 0.1$ is selected because no shear load is present. Since $N_x = 2N_y$, $\alpha = 0.3$ and $\beta = 0.6$ will provide approximately twice the strength in the hoop direction. In Figure 7.13, for $\alpha = 0.3$, $F_x = 320\ MPa$. From the same plot but interchanging the values β and α, results in $F_y = 670\ MPa$. Finally

$$t = \frac{3.6}{(0.8)\ 320} = 14.0\ mm$$
$$t = \frac{7.2}{(0.8)\ 670} = 13.5\ mm$$

An E-glass–polyester laminate with $V_f = 0.5$, $t = 14\ mm$, $\alpha = 0.3$, $\beta = 0.6$, and $\gamma = 0.1$ satisfies the requirements.

7.5 Stress Concentrations (*)[11]

In the stress and failure analysis carried out so far, it was assumed that the laminate had no irregularities such as abrupt changes in thickness, holes, notches, etc. Any irregularity in a laminate changes the stress values, usually raising them above the values computed far from the irregularity. This problem is accounted for in design by using stress concentration factors (SCF).

The theoretical SCF relates the maximum stress at the discontinuity (notch or hole) to the nominal stress σ_n computed assuming the discontinuity is not there (Figure 7.16)

$$K_t = \frac{\sigma_{\max}}{\sigma_n} \tag{7.59}$$

[11]Sections marked with (*) can be omitted during the first reading in an introductory course but are recommended for further study and reference.

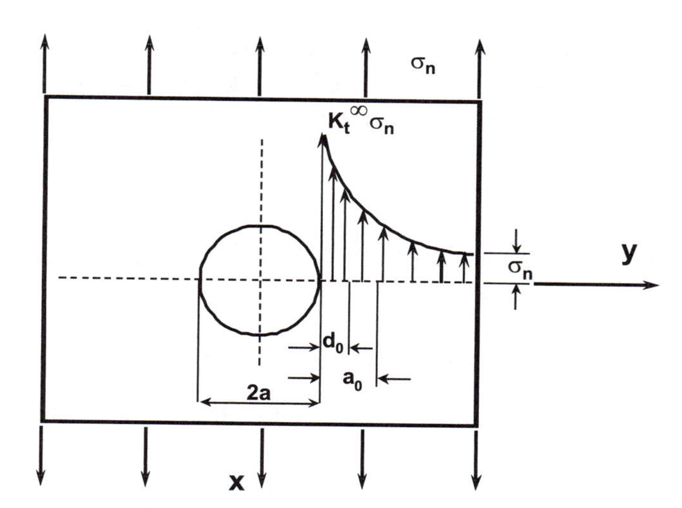

Figure 7.16: Distribution of stress near a hole in an infinite plate.

A brittle material will fail when the maximum stress $\sigma_{\max}$ reaches the strength of the material. For example, a notched unidirectional composite loaded transversely to the fibers, most likely will fail when the nominal stress reaches the value of the transverse tensile strength divided by the theoretical stress concentration factor

$$\sigma_{n,fail} = \frac{F_{2t}}{K_t} \tag{7.60}$$

In this case the material is said to be fully sensitive to notches. Multilayered composites, such as quasi-isotropic laminates, are able to withstand higher nominal stresses than that predicted by (7.60). This is explained using the concept of notch sensitivity, which is also used in metal design. The high values of stress $\sigma_{\max}$ predicted by (7.59) and illustrated in Figure 7.16 cause damage, early on during loading, in the vicinity of the discontinuity. The damaged material has lower stiffness, which contributes to a reduction of the maximum stress. Although stress concentration still occurs, its value is lower. The effective stress concentration factor K_e is lower than K_t because of a reduced sensitivity of the material to notches. The notch sensitivity is defined as

$$q = \frac{K_e - 1}{K_t - 1} \tag{7.61}$$

The notch sensitivity ranges from zero to one. If $q = 0$, then $K_e = 1$ and the material has no sensitivity to notches. If $q = 1$, then $K_e = K_t$ and the material is fully sensitive to notches (or brittle). The notch sensitivity of composites depends on the notch size as well as the material and laminate type.

Theoretical stress concentration factors K_t for laminated composites can be found in [190–192]. Then, the effective stress concentration factor is computed as

$$K_e = 1 + q(K_t - 1) \tag{7.62}$$

and failure is predicted when the nominal stress (load over unnotched area) reaches the strength of the unnotched material F_0 divided by the effective stress concentration factor K_e. The nominal stress at failure

$$F_N = \frac{F_0}{K_e} \tag{7.63}$$

is called *notched strength* because it represents the apparent strength of the notched material.

7.5.1 Notched Plate Under In-plane Load

Under in-plane loads only, a laminate can be modeled as an orthotropic plate if the following conditions are satisfied

$$A_{16} = A_{26} = B_{ij} = 0 \tag{7.64}$$

Stress concentration factors (SCF) for orthotropic plates can be found in [190–192]. For an infinitely wide orthotropic plate loaded in the x-direction (Figure 7.16) the SCF is

$$K_t^{\infty} = 1 + \sqrt{\frac{2}{A_{22}}\left(\sqrt{A_{11}A_{22}} - A_{12} + \frac{A_{11}A_{22} - A_{12}^2}{2A_{66}}\right)} \tag{7.65}$$

where A_{ij} are terms of the $[A]$ matrix for the laminate (see Section 6.4). In terms of equivalent laminate moduli (6.35)

$$K_t^{\infty} = 1 + \sqrt{2\left(\sqrt{\frac{E_x}{E_y}} - \nu_{xy} + \frac{E_x}{2G_{xy}}\right)} \tag{7.66}$$

Equation 7.65 is valid for balanced symmetric laminates with uniaxial in-plane loading. If bending or shear is introduced, or the laminate is not balanced (i.e., A_{16} and A_{26} are not zero) or the laminate is unsymmetrical, the $[B]$ and $[D]$ matrices (see Section 6.1.4) will have to be introduced and the theory presented next is no longer valid.

Equation 7.65 does not predict any influence of the hole diameter on the SCF. However, experimental data indicates that the strength reduces with increasing hole diameter. This is attributed to the fact that a damage zone is created near the edge of the hole. When the hole is small, the damage zone is small and no significant reduction of strength occurs. On the other hand, a large damage zone appears near a large hole, weakening the material.

The stress distribution shown in Figure 7.16 can be approximated as

$$\sigma_x \approx \left[1 + \frac{1}{2}\xi^2 + \frac{3}{2}\xi^4 - \frac{(K_t^{\infty} - 3)}{2}\left(5\xi^6 - 7\xi^8\right)\right]\sigma_n f_w \tag{7.67}$$

with $\xi = a/y$ where a is the radius of the hole, y is the distance from the center of the hole (see Figure 7.16), K_t^{∞} is given by (7.65), and f_w is a correction factor used when the plate is of finite width. This formula closely matches the exact solution with errors of less than ten percent in most cases [193–195]. When $2a/W \leq 1/3$ (Figure 7.17) the finite width correction factor can be approximated as [190, 196]

$$f_w = \frac{2 + (1 - 2a/W)^3}{3(1 - 2a/W)} \tag{7.68}$$

where a and W are the radius of the hole and width of the plate, respectively.

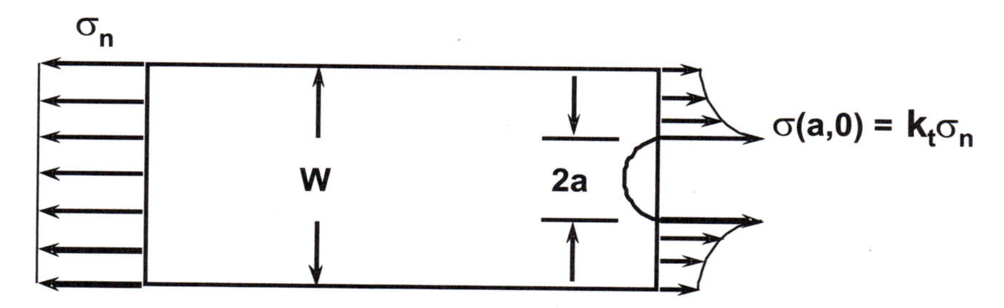

Figure 7.17: Notched plate of finite width showing the actual stress concentration.

According to the point stress criterion [194, 195], failure occurs when the stress predicted with (7.67) at a distance d_0 from the edge of the discontinuity is equal to the unnotched strength F_0. That is $\sigma_y = F_0$ at $y = a + d_0$, or

$$\left[1 + \frac{1}{2}\xi^2 + \frac{3}{2}\xi^4 - \frac{(K_t^\infty - 3)}{2}(5\xi^6 - 7\xi^8)\right] f_w F_N = F_0 \tag{7.69}$$

with $\xi = a/(a + d_0)$ where σ_n in (7.67) has been replaced by F_N to indicate the nominal stress at failure, also called *notched strength.* Using (7.63), the effective stress concentration factor is

$$K_e = \frac{F_0}{F_N} = f_w \left[1 + \frac{1}{2}\xi^2 + \frac{3}{2}\xi^4 - \frac{(K_t^\infty - 3)}{2}(5\xi^6 - 7\xi^8)\right] \tag{7.70}$$

with $\xi = a/(a + d_0)$ where a is the radius of the hole, a_0 is the characteristic length, and K_t^∞ is given by (7.65).

The characteristic distance d_0 is considered a material property for a given laminate, independent of the size of the hole. Finding the characteristic length requires a comparison with experimental data for the laminate in question, but generally choosing d_0 to be between 1 and 2 mm seems to give good results [194, 195, 197]. The square bracket in (7.70) is the effective stress concentration factor of an infinitely wide plate with a hole K_e^∞. The factor f_w is a reduction factor to account for the finite width of the plate

$$K_e = f_w K_e^\infty \tag{7.71}$$

The effective stress concentration factor can be obtained directly from (7.70), or using the notch sensitivity plot (Figure 7.18) as follows. First, compute the

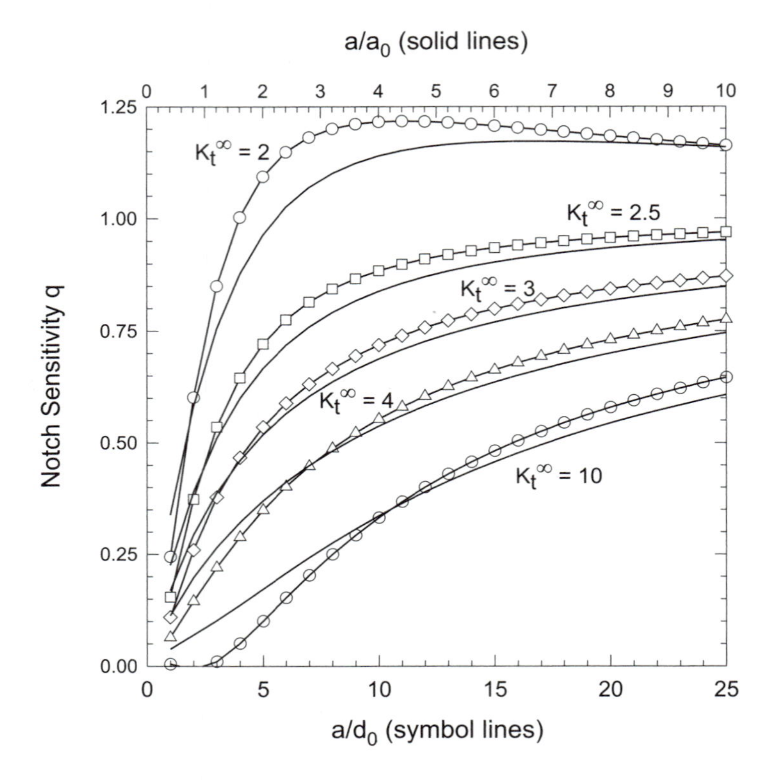

Figure 7.18: Notch sensitivity as a function of hole radius a. Lines with symbols correspond to the bottom axis and lines without symbols correspond to the top axis.

theoretical SCF using (7.65) or (7.66). Then, obtain the notch sensitivity from Figure 7.18 and the width correction factor f_w from (7.68). Finally,

$$K_e = f_w[1 + q(K_t - 1)] \tag{7.72}$$

Another interpretation of failure is given by the average strength criterion [194, 195]. According to this criterion, failure occurs when the average stress over a region of length a_0 measured from the edge of the hole equals the unnotched strength of the laminate F_0, or

$$\frac{1}{a_0} \int_a^{a+a_0} \sigma_x dy = F_0 \tag{7.73}$$

where F_0 is the unnotched strength (For a unidirectional lamina oriented along the y-direction $F_0 = F_{2t}$). Using (7.63) and 7.67

$$K_e = \frac{F_0}{F_N} = f_w \left[\frac{2 - \xi^2 - \xi^4 + (K_t^\infty - 3)\left(\xi^6 - \xi^8\right)}{2\left(1 - \xi\right)} \right] \tag{7.74}$$

with $\xi = a/(a + a_0)$.

The characteristic distance a_0 is an empirical constant. Values of a_0 between 2.5 and 5 mm seem to produce approximations of notched strength in good agreement with experimental data [195, 197]. While different material systems should be checked before accepting these values as universal constants, these values can be used as a first approximation during preliminary design. For very large holes, ξ approaches the value $\xi = 1$ and K_e approaches K_t. For very small holes, as $a \to 0$ and $K_e \to 1$, meaning that very small holes do not produce significant reductions in strength.

On a riveted joint, the notch strength is compared to the applied stress computed without reducing the area due to the holes and taking W as the hole pitch. The reduction of area is taken into account by the correction factor f_w.

The point stress criterion and average stress criterion can be used to develop expressions for the notched strength for other discontinuity besides circular holes, as long as an analytical solution, similar to (7.67) exists for the stresses near the discontinuity. For a crack of length $2a$ in an infinitely wide plate subject to stress perpendicular to the crack (Figure 7.19), the average stress criterion yields [194, 195]

$$\frac{F_N}{F_0} = \sqrt{\frac{a_0}{a_0 + 2a}} \tag{7.75}$$

and the point stress criterion

$$\frac{F_N}{F_0} = \sqrt{1 - \left(\frac{a}{a + d_0}\right)^2} \tag{7.76}$$

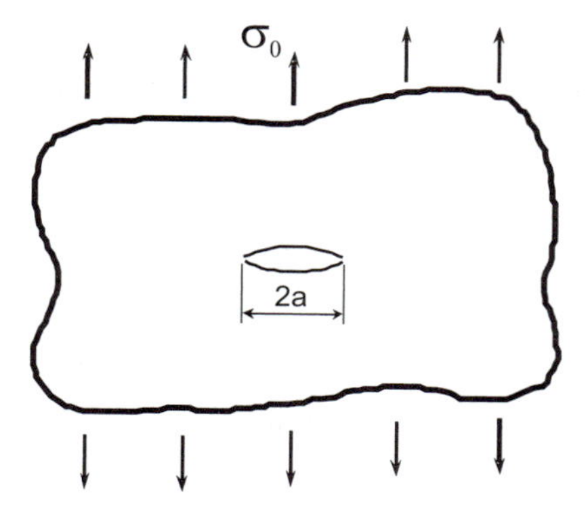

Figure 7.19: A crack in an infinite plate subjected to uniform stress far from the crack.

Sharp cracks are better analyzed using fracture mechanics [30, 198]. In that context, the energy release rate G_I is computed for a given crack in the structure being analyzed and its value is compared to the fracture toughness of the material G_{Ic}. For a crack of length $2a$ in an infinitely wide plate, the average stress criterion [194, 195] leads to a fracture toughness (in N/m)

$$G_{Ic} = \frac{F_0^2}{E'} \left[\frac{\pi a a_0}{a_0 + 2a} \right] \tag{7.77}$$

and the point stress criterion yields

$$G_{Ic} = \frac{F_0^2}{E'} \pi a \left[1 - \left(\frac{a}{a + d_0} \right)^2 \right] \tag{7.78}$$

where E' is a suitable value of laminate stiffness, which is necessary to transform from the traditional definition of fracture toughness K_{Ic} (in $N\sqrt{m}/m^2$) to the fracture toughness G_{Ic} (in $N/m = J/m^2$) used for the analysis composite materials, using the equation $G_{Ic} = K_{Ic}^2/E'$ when the laminate is in a state of plane stress.

For long cracks, the above formulas approach the values $G_{Ic} = F_0^2(\pi a_0/2)/E'$ and $G_{Ic} = F_0^2(2\pi d_0)/E'$, respectively, which can be used as a conservative estimate for all crack lengths. However, it must be noted that these approximations of fracture toughness apply only to a crack in a large plate (infinite in theory) that is subject to a uniform state of normal stress perpendicular to the crack.

Example 7.12 *Estimate the strength of a plate as shown in Figure 7.17, with* $W = 30\ mm$ *and* $a = 5\ mm$. *The laminate is* $[0/\pm 45/90]_{2s}$, *made of graphite–epoxy T300–N5208. The equivalent in-plane moduli of this quasi-isotropic laminate are* $E = 69.68\ GPa$, $G = 26.88\ GPa$, *and* $\nu = 0.3$. *The unnotched tensile strength of the laminate is* $F_0 = 494\ MPa$.

Solution to Example 7.12 *The theoretical SCF of the infinite plate is given by (7.66). Since the laminate is quasi-isotropic* $E_x = E_y = E$, $G_{xy} = G$, $\nu_{xy} = \nu$. *Therefore*

$$K_t^\infty = 1 + \sqrt{2\left(1 - 0.3 + \frac{69.68}{2(26.88)}\right)} \simeq 3$$

Selecting $d_0 = 1\ mm$, $a/d_0 = 5$, *and entering Figure 7.18 through the bottom of the graph, read* $q = 0.53$ *on the curve corresponding to* $K_t^\infty = 3$ *(labeled with diamonds). Next, use (7.62) to find the effective SCF of the infinite plate*

$$K_e^\infty = 1 + 0.53(3 - 1) = 2.06$$

where the superscript ∞ *has been added to indicate that* K_e^∞ *corresponds to the infinitely large plate. Since the plate in Figure 7.18 has finite width, the correct factor is computed with (7.68)*

$$f_w = \frac{2 + [1 - 2(5/30)]^3}{3[1 - 2(5/30)]} = 1.148$$

Therefore, the effective SCF of the finite width plate is

$$K_e = f_w K_e^\infty = 1.148(2.06) = 2.36$$

Finally, the strength of the notched plate is

$$F_N = \frac{F_0}{K_e} = \frac{494}{2.36} = 209\ MPa$$

In other words, the plate is predicted to fail when the stress, computed assuming the hole is not there, reaches 209 MPa. *Note that the result depends on the assumed value of* d_0. *Using* a_0 *instead of* d_0 *leads to slightly different results. Taking* $a_0 = 3.8$ *mm*, $a/a_0 = 1.32$, *and entering Figure 7.18 through the top of the graph, a value* $q = 0.42$ *is obtained. The rest of the procedure is identical, and the estimated notched strength is* $F_N = 234\ MPa$. *The two extreme values usually recommended for* a_0, *namely* 2.5 *and* 5 *mm lead to values of notched strength of* 211 MPa *and* 256 MPa, *respectively.*

Exercises

Exercise 7.1 *A laminate of undisclosed configuration subjected to undisclosed load has a reported first ply failure (FPF) strength ratio* $R = 0.5$. *Note: the material, laminate stacking sequence, and the loads are not known.*

(a) *How much does the load need to be increased/decreased to avoid first ply failure, assuming all components are increased/decreased proportionally?*

(b) *or, How much does the thickness of the laminate need to be increased/decreased to avoid first ply failure, assuming all laminae thicknesses are increased/decreased proportionally?*

(c) *or, How much does the strength of the material need to be increased/decreased to avoid first ply failure, assuming all strength components are increased/decreased proportionally?*

Exercise 7.2 *Consider a laminate* $[\pm 45]$ *of the following material:*

$$E_1 = 137.8\ GPa \quad E_2 = 10\ GPa \quad G_{12} = 5.236\ GPa \quad \nu_{12} = 0.3$$
$$F_{1t} = 826.8\ MPa \quad F_{1c} = 482.3\ MPa \quad F_{2t} = F_{2c} = 68.9\ MPa \quad F_6 = 241.15\ MPa$$

and thickness $t_k = 0.127$ *mm, subject to* $N_x = N_y = 1.0\ N/mm$. *Determine the strength ratio* R *at the bottom surface of the laminate, using the following strength criteria: (a) Maximum stress, (b) Maximum strain, (c) Interacting failure criterion. Do not adjust for in-situ strength. You are allowed to use a computer program to find the stresses but you must present detailed computations of the* R *values.*

Exercise 7.3 *Recompute Exercise 7.2 adjusting for in-situ strength, assuming the transition thickness for this material is* $t_t = 0.6$ *mm.*

Exercise 7.4 *Consider a* $[0/45/-45]_S$ *laminate, subject to a load* $N_x = N_y = 1000\ N/mm$. *Compute the [A], and [D] matrices (neglect shear deformations) for the intact laminate. Each lamina is* 2.5 *mm thick. Do not adjust for in-situ strength. The material properties are:*

$$E_1 = 204.6\ GPa \quad E_2 = 20.46\ GPa \quad G_{12} = 6.89\ MPa \quad \nu_{12} = 0.3$$
$$F_{1t} = 826.8\ MPa \quad F_{1c} = 482.3\ MPa \quad F_{2t} = F_{2c} = 68.9\ MPa \quad F_6 = 241.15\ MPa$$

Exercise 7.5 *Using the results for the intact laminate in Exercise 7.4, compute: (a) the strains* $\epsilon_x, \epsilon_y, \gamma_{xy}$ *at the interface between laminae, (b) the stresses* $\sigma_x, \sigma_y, \tau_{xy}$ *on all laminae, right next to the interface, (c) the strains* $\epsilon_1, \epsilon_2, \gamma_6$ *by using the transformation equations, (d) the stresses in material coordinates* $\sigma_1, \sigma_2, \sigma_6$.

Exercise 7.6 *Using the stresses found in Exercise 7.5, evaluate the strength ratio* R_{FPF} *on top and bottom of each lamina using the following failure criteria: (a) Interacting criterion, (b) Maximum strain criterion, and (c) Maximum stress criterion. In each case, the minimum of the values of R computed through the thickness corresponds to the first ply failure (FPF) strength ratio. Note that the FPF values computed with the various criteria are, in general, different from each other. Only experience or multiaxial strength test data can tell which criterion is the best for a particular material and stress state.*

Exercise 7.7 *Recompute Exercise 7.6 for loading beyond first ply failure (FPF). Simulate the damaged material by using a damage factor* $d = 0.999$ *(ply discount). The minimum value of strength ratio R thus obtained corresponds to the last ply failure (LPF), i.e.,* R_{LPF}.

Exercise 7.8 *Consider a* $[\pm 45]$ *laminate with the same material properties of Exercise 7.4. Determine the value of the forces* $N_x, N_y, N_{xy}, M_x, M_y, M_{xy}$ *required to produce a curvature* $\kappa_x = 0.00545\ mm^{-1}$, $\kappa_y = -0.00486\ mm^{-1}$. *Neglect small numbers and shear deformation.*

Exercise 7.9 *If all the forces determined in Exercise 7.8 are increased proportionally, use the interacting criterion to determine (a) the first ply failure load, and (b) the last ply failure load assuming* $d = 0.999$.

Exercise 7.10 *Consider a quasi-isotropic laminate* $[0/90/\pm 45]_S$, *8 mm thick, made of carbon–epoxy AS4–3501-6. Use the truncated-maximum-strain failure criterion without matrix cutoff. Estimate the strains-to-failure using the data in Table 1.3–1.4 and equation (7.8). Explain the mode of failure under the following loading conditions:*

1. $N_x = 1000$ *N/mm all other zero.*
2. $N_x = -1000$ *N/mm all other zero. Why is the strength ratio different from case (a)?*
3. $N_{xy} = 1000$ *N/mm all other zero.*
4. $N_{xy} = -1000$ *N/mm all other zero. Why is this different from case (c)?*

Exercise 7.11 *Repeat Exercise 7.10 using a matrix cutoff. Estimate the strains-to-failure using the data in Table 1.3–1.4 and equation (7.8)*

Exercise 7.12 *Use the material properties described in Exercise 4.10 (p. 141) and Example 7.1 (p. 223) to verify that the experimental values of strain to failure (listed below) are higher than those computed assuming linear elastic behavior (7.8). Comment your findings.*

$$\epsilon_{1t} = 11,900(10^{-6}),\ \epsilon_{1c} = 8,180(10^{-6})$$
$$\epsilon_{2t} = 2,480(10^{-6}),\ \epsilon_{2c} = 22,100(10^{-6})$$
$$\gamma_{6u} = 30,000(10^{-6})$$

Exercise 7.13 *Recompute Example 7.11 (p. 256) when the lamina is subjected to* $\sigma_2 = -90\ MPa$, $\sigma_6 = 90\ MPa$.

Exercise 7.14 *Show analytically that* $(f_{11}\sigma_1^2)R_1^2 + (f_1\sigma_1)R_1 - 1 = 0$, *with* $f_{11} = 1/(F_{1t}F_{1c})$ *and* $f_1 = 1/F_{1t} - 1/F_{1c}$ *is equivalent to (7.18). Specifically, show that when* $\sigma_1 > 0$, *the equations above reduce to* $R_1 = F_{1t}/\sigma_1$ *and when* $\sigma_1 < 0$ *it reduces to* $R_1 = -F_{1c}/\sigma_1$. *Remember that* F_{1t} *and* F_{1c} *are both positive.*

Exercise 7.15 *Recompute Example 7.7 (p. 242) using a* $[(\pm 55)_n]$ *laminate and ply thickness* $t_k = 0.127\ mm$.

Chapter 8

Damage

As we have seen in Chapter 7, first ply failure (FPF) is the load at which the first crack appears in a lamina, and last ply failure (LPF) is an approximation to the load for which the laminate is no longer able to sustain any load. In Chapter 7, predictions of FPF and LPF are done with *strength failure criteria* (Sections 7.1, 7.2, and 7.3). Strength failure criteria have at least three limitations, as follows:

First, some of the strength values used in strength failure criteria need to be adjusted for ply thickness and laminate stacking sequence (LSS). While the thickness dependency can be accounted for by using in-situ values (Section 7.2.1), the dependency on LSS is not easily accounted for. Even then, the criteria predict the appearance of the first matrix crack, not the saturation of matrix cracks that corresponds to total degradation of a lamina.

Second, strength failure criteria provide no indication about what happens after the first crack. For a unidirectional lamina (UD) lamina, the first crack coincides with the collapse of the lamina because there is nothing to prevent the lamina from fracturing into two pieces. Once the lamina is inside a laminate, the rest of the laminate will likely hold the lamina together. Furthermore, cracks cause reduction of stiffness of the lamina, which lead to stress redistribution to other laminae that eventually cause LPF. Strength failure criteria provide no prediction of this gradual process of damage accumulation, stiffness reduction, and stress redistribution.

Finally, equating laminate strength to the load at which the first fiber failure is detected in one lamina is only an approximation. Therefore, an alternative to strength failure criteria is developed in this chapter, inspired from continuum damage mechanics [2, Chapter 8].

8.1 Continuum Damage Mechanics

The applied (or apparent) stress σ is the stress calculated as load per unit area $\widetilde{A}$, which is the cross-sectional area known to the designer (Figure 8.1). The effective stress $\widetilde{\sigma}$ is the stress acting over the undamaged area of the material A. The undamaged (remaining) area is $A = (1 - D)\widetilde{A}$, were $0 \leq D < 1$ is the damage variable. Therefore, when the material is damaged, the effective stress is higher than the

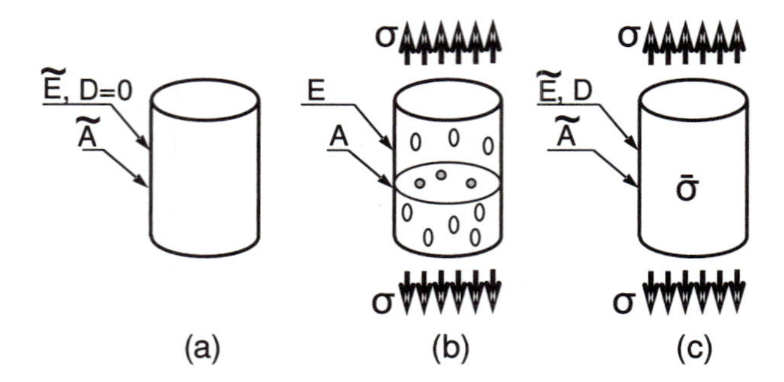

Figure 8.1: (a) Unstressed material, (b) stressed material, (c) effective configuration [199].

apparent stress as follows

$$\widetilde{\sigma} = \frac{\sigma}{(1-D)} \tag{8.1}$$

In this chapter, a *tilde* is used to indicate *effective* quantities such as effective stress $\widetilde{\sigma}$ and effective strain $\widetilde{\epsilon}$. The effective quantities are conjugate of *undamaged* properties such as the undamaged modulus $\widetilde{E}$ through constitutive equations such as $\widetilde{\sigma} = \widetilde{E}\,\widetilde{\epsilon}$.

Similarly, the undamaged stiffness matrix $\widetilde{C}$ and undamaged reduced stiffness matrix $\widetilde{Q}$ are indicated with a tilde. The effective stress and strain are conjugate to *undamaged* entities such as the undamaged reduced stiffness $\widetilde{Q}$ through constitutive equations such as $\{\widetilde{\sigma}\} = [\widetilde{Q}]\{\widetilde{\epsilon}\}$, where [ ] indicates a matrix and { } indicates a column array[1].

Damaged quantities such as damaged stress σ and damaged strain ϵ have no decoration and are conjugate to *damaged* properties such as damaged stiffness E, damaged stiffness matrix C, damaged reduced stiffness matrix Q, and so on, through constitutive equations such as $\sigma = E\epsilon$. Having the undamaged moduli $\widetilde{E}_1, \widetilde{E}_2, \widetilde{G}_{12}, \widetilde{\nu}_{12}$ and so on labeled with an overline may seem uncomfortable at first after so many years seeing the modulus of elasticity as simply E, but there is a need in continuum damage mechanics to identify clearly the unchanging value of the undamaged property as $\widetilde{E}$ to distinguish it from the changing value of damaged E. If this argument does not convince you, consider the following. Every time in college when we used the modulus of elasticity E, it did not occur to us to question whether or not the material had some intrinsic damage at the time a test was done to measure it. So, in a general sense, we were already using the *damaged* modulus, but we took it as constant because our analysis did not consider additional damage. Our analysis

[1]This notation is similar to the notation for undamaged moduli, stiffness, and compliance used in [2, Chapter 8] with the difference that an *overline* was used in [2, Chapter 8], while a *tilde* is used here.

prowess is now far superior and we can indeed consider additional damage on our old friend E.

The stiffness and compliance matrices C, S and their reduced counterparts Q, S with no decorations are computed in material coordinates, the 1-direction being aligned with the fiber direction, the 3-direction normal to the midsurface of the lamina, and the 2-direction obtained by the cross-product ($\overrightarrow{x}_2 = \overrightarrow{x}_3 \times \overrightarrow{x}_1$) of the former, to yield a right-hand sided coordinate system $1, 2, 3$. When these matrices are transformed to a different coordinate system, such as the laminate system x, y, z, their symbols are adorned by an *overline*, such as in $\overline{C}, \overline{S}, \overline{Q}$, which means *transformed*. In those rare cases when we might need to display *transformed effective* stiffness or *transformed effective* compliance matrices, we use both an overline to indicate *transformed* and a tilde line to indicate *effective*, such as in $[\overline{\widetilde{Q}}] = [\widetilde{\overline{Q}}]$.

Furthermore, for a composite lamina or composite tow, continuum damage mechanics requires the following variables to be defined:

1. A state variable to keep track of the amount of damage. The state variable may be physically based, such as the crack density λ in Section 8.4, or a phenomenological variable D, such as in (8.1).

2. An independent variable, thermodynamically conjugate to the state variable. The independent variable, also called *thermodynamics force*, is the physical quantity that *drives* the damage process. It could be the stress, the strain, the energy release rate, or other suitable defined thermodynamic force.

3. A damage activation function g that separates the damaging states from the undamaging ones.

4. A damage threshold used to control the initial size of the damage activation function.

5. A damage hardening function, which updates the damage threshold.

6. A damage evolution function, which is used to predict damage increments.

7. The value of critical damage when the lamina fractures.

8.2 Longitudinal Tensile Damage

In this section we describe how to use two material properties, the Weibull shape parameter m and the longitudinal tensile strength of the composite F_{1t}, to predict the damaging behavior and longitudinal tensile failure of a unidirectional fiber reinforced lamina.

The physics of damage under longitudinal tension stress is as follows [153]. Fiber strength varies from fiber to fiber according to a Weibull probability density function that is fully characterized by two parameters, the *Weibull shape parameter* m and the *Weibull scale parameter* σ_0 (see Appendix B). The shape parameter represents

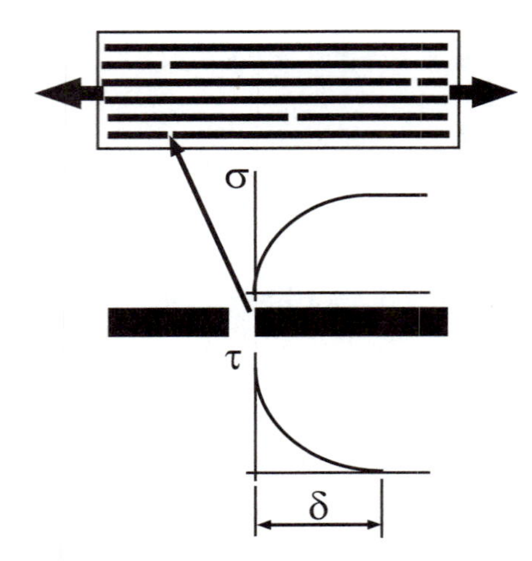

Figure 8.2: Shear lag near the end of a broken fiber.

the dispersion of the values of strength; the higher the m-number, the less dispersion (see Figure B.1 in Appendix B).

Upon application of longitudinal tensile stress, weaker fibers fail and their share of the load is redistributed to neighboring fibers. In addition, as depicted in Figure 8.2, stress recovers back into the broken fiber at a distance δ along the fiber direction and away from the break because the matrix transfers back the longitudinal stress through shear action onto the broken fiber (8.2). This phenomenon is called *shear lag* and the distance δ is called *ineffective length.* All this was nicely put together by Rosen [155] into a model that predicts the longitudinal tensile strength of fiber reinforced unidirectional composites in terms of the Weibull parameters. Later, Barbero and Kelly [153] recasted all this in the framework of continuum damage mechanics, so that the progressive failing of fibers could be interpreted as distributed damage. Those ideas are further developed here into a fully-functional damage model that only requires two parameters, the Weibull shape parameter m and the longitudinal tensile strength of the unidirectional composite F_{1t}. Both values are available in the literature for use in design or can be measured by standard tests (see Tables 2.3–2.4).

Under applied longitudinal tensile stress σ_1, the probability of fiber failure is given by the Weibull probability density. Loading a specimen up to a stress σ_1, the ratio of number of fibers broken to number of fibers in the cross-section can be calculated by the cumulative distribution function (CDF, B.4). Such ratio is also the ratio of damaged fiber area to initial fiber area. That is, the longitudinal tensile damage D_{1t} is given by [153]

$$D_{1t} = F(\widetilde{\sigma}_1) = 1 - exp\left[-\alpha\delta\widetilde{\sigma}_1^m\right] \tag{8.2}$$

where $\widetilde{\sigma}_1$ is the effective stress in the composite along the fiber direction, related to the applied stress σ_1 by

$$\sigma_1 = \widetilde{\sigma}_1 (1 - D_{1t}) \tag{8.3}$$

with δ being the ineffective length due to shear lag and $\alpha = (L_0 \sigma_0)^{-1}$ is a parameter related to the scale parameter of the fibers and the gauge length L_0 used in its experimental determination (see Section 4.4.1 and [2, Section 8.1.6]). Since neither σ_0 or L_0 are known, the product $\alpha\delta$, is calculated as follows. Substituting (8.2) into (8.3) we have

$$\sigma_1 = \widetilde{\sigma}_1 \exp(-\alpha\delta\widetilde{\sigma}_1^m) \tag{8.4}$$

Equation (8.4) has a maximum when $\partial\sigma_1/\partial\widetilde{\sigma}_1 = 0$, which leads to

$$\widetilde{\sigma}_{1,max} = (\alpha\delta\ m)^{-1/m} \tag{8.5}$$

$$F_{1t} = (\alpha\delta\ m\ e)^{-1/m} \tag{8.6}$$

$$\alpha\delta = (m\ e\ F_{1t}^m)^{-1} \tag{8.7}$$

where e is the basis of the natural logarithms. Substituting (8.7) into (8.2), the final explicit expression for the damage in the composite under tensile load along the fiber direction is

$$D_{1t} = F(\widetilde{\sigma}_1) = 1 - exp\left[-\frac{1}{m\ e}\left(\frac{\widetilde{\sigma}_1}{F_{1t}}\right)^m\right] \tag{8.8}$$

which can be used as long as $\widetilde{\sigma}_1 \geq 0$. In this way, modeling longitudinal fiber damage requires only to know one additional parameter, the Weibull shape parameter m of the fiber, for which values are available in the literature (see Tables 2.3, 2.4). The model predicts exactly the tensile strength F_{1t} as shown in Figure 8.3 for AS4–3501-6 (Table 1.4.) The evolution of damage is shown in Figure 8.3, where it can be seen that the tensile strength is reached for the critical value of fiber damage [2, (8.42)],

$$D_{1t}^{cr} = 1 - exp(-1/m) \tag{8.9}$$

which agrees well with experimental observations. To maintain damage constant during unloading, the threshold g_{1t} is set to the highest value of effective stress seen by the material, and damage is not recalculated unless $\widetilde{\sigma}_1 > g_{1t}$.

Summary

The basic ingredients of a continuum damage mechanics model to longitudinal tension damage are listed below.

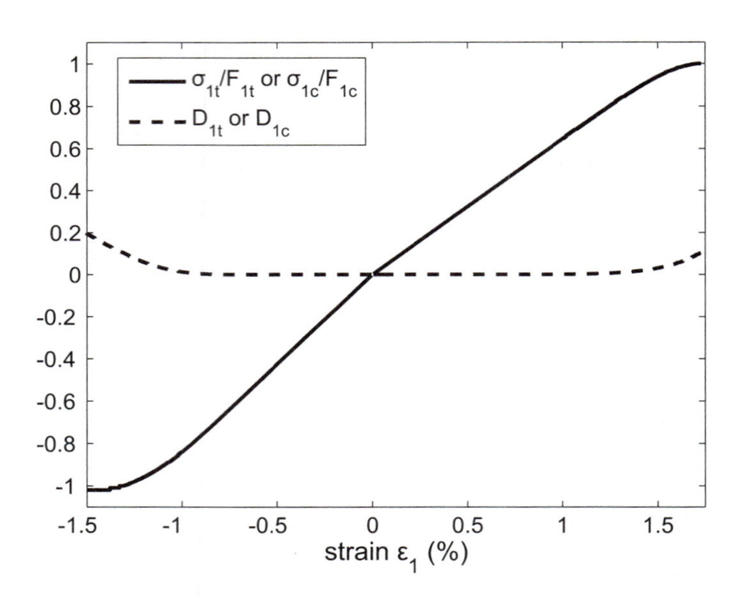

Figure 8.3: Longitudinal stress and damage $\sigma_{1t}/F_{1t}, D_{1t}$, (first quadrant) and $\sigma_{1c}/F_{1c}, D_{1c}$, (fourth quadrant) in a unidirectional laminate subject to longitudinal tension/compression.

1. The damage variable D_{1t} is the ratio of broken fibers to total number of fibers over a unit width of lamina (or a composite tow).

2. The independent variable is the effective stress in the longitudinal direction $\widetilde{\sigma}_1$ (8.3).

3. The damage activation function is

$$g = g_{1t} - \widetilde{\sigma}_1 \leq 0 \tag{8.10}$$

4. The damage threshold g_{1t} is initially set to zero because a small number of fibers break at very low stress. The Weibull probability density goes to zero for zero stress (see Figure B.1 in Appendix B).

5. The damage hardening function is obtained by updating the threshold g_{1t} to be the last value of effective stress; that is, $g_{1t} = \widetilde{\sigma}_1$. No more fibers break unless the previous value of stress is exceeded.

6. The damage evolution function is given by (8.8).

7. The critical damage when a lamina fails is given by (8.9).

Longitudinal tensile damage is not affected by other components of stress, such as transverse tensile, compressive, and in-plane shear, because the very high fiber strength dominates the behavior.

8.3 Longitudinal Compressive Damage

In this section we describe how to use one material property, the longitudinal compressive strength of the composite F_{1c}, to predict the damaging behavior leading to longitudinal compression failure of a unidirectional fiber reinforced lamina.

Rosen [159] calculated the buckling stress of perfectly straight fibers embedded in a matrix (see also Section 4.4.3). His formula, $F_{1c} = G_{12}$, overpredicts the measured compressive strength by as much as 100% because the fibers are not perfectly straight but wavy. The waviness of the fibers is represented by the misalignment angle α they form with the nominal fiber direction, and this waviness varies from fiber to fiber. Barbero and Tomblin [158] have shown experimentally that the misalignment angle in unidirectional fiber reinforced composites follows a Gaussian probability density distribution [156, (4),(5)]

$$f(\alpha) = \frac{1}{\alpha_\sigma \sqrt{2\pi}} \exp\left(-\frac{\alpha^2}{2\alpha_\sigma^2}\right) \tag{8.11}$$

The Gaussian distribution is centered at the origin $\alpha = 0$ and thus is completely characterized by a single parameter, the standard deviation of fiber misalignment α_σ. An simple, explicit formula to predict the compressive strength was developed in [156, p. 495], which is reproduced here (see also (4.85)–(4.86))

$$F_{1c} = G_{12} \left(1 + 4.76\chi\right)^{-0.69} \tag{8.12}$$

$$\chi = \frac{G_{12}\alpha_\sigma}{F_6} \tag{8.13}$$

where α_σ is the standard deviation of fiber misalignment, G_{12} is the tangent shear modulus at the origin ($\gamma_6 = 0$), and F_6 is the shear strength. Both G_{12} and F_6 can be measured by a shear test such as [118, ASTM D 5379], which recommends using a $[0/90]_S$ sample to yield lower scatter in the results.

The standard deviation of fiber misalignment α_σ, which is affected by processing conditions, can be measured by optical methods [157,158] but doing so is impractical for a designer. In this book, we propose to measure the compressive strength F_{1c}, the in-plane shear modulus G_{12} and strength F_6, then calculate the underlying standard deviation α_σ to satisfy (8.12), as follows

$$\alpha_\sigma = \frac{F_6}{4.76\ G_{12}} \left[\left(\frac{G_{12}}{F_{1c}}\right)^{1.45} - 1\right] \tag{8.14}$$

This approach is advantageous because the parameter α_σ is quite insensitive to laminate stacking sequence (LSS) and other design changes as long as the fiber type and processing conditions remain unchanged. This approach is customary practice in the composites industry; when these parameters are adjusted to industry's specific

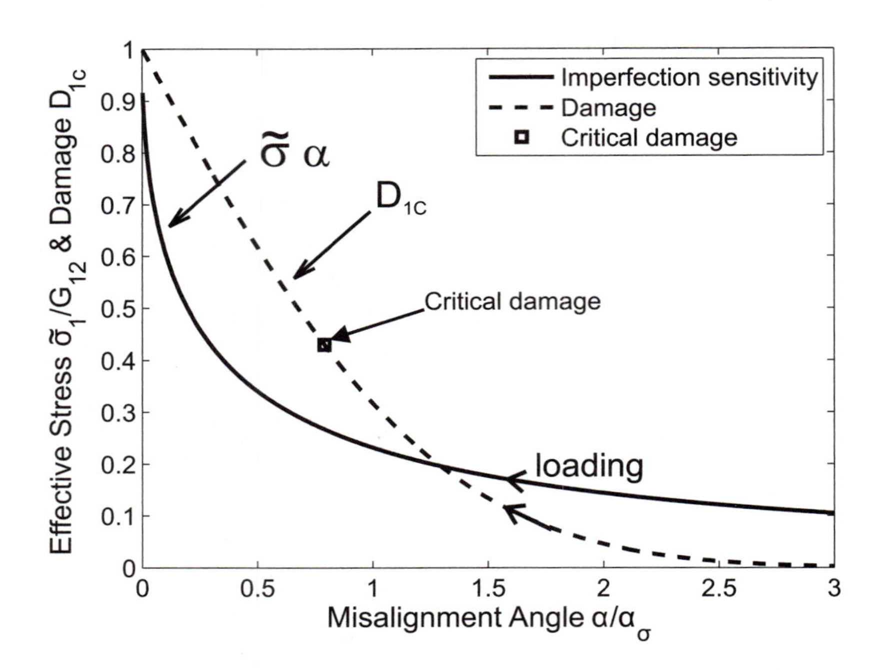

Figure 8.4: Imperfection sensitivity (8.15) for AS4–3501-6, Table 1.4.

manufacturing process, they become trade secrets for a given organization. The compressive strength F_{1c} can be measured by using standard test methods [41, SACMA SRM 1-94].

The physics behind compressive strength is as follows [156]. Upon application of a compressive stress, the fibers with larger misalignment buckle and the stress is redistributed to those that have not buckled. Fibers buckle at a stress[2] given by the imperfection sensitivity equation, derived in [156, (17)–(20)], and reproduced here using the notation of this book, as follows[3]

$$\widetilde{\sigma}_1(\alpha) = G_{12} + \frac{16C_2\alpha}{3\pi} - \frac{4\sqrt{2C_2\alpha\,(8C_2\alpha + 3\pi G_{12})}}{3\pi} \qquad (8.15)$$

$$C_2 = \frac{G_{12}^2}{4F_6}$$

and shown in Figure 8.4.

Each value of effective stress $\widetilde{\sigma}_1$ causes all fibers with angles larger than α to buckle. Damage represents the ratio of fibers that have buckled to the total in the cross-section. Since fibers will buckle the same regardless of the sign of the misalignment angle, damage is calculated as twice the area under the Gaussian

[2] Adjusted for fiber volume fraction so that $\widetilde{\sigma}_1$ is a composite stress, i.e., the stress averaged over a fiber and its surrounding matrix.

[3] Note the redefinition of C_2 with respect to [156, (17)].

probability density curve for angles outside the interval $|\alpha|$. Since the total area under the Gaussian probability density curve (8.11) is equal to one, it is easier to calculate damage as one minus the area in the interval $[-\alpha, \alpha]$, as follows

$$D_{1c} = 1 - \int_{-\alpha}^{\alpha} f(\alpha')d\alpha' = 1 - A(\alpha) \tag{8.16}$$

with

$$A(\alpha) = \int_{-\alpha}^{\alpha} f(\alpha')d\alpha' = erf(z) \tag{8.17}$$

with $f(\alpha')$ given by (8.11) and $erf(z)$ given in Appendix A.

Each value of applied stress σ_1 results in a value of effective stress $\widetilde{\sigma}_1$ according to (8.1) and each value of effective stress corresponds to a unique value of misalignment $\alpha(\widetilde{\sigma}_1)$ in the imperfection sensitivity plot (Figure 8.4). Loading proceeds from left to right in (Figure 8.4).

Since all the fibers with misalignment greater than the current one have already buckled, the damage is calculated simply as the integral of both tails of the Gaussian distribution function, i.e., (8.16). From (8.1) and (8.16), the apparent stress σ_1 in the tow is

$$\sigma_1 = \widetilde{\sigma}_1(1 - D_{1c}) = \widetilde{\sigma}_1 A(\alpha) \tag{8.18}$$

The apparent stress (8.18) has a maximum for which the tow cannot longer carry the applied load, that is, a failure strength given by (8.12).

The compressive strength is reached for a critical value of misalignment α_{cr}, obtained by setting the product of the effective stress in the unbuckled fibres (8.15) times the remaining area $A(\alpha)$ equal to the experimental value of compressive strength F_{1c} and solving for α_{cr} [156, (13)]

$$\widetilde{\sigma}_1(\alpha)A(\alpha) = F_{1c} \quad ; \quad \text{solve for } \alpha_{cr} \tag{8.19}$$

This in turn defines the critical value of damage [2, (8.60)]

$$D_{1c}^{cr} = 1 - A(\alpha_{cr}) \tag{8.20}$$

On a computer implementation, the effective stress can be calculated from the effective strain as

$$\{\widetilde{\sigma}\} = \left[\widetilde{S}\right]\{\widetilde{\epsilon}\} \tag{8.21}$$

where $\widetilde{\sigma}$, $\widetilde{\epsilon}$, $\widetilde{S}$, are the effective stress, effective strain, and undamaged compliance, respectively.

The angle α of the fiber that is buckling at that value of stress is calculated by solving numerically for α in (8.15). The value α_σ calculated by using (8.14) provides a convenient starting point for numerical solvers that need an initial guess. The solution is unique and convergence is achieved in a few iterations. Then, (8.16) yields the damage D_{1c} (Figure 8.4).

To maintain damage constant during unloading, the threshold g_{1c} is set to the highest (negative) value of longitudinal compressive stress seen by the material, and damage is not recalculated unless $\sigma < g_{1c}$.

Predicted longitudinal stress-strain response is shown in Figure 8.3. The compressive stress σ slightly exceeds the value of F_{1c} calculated with (8.12) because the imperfection sensitivity curve (8.15), used here for convenience, is an approximation to the exact one shown in [156, (9)] and [2, (8.53)].

Continuum damage mechanics ceases to represent the behavior near fracture, when sufficient density of damage effects coalesce to form a crack. From then on, only fracture mechanics can truly represent the material behavior. In the case of longitudinal compression, when the applied stress σ_1 reaches the longitudinal compressive strength given by (8.12), a kink band forms and the material fails catastrophically if subjected to load control. Damage suddenly becomes $D_{1c} = 1$ and the longitudinal stiffness of the affected lamina is completely lost.

At each point in the structure, the longitudinal direction may accumulate both tensile and compressive damage if it has experienced both tension and compression. Therefore, in the linear elastic domain (damage not being recalculated), one must decide about which damage, tension or compression, determines the modulus of the material. The decision is simple, as it only involves considering the sign of the current applied stress.

Summary

The basic ingredients of a continuum damage mechanics model for longitudinal compressive damage are listed below.

1. The damage variable is D_{1c}, which is the ratio of buckled fibers to total number of fibers in a unit width of lamina.

2. The independent variable is the effective stress in the longitudinal direction $\widetilde{\sigma}_1$ (8.18).

3. The damage activation function is

$$g = g_{1c} - \widetilde{\sigma}_1 \leq 0 \tag{8.22}$$

4. The damage threshold is g_{1c}, initially is zero because a small number of fibers buckle at very low stress. Since the Gaussian distribution of fiber misalignment has a tail that goes to infinity, there are a few fibers with very large misalignment, and these will buckle with stress approaching zero.

5. The damage hardening function is simply given by the fact that the threshold g_{1c} is always updated to be the last value of effective stress, as $g_{1c} = \widetilde{\sigma}_1$. No more fibers will buckle unless the effective stress increases.

6. The damage evolution function is defined by (8.16).

7. The critical damage when a lamina fails is given by (8.20).

8.4 Transverse Tension and In-plane Shear

In this section we describe how to use two material properties, the fracture toughness in modes I and II, G_{Ic}, G_{IIc}, to predict the damaging behavior and transverse tensile failure of a unidirectional fiber reinforced lamina embedded in a laminate. The constraining effect of adjacent laminae is taken into account, leading to apparent transverse strength F_{2t} being a function of ply thickness. The crack initiation strain, crack density evolution as a function of stress (strain) up to crack saturation, and stress redistribution to adjacent laminae is predicted accurately.

The physics of matrix cracking under transverse tension and in-plane shear is as follows. No matter how much care is taken during the production process, there are always defects in the material. These defects may be voids, microcracks, fiber–matrix debonding, and so on, but all of them can be represented by a typical matrix crack of representative size $2a_0$, as shown in Figure 7.8.

When subject to load, matrix cracks grow parallel to the fiber orientation, as shown in Figure 7.7, where it can be seen that cracks are aligned with the fiber direction in the 90° and 30° laminae. These sets of parallel cracks reduce the stiffness of the cracked lamina, which then sheds its share of the load onto the remaining laminae. In each lamina, the damage caused by this set of parallel cracks is represented by the crack density, defined as the inverse of the distance between two adjacent cracks $\lambda = 1/(2l)$, as shown in Figure 8.5. Therefore, the crack density is the only state variable needed to represent the state of damage in the cracked lamina. Note that the actual, discrete cracks are modeled by the theory, which is thus named *discrete damage mechanics* (DDM).

The basic ingredients of the DDM model for transverse tension and in-plane shear damage are listed below:

1. In each lamina, the state variable is the crack density λ. Three damage variables $D_{22}(\lambda), D_{66}(\lambda)$, and $D_{12}(\lambda)$ are defined for convenience but they are computed in terms of the crack density.

2. The independent variable is the midsurface[4] strain $\epsilon = \{\epsilon_1, \epsilon_2, \gamma_6\}^T$.

[4]The analysis presented in this section is for symmetric laminates under membrane forces. A generalization for unsymmetric laminates and/or laminates under bending is being developed.

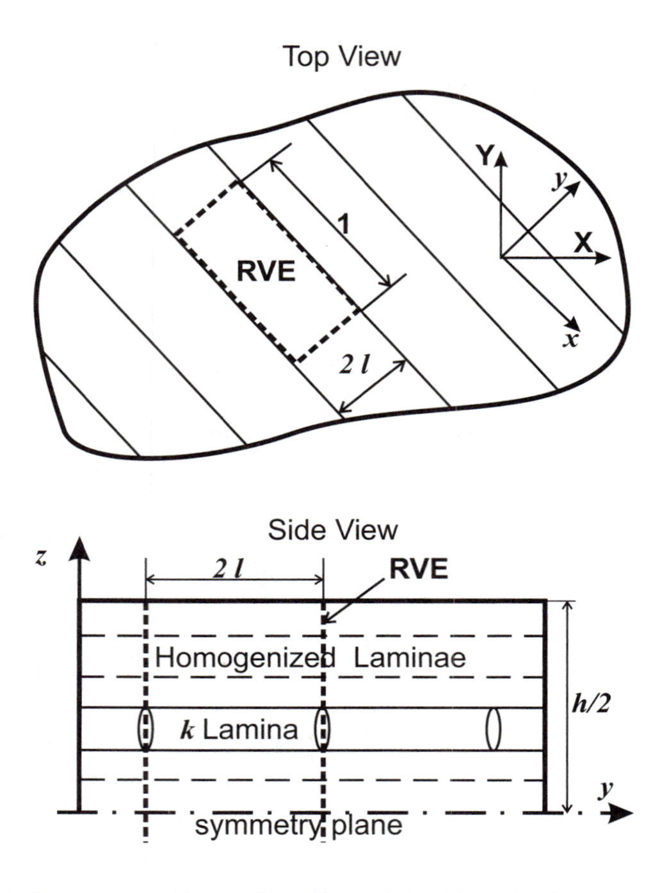

Figure 8.5: Representative unit cell used in *discrete damage mechanics*.

3. The damage activation function, which separates the damaging states from the undamaging ones, is

$$g = (1-r)\sqrt{\frac{G_I(\lambda,\epsilon)}{G_{Ic}}} + r\frac{G_I(\lambda,\epsilon)}{G_{Ic}} + \frac{G_{II}(\lambda,\epsilon)}{G_{IIc}} - 1 \leq 0$$

$$r = \frac{G_{Ic}}{G_{IIc}} \tag{8.23}$$

4. The damage threshold is embedded into g, and represented by the (invariant) material properties G_{Ic}, G_{IIc} (see Section 1.5). Before damage starts, $\lambda = 0$ and (8.23) is a damage initiation criterion [200]. With $\lambda = 0$, the strain for which $g = 0$ is the strain for crack initiation. Once damage starts, (8.23) becomes a damage activation function by virtue of the automatic hardening described below.

5. The hardening function is embedded into the damage activation function g. For a given value of strain, the calculated values of energy release rate

$G_I(\lambda), G_{II}(\lambda)$ (see Section 1.5) are monotonically decreasing functions of λ. Therefore, as soon as λ grows, $G_I(\lambda), G_{II}(\lambda)$ decrease, making $g < 0$ and thus stopping further damage until the *driving force*, the strain, is increased by the application of additional load.

6. No damage evolution function need to be postulated, with the advantage that no new empirical parameters are needed. Simply the crack density λ adjusts itself to a value that will set the laminate in equilibrium with the external loads for the current strain while satisfying $g = 0$. A return mapping algorithm [2, Chapter 8] achieves this by iterating until $g = 0$ by updating the crack density with iterative increments calculated as $\Delta\lambda = -g/\frac{\partial g}{\partial \lambda}$.

7. The crack density grows until the lamina is saturated with cracks ($\lambda \to \infty$). At that point the lamina looses all of its transverse and shear stiffness ($D_2 \approx 1, D_6 \approx 1, D_{12} \approx 1$), at which point all of the load is already transferred to the remaining laminae in the laminate. The analysis of the cracked lamina is stopped when the crack density reaches $\lambda_{lim} = 1/h^k$, where h^k is the thickness of lamina k; this means that cracks are closely spaced at a distance equal to the lamina thickness.

Having described the ingredients of the model, it now remains to show how to calculate the various quantities. The solution begins by calculating the reduced stiffness of the laminate $Q = A/h$ for a given crack density λ_k in a cracked lamina k, where A is the in-plane laminate stiffness matrix, and h is the thickness of the laminate.

The following conventions are used in this section:

- the superscript/subscript i denotes any laminae in the laminate.
- the superscript/subscript k denotes the cracking lamina.
- the superscript/subscript m denotes any laminae other than the cracking one ($m \neq k$).

8.4.1 Limitations

Most practical laminates are symmetric and the most efficient use of them is by designing the structure to be loaded predominantly with membrane loads (see Chapter 12). Therefore, the solution presented here is for a symmetric laminate under membrane loads. In this case,

$$\frac{\partial w^i}{\partial x} = \frac{\partial w^i}{\partial y} = 0 \tag{8.24}$$

where $u(x, y, z), v(x, y, z), w(x, y, z)$, are the displacements of a point in lamina i as a function of the coordinates x, y, z.

Furthermore, the thickness h^i of the laminae are assumed to be small, so that

$$\sigma_z^i = 0 \tag{8.25}$$

As it has been mentioned before, all cracks are assumed to be parallel to the fiber direction. Furthermore, practical designs avoid thick laminae. For thin laminae the cracks occupy the entire thickness of the lamina.

8.4.2 Approximations

Since the objective is to calculate the laminate stiffness reduction due to cracks, it suffices to work with thickness averages of the variables. A thickness average is denoted by

$$\hat{\phi}_i(x,y) = \frac{1}{h^i}\int_{h^i} \phi_i(x,y,z)dz \tag{8.26}$$

where h^i denotes the thickness of lamina i.

Specifically,

- $\hat{u}^i(x,y), \hat{v}^i(x,y), \hat{w}^i(x,y)$, are the thickness-average displacements in lamina i as a function of the in-plane coordinates x, y.
- $\hat{\epsilon}_1^i(x,y), \hat{\epsilon}_2^i(x,y), \hat{\gamma}_6^i(x,y)$, are the thickness-average strains in lamina i.
- $\hat{\sigma}_1^i(x,y), \hat{\sigma}_2^i(x,y), \hat{\sigma}_6^i(x,y)$, are the thickness-average stress in lamina i.

The same averaging is used to compute the average over the thickness of the laminate h. Out-of-plane (intralaminar) shear stress components appear due to the the perturbation of the displacement field caused by the crack. These are approximated by linear functions through the thickness of the lamina i, as follows,

$$\begin{aligned}\sigma_4^i(z) &= \sigma_4(z_{i-1}) + [\sigma_4(z_i) - \sigma_4(z_{i-1})]\,\frac{z - z_{i-1}}{h^i}\\ \sigma_5^i(z) &= \sigma_5(z_{i-1}) + [\sigma_5(z_i) - \sigma_5(z_{i-1})]\,\frac{z - z_{i-1}}{h^i}\end{aligned} \tag{8.27}$$

where z_{i-1} and z_i are the thickness coordinate at the bottom and top of lamina i, respectively, as shown in Figure 6.6.

8.4.3 Displacement Field

The objective now is to solve for the average displacements $\hat{u}^i(x,y), \hat{v}^i(x,y)$, in all laminae i for a given crack density λ and applied strain ϵ. To do this, the in-plane and intralaminar equilibrium equations, the kinematic equations, and the constitutive equations are invoked (for details see [201]). This leads to a system of $2n$ partial differential equations in $\hat{u}^i(x,y), \hat{v}^i(x,y)$. The system of partial differential equations (PDE) has a particular solution of the form

$$
\begin{aligned}
\hat{u}^i &= a^i \sinh \lambda y + ax + by \\
\hat{v}^i &= b^i \sinh \lambda y + bx + cy
\end{aligned} \tag{8.28}
$$

which substituted into the PDE system leads to an eigenvalue problem

$$
([K] - \eta_e^2[M]) \left\{ \begin{array}{c} a^i \\ b^i \end{array} \right\}_e = \{0\} \tag{8.29}
$$

where η are the $2n$ eigenvalues and $\{a^i, b^i\}_e^T$ are the $2n$ eigenvectors. It turns out that always two of the eigenvalues are zero, which can be taken to be the last two in the set, so that there remain only $2n-2$ independent solutions. Then, the general solution of the PDE system is built as the linear combination of the $2n-2$ independent solutions as follows

$$
\left\{ \begin{array}{c} \hat{u}^i \\ \hat{v}^i \end{array} \right\} = \sum_{e=1}^{2n-2} \alpha_e \left\{ \begin{array}{c} a^i \\ b^i \end{array} \right\}_e \sinh \eta_e y + \left\{ \begin{array}{c} a \\ b \end{array} \right\} x + \left\{ \begin{array}{c} b \\ c \end{array} \right\} y \tag{8.30}
$$

where α_e are unknown coefficients in the linear combination. It can be seen that the general solution contains $2n+1$ unknown coefficients, including a, b, c, and α_e with $e = 1 ... 2n-2$. To determine these coefficients, one needs $2n+1$ boundary conditions on the boundary of the representative volume element (RVE) in Figure 8.5. Note the that RVE spans a unit length along the fiber direction x, a distance $2l$ between successive cracks (along y) and the whole thickness h of the symmetric laminate.

Two very important parameters are introduced through the boundary conditions, namely the crack density λ and the stress $\hat{\sigma} = N/h$ applied to the laminate, where N is the in-plane force per unit length. The crack density enters through the dimension of the RVE, which has a width of $2L = 1/\lambda$. The applied stress (or strain) enters through the force equilibrium on the RVE. In summary, there are $2n+1$ boundary conditions that lead to a system of $2n+1$ algebraic equations that can be solved for the $2n+1$ coefficients in (8.30). Therefore, the average displacements in all laminae are now known from (8.30) for given values of crack density λ and applied load $\hat{\sigma} = N/h$.

8.4.4 Strain Field

The thickness-averaged strain field in all laminae can now be obtained by using the kinematic equations [145], namely by directly differentiating (8.30) as follows

$$\hat{\epsilon}_x^i = \frac{\partial \hat{u}^i}{\partial x}$$
$$\hat{\epsilon}_y^i = \frac{\partial \hat{v}^i}{\partial y}$$
$$\hat{\gamma}_{xy}^i = \frac{\partial \hat{u}^i}{\partial y} + \frac{\partial \hat{v}^i}{\partial x} \tag{8.31}$$

8.4.5 Laminate Reduced Stiffness

The compliance of the laminate $\overline{S}$ in the coordinate system of lamina k can be calculated one column at a time by solving for the strains (8.31) for three load cases, a, b, and c, as follows

$$^a\hat{\sigma} = \begin{Bmatrix} 1 \\ 0 \\ 0 \end{Bmatrix}; \quad ^b\hat{\sigma} = \begin{Bmatrix} 0 \\ 1 \\ 0 \end{Bmatrix}; \quad ^c\hat{\sigma} = \begin{Bmatrix} 0 \\ 0 \\ 1 \end{Bmatrix} \tag{8.32}$$

Since the three applied stress states are unit values, for each case, a, b, c, the average strain calculated with (8.31) represents one column in the laminate compliance matrix,

$$\overline{S} = \begin{bmatrix} ^a\epsilon_x & ^b\epsilon_x & ^c\epsilon_x \\ ^a\epsilon_y & ^b\epsilon_y & ^c\epsilon_y \\ ^a\gamma_{xy} & ^b\gamma_{xy} & ^c\gamma_{xy} \end{bmatrix} \tag{8.33}$$

where x, y, are the coordinates of lamina k (Figure 8.5). Next, the laminate stiffness in the coordinate system of lamina k is

$$\overline{C} = \overline{S}^{-1} \tag{8.34}$$

8.4.6 Lamina Reduced Stiffness

The stiffness of lamina m, with $m \neq k$, in the coordinate system of lamina k (see Figure 8.5) is given by (5.48) as

$$\overline{Q}^m = \begin{bmatrix} \overline{Q}_{11} & \overline{Q}_{12} & \overline{Q}_{16} \\ \overline{Q}_{12} & \overline{Q}_{22} & \overline{Q}_{26} \\ \overline{Q}_{16} & \overline{Q}_{26} & \overline{Q}_{66} \end{bmatrix} \tag{8.35}$$

where *overline* denotes a quantity transformed to the coordinate system of lamina k. The stiffness of the cracking lamina $Q^k = \overline{Q}^k$ is yet unknown. Note that $Q^k = \overline{Q}^k$ because all quantities are rotated to the coordinate system of lamina k.

In principle, the laminate stiffness is defined by the contribution of the cracking lamina k plus the contribution of the remaining $n-1$ laminae, as follows

$$\overline{C} = Q^k \frac{h^k}{h} + \sum_{m=1}^{n} (1 - \delta_{mk}) \overline{Q}^m \frac{h^m}{h} \tag{8.36}$$

where the *delta* Dirac is defined as $\delta_{mk} = 1$ if $m = k$, otherwise 0. The left-hand side (LHS) of (8.36) is known from (8.34) and all values of $\overline{Q}^m$ can be easily calculated since the m laminae are not cracking at the moment. Therefore, one can calculate the reduced stiffness Q^k of lamina k as follows

$$Q^k = \frac{h}{h^k} \left[\overline{C} - \sum_{m=1}^{n} (1 - \delta_{mk}) \overline{Q}^m \frac{h^m}{h} \right] \tag{8.37}$$

To facilitate later calculations, the stiffness Q^k can be written using concepts of continuum damage mechanics [2, (8.63)–(8.67)], in terms of the stiffness of the undamaged lamina and damage variables $D^k_{22}, D^k_{66}, D^k_{12}$, as follows

$$Q^k = \begin{bmatrix} \widetilde{Q}^k_{11} & (1 - D_{12})\, \widetilde{Q}^k_{12} & 0 \\ (1 - D_{12})\, \widetilde{Q}^k_{12} & (1 - D_{22})\, \widetilde{Q}^k_{22} & 0 \\ 0 & 0 & (1 - D_{66})\, \widetilde{Q}^k_{66} \end{bmatrix} \tag{8.38}$$

with D^k_{ij}, $(i, j = 1, 2, 6)$, calculated for a given crack density λ^k and applied strain ϵ^0, as follows

$$D^k_{ij}(\lambda^k, \epsilon^0) = 1 - Q^k_{ij} / \widetilde{Q}^k_{ij} \tag{8.39}$$

where $\widetilde{Q}^k$ is the original value of the undamaged property in and Q^k is the damaged value computed in (8.37), both computed in the coordinate system of lamina k.

8.4.7 Fracture Energy

Under displacement control, the energy release rate (ERR) is defined as the partial derivative of the strain energy U with respect to the crack area A (see Section 1.5). According to experimental observations on laminated, brittle matrix composites (e.g., using most toughened epoxy matrices), cracks develop suddenly over a finite length, and thus are not infinitesimal. Then, Griffith's energy principle is applied on its discrete (finite) form in order to describe the observed, discrete (finite) behavior of crack growth, as follows

$$\begin{aligned} G_I &= -\frac{\Delta U_I}{\Delta A} \\ G_{II} &= -\frac{\Delta U_{II}}{\Delta A} \end{aligned} \tag{8.40}$$

where $\Delta U_I, \Delta U_{II}$, are the change in laminate strain energy during mode I and mode II finite crack growth, respectively; and ΔA is the is the newly created (finite) crack area, which is one half of the new crack surface. Counting crack area as one-half of crack surface is consistent with the classical fracture mechanics convention adopted in Chapter 1, for which fracture toughness G_c is twice of Griffith's surface energy γ_c.

The strain energy can be written as one-half the integral over the volume of the product of stress and strain. Furthermore, the average strain theorem [2, (6.16)] for an RVE states that the volume integral of the strain $\widehat{\epsilon}(x, y)$ in the RVE is equal to the applied strain ϵ, the later being constant over the RVE. Therefore,

$$\begin{aligned} G_I &= -\frac{1}{2}\int_{RVE} \frac{\Delta(\hat{\sigma}_x\hat{\epsilon}_x)}{\Delta A} + \frac{\Delta(\hat{\sigma}_y\hat{\epsilon}_y)}{\Delta A} = -\frac{V_{RVE}}{2}\left[\frac{\Delta(\sigma_x\epsilon_x)}{\Delta A} + \frac{\Delta(\sigma_y\epsilon_y)}{\Delta A}\right] \\ G_{II} &= -\frac{1}{2}\int_{RVE} \frac{\Delta(\hat{\sigma}_{xy}\hat{\gamma}_{xy})}{\Delta A} = -\frac{V_{RVE}}{2}\frac{\Delta(\sigma_{xy}\gamma_{xy})}{\Delta A} \end{aligned} \tag{8.41}$$

The constitutive equation for the laminate is

$$\begin{aligned} \sigma_x &= \overline{C}_{11}\epsilon_x + \overline{C}_{12}\epsilon_y + \overline{C}_{16}\gamma_{xy} \\ \sigma_y &= \overline{C}_{12}\epsilon_x + \overline{C}_{22}\epsilon_y + \overline{C}_{26}\gamma_{xy} \\ \sigma_{xy} &= \overline{C}_{16}\epsilon_x + \overline{C}_{26}\epsilon_y + \overline{C}_{66}\gamma_{xy} \end{aligned} \tag{8.42}$$

where $\overline{C} = \overline{S}^{-1}$ (8.34).

For the current damage state i, characterized by crack density (state variable) $\lambda_i = 1/(2l)$, the RVE in Figure 8.5 has thickness h, a unit length along the crack direction x, and width $2l$ (between cracks), resulting in a volume $V_{RVE} = h/\lambda$. Every new crack appears between two existing cracks, yielding new crack spacing l, new crack density $\lambda_{i+1} = 2\lambda_i$ and newly created area $\Delta A = 2t_k$. Note that two symmetric cracks, each with area $t_k \times 1$, appear simultaneously due to the symmetry of the LSS. Therefore, the ERR for the RVE encompassing the whole thickness of the symmetric laminate is given by

$$\begin{aligned} G_I(\lambda, \epsilon) &= -\frac{h}{2\lambda}\left\{\left(\frac{\Delta\overline{C}_{11}}{\Delta A}\epsilon_x + \frac{\Delta\overline{C}_{12}}{\Delta A}\epsilon_y + \frac{\Delta\overline{C}_{16}}{\Delta A}\gamma_{xy}\right)\epsilon_x \right. \\ &\quad \left. + \left(\frac{\Delta\overline{C}_{12}}{\Delta A}\epsilon_x + \frac{\Delta\overline{C}_{22}}{\Delta A}\epsilon_y + \frac{\Delta\overline{C}_{26}}{\Delta A}\gamma_{xy}\right)\epsilon_y\right\} \\ G_{II}(\lambda, \epsilon) &= -\frac{h}{2\lambda}\left\{\left(\frac{\Delta\overline{C}_{16}}{\Delta A}\epsilon_x + \frac{\Delta\overline{C}_{26}}{\Delta A}\epsilon_y + \frac{\Delta\overline{C}_{66}}{\Delta A}\gamma_{xy}\right)\gamma_{xy}\right\} \end{aligned} \tag{8.43}$$

The damage activation function (8.23) can now be calculated for any value of λ and applied strain $\epsilon_x, \epsilon_y, \gamma_{xy}$. Note that the computation of the ERR components derives directly from the displacement solution (8.30) for a discrete crack (Figure 8.5). When this formulation is used along with the finite element method (FEM), it does not display mesh dependency on the solution and does not require the arbitrary specification of a characteristic length [202], in contrast to formulations based on smeared crack approximations. The effect of residual thermal stresses is incorporated into this formulation in [202]. A shell element capable of utilizing this formulation is implemented as a user element into ANSYS [203].

8.4.8 Solution Algorithm

The solution algorithm consists of (a) strain steps, (b) laminate-iterations, and (c) lamina-iterations. The state variables for the laminate are the array of crack densities for all laminae i and the membrane strain ϵ. At each load (strain) step, the strain on the laminate is increased and the laminae are checked for damage modes by evaluating the damage activation functions of all possible modes of damage, including:

1. Longitudinal tension (Section 8.2).
2. Longitudinal compression (Section 8.3).
3. Transverse tension and/or shear (this section).
4. Transverse compression (Section 4.4.7).

8.4.9 Lamina Iterations

When matrix cracking is detected (mode 3) in lamina k, a return mapping algorithm (RMA) [2, Chapter 8] is invoked to iterate and adjust the crack density λ^k in lamina k in such a way that g^k returns to zero while maintaining equilibrium between the external forces and the internal forces in the laminae. The iterative procedure works as follows. At a given strain level ϵ for the laminate and given λ^k for lamina k, calculate the value of the damage activation function g^k and the damage variables, which are both functions of λ^k. The RMA calculates the increment (decrement) of crack density as [2, Chapter 8],

$$\Delta\lambda^k = -g^k / \frac{\partial g^k}{\partial \lambda} \tag{8.44}$$

until $g^k = 0$ is satisfied within a given tolerance, for all $k = 1...n$, where n is the number of laminae in the laminate. The analysis starts with a negligible value of crack density present in all laminae ($\lambda = 0.01$ cracks/mm were used in the examples).

8.4.10 Laminate Iterations

To calculate the stiffness reduction of a cracked lamina (k-lamina), all of the other laminae (m-laminae) in the laminate are considered not damaging during the course of lamina-iterations in lamina k, but with damaged properties calculated according to the current values of their damage variables D_{ij}^{m}. Given a trial value of λ^k, the analytical solution provides g^k, D_{ij}^k, for lamina k assuming all other laminae do not damage while performing lamina iterations in lamina k. Since the solution for lamina k depends on the stiffness of the remaining laminae, a converged iteration for lamina k does not guarantee convergence for the same lamina once the damage in the remaining laminae is updated. In other words, within a given strain step, the stiffness and damage of all the laminae are interrelated and they must all converge. This can be accomplished by laminate-iterations; that is, looping over all laminae repeatedly until all laminae converge to $g = 0$ for all k. Unlike classical RMA set up for plasticity, where the hardening parameter is monotonically increasing [15, Chapter 8], the crack density λ^k must be allowed to decrease if in the course of laminate iterations, other laminae sustain additional damage that makes the laminate more compliant and thus requires a reduction of λ^k.

Example 8.1 *Using both, the maximum stress criterion with calculated in-situ values and the DDM formulation, compute and compare the crack initiation strain for two laminates: a)* $[0/90_3/0]_S$ *and b)* $[0_2/90_3]_S$*, both made of glass–epoxy with ply thickness* $t_k = 0.125$ *mm and properties given below.*

E_1	E_2	G_{12}	ν_{12}	ν_{23}		
MPa	*MPa*	*MPa*				
44700	*12700*	*5800*	*0.297*	*0.41*		
F_{1t}	F_{1c}	F_{2t}	F_{2c}	F_6	G_{Ic}	G_{IIc}
MPa	*MPa*	*MPa*	*MPa*	*MPa*	kJ/m^2	kJ/m^2
-	-	*43.4*	-	*135.8*	*0.175*	*1.5*

Solution to Example 8.1 *Assuming the transition thickness of glass–epoxy to be* $t_t = 0.6$ *mm, the in-situ values for the* $90°$*-lamina calculated with (7.42) are reported below. With these in-situ values, the maximum stress criterion (max.* σ*) yields the strain at crack initiation* ϵ_x *and first ply failure* N_{FPF} *are reported below. A computer program implementing the DDM formulation is available on the Web site [1]. The crack initiation strain and FPF obtained with the DDM model are reported below.*

				Maximum Stress		DDM	
LSS	t_{90}	F_{2t}^{is}	F_6^{is}	ϵ_x	N_{FPF}	ϵ_x	N_{FPF}
	mm	*MPa*	*MPa*	%	*N/mm*	%	*N/mm*
$[0/90_3/0]_S$	*0.375*	*87*	*243*	*0.69*	*222*	*0.73*	*234*
$[0_2/90_3]_S$	*0.750*	*69*	*192*	*0.55*	*176*	*0.53*	*185*

Exercises

Exercise 8.1 *Consider a* $[0_{10}/90_{10}]_S$ *with ply thickness* $t_k = 0.125$ *mm. The material is E-glass–polyester (Table:1:1) with fiber's Weibull shape parameter* $m = 6.2$*. Using the notation of carpet plots, this laminate has* $\alpha = 0.5, \beta = 0.5, \gamma = 0$*. Using the methodology presented in this chapter, compute:*

1. *the laminate FPF strength F_{xt}. Which lamina cracks first, thus defining FPF?*
2. *the laminate FPF strength F_{yt}. Which lamina cracks first, thus defining FPF?*
3. *recompute using in-situ strengths with transition thickness $t_t = 1.2\,mm$.*
4. *Explain why $F_{yt} = F_{xt}$.*

Exercise 8.2 *Consider a $[0_8/90_8/45_2/-45_2]_S$ with ply thickness $t_k = 0.125$ mm. The material is E-glass–polyester (Table 1.3) with fiber's Weibull shape parameter $m = 6.2$. Using the notation of carpet plots, this laminate has $\alpha = 0.4, \beta = 0.4, \gamma = 0.2$. Using the methodology presented in this chapter, compute:*

1. *the laminate FPF strength F_{xt}. Note that the 90_4 lamina is the first to crack, thus defining FPF.*
2. *the laminate FPF strength F_{yt}. Note that the exterior 0_4 lamina is the first to crack, thus defining FPF.*
3. *recompute using in-situ strengths with transition thickness $t_t = 1.2\,mm$.*
4. *Explain why $F_{yt} < F_{xt}$.*

Exercise 8.3 *Re-design the structure in Example 7.11 (p. 256) avoiding thick clusters of plies and taking into account in-situ strength. Use $t_t = 1.2\,mm$ and $\nu_{23} = 0.38$.*

Chapter 9

Fabric-reinforced Composites

Fabric reinforcement offers some advantages with respect to laminates made of unidirectional composites, including ease of lay-up, superior damage tolerance, and impact resistance of the resulting composite. The vast majority of the literature focuses on plain weave fabrics but twill, satin, and basket weaves (Figure 2.3) are becoming popular because of their superior drapeability. An introduction to fabric (textile) reinforcements and the nomenclature used to describe them is given in Sections 2.4.2–2.4.3.

Analysis methods for fabric-reinforced composites can be grouped into two categories: numerical methods involving finite element analysis (FEA, [2]), and closed-form (analytical) methods. While FEA models provide higher accuracy using less approximations, they also require a unique discretization (mesh) for each specific case, making preliminary design and parametric studies cumbersome. On the other hand, analytical methods, such as the one presented in this chapter, are able to provide good agreement with experimental data if they incorporate enough features to model the material behavior.

A number of analytical models for woven fabric-reinforced composites have been proposed in the literature, each having some valuable features as well as some limitations [204–211]. A generalization and refinement of these models is presented in the remainder of this chapter, resulting in a formulation that can deal with multi-directional in-plane loading of plain weave as well as general twill and satin weaves. Further, the undulation and cross-section geometry of both the warp and fill tows are modeled accurately, as well as the presence of a gap between the tows. All failure events are accounted for, including matrix cracking, transverse tow damage, and longitudinal tow failure, leading to accurate prediction of stiffness as well as strength.

9.1 Weave Pattern Description

A *laminate* is a collection of *laminae*, where each *lamina* is reinforced with a layer of fabric. Fabric-reinforced composites may be *laminated* or not, but even a single lamina reinforced with a fabric displays many of the features of laminated compos-

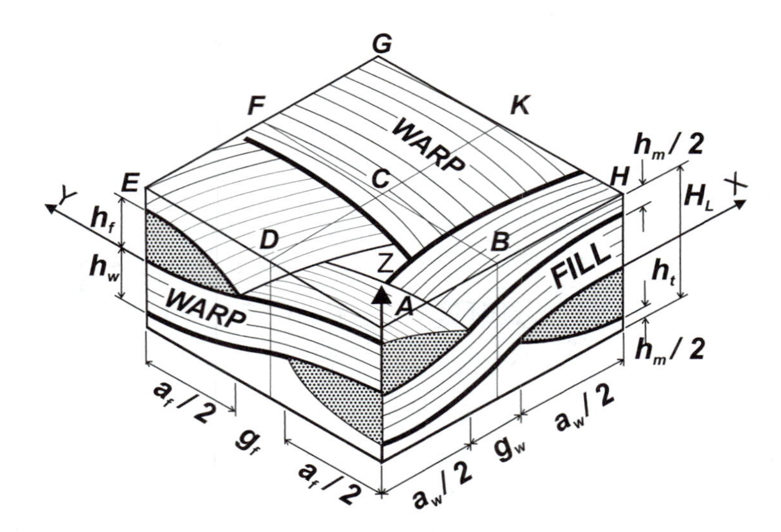

Figure 9.1: One-quarter of the repetitive unit cell (RUC) of a plain weave fabric [212].

ites, such as high stiffness and strength in two perpendicular directions and so on. Therefore, in this chapter, the term laminate is used when the discussion applies indistinctly to a single fabric-reinforced lamina or to a laminated fabric-reinforced composite.

A *fabric* is a collection of tows woven in a given pattern. Both fiber and matrix are responsible for bearing the mechanical loads while the matrix protects the fibers from environmental attack. Fabrics are classified as woven, nonwoven, knitted, braided, two-dimensional (2D), and three-dimensional (3D).

The woven fabrics considered in this chapter are planar, orthogonal, square weaves such as plain weave, twill, and satin [204, 208, 209]. A fabric is woven by interlacing tows, roving, strands, or yarns. While these terms are defined in Section 2.4, the term *tow* will be used in this chapter to denote a group of fibers used as a unit to weave a fabric. Unlike stitched fabrics, a planar weave has no through-the-thickness features other than the undulation resulting from weaving. An orthogonal weave has tows woven at right angles in the plane of the fabric. In this case, the tows laid down in one direction (usually the lamina x-direction) are called *fill* and the perpendicular tows are called *warp* or weft, the later being aligned with the length of the roll of fabric. The pitch is the width of a tow a_i (Figure 9.1) plus the separation between tows g_i, if any; where the subscript $i = f, w$, refers to fill and warp, respectively. The weave is said to be square if the pitch of the fill is equal to that of the warp $(a_f + g_f = a_w + g_w)$.

The integrity of the dry fabric is maintained by the mechanical interlocking of the tows. Fabric properties such as drapeability are controlled primarily by the weave style. Many weave patterns, or styles, exist; the most common being plain weave, satin weave, and twill weave.

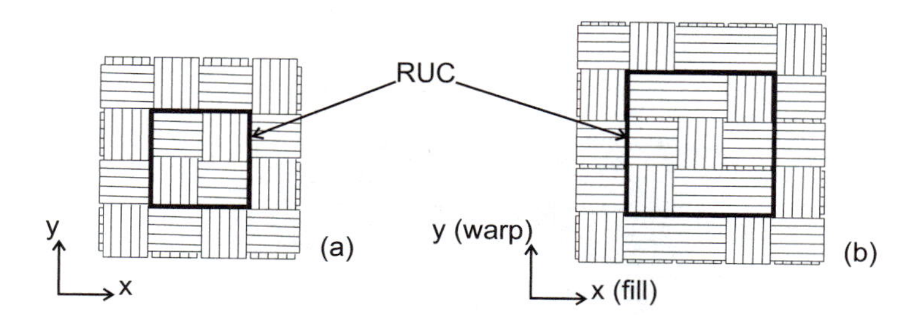

Figure 9.2: (a) plain weave fabric 2/1/1. (b) twill weave fabric 3/1/1.

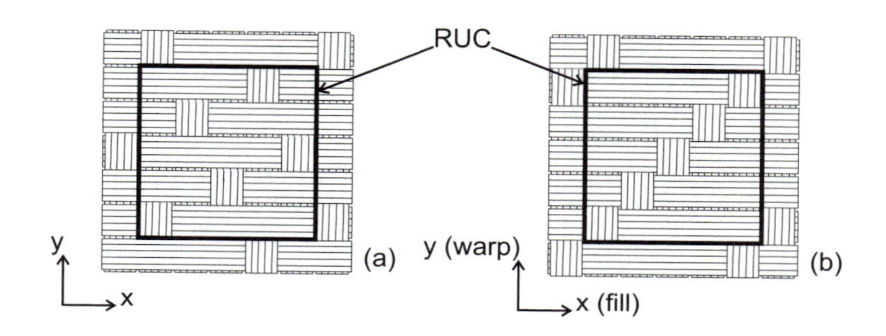

Figure 9.3: (a) satin 5/2/1. (b) twill 5/1/1.

Plain weave is the simplest, most commonly used style. The fill and warp are interlaced to produce a checkerboard effect (Figure 9.2a) with about 50% of fill and 50% of warp exposed on each face of the fabric. Plain weave fabrics are strong, hard wearing, and snag resistant, but they are not very drapeable and tend to wrinkle when trying to conform to contoured molds.

Twill weave is recognized by its characteristic diagonal ribs easily seen on the faces (Figure 9.2b). A twill is made by weaving the fill under the warp ($n_i \geq 1$), and then over two or more warps repeatedly with a *shift* n_s between rows that creates the characteristic diagonal pattern. Twill weaves are more drapeable than plain weave but less drapeable than satin.

Satin weave has interlacing tows arranged in such a way that on one face of the fabric, most of the exposed tows are fill, while on the other face most of the exposed tows are warp. No twill diagonal line can be identified. The satin is made by floating warp tows across a number of fill tows to bring the former to one face and the later to the opposite face of the fabric. The interlacing floats three or more tows between consecutive interlacing (Figure 9.3a). Satin weaves drape and slide extremely well and thus can follow complicated contours in the mold. Long floats such as 7/1/1 or longer may easily snag and are poorly resistant to abrasion. However, shorter floats such as 4/1/1 may be tough and durable.

To describe a given fabric, look at the fabric in such a way that the fill tows are oriented from left to right along the x-direction (fill-direction). First, determine the *harness* n_g; that is, the *smallest* number of adjacent tows needed to define a

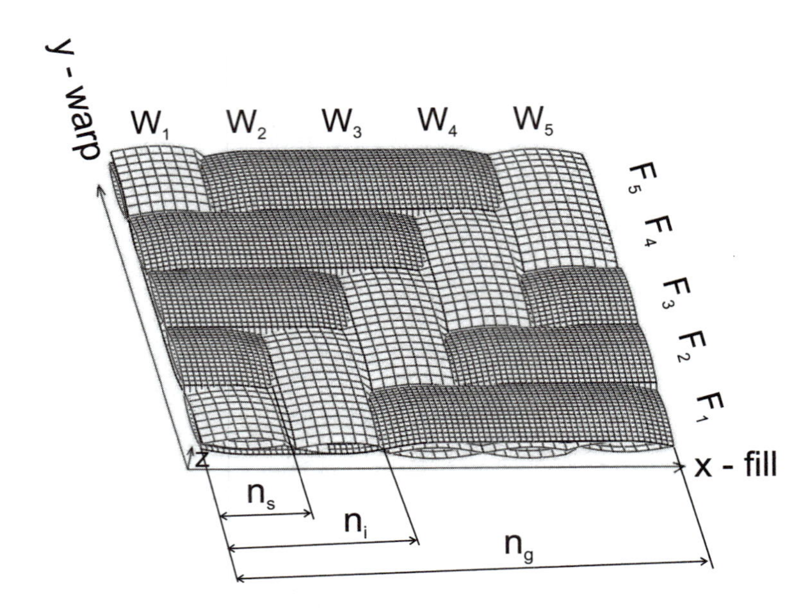

Figure 9.4: Weave parameters n_g, n_s, n_i shown on a 5/1/2 pattern.

repetitive unit cell (RUC). Then, look at the fill tows floating over/under warps. Looking at a *single* fill, say F_1, and counting from left to right, the number of adjacent warps that consecutively appear *on top* of the fill is the *interlacing* n_i (Figure 9.4). Fill F_2 floats *under* n_i warps, but the float is shifted to the right by a number of warps equal to the *shift* n_s. Next, select a representative unit cell (RUC) in such a way that a fill floating *under* n_i warps starts at the left bottom corner of the RUC, as shown in Figure 9.4. In fact, all figures depicting fabrics in this chapter have been drawn following this convention.

The *repetitive unit cell* (RUC) is the smallest portion of the fabric-reinforced composite that contains all the features that repeat periodically on the face of the fabric, as shown in Figures 9.2–9.5.

Three geometric parameters are needed to completely describe the weave pattern in a fashion amenable for modeling and computation. Although the three parameters proposed herein, n_g, n_s, n_i, can be identified along both the fill and warp directions, it helps to look at one direction only when counting geometrical features used to define the parameters (see Figures 9.2–9.5). The three parameters are:

- Harness n_g is the number of subcells along one direction of the RUC ($n_g = 5$ in Figure 9.4).

- Shift n_s is number of subcells between consecutive interlacing regions ($n_s = 1$ in Figure 9.4).

- Interlacing n_i is the number of subcells in the interlacing region ($n_i = 2$ in Figure 9.4).

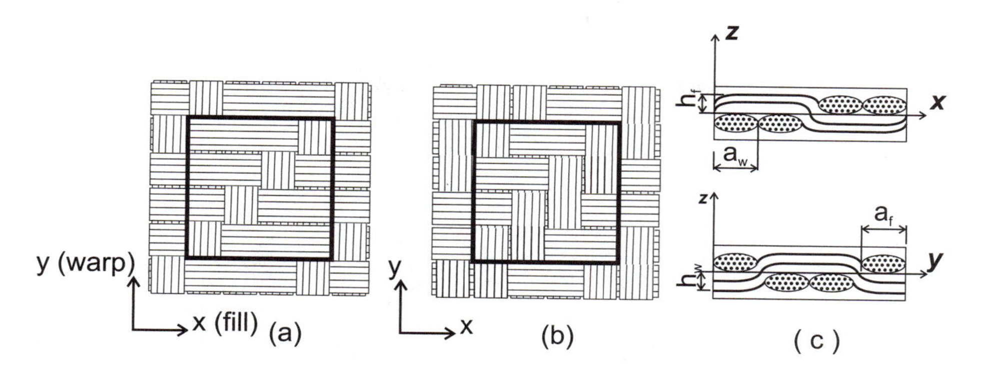

Figure 9.5: (a) twill 4/1/1, (b) twill 4/1/2 (top view), (c) twill 4/1/2 (side view).

A plain weave is described as $n_g/n_s/n_i = 2/1/1$. A subset of twill weaves is obtained when $n_g = 3$, including 3/1/1 (shown in Figure 9.2b) and 3/2/1 (not shown). Fabrics with $n_g \geq 4$ generate either twill or satin depending on the other two parameters. While 4/1/1 (Figure 9.5a) and 4/1/2 (Figure 9.5b) are both twills, and 5/1/1 in (Figure 9.3b) is also a twill, 5/2/1 in (Figure 9.3a) is a satin.

Note that the n_s, n_i, parameters are defined counting along the arbitrary x-direction but they could also be defined counting along the x-direction of the same fabric but rotated by 90° (the former y-direction), which would not necessarily give the same numbers. Furthermore, the counting was done on one face of the fabric but could be done on the opposite (reverse) face as well. Finally, there are several ways to draw valid repetitive unit cells, all with the same hardness n_g but different arrangements of floats (also called *overlaps* or *interlacing*). Any of these definitions would accurately describe the weave. For example, twill weaves can be described indistinctly by $n_g/1/n_i$, or by $n_g/n_g - 1/n_g - n_i$ on the 90° rotated fabric, or by $n_g/1/n_g - n_i$ on the opposite face, or by $n_g/n_g - 1/n_i$ on the opposite, rotated face (see Exercise 9.1).

Also, satin weaves can be described indistinctly by various $n_g/n_s/n_i$ combinations (see Exercise 9.2). Any of these multiple codifications for the same weaving pattern can be reduced to $n_g/1/n_i$ for twills or to $n_g/n_s/1$ for satins ($n_s > 1$). The formulation presented in Section 9.8 assumes that twill patterns are described in such a way that the second parameter is equal to 1 and that satin patterns are described in such a way that the third parameter is equal to 1. For this reason, it is recommended, but not mandatory, to choose the face and orientation of the weaving pattern that displays the minimum length of fill-float-under-warp n_i and the minimum skip n_s.

9.2 Analysis

For sake of pedagogy, the analysis methodology is developed first for the case of plain weave fabric and later generalized for other weave patterns. The methodology

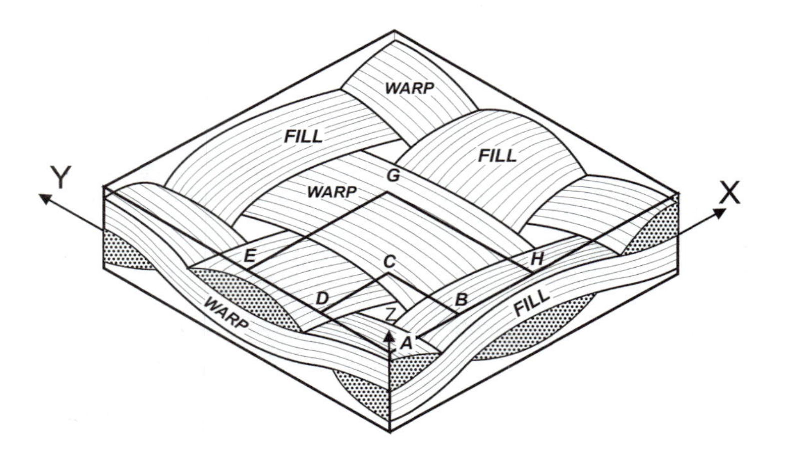

Figure 9.6: Repetitive unit cell (RUC) of a plain weave fabric-reinforced composite [212].

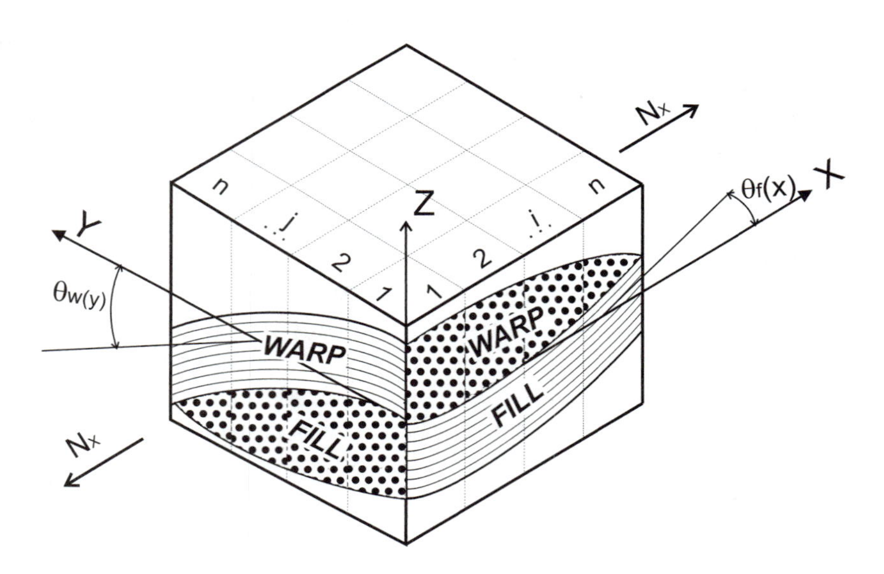

Figure 9.7: Plain weave RVE displaying the 2D mesh on the top surface with elements in positions ij, with $i, j = 1...n$.

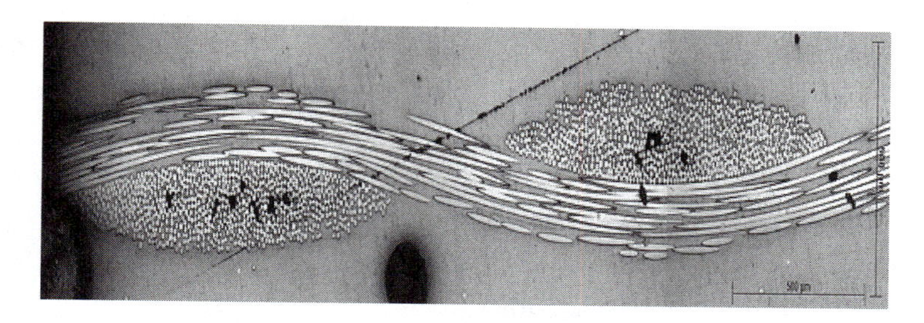

Figure 9.8: Photomicrograph of a fabric-reinforced composite lamina. The undulation of the fill and the cross-section of the warp are shown.

takes into account the undulation of both, fill and warp. The mechanical behavior of the RUC represents the mechanical behavior of the whole composite and thus only the RUC needs to be analyzed (Figure 9.6).

Due to geometric and material symmetry, 1/4 of the RUC (A-H-G-E in Figure 9.1) suffices to analyze a plain weave fabric. Furthermore, using antisymmetry conditions, the behavior of the RUC can be approximated by using a *representative volume element* (RVE) encompassing only 1/16 of the RUC (A-B-C-D in Figure 9.1, see also Figure 9.7). For weaving patterns other than plain weave, no smaller representative volume than the repetitive unit cell can be selected to describe all the material features. In this case, the RVE is the RUC itself. For this reason, the term RVE is used throughout the rest of this chapter.

To define the geometry of the RVE, the following geometrical parameters are needed:

- h_f, h_w, h_m are the fill, warp, and neat-matrix thicknesses, respectively (see Figure 9.1).
- a_f, a_w are the fill and warp widths, respectively (see Figure 9.1).
- g_f, g_w are the gaps between tows (see Figure 9.1).

These geometric parameters can be measured from photomicrographs of composite samples (see Figure 9.8 and [213–215]), or they can be *estimated* from manufacturing specifications, as explained in Section 9.3. Most woven fabrics used for composite reinforcement are tightly woven in order to maximize the fiber volume fraction. Therefore, g is usually negligible and it is sometimes neglected as a first approximation. In Section 9.7, the model will be refined to include the gap. All geometrical parameters for eleven different samples of plain weave fabric, for which experimental data is available, are given in Table 9.1.

At each location x, y, each of the constituents (fill, warp, and neat-matrix) has a thickness (Figures 9.1, 9.7). The z-coordinate of the midpoint through the thickness of each constituent is used to describe the undulation of the tows. Sinusoidal functions are used to describe the undulation of the midpoint of the fill $z_f(x)$ and warp $z_w(y)$; that is, the function that describes the position z of the midpoint along the tow, is defined [207] as follows

$$
\begin{aligned}
z_f(x) &= -\frac{h_f}{2} cos \frac{\pi x}{a_w} \\
z_w(y) &= \frac{h_w}{2} cos \frac{\pi y}{a_f}
\end{aligned}
\tag{9.1}
$$

where the subscripts f, w, refer to fill and warp, respectively. The thickness of cross-section of fill and warp are defined by the following thickness functions[1]

$$
\begin{aligned}
e_f(y) &= |h_f cos \frac{\pi y}{a_f}| \\
e_w(x) &= |h_w cos \frac{\pi x}{a_w}|
\end{aligned}
\tag{9.2}
$$

The top and bottom surfaces bounding the fill and warp tows are obtained simply as

$$
\begin{aligned}
z_f^{top}(x, y) &= z_f(x) + \frac{1}{2}\ e_f(y) \\
z_f^{bot}(x, y) &= z_f(x) - \frac{1}{2}\ e_f(y) \\
z_w^{top}(x, y) &= z_w(y) + \frac{1}{2}\ e_w(x) \\
z_w^{bot}(x, y) &= z_w(y) - \frac{1}{2}\ e_w(x)
\end{aligned}
\tag{9.3}
$$

as shown in Figure 9.4. In the figure, the surface of the warp is represented by the coarse grid and the surface of the fill by the finer grid. The convention in this chapter is that the warp is woven along the y-direction and the fill along the x-direction.

Next, the tow undulation angle (Figure 9.7) can be calculated from the undulation functions as follows

$$
\begin{aligned}
\theta_f(x) &= \left| arctan \left(\frac{\partial}{dx} z_f(x) \right) \right| \\
\theta_w(y) &= \left| arctan \left(\frac{\partial}{dy} z_w(y) \right) \right|
\end{aligned}
\tag{9.4}
$$

Note that the z-coordinate of the midpoint through the thickness of the warp is the same for all values of x and that the z-coordinate of the midpoint through the thickness of the fill is the same for all values of y.

[1] || denotes absolute value.

9.3 Tow Properties

The hygrothermo-mechanical properties of the composite are derived from the properties of the neat matrix regions and those of the fill and warp tows. The properties of the tows (e.g., $E_1, F_{1t}, ...$) are calculated in terms of the thermo-mechanical properties and apparent strength[2] of the constituents (fiber and matrix, Table 9.2) and their respective volume fractions (Table 9.1), as explained in this section. Therefore, the fiber volume fraction inside the tow is needed, but only the overall fiber volume fraction V^o_{fiber} of the lamina is known. The latter can be measured experimentally by acid digestion [118, D 3171] or burn-out tests [118, D 2854]. If candidate materials considered in the design have not been fabricated yet, the overall fiber volume fraction V^o_{fiber} cannot be measured experimentally, but it can be roughly estimated as

$$V^o_{fiber} = \frac{w}{\rho_f h} \tag{9.5}$$

where w is the weight per unit area of the fabric, ρ_f is the density of the fiber material (e.g., glass, carbon, aramid, and so on), and h is the thickness of the fabric-reinforced composite lamina. Again, at the design stage when candidate material samples are not yet available, the thickness of the lamina can be estimated as the thickness of the dry fabric.

The mesoscale volume fraction V_{meso} is the volume of impregnated warp and fill tows $v_w + v_f$ per unit volume of laminate v_{RVE}, calculated as follows

$$V_{meso} = \frac{v_w + v_f}{v_{RVE}} = \frac{4(v_w + v_f)}{h(a_f + g_f)(a_w + g_w)} \tag{9.6}$$

Note that starting with (9.6), the terms g_f, g_w, appear in most equations, even if it was previously stated that the zero gap case is treated first. Generality is achieved in this way, with the equations covering the cases of plain weave fabrics with and without gap, and thus avoiding rewriting these same equations when the effect of the gap is addressed in Section 9.7.

For a plain weave fabric, the fiber volume fraction in the fill V^f_{fiber} and warp V^w_{fiber} are identical, so they are determined as follows

$$V^f_{fiber} = V^w_{fiber} = V^o_{fiber}/V_{meso} \tag{9.7}$$

The volume of the fill and warp tows are calculated as

[2]Apparent strength values are back-calculated from experimental data of unidirectional laminae through the use of micromechanics formulae such as those presented in Section 4.4. Specifically (4.82) and (7.40) have been used in the examples in this chapter. The use of knock-down factors is also common practice, see Table 2.5.

$$
\begin{aligned}
v_f &= A_f \; L_f \\
v_w &= A_w \; L_w
\end{aligned}
\tag{9.8}
$$

The cross-sectional area of the tows in the RVE (Figure 9.7) are calculated using the description of the geometry of the cross-section of fill and warp, as follows

$$
\begin{aligned}
A_f &= \int_0^{(a_f+g_f)/2} e_f(y)dy = \frac{ha_f}{2\pi} \\
A_w &= \int_0^{(a_w+g_w)/2} e_w(x)dx = \frac{ha_w}{2\pi}
\end{aligned}
\tag{9.9}
$$

where the pitch can be either measured from photomicrographs of samples (Figure 9.8) or calculated as the inverse of the tow count, $a_f = 1/N_f$, and $a_w = 1/N_w$, where N_w, N_f are the number of fill and warp tows per unit length, respectively.

The length of the tows in the RVE are calculated using the description of the undulation as follows

$$
\begin{aligned}
L_f &= \int_0^{(a_w+g_w)/2} \sqrt{1+\left(\frac{d}{dx}z_f(x)\right)^2}\, dx \\
L_w &= \int_0^{(a_f+g_f)/2} \sqrt{1+\left(\frac{d}{dy}z_w(y)\right)^2}\, dy
\end{aligned}
\tag{9.10}
$$

Using (9.1) if there is no gap, or (9.47) with gap, and approximating the derivatives in (9.10) using Taylor series expansions, i.e., $\sin(\pi\ x)/(a_w+g_w) \simeq (\pi\ x)/(a_w+g_w)$ and $\sin(\pi\ y)/(a_f+g_f) \simeq (\pi\ y)/(a_f+g_f)$, then (9.10) can be approximated by the following closed-form formulas

$$
\begin{aligned}
L_f &= \frac{1}{2c_f}\left\{\frac{a_w+g_w}{2}\sqrt{c_f^2+\left(\frac{a_w+g_w}{2}\right)^2}\right. \\
&\left. +c_f^2 \ln\left[\frac{1}{c_f}\left(\frac{a_w+g_w}{2}+\sqrt{c_f^2+\left(\frac{a_w+g_w}{2}\right)^2}\right)\right]\right\} \\
L_w &= \frac{1}{2c_w}\left\{\frac{a_f+g_f}{2}\sqrt{c_w^2+\left(\frac{a_f+g_f}{2}\right)^2}\right. \\
&\left. +c_w^2 \ln\left[\frac{1}{c_w}\left(\frac{a_f+g_f}{2}+\sqrt{c_w^2+\left(\frac{a_f+g_f}{2}\right)^2}\right)\right]\right\}
\end{aligned}
\tag{9.11}
$$

$$
\text{with} \quad c_f = \frac{2(a_w+g_w)^2}{\pi^2 h_f} \quad ; \quad c_w = \frac{2(a_f+g_f)^2}{\pi^2 h_w}
$$

Regarding the estimation of composite material geometrical properties (e.g., tow width, tow thickness, the gap between tows) from the dry fabric material properties in the design stage, it has to be noted that, according with the manufacturing process of the composite material, the geometry of the fabric embedded in the composite can be different from the geometry of the dry fabric. This is because of compaction (tows compression in consolidation) that takes place during composite material manufacturing that causes both tow geometry changes and ply nesting (a shift and interlocking between adjacent plies) [216]. The severity of these effects depends on the type of manufacturing technology. For example, it takes place during vacuum bagging but it is more noticeable in resin transfer molding when dry compaction of woven plies is applied prior to resin injection. Therefore, it is more accurate to use the tow and ply geometrical input data as obtained from photomicrographic observation of composite samples.

Once the fiber volume fraction of the tow is calculated by using (9.7), micromechanics formulas are used to calculate the properties of the tow (Tables 9.3–9.4). Properties of interest are the elastic moduli and the strength values for a transversely isotropic material. These can be calculated as shown in Chapter 4.

Applying a micromechanics model, the properties of the tow in its own material coordinate system $(1, 2, 3)$ (Figure 9.9) are calculated and the transversely isotropic compliance matrix of the tow is built [2, (1.91)] as follows

$$S = \begin{pmatrix} 1/E_{11} & -\nu_{21}/E_{22} & -\nu_{31}/E_{33} & 0 & 0 & 0 \\ -\nu_{12}/E_{11} & 1/E_{22} & -\nu_{32}/E_{33} & 0 & 0 & 0 \\ -\nu_{13}/E_{11} & -\nu_{23}/E_{22} & 1/E_{33} & 0 & 0 & 0 \\ 0 & 0 & 0 & 1/G_{23} & 0 & 0 \\ 0 & 0 & 0 & 0 & 1/G_{13} & 0 \\ 0 & 0 & 0 & 0 & 0 & 1/G_{12} \end{pmatrix} \tag{9.12}$$

Since the tows are undulated, the transformation matrices from the laminate coordinate system (x, y, z) to the material coordinate system $(1, 2, 3)$ in Figure 9.9 are calculated [2]. The fill tow undergoes a clockwise rotation $\theta_f(x)$ about the $2f$ axis, while the warp tow undergoes two rotations: one 90° counterclockwise rotation about z-axis, and one counterclockwise rotation $\theta_w(y)$ about the $2w$-axis. Considering this, the fill and warp rotation matrices from (x, y, z) to $(1, 2, 3)$ coordinate system can be written [2, (1.18), Section 1.10] as

$$\Theta_f(x) = \begin{pmatrix} l_1 & m_1 & n_1 \\ l_2 & m_2 & n_2 \\ l_3 & m_3 & n_3 \end{pmatrix} = \begin{pmatrix} \cos\theta_f & 0 & \sin\theta_f \\ 0 & 1 & 0 \\ -\sin\theta_f & 0 & \cos\theta_f \end{pmatrix} \tag{9.13}$$

$$\Theta_w(y) = \begin{pmatrix} l_1 & m_1 & n_1 \\ l_2 & m_2 & n_2 \\ l_3 & m_3 & n_3 \end{pmatrix} = \begin{pmatrix} 0 & \cos\theta_w & \sin\theta_w \\ -1 & 0 & 0 \\ 0 & -\sin\theta_w & \cos\theta_w \end{pmatrix} \tag{9.14}$$

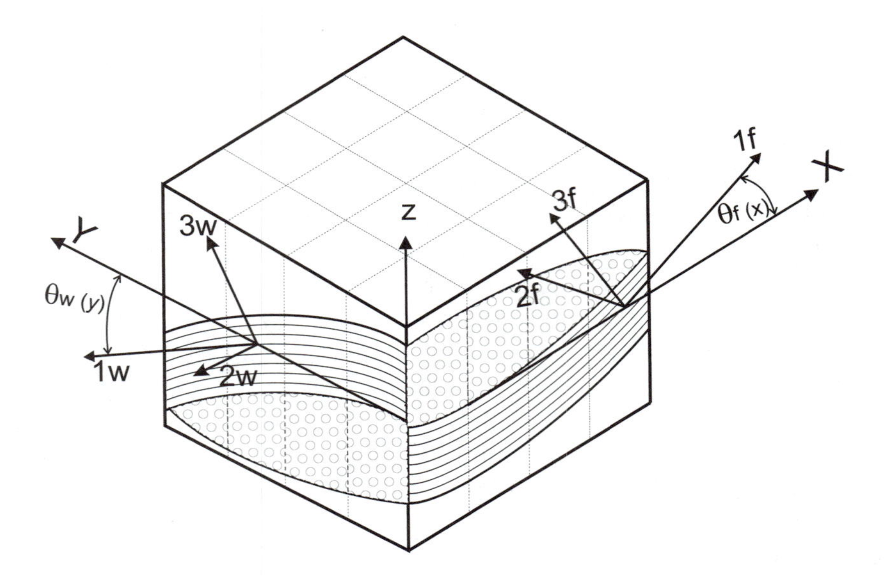

Figure 9.9: The tow material coordinate system $(1, 2, 3)$ vs. the RVE laminate coordinate system (x, y, z).

where f, w, indicates fill and warp, respectively. Note that the direction cosines l, m, n of the fill are functions of x, and the direction cosines of the warp are functions of y, according with the $\theta_f(x)$ and $\theta_w(y)$ rotations. Further, the transformation matrices *from laminate to material coordinate system*, for fill and warp can be calculated [2, (1.33)] as

$$T_{(f,w)}(x) = \begin{pmatrix} l_1^2 & m_1^2 & n_1^2 & 2m_1 n_1 & 2l_1 n_1 & 2l_1 m_1 \\ l_2^2 & m_2^2 & n_2^2 & 2m_2 n_2 & 2l_2 n_2 & 2l_2 m_2 \\ l_3^2 & m_3^2 & n_3^2 & 2m_3 n_3 & 2l_3 n_3 & 2l_3 m_3 \\ l_2 l_3 & m_2 m_3 & n_2 n_3 & m_2 n_3 + n_2 m_3 & l_2 n_3 + n_2 l_3 & l_2 m_3 + m_2 l_3 \\ l_1 l_3 & m_1 m_3 & n_1 n_3 & m_1 n_3 + n_1 m_3 & l_1 n_3 + n_1 l_3 & l_1 m_3 + m_1 l_3 \\ l_1 l_2 & m_1 m_2 & n_1 n_2 & m_1 n_2 + n_1 m_2 & l_1 n_2 + n_1 l_2 & l_1 m_2 + m_1 l_2 \end{pmatrix} \tag{9.15}$$

Equation 9.15 for the fill reduces to

$$T_f(x) = \begin{bmatrix} c_f^2 & 0 & s_f^2 & 0 & 2c_f s_f & 0 \\ 0 & 1 & 0 & 0 & 0 & 0 \\ s_f^2 & 0 & c_f^2 & 0 & -2c_f s_f & 0 \\ 0 & 0 & 0 & c_f & 0 & -s_f \\ -c_f s_f & 0 & c_f s_f & 0 & c_f^2 - s_f^2 & 0 \\ 0 & 0 & 0 & s_f & 0 & c_f \end{bmatrix} \tag{9.16}$$

and for the warp to

$$T_w(y) = \begin{bmatrix} 0 & c_w^2 & s_w^2 & -2c_w s_w & 0 & 0 \\ 1 & 0 & 0 & 0 & 0 & 0 \\ 0 & s_w^2 & c_w^2 & 2c_w s_w & 0 & 0 \\ 0 & 0 & 0 & 0 & -c_w & -s_w \\ 0 & c_w s_w & -c_w s_w & c_w^2 - s_w^2 & 0 & 0 \\ 0 & 0 & 0 & 0 & s_w & -c_w \end{bmatrix} \tag{9.17}$$

where

$$\begin{aligned} s_f(x) &= sin\ \theta_f(x) \\ c_f(x) &= cos\ \theta_f(x) \\ s_w(y) &= sin\ \theta_w(y) \\ c_w(y) &= cos\ \theta_w(y) \end{aligned} \tag{9.18}$$

The transformed compliance matrices $\overline{S}_f(x), \overline{S}_w(y)$, in laminate coordinate system (x, y, z) can now be calculated as

$$\begin{aligned} \overline{S}_f(x) &= (T_f(x))^T \cdot S \cdot T_f(x) \\ \overline{S}_w(y) &= (T_w(y))^T \cdot S \cdot T_w(y) \end{aligned} \tag{9.19}$$

where $\overline{S}_f(x)$ and $\overline{S}_w(y)$ are 3D compliance matrices. Because of the plane stress assumption, only the $\overline{S}_{i,j}$ terms with $i, j = 1, 2, 6$ are considered and the reduced stiffness matrices in laminate coordinate system have the form

$$^i\overline{Q} = \begin{bmatrix} ^i\overline{Q}_{11} & ^i\overline{Q}_{12} & 0 \\ ^i\overline{Q}_{21} & ^i\overline{Q}_{22} & 0 \\ 0 & 0 & ^i\overline{Q}_{66} \end{bmatrix} = \begin{bmatrix} ^i\overline{S}_{11} & ^i\overline{S}_{12} & 0 \\ ^i\overline{S}_{21} & ^i\overline{S}_{22} & 0 \\ 0 & 0 & ^i\overline{S}_{66} \end{bmatrix}^{-1} \tag{9.20}$$

where $i = f, w$, indicates fill and warp, respectively. Unlike for classical laminates, the 16 and 26 terms are zero for fabric-reinforced composites because there are no rotations around the z-axis; the only rotations are due to the undulation of the fill and warp.

For calculating thermal and moisture expansion coefficients α and β of the woven fabric-reinforced laminate, a similar approach to that used for elastic moduli is proposed. A micromechanics model (Chapter 4) is applied to calculate the coefficients α, β, for fill and warp tows, taken as unidirectional laminae, at their corresponding tow fiber volume fraction, from the transversely isotropic fiber and isotropic matrix constituents properties that are presented in Table 9.2.

9.4 Element Stiffness and Constitutive Relationship

The RVE is meshed on the surface by a 2D mesh as shown in Figure 9.7 and the volume defined as the area of each 2D element times the thickness of the lamina

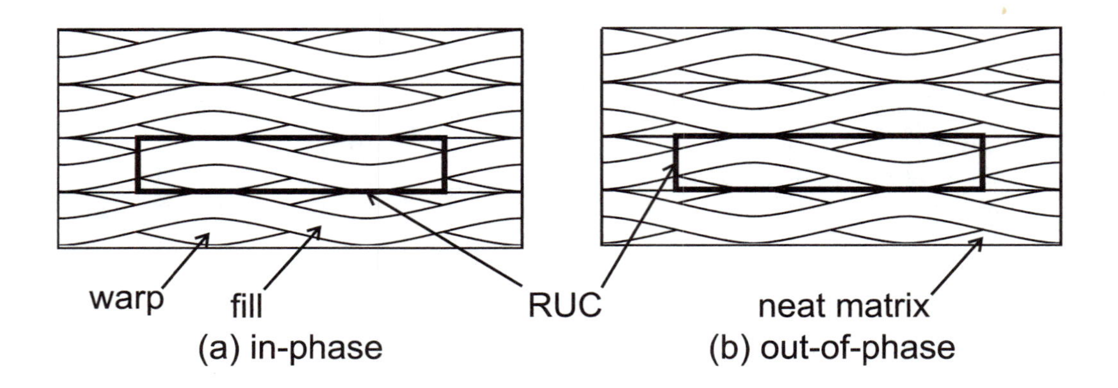

Figure 9.10: Fabric-reinforced laminates: (a) in-phase configuration, and (b) out-of-phase configuration.

is analyzed using classical lamination theory (CLT) in the usual way (Chapter 6). Each element is an asymmetric laminate with four layers ($k = 1..4$) and laminate stacking sequence (LSS) $[m, f, w, m]$ or $[m, w, f, m]$, depending on the x, y position, where m, f, w indicate neat-matrix, fill, and warp, respectively.

Since the element LSS is in general asymmetric, bending-extension coupling occurs and the local stiffness coefficients for each point of RVE can be calculated according to classical lamination theory, as follows

$$A_{i,j}(x, y), B_{i,j}(x, y), D_{i,j}(x, y) = \int_{-h/2}^{h/2} (1, z, z^2) \cdot \overline{Q}_{i,j}(x, y)\, dz \tag{9.21}$$

where $i, j = 1, 2, 6$.

When the fabric-reinforced laminate is loaded by in-plane forces, local rotations appear due to the undulation and the interlacing of the tows. Regarding the bending-extension coupling, two models can be used: bending-restrained and bending-allowed.

9.4.1 Bending-Restrained Model

This model is especially appropriate for plain weave pattern. In this case, the average curvature of the RUC in the single lamina case is zero because of the sinusoidal nature of the undulation; that is, positive and negative curvatures from two adjacent yarns cancel out each other because the stacking sequence is inverted. Moreover, in the case of the laminate made of several fabric-reinforced laminae, two extreme cases are identified depending on the relative position of the maxima of the undulations in the various laminae.

If the maxima of the undulation profile on one lamina is aligned with the minima of the profile of the adjacent lamina, the laminate is said to be out-of-phase because the sinusoidal curves are exactly out-of-phase (Figure 9.10b). In this case the overall laminate stacking sequence is symmetric, even if individual fabric-reinforced laminae are asymmetric. Therefore, the local laminate bending-extension coefficients $B_{i,j}(x, y)$ are zero and no local curvatures are induced under in-plane load.

If the maxima of the undulation profile are aligned at the same x, y location for all laminae, the laminate is said to be in-phase (Figure 9.10a) because the phase of the sinusoidal curves coincide for all laminae. In this case the overall laminate stacking sequence is no longer symmetric, but it can be shown (see Exercise 6.9) that the bending-extension coupling coefficients become negligible as the number of laminae increases.

In the case of bending-restrained model, the material represented by the RVE has a constitutive relationship given by

$$\begin{Bmatrix} N_x \\ N_y \\ N_{xy} \end{Bmatrix} = [A_{i,j}(x, y)] \begin{Bmatrix} \epsilon_x^0 \\ \epsilon_y^0 \\ \gamma_{xy}^0 \end{Bmatrix} \tag{9.22}$$

where $\epsilon_x^0, \epsilon_y^0, \gamma_{xy}^0$ are the midplane strains (constant over the lamina thickness), calculated for each (x, y) point. Next, the local strains are calculated as

$$\begin{Bmatrix} \epsilon_x^0(x, y) \\ \epsilon_y^0(x, y) \\ \gamma_{xy}^0(x, y) \end{Bmatrix} = [A_{i,j}(x, y)]^{-1} \begin{Bmatrix} N_x \\ N_y \\ N_{xy} \end{Bmatrix} \tag{9.23}$$

The bending-restrained model works particularly well for plain weave pattern without damage.

9.4.2 Bending-Allowed Model

In this case the local nonzero terms $B_{i,j}(x, y)$ and $D_{i,j}(x, y)$ are considered. Each infinitesimal element is regarded as an asymmetric laminate, having the stacking sequence matrix–fill–warp-matrix (from bottom to top for the considered RVE, Figure 9.7). Because of this asymmetric laminate structure, bending-extension coupling appears. In this case, the constitutive equation for woven fabric lamina under in-plane loading becomes

$$\begin{Bmatrix} N_x \\ N_y \\ N_{xy} \\ M_x \\ M_y \\ M_{xy} \end{Bmatrix} = \begin{bmatrix} [A_{i,j}(x, y)] & [B_{i,j}(x, y)] \\ [B_{i,j}(x, y)] & [D_{i,j}(x, y)] \end{bmatrix} \begin{Bmatrix} \epsilon_{x0} \\ \epsilon_{y0} \\ \gamma_{xy0} \\ k_x \\ k_y \\ k_{xy} \end{Bmatrix} \tag{9.24}$$

and the local midplane strains and curvatures are

$$\begin{Bmatrix} \epsilon_x^0 \\ \epsilon_y^0 \\ \gamma_{xy}^0 \\ k_x \\ k_y \\ k_{xy} \end{Bmatrix} (x, y) = \begin{bmatrix} [A_{i,j}(x, y)] & [B_{i,j}(x, y)] \\ [B_{i,j}(x, y)] & [D_{i,j}(x, y)] \end{bmatrix}^{-1} \begin{Bmatrix} N_x \\ N_y \\ N_{xy} \\ M_x \\ M_y \\ M_{xy} \end{Bmatrix} \tag{9.25}$$

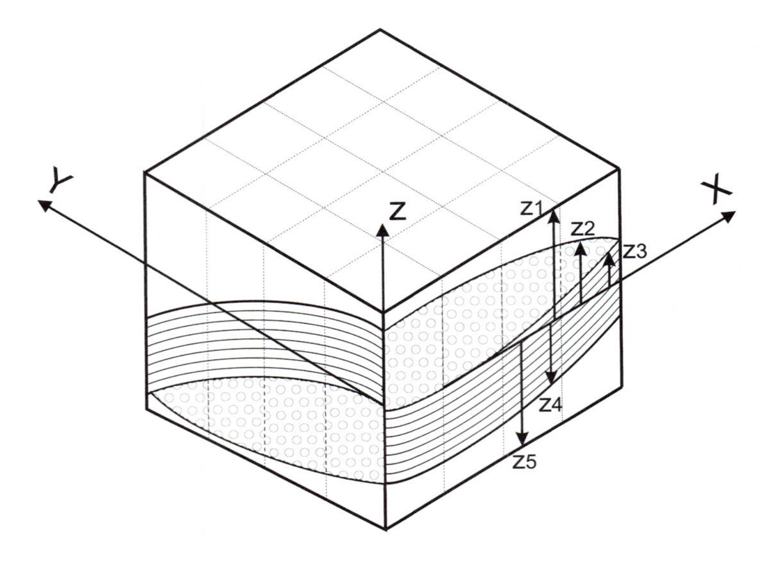

Figure 9.11: The asymmetric stacking sequence of woven fabric-reinforced lamina.

After that, the strain distribution through the thickness for each grid element (Figure 9.11) can be calculated as

$$\begin{Bmatrix} \epsilon_x \\ \epsilon_y \\ \gamma_{xy} \end{Bmatrix} (x, y, z) = \begin{Bmatrix} \epsilon_{x0} \\ \epsilon_{y0} \\ \gamma_{xy0} \end{Bmatrix} (x, y) + z \begin{Bmatrix} k_x \\ k_y \\ k_{xy} \end{Bmatrix} (x, y) \tag{9.26}$$

In this case, a strain (and consequently stress) distribution over the thickness (along z-coordinate) is obtained, complicating the computation of the stress-strain response (Figure 9.11). For each fill–warp–matrix constituent, the strains and stresses have to be calculated at the top and the bottom of each constituent, according with the considered asymmetric LSS over the RVE, and the maximum value of strain–stress has to be considered for failure analysis.

In real situations, the laminae are placed randomly on top of each other and the situation is somewhere in between the two extreme cases already discussed. As long as the stress distribution and the material behavior up to initial damage is regarded, the local bending constraint model offers good agreement with experimental data, and it also offers the advantage of simplicity. But if damage is considered, attention should be paid to this assumption, because after matrix damage initiation the constraint effect between adjacent tows and layers is progressively lost; individual layers are not bending-constrained any more and the bending-extension coupling effect might not be negligible.

9.5 Laminate Properties

9.5.1 Elastic Constants

The stiffness $E_x, E_y, G_{xy}, \nu_{xy}$ of the laminate, or of its surrogate the RVE, are obtained by assembling the stiffness of all the elements in the RVE. This can be accomplished in a variety of ways, which are all instances of homogenization theory. The simplest approach is to use an isostrain assumption [217]; that is, to assume that all the elements are subjected to the same in-plane strain $\{\epsilon_x, \epsilon_y, \gamma_{xy}\}^T$ and to use the strain energy of RVE for deriving the equivalent laminate elastic constants from the elastic properties of the constituent materials (matrix and tows). This is an approximation because the RVE is inhomogeneous and thus certain parts of it do deform more than others when a load $\{N_x, N_y, N_{xy}\}^T$ is applied on its boundary. This is the reason why the isostrain assumption is used only for evaluating the material elastic constants, and it is not applicable for stress analysis and failure. Better approximations exist [213] or are currently being developed. For example, micromechanics of composites with periodic microstructure yield excellent predictions of moduli but it is somewhat computationally expensive [213]. For the sake of simplicity and pedagogy, the isostrain assumption [204, 209, 216] is used herein to compute the extensional stiffness matrix of the laminate, simply as follows

$$
\begin{aligned}
[A] &= \frac{4}{(a_f + g_f)(a_w + g_w)} \int_0^{(a_w+g_w)/2} \int_0^{(a_f+g_f)/2} [A(x,y)] dx dy \\
&= \frac{4}{(a_f + g_f)(a_w + g_w)} \sum_0^{n^2} [A]^{(e)} S^e
\end{aligned}
\tag{9.27}
$$

where $[A]^{(e)}, S^e$ is the extensional stiffness matrix of the e-th element and its area, respectively, and n^2 is the number of elements used to discretize the RVE (Figure 9.7). The laminate stiffness can be inverted to yield the laminate compliance $[\alpha] = [A]^{-1}$, and the laminate moduli listed in Table 9.5 are obtained simply as

$$
\begin{aligned}
E_x &= (\alpha_{11} h)^{-1} \\
E_y &= (\alpha_{22} h)^{-1} \\
G_{xy} &= (\alpha_{66} h)^{-1} \\
\nu_{xy} &= -\frac{\alpha_{12}}{\alpha_{11}}
\end{aligned}
\tag{9.28}
$$

where h is the thickness of the laminate. Comparison between predicted and experimental values is shown in Table 9.6.

9.5.2 Thermal and Moisture Expansion Coefficients

The mechanical analysis of polymer matrix composite materials needs to consider not only the effect of mechanical applied loads, but also the effect of temperature and moisture concentration change. The resulting thermal and moisture expansion coefficients for unidirectional fill and warp tows, regarded in material $(1,2,3)$ coordinate system, are presented in Table 9.3.

The unidirectional tow properties are affected by the fill and warp undulation angles, $\theta_f(x)$ and $\theta_w(y)$ from (9.4), which are used for calculating the corresponding local off-axis thermal and moisture expansion coefficients in the laminate coordinate system (x,y,z). Since $\beta\Delta m$ and $\alpha\Delta T$ are strains, the strain transformation relation from laminate to material coordinate system $\{\epsilon'\}^{T,M} = [T]\{\epsilon\}^{T,M}$ is used, where $\{\epsilon'\}^{T,M}$ and $\{\epsilon\}^{T,M}$ are the tensor strains due to thermal and moisture expansion effect in material $(1,2,3)$ and laminate (x,y,z) coordinate system, respectively [2, (1.35)]. The transformation matrix $[T]$ is given by (9.16) and (9.17) for fill and warp tows. In the case of thermal and moisture expansion coefficients, the inverse transformation, from material to laminate coordinate system is required $\{\epsilon\}^{T,M} = [T]^{-1}\{\epsilon'\}^{T,M}$. Next, the coordinate transformation for the thermal and moisture expansion coefficients can be derived as

$$\left\{\begin{matrix} \alpha_x(x,y) \\ \alpha_y(x,y) \\ \alpha_z(x,y) \\ \frac{1}{2}\alpha_{yz}(x,y) \\ \frac{1}{2}\alpha_{xz}(x,y) \\ \frac{1}{2}\alpha_{xy}(x,y) \end{matrix}\right\}_{f,w} = [T]^{-1}_{f,w}(x,y)\left\{\begin{matrix} \alpha_1 \\ \alpha_2 \\ \alpha_3 \\ 0 \\ 0 \\ 0 \end{matrix}\right\}_{f,w} \tag{9.29}$$

$$\left\{\begin{matrix} \beta_x(x,y) \\ \beta_y(x,y) \\ \beta_z(x,y) \\ \frac{1}{2}\beta_{yz}(x,y) \\ \frac{1}{2}\beta_{xz}(x,y) \\ \frac{1}{2}\beta_{xy}(x,y) \end{matrix}\right\}_{f,w} = [T]^{-1}_{f,w}(x,y)\left\{\begin{matrix} \beta_1 \\ \beta_2 \\ \beta_3 \\ 0 \\ 0 \\ 0 \end{matrix}\right\}_{f,w} \tag{9.30}$$

where $\alpha_1, \alpha_2, \alpha_3$ and $\beta_1, \beta_2, \beta_3$ are given in (4.72–4.73) and (4.76), respectively.

The matrix inversion $[T]^{-1}_{f,w}$ can be performed easily taking into account that $[T]^{-1}_{f,w} = [\widehat{T}]^T_{f,w}$, where $[\widehat{T}] = [R][T][R]^{-1}$, with $[R]$ being the Reuter matrix, defined as $R_{ij} = 1$ for $i = j = 1\cdots3$, $R_{ij} = 2$ for $i = j = 4\cdots6$, and $R_{ij} = 0$ for $i \neq j$ (See [2, (1.47)]). Selecting the in-plane (x,y) terms of interest, the local fill and warp thermal and moisture expansion coefficients in the (x,y) plane are obtained from (9.30) as

$$
\begin{array}{lll}
\alpha_x^f(x) = \alpha_1 c_f^2 + \alpha_2 s_f^2 & \alpha_y^f(x) = \alpha_2 & \alpha_{xy}^f = 0 \\
\alpha_x^w(y) = \alpha_2 & \alpha_y^w(y) = \alpha_1 c_w^2 + \alpha_2 s_w^2 & \alpha_{xy}^w = 0 \\
\beta_x^f(x) = \beta_1 c_f^2 + \beta_2 s_f^2 & \beta_y^f(x) = \beta_2 & \beta_{xy}^f = 0 \\
\beta_x^w(y) = \beta_2 & \beta_y^w(y) = \beta_1 c_w^2 + \beta_2 s_w^2 & \beta_{xy}^w = 0
\end{array}
\tag{9.31}
$$

where c_f, s_f, c_w, s_w are given by (9.18). The local hygrothermal forces and moments per unit length are calculated according with CLT (Section 6.6)

$$
\begin{aligned}
\begin{Bmatrix} N_x^T \\ N_y^T \\ N_{xy}^T \end{Bmatrix}(x,y) &= \Delta T \int_{-h/2}^{h/2} [\overline{Q}(x,y)] \cdot \begin{Bmatrix} \alpha_x(x,y) \\ \alpha_y(x,y) \\ \alpha_{xy}(x,y) \end{Bmatrix} dz \\
\begin{Bmatrix} N_x^M \\ N_y^M \\ N_{xy}^M \end{Bmatrix}(x,y) &= \Delta m \int_{-h/2}^{h/2} [\overline{Q}(x,y)] \cdot \begin{Bmatrix} \beta_x(x,y) \\ \beta_y(x,y) \\ \beta_{xy}(x,y) \end{Bmatrix} dz \\
\begin{Bmatrix} M_x^T \\ M_y^T \\ M_{xy}^T \end{Bmatrix}(x,y) &= \Delta T \int_{-h/2}^{h/2} [\overline{Q}(x,y)] \cdot \begin{Bmatrix} \alpha_x(x,y) \\ \alpha_y(x,y) \\ \alpha_{xy}(x,y) \end{Bmatrix} z dz \\
\begin{Bmatrix} M_x^M \\ M_y^M \\ M_{xy}^M \end{Bmatrix}(x,y) &= \Delta m \int_{-h/2}^{h/2} [\overline{Q}(x,y)] \cdot \begin{Bmatrix} \beta_x(x,y) \\ \beta_y(x,y) \\ \beta_{xy}(x,y) \end{Bmatrix} z dz
\end{aligned}
\tag{9.32}
$$

where the superscripts T, M indicate the effect of temperature variation ΔT, and moisture content variation Δm, respectively.

As in the case of elastic constants for woven fabric-reinforced laminate, a homogenization technique is required, similarly to the isostrain method used in Section 9.5.1. According with the isostrain method, the average in-plane load over the RVE due to temperature and moisture content change becomes

$$
\{N\}^{T,M} = \frac{4}{(a_f+g_f)(a_w+g_w)} \int_0^{(a_w+g_w)/2} \int_0^{(a_f+g_f)/2} \{N^{T,M}(x,y)\}\, dxdy \tag{9.33}
$$

and the linear thermal and moisture expansion coefficients become

$$
\begin{Bmatrix} \alpha_x \\ \alpha_y \\ \alpha_{xy} \end{Bmatrix} = [A]^{-1} \begin{Bmatrix} N_x^T \\ N_y^T \\ N_{xy}^T \end{Bmatrix} \quad ; \quad \begin{Bmatrix} \beta_x \\ \beta_y \\ \beta_{xy} \end{Bmatrix} = [A]^{-1} \begin{Bmatrix} N_x^M \\ N_y^M \\ N_{xy}^M \end{Bmatrix} \tag{9.34}
$$

Calculated linear thermal and moisture expansion coefficients are presented in Table 9.5. Comparison between predicted and experimental values is shown in Table 9.6.

9.6 Failure Analysis

Failure analysis is carried out by computing the stresses in each constituent (f, w, m) at each coordinate (x, y) and using both a *damage activation function* (failure criterion) and a *damage evolution function* to predict initiation and evolution of damage up to laminate collapse. Again for the sake of simplicity and pedagogy, the *maximum stress criterion* (Section 7.1.2) and a *damage factor* d_f (Section 7.3.1) are used for damage initiation and evolution, respectively. These choices, while expedient, leave much potential for improvement.

9.6.1 Stress Analysis

At each increment of the applied load, strains and stresses are computed at the mid-point of each constituent (neat-matrix, fill, and warp; Figure 9.7) of each element. Using a plane stress assumption and distributing the load N_x equally among all divisions in the RVE in Figure 9.7, the local in-plane strain in each element in laminate (x, y, z) coordinate system is computed using either (9.23) or (9.26), according with the bending-restrained or bending-allowed model. Then, the local stresses for each constituent (fill, warp, matrix) of each element in the laminate coordinate system (x, y, z) are calculated as

$$\left\{ \begin{array}{c} \sigma_x \\ \sigma_y \\ \sigma_{xy} \end{array} \right\}^{f,w,m} (x, y, z) = [\overline{Q}(x, y)]^{f,w,m} \left\{ \begin{array}{c} \epsilon_x \\ \epsilon_y \\ \gamma_{xy} \end{array} \right\} (x, y, z) \tag{9.35}$$

where $[\overline{Q}]^{f,w,m}$ is the reduced stiffness matrix in laminate coordinates (x, y, z) of each fill, warp, and matrix constituent at the (x, y) position considered. Note that the reduced stiffness matrix of each phase is different in each element because the undulation angle and the thickness of the phases vary as a function of x and y.

Regarding the use of the plane stress assumption (9.20), classical lamination theory ((9.23) or (9.26), (9.35)), and in light of the 3D features of woven fabric-reinforced composites, the following observation is made: the 3D transformed compliance matrices in laminate coordinates (x, y, z) for fill and warp tows (9.19) have the following form at any point (x, y) inside the RVE

$$\overline{S}_f(x) = \begin{bmatrix} S_{11}^f & S_{12}^f & S_{13}^f & 0 & S_{15}^f & 0 \\ S_{21}^f & S_{22}^f & S_{23}^f & 0 & S_{25}^f & 0 \\ S_{31}^f & S_{32}^f & S_{33}^f & 0 & S_{35}^f & 0 \\ 0 & 0 & 0 & S_{44}^f & 0 & S_{46}^f \\ S_{51}^f & S_{52}^f & S_{53}^f & 0 & S_{55}^f & 0 \\ 0 & 0 & 0 & S_{64}^f & 0 & S_{66}^f \end{bmatrix}$$

$$\overline{S}_w(y) = \begin{bmatrix} S_{11}^w & S_{12}^w & S_{13}^w & S_{14}^w & 0 & 0 \\ S_{21}^w & S_{22}^w & S_{23}^w & S_{24}^w & 0 & 0 \\ S_{31}^w & S_{32}^w & S_{33}^w & S_{34}^w & 0 & 0 \\ S_{41}^w & S_{42}^w & S_{43}^w & S_{44}^w & 0 & 0 \\ 0 & 0 & 0 & 0 & S_{55}^w & S_{56}^w \\ 0 & 0 & 0 & 0 & S_{65}^w & S_{66}^w \end{bmatrix} \qquad (9.36)$$

where the terms $S_{ij}^{f,w}$ are generally nonzero. Applying the plane stress assumption (9.20), some of the nonzero terms are dropped. In this way, some coupling effects are neglected during stress calculation. For example, in the case of fill tow, dropping the S_{51}^f term, the shear-extension coupling σ_x–γ_{xz} is neglected, even though a shear strain γ_{xz} could be calculated as a result of an applied σ_x stress. The same applies for the shear-extension coupling $\sigma_y - \gamma_{xz}$ neglected by dropping the S_{52}^f term and the shear-shear coupling $\tau_{xy} - \gamma_{yz}$ neglected by dropping the S_{46}^f term. In the case of warp tow, the neglected couplings are shear-extension $\sigma_y - \gamma_{yz}$ by dropping the S_{42}^w term, shear-extension $\sigma_x - \gamma_{yz}$ by dropping the S_{41}^w term, and shear-shear $\tau_{xy} - \gamma_{xz}$ by dropping the S_{46}^w term.

In order to apply a failure criterion, the stress is transformed from the laminate coordinate system to the material coordinates system of the fill and warp as follows

$$\{\sigma_1, \sigma_2, \sigma_3, \sigma_4, \sigma_5, \sigma_6\}_{f,w}^T = [T]_{f,w} \{\sigma_x, \sigma_y, 0, 0, 0, \sigma_{xy}\}_{f,w}^T \qquad (9.37)$$

where the $[T]$ represents the transformation matrix from laminate (x, y, z) to material $(1, 2, 3)$ coordinate system for each of fill and warp phase (9.16)–(9.17).

The RVE stiffness matrices $[A]$, $[B]$, $[D]$ can be calculated by assembling the local element matrices $[A(x, y)]$, $[B(x, y)]$, $[D(x, y)]$ using any homogenization method. In this case, the isostrain assumption is used for simplicity (9.27). The average strain over the RVE is then calculated accordingly with the bending-restrained model as

$$\{\epsilon_x^0, \epsilon_y^0, \gamma_{xy}^0\}^T = [A_{i,j}]^{-1} \{N_x, N_y, N_{xy}\}^T \qquad (9.38)$$

or with the bending-allowed model

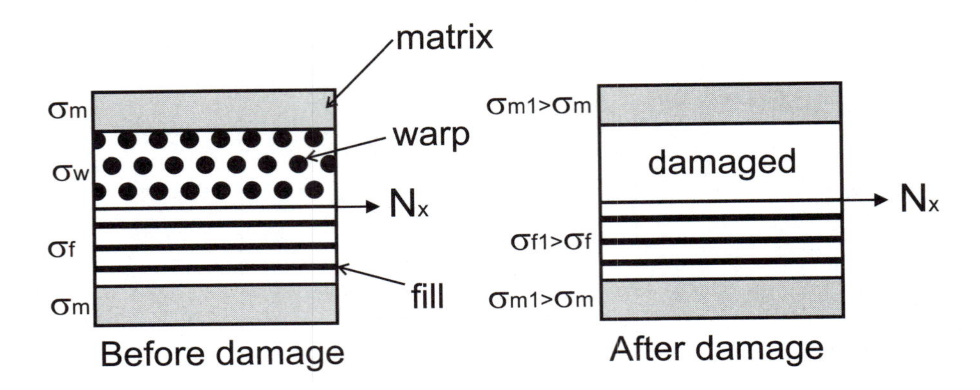

Figure 9.12: Stress redistribution over the cell thickness as a result of stiffness reduction due to phase damage (matrix, warp, fill).

$$\begin{Bmatrix} \epsilon_x^0 \\ \epsilon_y^0 \\ \gamma_{xy}^0 \\ k_x \\ k_y \\ k_{xy} \end{Bmatrix} = \begin{bmatrix} [A_{i,j}] & [B_{i,j}] \\ [B_{i,j}] & [D_{i,j}] \end{bmatrix}^{-1} \begin{Bmatrix} N_x \\ N_y \\ N_{xy} \\ M_x \\ M_y \\ M_{xy} \end{Bmatrix} \tag{9.39}$$

where $i, j = 1, 2, 6$. The average in-plane stress is calculated as

$$\{\sigma_x, \sigma_y, \sigma_{xy}\}^T = \{N_x/h, N_y/h, N_{xy}/h\}^T \tag{9.40}$$

where h is the laminate thickness.

9.6.2 Damage Initiation, Evolution, and Fracture

Additional modeling features are proposed in this section in order to describe the material behavior before and after failure initiation. A bending-restrained model (9.22)–(9.23) is used as it yields results in closer agreement to experimental data than a bending-allowed model.

a. Ineffective Longitudinal Stress Recovery

In the case of tensile loading, lamina failure is controlled by longitudinal tow failure. Due to fully 3D undulation featured by the present model and the oblong geometry of the tow cross-section, longitudinal tow failure initiates where the tow thickness is smallest, at the edges of the cross-section of the tow (elements in row n in Figure 9.7), and progresses towards the thicker middle point (elements in row 1 in Figure 9.7). Furthermore, longitudinal tow failure is influenced by the the local tow angle $\theta^{f,w}$,

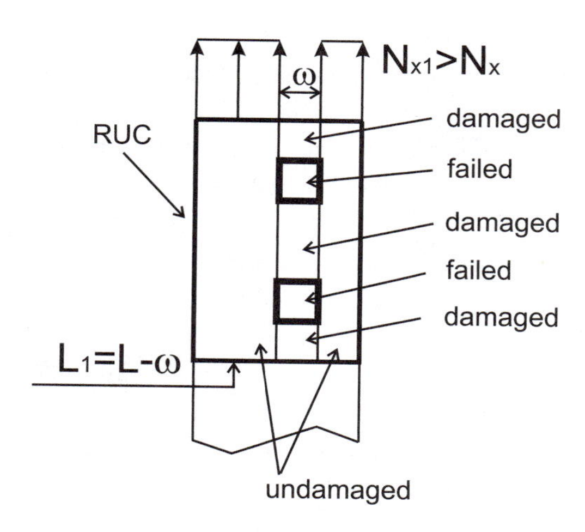

Figure 9.13: Load redistribution as a result of longitudinal fill failure in one cell.

this angle playing a double role. On one hand it counts for stiffness reduction (9.19) of fill and warp tows, and on the other hand it counts for stress transformation from laminate (x, y, z) to material $(1, 2, 3)$ coordinate system (9.37), influencing in this way the value of stress along fiber direction (longitudinal direction) and perpendicular to fiber direction (transverse direction) in material coordinate system.

Due to the periodic behavior of the RVE, the tow fails periodically on every RVE along the load direction. Since the dimensions of the RVE are of the same order of magnitude of the dimensions of the cross-section of the tow, the tow cannot recover longitudinal stress through shear before it encounters the next closest periodic point of failure. In other words, the ineffective length [153, 154] is longer than the dimensions of the RVE. Because of this, once a portion of the cross-section of a tow fails in the longitudinal direction, the entire length of that portion of the tow is assumed to have failed, and therefore degraded. This phenomenon does not occur in the case of shear failure due to applied shear load N_{xy}.

b. Internal Load Redistribution

There are two types of internal load redistribution. The first one is called *local redistribution* because the local phase stiffness is reduced when damage occurs in a phase (matrix, warp, fill). The effect of this is a local stress redistribution through the thickness at the same level of applied load. This kind of stress redistribution is illustrated in Figure 9.12 for the case of tensile load along the fill N_x. After the transverse warp failure in a cell, and consequently after warp transverse stiffness reduction in that same cell, the stress in the remaining phases is increased: $\sigma_{f1} > \sigma_f$ and $\sigma_{m1} > \sigma_m$, at the same level of N_x load. This stress redistribution happens inherently as a result of the stiffness reduction process, and applies for any load type (N_x, N_y or N_{xy}).

The second type of load redistribution, called *global redistribution*, takes into account the overloading of the remaining cells when the fill or warp phase in one cell fails in the longitudinal direction as a result of tensile load. This kind of load redistribution is illustrated in Figure 9.13. A longitudinal fill failure in one cell (failed) is followed by degrading (damaging) the whole cell line as a result of *ineffective longitudinal stress recovery* (explained in Section 9.6.2a). Then, the load increases in the remaining undegraded (undamaged) cells proportionally to the width ω of the degraded cell. In this way, at a certain level of applied load N_x corresponding to longitudinal fill failure in one cell (failed), the load in the remaining cells becomes $N_{x1} = N_x L/L_1 > N_x$, at the same level of applied N_x, where L represents the transverse dimension of load carrying part of RVE right before the local longitudinal fill failure, and L_1 represents the transverse dimension of load carrying part of RVE right after the local longitudinal fill failure. This kind of load redistribution happens only in the case of tensile loading N_x or N_y.

Damage initiation and evolution is approached as follows. Stresses are transformed from *laminate (x,y,z)* to *material (1,2,3)* coordinate system and the maximum stress damage activation criteria (Section 7.1.2) is used to predict damage initiation in the tow region. Due to the 3D nature of fabric-reinforced materials, a complex state of stress is obtained for each fill and warp constituent (according with the external applied load in the x, y plane). The stresses in material (tow) coordinate system are noted as $[\sigma_1, \sigma_2, \sigma_3, \sigma_4, \sigma_5, \sigma_6]^T$. The corresponding strengths on material directions are $[F_{1t}, F_{2t}, F_{3t}, F_4, F_5, F_6]^T$ and the corresponding failure index (in the case of maximum stress damage activation criterion) are

$$[I_1, I_2, I_3, I_4, I_5, I_6]^T = [\sigma_1/F_{1t}, \sigma_2/F_{2t}, \sigma_3/F_{3t}, \sigma_4/F_4, \sigma_5/F_5, \sigma_6/F_6]^T \qquad (9.41)$$

The following damage modes for fill and warp (regarded as unidirectional laminae in their own $(1, 2, 3)$ material coordinate system) are considered:

- *mode1*: longitudinal damage, due to tensile stress along the *1*-direction (fiber direction), $I_1 \geq 1$
- *mode2*: transverse damage, due to tensile stress along the *2*-direction (perpendicular to fiber direction), $I_2 \geq 1$
- *mode3*: transverse damage, due to tensile stress along the *3*-direction (perpendicular to fiber direction), $I_3 \geq 1$
- *mode4*: transverse shear damage, due to shear stress in 2–3 plane, $I_4 \geq 1$
- *mode5*: transverse shear damage, due to shear stress in 1–3 plane, $I_5 \geq 1$
- *mode6*: in-plane shear damage, due to shear stress in 1–2 plane, $I_6 \geq 1$

When damage corresponding to any mode is detected ($I_i = \sigma_i/F_i \geq 1$, where $i = 1 \cdots 6$), the 3D stiffness matrix $C = S^{-1}$ (in material $(1, 2, 3)$ coordinate system) of the affected fill or warp (9.12) is degraded using a *second-order damage tensor*

$\mathbf{D}$ [2, Chapter 8] to represent damage of fill and warp tows. The tows are regarded as orthotropic fiber-reinforced unidirectional laminae. The second-order damage tensor $\mathbf{D}$ can be represented as a diagonal tensor $D_{ij} = d_i \delta_{ij}$ (no sum on i) in a coordinate system coinciding with the principal directions of $\mathbf{D}$, which coincides with the $(1,2,3)$ material directions of fill and warp; d_i are the eigenvalues of the damage tensor, which represent the damage ratio along these directions, and δ_{ij} is the Kronecker delta.[3] The dual variable of the damage tensor is the *integrity tensor* $\mathbf{\Omega} = \sqrt{\mathbf{I} - \mathbf{D}}$, which represents the undamaged ratio [2, Chapter 8]. The *second-order damage tensor* $\mathbf{D}$ and the *integrity tensor* $\mathbf{\Omega}$ are diagonal and have the following explicit forms

$$D_{ij}^{f,w} = \begin{bmatrix} d_1 & 0 & 0 \\ 0 & d_2 & 0 \\ 0 & 0 & d_3 \end{bmatrix} \tag{9.42}$$

$$\Omega_{ij}^{f,w} = \begin{bmatrix} \Omega_1 & 0 & 0 \\ 0 & \Omega_2 & 0 \\ 0 & 0 & \Omega_3 \end{bmatrix} = \begin{bmatrix} \sqrt{1-d_1} & 0 & 0 \\ 0 & \sqrt{1-d_2} & 0 \\ 0 & 0 & \sqrt{1-d_3} \end{bmatrix} \tag{9.43}$$

where f, w represents fill and warp phases, and $i, j = 1, 2, 3$.

The damaged stiffness tensor $\mathbf{C}_d^{f,w}$ can be written in explicit contracted notation for an orthotropic material as a 6 x 6 array in terms of the undamaged stiffness tensor $\mathbf{C}$ as follows [2, Appendix C, Eq. C.6]

$$C_d^{f,w} = \begin{bmatrix} C_{11}\Omega_1^4 & C_{12}\Omega_1^2\Omega_2^2 & C_{13}\Omega_1^2\Omega_3^2 & 0 & 0 & 0 \\ C_{12}\Omega_1^2\Omega_2^2 & C_{22}\Omega_2^4 & C_{23}\Omega_2^2\Omega_3^2 & 0 & 0 & 0 \\ C_{13}\Omega_1^2\Omega_3^2 & C_{23}\Omega_2^2\Omega_3^2 & C_{33}\Omega_3^4 & 0 & 0 & 0 \\ 0 & 0 & 0 & C_{44}\Omega_2^2\Omega_3^2 & 0 & 0 \\ 0 & 0 & 0 & 0 & C_{55}\Omega_1^2\Omega_3^2 & 0 \\ 0 & 0 & 0 & 0 & 0 & C_{66}\Omega_1^2\Omega_2^2 \end{bmatrix} \tag{9.44}$$

In the numerical examples in this chapter, $d_1 = 0.99$ and $d_2 = d_3 = 0.9$ are used for stiffness reduction along the material directions. This means that while longitudinal failure is represented by a 99% reduction of stiffness, transverse failure is represented by a 90% reduction. Numerical values are chosen to obtain the best possible agreement with experimental data. For a better approach the reader is encouraged to consult Chapter 8.

Therefore, when the stress level in any of fill or warp reaches the level corresponding to $I_i \geq 1$, the d_1, d_2, d_3 stiffness reduction coefficients are applied as follows:

- if the detected damage is in *mode 1*, set $d_1 = 0.99, d_2 = d_3 = 0$

[3] $\delta_{ij} = 1$ if $i = j$, zero otherwise.

- if the detected damage is in *mode 2* or *mode 6*, set $d_1 = 0, d_2 = 0.9, d_3 = 0$
- if the detected damage is in *mode 3* or *mode 5*, set $d_1 = d_2 = 0, d_3 = 0.9$
- if the detected damage is in *mode 4*, set $d_1 = 0, d_2 = d_3 = 0.9$

At high stress levels, it is possible that these individual damage modes follow each other. In that case, the corresponding second-order damage tensors are superposed.

In the case of tension load along fill direction, it is possible that not only transverse warp damage due to N_x load occurs, but also other types of early fill and warp damage may occur. In this way, the analysis methodology proposed is quite general and the 3D features of fabric-reinforced materials are preserved.

In the case of neat-matrix isotropic regions, the von Misses yield criterion is used to predict damage initiation. The elastic nondamaging region in principal stress $\sigma_I, \sigma_{II}, \sigma_{III}$ space is given by

$$\frac{1}{\sqrt{2}}\left[(\sigma_I - \sigma_{II})^2 + (\sigma_{II} - \sigma_{III})^2 + (\sigma_I - \sigma_{III})^2\right]^{1/2} - S_y \leq 0 \tag{9.45}$$

That is, damage occurs when the equality holds and no damage occurs when the inequality holds. Upon damage initiation, the *second-order damage tensor* for the isotropic case becomes

$$D_{ij}^m = \begin{bmatrix} d & 0 & 0 \\ 0 & d & 0 \\ 0 & 0 & d \end{bmatrix} \tag{9.46}$$

with $d = 0.99$, and the isotropic matrix stiffness C^m in that cell is degraded accordingly. This means that the stiffness of the neat matrix is degraded to 1% of its original value.

Before proceeding to the next load step, stresses and strains are recalculated in the element to account for *local and global* redistribution of stress (Section 9.6.2) and possible additional damage. Then, the load is incremented and the process repeated, thus obtaining a stress–strain plot as shown in Figure 9.14. Comparison between predicted and experimental results is shown in Table 9.7.

9.6.3 Cross-ply Approximation

An expedient but not always accurate way to estimate the fabric-reinforced material moduli and strength is to consider a symmetric cross-ply laminate with the same overall fiber volume fraction V_{fiber}^o of the fabric-reinforced material. Even if this method is easier to apply and less expensive from the computational effort point of view, it may produce inaccurate results. This is exemplified in Table 9.8, where it can be observed that the equivalent cross-ply laminate model is not able to describe

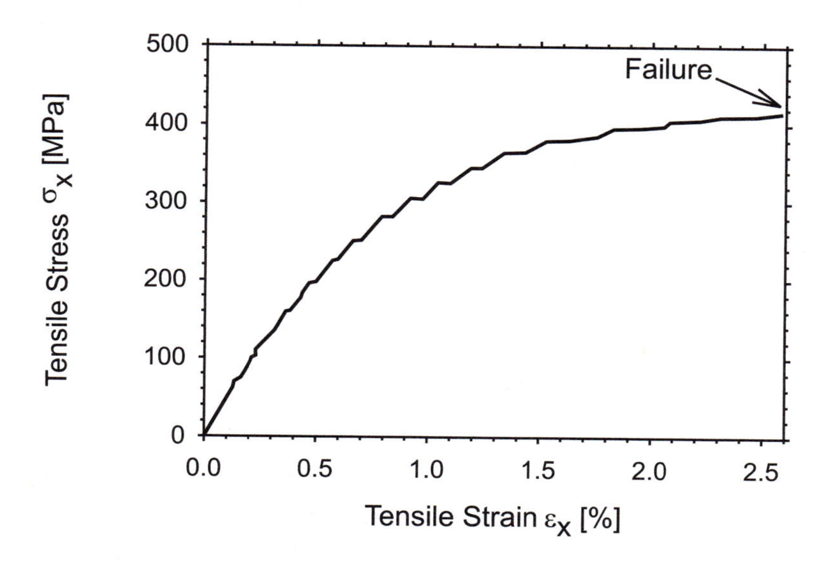

Figure 9.14: Stress–strain plot for plain weave carbon–epoxy (sample type #10 in Table 9.1), subjected to $N_x \neq 0, N_y = N_{xy} = 0$.

the fabric-reinforced material. This is because the cross-ply model does not take into account the tow geometry.

Example 9.1 *Estimate the elastic moduli $E_x, E_y, G_{xy}, \nu_{xy}$ of the eleven plain weave E-glass–epoxy and T300 carbon–epoxy material systems with geometric properties given in Table 9.1 [211, 218]. Constituent materials properties are given in Table 9.2 [211, 218].*

Solution to Example 9.1 *First calculate the tow fiber volume fraction for all eleven cases using (9.7). The results are given on the last column of Table 9.1. Next, use PMM (Section 4.2.8) to calculate the tow properties. The calculated values are shown in Table 9.3. Then, calculate the laminate moduli using (9.28). The calculated values of moduli are shown in Table 9.5 and compared to experimental values [218] in Table 9.6.*

Example 9.2 *Calculate the laminate tensile strength F_{xt} of the laminates considered in Example 9.1.*

Solution to Example 9.2 *Using a computer program implementing the damage activation and damage evolution methodology described in this chapter [1], perform the strength analysis of plain weave material systems with geometric parameters given Table 9.1. The predicted values of strength are shown in Table 9.5 and compared to experimental values [211] in Table 9.7.*

Example 9.3 *Estimate the lamina shear strength F_s of plain weave E-glass–epoxy and T300 carbon–epoxy material systems with geometric properties [211,218] given in Table 9.1 and constituent materials properties [211,218] given in Table 9.2.*

Solution to Example 9.3 *The solution procedure is identical to that of Example 9.2 [1] but the load is $N_{xy} \neq 0, N_x = N_y = 0$. The results are shown in Table 9.5, and compared to experimental values [218] in Table 9.7.*

9.7 Woven Fabrics with Gap

If the gap region between adjacent fill or/and warp tows is appreciable, it has to be considered in the model. In this case a new geometrical description of fill and warp tows is necessary. Therefore, the previous equations (9.1–9.2) that did not consider the gap are extended for the case of plain weave fabrics with gap, as follows

$$z_f(x,y) = \begin{cases} -\frac{h_f}{2} \cdot cos\frac{\pi x}{a_w + g_w} & \text{if} \quad y \in \left[0, \frac{a_f}{2}\right] \\ 0 & \text{if} \quad y \in \left[\frac{a_f}{2}, \frac{a_f + g_f}{2}\right] \end{cases}$$
$$z_w(x,y) = \begin{cases} \frac{h_w}{2} \cdot cos\frac{\pi y}{a_f + g_f} & \text{if} \quad x \in \left[0, \frac{a_w}{2}\right] \\ 0 & \text{if} \quad x \in \left[\frac{a_w}{2}, \frac{a_w + g_w}{2}\right] \end{cases} \tag{9.47}$$

and

$$e_f(y) = \begin{cases} |h_f \cdot cos\frac{\pi y}{a_f}| & \text{if} \quad y \in \left[0, \frac{a_f}{2}\right] \\ 0 & \text{if} \quad y \in \left[\frac{a_f}{2}, \frac{a_f + g_f}{2}\right] \end{cases}$$
$$e_w(x) = \begin{cases} |h_w \cdot cos\frac{\pi x}{a_w}| & \text{if} \quad x \in \left[0, \frac{a_w}{2}\right] \\ 0 & \text{if} \quad x \in \left[\frac{a_w}{2}, \frac{a_w + g_w}{2}\right] \end{cases} \tag{9.48}$$

A close analysis of these new shape functions reveals the RVE configuration depicted in Figure 9.15. It can be seen that due to the geometrical description used, a thin matrix layer is simulated between fill and warp tows. The thickness of this matrix layer is small and it depends on the geometrical parameters of the RVE (gap dimension, fill and warp width, fill and warp thickness). The analysis follows the same steps as described for the case of no gap in Sections 9.2–9.6. The computational effort increases due to different LSS that are formed in different regions of the RVE (see Figure 9.15):

- region A, the LSS is (from bottom to top): matrix/fill/intermediate matrix layer/warp/matrix.
- region B, the LSS is: matrix/fill/matrix.

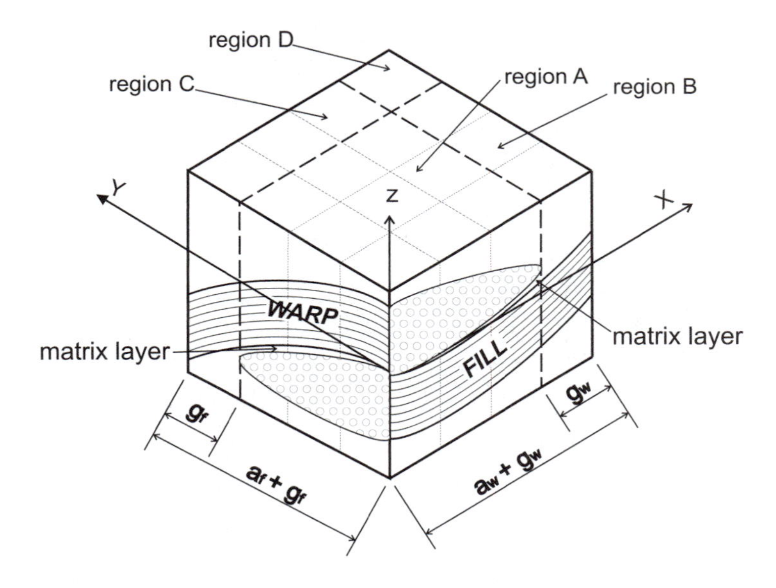

Figure 9.15: RVE considering the gap.

- region C, the LSS is: matrix/warp/matrix.
- region D, only one layer of pure neat matrix.

While equations (9.47)–(9.48) describe the geometry of plain weave fabric considering the gap between fill/warp tows, all the other relations necessary to run the analysis of this kind of fabric reinforcement are the same as those in Section 9.1–9.6. One only has to substitute equations (9.47)–(9.48) for equations (9.1)–(9.2), and run the analysis as described in Section 9.1–9.6.

Two basic parameters in woven fabric manufacturing process are the maximum thickness of tows h_f, h_w and the gap length g_f, g_w. The former has a direct influence on the maximum undulation angle of tows θ_f, θ_w while the latter has a direct influence on overall fiber volume fraction of composite. Both of these, the undulation angle and overall fiber volume fraction, play an important role in mechanical properties of the final composite material.

The following results are obtained for a E-glass–epoxy material system whose constituent properties are presented in Table 9.2. A fill and warp tow width $a_f = a_w = 2$ mm and a tow fiber volume fraction $V_{fiber}^{f,w} = 0.7$ were considered for a parametric study.

While the overall fiber volume fraction as a function of h/a is nearly constant (Figure 9.16), the modulus varies significantly as a function of h/a for small g/a ratios, because the undulation angle increases with h/a, thus reducing the modulus. The variation of modulus with h/a at high g/a ratios is not considerable, but the modulus is highly reduced at high g/a ratio.

The overall fiber volume fraction varies significantly as a function of g/a (Figure 9.17) for both low and high h/a ratios, and the modulus (Figure 9.17) follows the

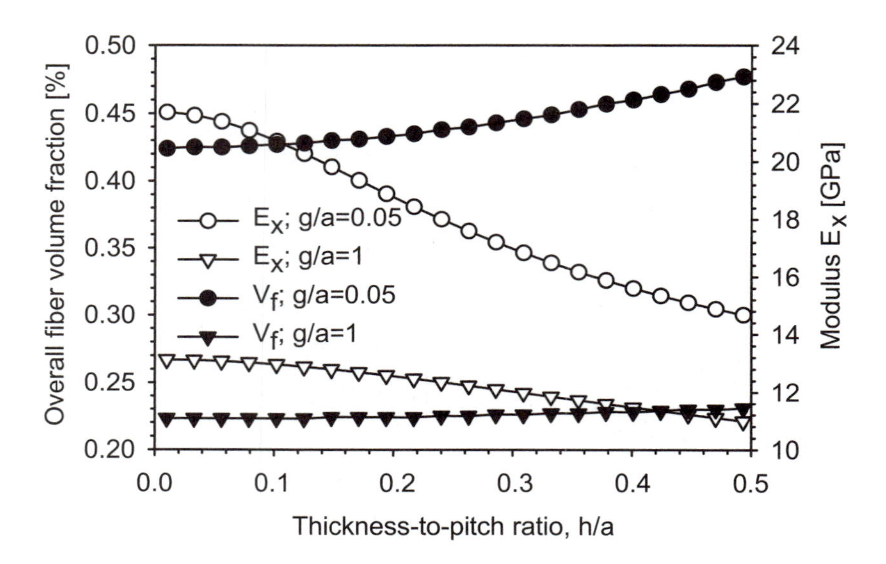

Figure 9.16: Overall fiber volume fraction V^o_{fiber} and modulus E_x as a function of thickness ratio h/a.

same trend, most significantly for small h/a ratios, and a bit less at high h/a ratios.

9.8 Twill and Satin

In order to extend the model described in the previous section to weaving patterns other than plain weave, a RVE has to be selected and the complete geometrical description of the fill and warp undulation in the RVE needs to be developed in terms of the weave parameters n_g, n_s, n_i. As it was noted in Section 9.2, in the case of twill and satin weaves the RVE is represented by the RUC itself. Three cases are considered next.

9.8.1 Twill Weave with $n_g > 2, n_s = 1, n_i = 1$

A weave pattern with $n_g > 2, n_s = 1, n_i = 1$ represents a subset of all the possible twill patterns (Figure 9.18). For $n_g = 2$, it reduces to a plain weave.

To describe the undulation when $n_g \neq 0$, two additional parameters $i = 1...n_g, j = 1...n_g$, need to be introduced, where n_g is the number of subelements along one direction of the RVE, as defined in Section 9.1. These parameters identify univocally the position of each subelement (i, j) of area $(a_f + g_f) \times (a_w + g_w)$ inside the RVE, as shown in Figure 9.18. Various regions, labeled I to VI in Figure 9.18 are described by different functions.

For $n_g > 2, n_s = 1, n_i = 1$, the undulation of the fill tow can be described with three functions spanning five regions in Figure 9.18; first the interlacing region I and adjacent regions II and III, second, the noninterlacing region IV, and third, the

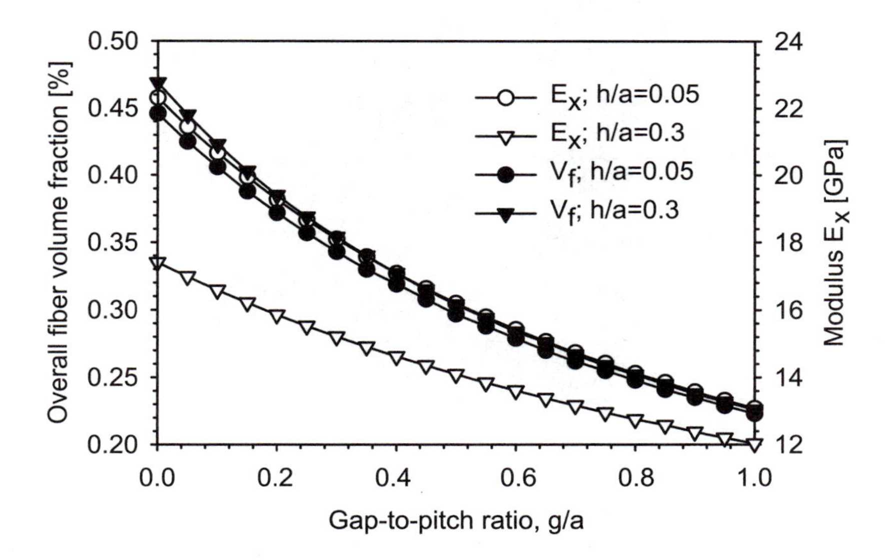

Figure 9.17: Overall fiber volume fraction V^o_{fiber} and modulus E_x as a function of gap ratio g/a.

transition regions V, and VI, as follows

$$
\begin{cases}
\text{if} \quad (j-1)w_f + \frac{g_f}{2} \le y \le jw_f - \frac{g_f}{2} \\
z_f(x,y,j) = \\
\begin{cases}
-\frac{h_f}{2} sin\left[\frac{\pi x}{w_w} + (j-1)\pi\right] & \text{if} \quad (j-\frac{3}{2})w_w \le x \le (j+\frac{1}{2})w_w \\
\frac{h_f}{2} & \text{if} \quad \begin{array}{l} (j+\frac{1}{2}-n_g)w_w < x < (j-\frac{3}{2})w_w \\ \text{or } (j+\frac{1}{2})w_w < x < (j-\frac{3}{2}+n_g)w_w \end{array} \\
(-1)^{n_g+1}\times & \\
\quad \frac{h_f}{2} sin\left[\frac{\pi x}{w_w} + (j-1)\pi\right] & \text{if} \quad \begin{array}{l} (j-\frac{3}{2}-n_g)w_w \le x \le (j+\frac{1}{2}-n_g)w_w \\ \text{or } (j-\frac{3}{2}+n_g)w_w \le x \le (j+\frac{1}{2}+n_g)w_w \end{array}
\end{cases} \\
\text{if} \quad (j-1)w_f \le y < (j-1)w_f + \frac{g_f}{2} \quad \text{or} \quad jw_f - \frac{g_f}{2} < y \le jw_f \\
z_f(x,y,j) = 0
\end{cases}
\tag{9.49}
$$

where $i = 1...n_g, j = 1...n_g$ indicate the position of the interlacing region inside the RVE (Figure 9.18). A similar function can be written for the warp tow as follows. In all cases when $n_s = n_i = 1$, the function that describes the undulation of the warp is minus the function describing the fill. One needs only to multiply by minus one, and substitute in order $y, x, i, w_w, w_f, g_w, h_w$ for $x, y, j, w_f, w_w, g_f, h_f$ in (9.49), as follows

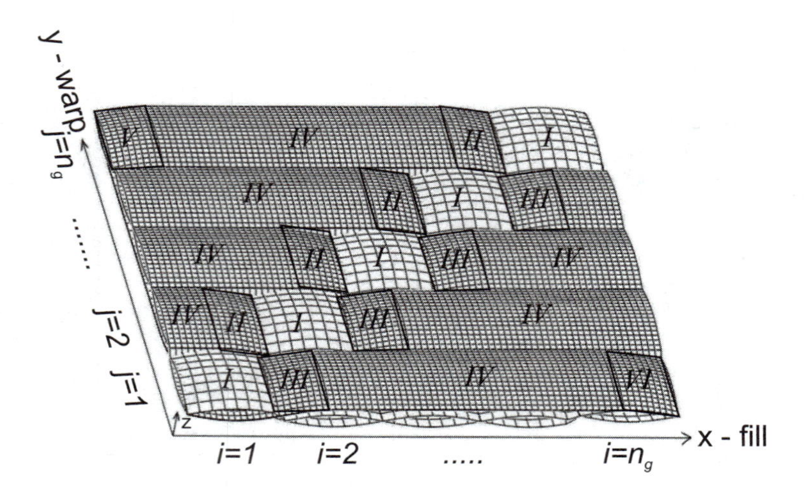

Figure 9.18: Twill 5/1/1 displaying the various sections (I to VI) used for modeling of the fill tow. The warp is shown by the coarse grid and the fill by the finer grid.

$$
\begin{cases}
\text{if } \ (i-1)w_w + \frac{g_w}{2} \le x \le i w_w - \frac{g_w}{2} \\
z_w(x,y,i) = \\
\quad \begin{cases}
\frac{h_w}{2} sin\left[\frac{\pi y}{w_f} + (i-1)\pi\right] & \text{if } \ (i-\frac{3}{2})w_f \le y \le (i+\frac{1}{2})w_f \\
-\frac{h_w}{2} & \text{if } \ \begin{array}{l} (i+\frac{1}{2}-n_g)w_f < y < (i-\frac{3}{2})w_f \\ \text{or } (i+\frac{1}{2})w_f < y < (i-\frac{3}{2}+n_g)w_f \end{array} \\
(-1)^{n_g} \times & \\
\frac{h_w}{2} sin\left[\frac{\pi y}{w_f} + (i-1)\pi\right] & \text{if } \ \begin{array}{l} (i-\frac{3}{2}-n_g)w_f \le y \le (i+\frac{1}{2}-n_g)w_f \\ \text{or } (i-\frac{3}{2}+n_g)w_f \le y \le (i+\frac{1}{2}+n_g)w_f \end{array}
\end{cases} \\
\text{if } \ (i-1)w_w \le x < (i-1)w_w + \frac{g_w}{2} \ \text{ or } \ i w_w - \frac{g_w}{2} < x \le i w_w \\
z_w(x,y,i) = 0
\end{cases}
\tag{9.50}
$$

It can be noticed a new notation used in (9.49), (9.50), $w_f = a_f + g_f, w_w = a_w + g_w$, the gap presence in the weaving pattern being accounted for.

As the equations (9.49), (9.50) represent a generalization of (9.47) to the general weaving pattern treated in this Section, a generalization of (9.48) is also required for the thickness of fill/warp cross-sections, as follows

$$e_f(y, j) = \qquad (9.51)$$
$$\begin{cases} \left| h_f sin \left[\frac{\pi(-(j-1)w_f + y - g_f/2)}{a_f} \right] \right| & \text{if} \quad (j-1)w_f + \frac{g_f}{2} \leq y \leq jw_f - \frac{g_f}{2} \\ 0 \quad \text{if} \quad (j-1)w_f \leq y < (j-1)w_f + \frac{g_f}{2} & \text{or} \quad jw_f - \frac{g_f}{2} < y \leq jw_f \end{cases}$$

$$e_w(x, i) =$$
$$\begin{cases} \left| h_w sin \left[\frac{\pi(-(i-1)w_w + x - g_w/2)}{a_w} \right] \right| & \text{if} \quad (i-1)w_w + \frac{g_w}{2} \leq x \leq iw_w - \frac{g_w}{2} \\ 0 \quad \text{if} \quad (i-1)w_w \leq x < (i-1)w_w + \frac{g_w}{2} & \text{or} \quad iw_w - \frac{g_w}{2} < x \leq iw_w \end{cases}$$

The same kind of change of variables relationship which connects $z_f(x, y, j) - z_w(x, y, i)$ also works in the case of $e_f(y, j) - e_w(x, i)$: one needs only to substitute in order x, i, w_w, g_w, h_w, a_w for y, j, w_f, g_f, h_f, a_f in $e_f(y, j)$ for obtaining $e_w(x, i)$ (the multiplication by minus one is not required in this case).

The undulation and thickness functions over the entire RVE are constructed as follows

$$\begin{aligned} &\text{when} \quad (j-1)\ w_f \leq y \leq j\ w_f \\ &\qquad z_f(x, y) = z_f(x, y, j) \quad , \quad e_f(y) = z_f(y, j) \\ &\text{when} \quad (i-1)\ w_w \leq x \leq i\ w_w \\ &\qquad z_w(x, y) = z_w(x, y, i) \quad , \quad e_w(x) = e_w(x, i) \end{aligned} \qquad (9.52)$$

9.8.2 Twill Weave with $n_s = 1$

A weave pattern with $n_s = 1$ and arbitrary values of both n_g and n_i represents a twill with fixed shift $n_s = 1$, which features a wider diagonal rib (Figure 9.19). The undulation functions for this case are more complex than for the previous case, as follows

$$
\begin{cases}
\text{if} \quad (j-1)w_f + \frac{g_f}{2} \le y \le j w_f - \frac{g_f}{2} \\
z_f(x,y,j) = \\
\quad \begin{cases}
\frac{-h_f}{2} sin\left[\frac{\pi x}{w_w} + (j-1)\pi\right] \\
\qquad \text{if } (j-\frac{3}{2})w_w \le x \le (j-\frac{1}{2})w_w \\
\\
\frac{-h_f}{2} \\
\qquad \text{if } (j-\frac{1}{2})w_w < x < \left(j-\frac{3}{2}+n_i\right) w_w \\
\\
(-1)^{n_i}\frac{h_f}{2} sin\left[\frac{\pi x}{w_w} + (j-1)\pi\right] \\
\qquad \text{if } \left(j-\frac{3}{2}+n_i\right) w_w \le x \le \left(j-\frac{1}{2}+n_i\right) w_w \\
\\
\frac{h_f}{2} \\
\qquad \text{if } \left(j-\frac{1}{2}-n_g+n_i\right) w_w < x < \left(j-\frac{3}{2}\right) w_w \\
\qquad \text{or } \left(j-\frac{1}{2}+n_i\right) w_w < x < \left(j-\frac{3}{2}+n_g\right) w_w \\
\\
(-1)^{(n_g+1)}\frac{h_f}{2} sin\left[\frac{\pi x}{w_w} + (j-1)\pi\right] \\
\qquad \text{if } \left(j-\frac{3}{2}-n_g\right) w_w \le x \le \left(j-\frac{1}{2}-n_g\right) w_w \\
\qquad \text{or } \left(j-\frac{3}{2}+n_g\right) w_w \le x \le \left(j-\frac{1}{2}+n_g\right) w_w \\
\\
(-1)^{(n_g+1)}(-1)^{(n_i+1)}\frac{h_f}{2} sin\left[\frac{\pi x}{w_w} + (j-1)\pi\right] \\
\qquad \text{if } \left(j-\frac{3}{2}-n_g+n_i\right) w_w \le x \le \left(j-\frac{1}{2}-n_g+n_i\right) w_w \\
\\
\frac{-h_f}{2} \\
\qquad \text{if } \left(j-\frac{1}{2}-n_g\right) w_w < x < \left(j-\frac{3}{2}-n_g+n_i\right) w_w
\end{cases} \\
\\
\text{if} \quad \left[(j-1)w_f \le y < (j-1)w_f + \frac{g_f}{2} \quad \text{or} \quad j w_f - \frac{g_f}{2} < y \le j w_f\right] \\
\qquad \text{and} \quad g_f \ne 0 \\
z_f(x,y,j) = 0
\end{cases}
\tag{9.53}
$$

The warp functions do not correspond one-to-one with the fill functions as it was the case in Section 9.8.1 but they are correlated in a particular way for each specific weave pattern. For example, in a 5/1/2 twill type (Figure 9.19), fill tows numbered $j = 1...5$ have identical undulation as warp tows numbered $i = 2, 3, 4, 5, 1$. This means that a mapping (correlation) vector has to be defined (in the previous example the mapping vector is $(2, 3, 4, 5, 1)$) in order to define the warp geometry based on fill geometry. Other than that, one needs only to multiply by minus one, and substitute in order $y, x, i, w_w, w_f, g_w, h_w$ for $x, y, j, w_f, w_w, g_f, h_f$ in in (9.53). The corresponding functions in terms of (x, y) are found by applying (9.52).

Regarding the fill/warp thickness functions for the preset case, they are identical to those defined in (9.51). The corresponding functions in terms of (x, y) are found

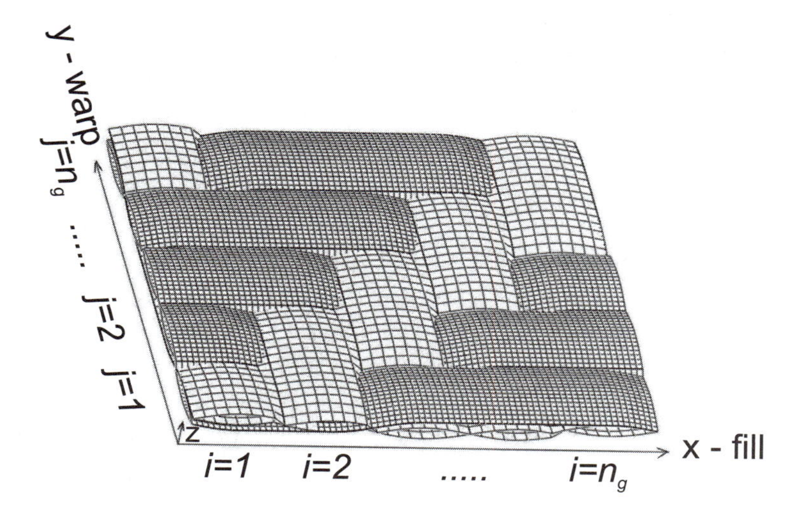

Figure 9.19: Twill 5/1/2.

by applying (9.52).

9.8.3 Satin Weave with $n_i = 1$

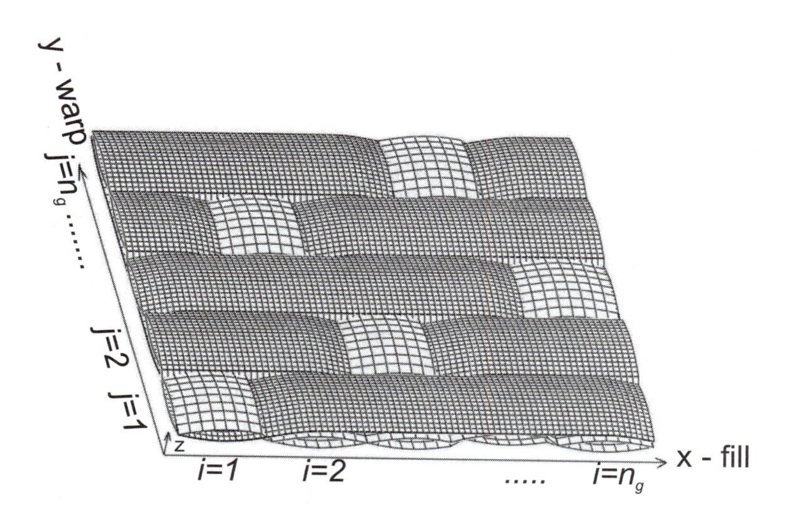

Figure 9.20: Satin 5/2/1.

A weave pattern with $n_i = 1$ and arbitrary values of both n_g and n_s represents all the satin weaves with interlacing over only one subcell $n_i = 1$ (Figure 9.20). The undulation function for the fill in this case is

$$
\begin{cases}
\text{if} \quad (j-1)w_f + \frac{g_f}{2} \le y \le jw_f - \frac{g_f}{2} \\
z_f(x,y,j) = \\
\quad \begin{cases}
\text{for } k \in 1 \ldots \text{trunc}\left[\frac{n_s(j-1)+1}{n_g}\right] + 1 \\
\quad \begin{cases}
(-1)^{(kn_g+1)} \frac{h_f}{2} sin\left[\frac{\pi x}{w_w} + n_s(j-1)\pi\right] \\
\qquad \text{if } \left[n_s(j-1) - \frac{1}{2} - kn_g\right] w_w \le x \le \left[n_s(j-1) + \frac{3}{2} - kn_g\right] w_w \\
\\
\frac{h_f}{2} \\
\qquad \text{if } (k-1) \ge 1 \\
\qquad \text{and } \left[n_s(j-1) + \frac{3}{2} - kn_g\right] w_w < x < \left[n_s(j-1) - \frac{1}{2} - (k-1)n_g\right] w_w
\end{cases} \\
\\
\frac{-h_f}{2} sin\left[\frac{\pi x}{w_w} + n_s(j-1)\pi\right] \\
\qquad \text{if } \left[n_s(j-1) - \frac{1}{2}\right] w_w \le x \le \left[n_s(j-1) + \frac{3}{2}\right] w_w \\
\\
(-1)^{(n_g+1)} \frac{h_f}{2} sin\left[\frac{\pi x}{w_w} + n_s(j-1)\pi\right] \\
\qquad \text{if } \left[n_s(j-1) - \frac{1}{2} + n_g\right] w_w \le x \le \left[n_s(j-1) + \frac{3}{2} + n_g\right] w_w \\
\\
\frac{h_f}{2} \\
\qquad \text{if } \left[n_s(j-1) + \frac{3}{2} - n_g\right] w_w < x < \left[n_s(j-1) - \frac{1}{2}\right] w_w \\
\qquad \text{or } \left[n_s(j-1) + \frac{3}{2}\right] w_w < x < \left[n_s(j-1) - \frac{1}{2} + n_g\right] w_w
\end{cases} \\
\\
\text{if} \quad \left[(j-1)w_f \le y < (j-1)w_f + \frac{g_f}{2} \quad \text{or} \quad jw_f - \frac{g_f}{2} < y \le jw_f\right] \\
\qquad \text{and} \quad g_f \ne 0 \\
z_f(x,y,j) = 0
\end{cases}
\tag{9.54}
$$

In this case, the warp functions do not correspond one-to-one with the fill functions as it was the case in Section 9.8.1 but they are correlated in a particular way for each specific weave pattern. For example, in a 5/2/1 satin type (Figure 9.20), fill tows numbered $j = 1...5$ have identical undulation as warp tows numbered $i = 1, 5, 4, 3, 2$. This means that a mapping (correlation) vector has to be defined (in the previous example the mapping vector is $(1, 5, 4, 3, 2)$) in order to define the warp geometry based on fill geometry. Other than that, one needs only to multiply by minus one, and substitute in order $y, x, i, w_w, w_f, g_w, h_w$ for $x, y, j, w_f, w_w, g_f, h_f$ in in (9.54). The corresponding functions in terms of (x, y) are found by applying (9.52).

Regarding the fill/warp thickness functions for the preset case, they are identical to those defined in (9.51). The corresponding functions in terms of x, y, are found by applying (9.52).

The most general case with both $n_s > 1$ and $n_i > 1$ does not have practical application and thus is not considered here.

9.8.4 Twill and Satin Thermo-elastic Properties

While the equations provided in Sections 9.8.1–9.8.3 deal only with the geometrical description of general twill and satin weaving patterns (which is basically a generalization of (9.1)–(9.2) and of their counterpart, (9.47)–(9.48) when the gap is considered), other equations throughout the Sections 9.2–9.5 need to be modified in order to provide a general level for the analysis of the geometries described in Sections 9.8.1–9.8.3, and to account for the fact that in these cases the RVE is the RUC itself (i.e., no smaller entity can be selected in order to reduce the size of the analysis). In the following, the equations from Sections 9.2–9.5 that need to be modified in order to run the thermo-elastic analysis of general twill/satin reinforcements are highlighted next.

– Equation (9.4) changes to

$$\theta_f(x,y) = arctan\left(\frac{\partial}{dx} z_f(x,y)\right)$$
$$\theta_w(x,y) = arctan\left(\frac{\partial}{dy} z_w(x,y)\right) \tag{9.55}$$

Consequently, all the equations that make use of undulation angles θ_f, θ_w, will be accordingly considered in terms the two space variables, x and y, namely $\theta_f(x,y), \theta_w(x,y)$.

– Equation (9.6) changes to

$$V_{meso} = \frac{v_w + v_f}{v_{RVE}} = \frac{v_w + v_f}{n_g^2 \; w_f \; w_w \; h} \tag{9.56}$$

– Equation (9.8) changes to

$$v_f = n_g \; A_f \; L_f$$
$$v_w = n_g \; A_w \; L_w \tag{9.57}$$

– Equation (9.9) changes to

$$A_f = \int_0^{w_f} e_f(y)dy$$
$$A_w = \int_0^{w_w} e_w(x)dx \tag{9.58}$$

– Equation (9.10) changes to

$$L_f = \int_0^{n_g \cdot w_w} \sqrt{1 + \left(\frac{d}{dx} z_f(x,y)\right)^2} dx$$
$$L_w = \int_0^{n_g \cdot w_f} \sqrt{1 + \left(\frac{d}{dy} z_w(x,y)\right)^2} dy \tag{9.59}$$

– While (9.16) does not change, (9.17) changes to

$$T_w(x,y) = \begin{bmatrix} 0 & c_w^2 & s_w^2 & 2c_w s_w & 0 & 0 \\ 1 & 0 & 0 & 0 & 0 & 0 \\ 0 & s_w^2 & c_w^2 & -2c_w s_w & 0 & 0 \\ 0 & 0 & 0 & 0 & -c_w & s_w \\ 0 & -c_w s_w & c_w s_w & c_w^2 - s_w^2 & 0 & 0 \\ 0 & 0 & 0 & 0 & -s_w & -c_w \end{bmatrix} \tag{9.60}$$

– Equation (9.27) changes to

$$[A] = \frac{1}{n_g^2\ w_f\ w_w} \sum_{j=1}^{n_g} \sum_{i=1}^{n_g} \int_{w_f(j-1)}^{w_f j} \int_{w_w(i-1)}^{w_w i} [A(x,y)] dx dy \tag{9.61}$$

– Equation (9.33) changes to

$$\{N\}^{T,M} = \frac{1}{n_g^2\ w_f\ w_w} \sum_{j=1}^{n_g} \sum_{i=1}^{n_g} \int_{w_f(j-1)}^{w_f j} \int_{w_w(i-1)}^{w_w i} \left\{ N^{T,M}(x,y) \right\} dx dy \tag{9.62}$$

9.9 Randomly Oriented Reinforcement

Randomly oriented reinforcements are used to produce low cost laminates with isotropic in-plane properties. Typical reinforcements include chopped fibers arranged randomly in a chopped strand mat, or continuous fibers placed on a swirl pattern to form a continuous strand mat (CSM, see Figure 2.3). Either way, fiber orientations are randomly distributed with a uniform probability density function, held together by a binder (Section 2.4). CSM is used to obtain bidirectional properties on pultrusion and other processes where continuous rovings constitute the main reinforcement. Since CSM is produced with continuous roving, it is more expensive than chopped strand mat, which is produced directly from the glass furnace, thus saving the intermediate step of making a roving. Furthermore, better quality E-glass is commonly used for CSM, while the chopped version is usually made of lower quality glass. Therefore, CSM has in general better mechanical properties than chopped strand mat, but hand lay-up production of not-so-critical structures (e.g., small boats) may use chopped strand mat. The interlaminar shear strength values F_4 and F_5 (Section 4.4.10) are usually lower in composites made with lower quality glass. If these properties are critical, E- or S-glass should be used.

9.9.1 Elastic Moduli

The elastic properties of both continuous strand mat (CSM) and chopped strand mat can be predicted assuming that they are random composites. A layer of composite with randomly oriented fibers can be idealized as a laminate with a large number of thin unidirectional layers, each with a different orientation from 0° to 180°. The properties of the random composite are the average properties of this fictitious laminate. Each fictitious unidirectional layer has a reduced stiffness matrix $[Q]$ given by (5.23). The orientation of each layer is accounted for by rotating the matrix $[Q]$ to a common coordinate system by using (5.48). Then, the average is computed by

$$[Q^{CSM}] = \frac{1}{\pi}\int_0^\pi [T]^{-1}[Q][T]^{-T}d\theta \tag{9.63}$$

which leads to

$$\begin{aligned}
Q_{11}^{CSM} &= \frac{3}{8}Q_{11} + \frac{1}{4}Q_{12} + \frac{3}{8}Q_{22} + \frac{1}{2}Q_{66} \\
Q_{12}^{CSM} &= \frac{1}{8}Q_{11} + \frac{3}{4}Q_{12} + \frac{1}{8}Q_{22} - \frac{1}{2}Q_{66} \\
Q_{66}^{CSM} &= \frac{1}{8}Q_{11} - \frac{1}{4}Q_{12} + \frac{1}{8}Q_{22} + \frac{1}{2}Q_{66} \\
Q_{16}^{CSM} &= Q_{26}^{CSM} = 0
\end{aligned} \tag{9.64}$$

Using (5.23) to write the coefficients of both $[Q]$ and $[Q^{CSM}]$, the isotropic properties E, G, and ν of the random composite can be obtained from the known properties of a unidirectional material with the same fiber volume fraction

$$\begin{aligned}
\frac{E}{1-\nu^2} &= \frac{3}{8\Delta}(E_1+E_2) + \frac{\nu_{12}E_2}{4\Delta} + \frac{G_{12}}{2} \\
\frac{\nu E}{1-\nu^2} &= \frac{E_1+E_2}{8\Delta} - \frac{G_{12}}{2} + \frac{3\nu_{12}E_2}{4\Delta} \\
G &= \frac{E_1+E_2}{8\Delta} + \frac{G_{12}}{2} - \frac{\nu_{12}E_2}{4\Delta}
\end{aligned} \tag{9.65}$$

with $\Delta = 1 - \nu_{12}\nu_{21}$. Solving the above three equations for E, G, and ν results in

$$\begin{aligned}
E &= \frac{(E_1+E_2)^2 + \Psi[E_1 + E_2(1+2\nu_{12})] - (2\nu_{12}E_2)^2}{\Delta\Phi} \\
G &= \frac{E_1 + E_2(1-2\nu_{12}) + \Psi}{8\Delta} \\
\nu &= \frac{E_1 + E_2(1+6\nu_{12}) - \Psi}{\Phi} \\
\Psi &= 4G_{12}\Delta \\
\Phi &= 3E_1 + (3+2\nu_{12})E_2 + \Psi
\end{aligned} \tag{9.66}$$

Note that the equation above satisfies the isotropic relationship $G = E/2(1+\nu)$. These expressions can be approximated [182] by

$$
\begin{aligned}
E &= \frac{3}{8}E_1 + \frac{5}{8}E_2 \\
G &= \frac{1}{8}E_1 + \frac{1}{4}E_2 \\
\nu &= \frac{E}{2G} - 1
\end{aligned}
\tag{9.67}
$$

where E_1 and E_2 are the longitudinal and transverse moduli of a fictitious unidirectional layer having the same volume fraction as the CSM layer. The approximation for G is obtained by assuming $\Delta = 1$, $\nu_{12} = 1/4$, and $G_{12} = 3/8\ E_2$. The approximation for E is obtained dividing the equation for E by $(1-\nu^2)$ and using the same assumptions.

Comparison between experimental and predicted modulus (using (4.24) and (4.31) to obtain E_1 and E_2) is presented in Table 9.9, for vinyl ester reinforced with E-glass CSM (Owens Corning 457.5 g/m^2). Predicted values are based on the following material properties: $E_{fiber} = 72.3\ GPa$, $E_m = 3.4\ GPa$, $\nu_{fiber} = 0.22$, $\nu_m = 0.38$, stress partition parameter $\eta_s = 2$.

From the experimental data it can be seen that the reinforcement is not truly random. The direction along the length of the CSM roll is called *longitudinal.* Along the transverse direction, the stiffness of the composite is about 13% lower because of a slightly preferential orientation along the length of the roll. The predicted values are higher than expected, perhaps because the predictions are based on the nominal weight of the CSM and a variability of up to ± 6% is possible.

9.9.2 Strength

Chopped strand mat and continuous strand mat are considered randomly oriented composites, even if in practice there is some preferential orientation of the fibers. The resulting composite is considered to be quasi-isotropic (see Section 6.3.5). That is, the properties are the same along any orientation on the surface of the layers. The tensile strength F_{csm-t} may be estimated by the empirical formulas taken from [219] and rewritten here as

$$
\alpha = \sqrt{\frac{F_{1t}\,F_{2t}}{F_6^2}} \quad ; \; F_{csm-t} = \begin{cases} \dfrac{4}{\pi} F_6\,(1+\ln\alpha) & \text{for } \alpha > 1 \\ \dfrac{4}{\pi} F_6\,\alpha & \text{for } \alpha \leqslant 1 \end{cases}
\tag{9.68}
$$

where F_{1t}, F_{2t}, and F_6 are the longitudinal tensile, transverse tensile, and in-plane shear strength of a fictitious unidirectional material containing the same fiber volume fraction as that of the CSM material. The formula compares well with experimental data for some PMC with fiber volume fraction up to 20% [219].

Exercises

Exercise 9.1 *Draw the top view repetitive unit cell (RUC) for* 5/1/2 *weave counting along the x-direction. Next, redefine the weave parameters, first by* 90° *rotating the weaving pattern, and second by considering the opposite face of the weaving pattern, on both initial orientation and* 90° *rotated orientation of the weaving pattern. Is this a twill or a satin weave?*

Exercise 9.2 *Repeat Exercise 9.1 for* 5/2/1 *weaving pattern.*

Exercise 9.3 *Draw the top view repetitive unit cell (RUC) for* 4/2/1 *weave counting along the x-direction. Would this weave pattern remain woven without loosing tows?*

Exercise 9.4 *Find the values of shift n_s in weaves defines by* $6/n_s/1$, $8/n_s/1$, *and* $9/n_s/1$, *that feature the same property as* 4/2/1 *in Exercise 9.3. Sketch these hypothetical weaves.*

Exercise 9.5 *Draw the top view repetitive unit cell (RUC) for* 5/1/3 *weave counting along the x-direction. Next, redefine the weave parameters by considering the opposite face of the weaving pattern. Is this a twill or sating weave? How does this weave compare to* 5/1/2 *in Exercise 9.1? (see Section 9.1)*

Exercise 9.6 *Use solid modeling software such as ProE(TM) to draw to scale the RUC's for sample type 2 in Table 9.1.*

Exercise 9.7 *Using any plotting software, draw to scale one quarter period of the weave pattern in both fill/warp directions for sample types 2 in Table 9.1, using the geometrical functions provided in 9.2 and 9.7. Consider the gap between fill/warp tows as being zero. Consider only the tow midpoint for plotting.*

Exercise 9.8 *Using any plotting software, draw to scale a full period of the weave pattern in both fill/warp directions for sample types 2 in Table 9.1, using the geometrical functions provided in 9.8. Consider only the tow midpoint for plotting.*

Exercise 9.9 *Using any plotting software,draw to scale a full period of the weave pattern in fill directions for sample types 2 in Table 9.1, using the geometrical functions provided in 9.8. Consider the bounding surfaces for both fill/warp tows, at different positions along warp directions, while plotting along fill direction.*

Exercise 9.10 *Redo the plots in 9.9 for the following gap values:* $g_f = g_w = 0.15\,mm$ *and* $g_f = g_w = 0\,mm$. *Comment on the model for woven fabric with gap.*

Exercise 9.11 *Estimate the weight per unit area w of the carbon fabrics in Table 9.1 based on the information given.*

Exercise 9.12 *Estimate the overall fiber volume fraction* V^0_{fiber} *of weave type 5 in Table 9.1 assuming that the weight the fabric is* $270\ g/m^2$.

Exercise 9.13 *Calculate the mesoscale tow volume fraction* V_{meso} *and the tow fiber volume fraction* $V^f_{fiber} = V^w_{fiber}$ *of the weave type 5 in Table 9.1, based on the information given in the table.*

Exercise 9.14 *Recalculate the tow elastic properties* $E_1, E_2, G_{12}, G_{23}, \nu_{12}, \nu_{23}$ *of weave type 1 in Table 9.3 using the values of* V^{tow}_{fiber} *calculated in Exercise 9.13.*

Exercise 9.15 *Recalculate the tow hygrothermal properties of weave type 1 in Table 9.3 using the values of* V_{fiber}^{tow} *calculated in Exercise 9.13.*

Exercise 9.16 *Calculate the compliance matrix* $[\overline{S}]$ *in laminate coordinates at* $x = a_w/4, y = a_f/2$, *for the fill and warp tows of sample type 5 in Table 9.1 using the information given in the tables in this chapter.*

Exercise 9.17 *Calculate the reduced stiffness matrix* $[\overline{Q}]$ *in laminate coordinates of the fill and warp tows in Exercise 9.16 at location* $x = a_w/4,\ y = a_f/2$. *Compare with the output of 9.16, and make comments.*

Exercise 9.18 *Calculate the* $[A]$ *matrix of the RVE in Exercise 9.17 at location* $x = a_w/4,\ y = a_f/2$.

Exercise 9.19 *Use the computer program available on the Web site to calculate the* $[A]$ *matrix and* $[\alpha]$ *matrix of sample type 5 in Table 9.1 using the information given in the tables in this chapter. From those values, compute the laminate properties* $E_x, E_y, G_{xy}, \nu_{xy}$.

Exercise 9.20 *For sample type 5 in Table 9.1, calculate the strain at position* $x = a_w/4,\ y = a_f/2$ *for the case of a longitudinal tensile load* $N_x = 5\,N/mm$ *using the bending-restrained model. Calculate the corresponding stress level in fill and warp, at the same location, in material coordinate system. Determine if any kind of failure takes place at this load level (using data from Table 9.5).*

Exercise 9.21 *Use the computer program available on the Web site to calculate the tensile and shear strength values* F_{xt}, F_{xy} *of sample type 5 in Table 9.1 using the information given in the tables in this chapter.*

Exercise 9.22 *Compare the results obtained in Exercise 9.21 to an equivalent symmetric cross-ply approximation using the analysis of Chapter 7, as implemented in CADEC [1].*

Table 9.1: Geometrical parameters for several plain weave fabrics.

Sample type	Tow width [mm]		Tow thickness [mm]		Gap [mm]		Matrix thickness [μm]	Fiber volume fraction Overall (exp.)	Fiber volume fraction Tow (calc.)
	a_f	a_w	h_f	h_w	g_f	g_w	h_m	V^o_{fiber}	$V^{f,w}_{fiber}$ [a]
E-glass fiber									
1	0.40	0.40	0.043	0.043	0.25	0.25	0.0	0.28	0.71
2	0.45	0.45	0.048	0.048	0.30	0.30	2.0	0.23	0.63
3	0.68	0.62	0.090	0.090	0.04	0.10	0.0	0.40	0.69
4	0.62	0.68	0.090	0.090	0.10	0.04	0.0	0.43	0.74
5	0.86	0.84	0.110	0.110	0.00	0.02	0.5	0.46	0.73
6	0.84	0.86	0.110	0.110	0.02	0.00	0.5	0.47	0.74
7	1.08	1.21	0.225	0.225	0.30	0.17	0.0	0.41	0.76
Carbon fiber									
8	1.80	1.45	0.100	0.100	0.65	1.00	0.0	0.27	0.64
9	1.45	1.80	0.100	0.100	1.00	0.65	0.0	0.27	0.64
10	0.96	1.10	0.080	0.080	0.18	0.04	1.0	0.44	0.77
11	1.10	0.96	0.080	0.080	0.04	0.18	1.0	0.44	0.77

[a] $V^f_{fiber} = V^w_{fiber}$

Table 9.2: Mechanical properties of fiber and matrix.

Property	Symbol	Units	Value
E-glass fiber			
Modulus	E_{fiber}	GPa	72.0
Poisson's Ratio	ν_{fiber}		0.3
Strength (manufacturer specification)	F_{fiber}	MPa	3450.0
Strength (apparent, see footnote 2)	F_{fa}	MPa	1995.0
Thermal Expansion Coefficient	α_{fiber}	$\mu\varepsilon/C$	5.4
Moisture Expansion Coefficient	β_{fiber}	ε	0.0
Density	ρ_{fiber}	g/cm^3	2.62
Carbon fiber			
Modulus	E_{1f}	GPa	230.0
	E_{2f}	GPa	40.0
Poisson's Ratio	ν_{12f}		0.26
	ν_{23f}		0.39
Strength (manufacturer specification)	F_{fiber}	MPa	3240.0
Strength (apparent, see footnote 2)	F_{fa}	MPa	2475.0
Thermal Expansion Coefficient	α_{1f}	$\mu\varepsilon/C$	-0.7
	α_{2f}	$\mu\varepsilon/C$	10.0
Moisture Expansion Coefficient	β_f	ε	0.0
Density	ρ_f	g/cm^3	1.76
Epoxy matrix			
Modulus	E_m	GPa	3.5
Poisson's Ratio	ν_m		0.35
Tensile Strength, Bulk (manufacturer specification)	F_{mt}	MPa	60.0
Tensile Strength, Apparent in E-glass–epoxy system	F_{mta}	MPa	36.6
Tensile Strength, Apparent in carbon–epoxy system	F_{mta}	MPa	64.4
Shear Strength, Bulk (manufacturer specification)	F_{ms}	MPa	100.0
Shear Strength, Apparent in E-glass–epoxy system	F_{msa}	MPa	43.0
Shear Strength, Apparent in carbon–epoxy system	F_{msa}	MPa	118.2
Thermal Expansion Coefficient	α_m	$\mu\varepsilon/C$	63
Moisture Expansion Coefficient	β_m	ε	0.33
Density	ρ_m	g/cm^3	1.17

Table 9.3: Hygrothermo-mechanical properties of the tow for various fabrics.

Sample type	Modulus [GPa]				Poisson's ratio		Thermal expansion [$\mu\varepsilon$/C]		Moisture expansion [$10^{-3}\varepsilon$]	
	E_1	E_2	G_{12}	G_{23}	ν_{12}	ν_{23}	α_1	α_2	β_1	β_2
1	52.33	16.43	6.20	6.31	0.31	0.30	6.50	23.70	6.34	53.03
2	46.36	13.15	4.78	4.96	0.32	0.33	7.04	29.69	9.33	72.56
3	50.72	15.45	5.76	5.91	0.32	0.31	6.63	25.29	7.07	58.06
4	54.26	17.71	6.83	6.84	0.31	0.29	6.35	21.80	5.51	47.14
5	53.31	17.06	6.51	6.57	0.31	0.30	6.42	22.74	5.91	50.03
6	54.39	17.80	6.87	6.88	0.31	0.29	6.34	21.68	5.46	46.76
7	55.82	18.87	7.41	7.32	0.31	0.29	6.26	20.29	4.89	42.53
8	148.17	12.25	4.83	4.72	0.30	0.30	-0.20	33.25	2.82	69.80
9	148.17	12.25	4.83	4.72	0.30	0.30	-0.20	33.25	2.82	69.80
10	178.39	16.37	7.34	6.60	0.29	0.24	-0.47	24.27	1.47	41.02
11	178.39	16.37	7.34	6.60	0.29	0.24	-0.47	24.27	1.47	41.02

Table 9.4: Predicted tow strength for various fabrics.

Sample type	Strength [MPa] Longitudinal Tensile F_{1t} Eq.(4.82)	Transverse Tensile F_{2t} Eq.(4.103)	Shear F_6 Eq.(4.115)
1	1449.9	32.0	37.6
2	1237.0	30.5	35.8
3	1405.4	31.7	37.2
4	1503.6	32.4	38.1
5	1477.9	32.2	37.9
6	1507.9	32.5	38.1
7	1546.7	32.8	38.5
8	1594.6	55.0	100.3
9	1594.6	55.0	100.3
10	1919.7	58.1	106.3
11	1919.7	58.1	106.3

Table 9.5: Hygrothermo-mechanical properties of a fabric-reinforced laminate.

Sample type	Modulus [GPa]			Poisson's ratio	Thermal expansion [$\mu\varepsilon$/C]		Moisture expansion [$10^{-3}\varepsilon$]		Strength [MPa] Tensile		Shear
	E_x	E_y	G_{xy}	ν_{xy}	α_x	α_y	β_x	β_y	F_{xt}	F_{yt}	F_{xy}
1	15.48	15.48	3.22	0.18	20.29	20.29	73.28	73.28	320	320	38
2	13.02	12.02	2.57	0.18	23.75	23.75	91.26	91.26	275	275	36
3	19.68	18.87	3.86	0.17	16.80	17.64	51.16	54.30	310	310	38
4	20.61	21.46	4.47	0.17	16.16	15.43	47.83	45.10	335	335	38
5	22.12	21.89	4.56	0.17	15.25	15.44	42.73	43.40	325	325	38
6	22.49	22.73	4.79	0.17	15.02	14.84	41.61	40.98	330	330	38
7	19.30	20.20	4.52	0.17	16.83	15.95	52.07	48.70	345	345	40
8	37.52	31.68	2.79	0.07	7.30	9.23	33.26	40.19	355	355	100
9	31.70	37.52	2.79	0.06	9.23	7.30	40.19	33.26	355	355	100
10	48.94	54.45	4.73	0.06	5.44	4.60	21.11	18.67	415	415	106
11	54.45	48.94	4.73	0.06	4.60	5.44	18.67	21.11	420	420	106

Table 9.6: Comparison of predicted and experimental thermo-mechanical properties.

Sample	E_x [GPa]			[%]	G_{xy} [GPa]			ν_{xy}		α_x [$\mu\varepsilon/C$]		
type	Calc.	Exp.	SD[a]	error	Calc.	Exp. $[10^o]^b$	Exp. $[\pm 45^o]^c$	Calc.	Exp.	Calc.	Exp. SG[d]	Exp. DM[e]
1	15.48	17.5	1.0	11.51	3.22	-	-	0.18	-	20.29	20.86	21.12
2	13.02	20.0	0.8	34.88	2.57	6.25	2.94	0.18	-	23.75	19.58	19.51
3	19.68	21.5	1.0	8.64	3.86	6.89	3.57	0.17	-	16.80	13.69	15.45
4	20.61	20.8	0.7	0.91	4.47	-	-	0.17	-	16.16	-	-
5	22.19	22.8	1.2	2.99	4.56	8.33	5.50	0.17	-	15.25	12.88	14.06
6	22.49	22.4	1.0	0.42	4.79	-	-	0.17	-	15.02	-	-
7	19.30	19.6	1.3	1.55	4.52	-	-	0.17	-	16.83	15.20	18.60
8	37.52	62.5	3.2	39.97	2.79	10.0	5.0	0.07	0.1	7.30	5.31	5.68
9	31.68	-	-	-	2.79	-	-	0.06	-	9.23	6.04	8.33
10	48.94	49.3	1.9	0.74	4.73	-	-	0.06	-	5.44	-	-
11	54.45	60.3	2.1	9.7	4.73	-	-	0.06	-	4.60	-	-

[a] Standard deviation of experimental results.

[b] Shear modulus or strength of a woven fabric-reinforced laminate, obtained by longitudinal off-axis tension test at 10° with respect to the tow direction.

[c] Shear modulus or strength of a woven fabric-reinforced laminate, obtained by longitudinal tension test at $\pm 45^o$ with respect to the tow direction.

[d] Strain gauge method.

[e] Dilatometer method.

Table 9.7: Comparison between predicted and experimental strength values.

Sample type	F_{xt} [MPa] Calc.	Exp.	[%] SD[a]	error	F_{xy} [MPa] Calc.	Exp. D5379[b]	Exp. [10°][a]	Exp. [±45°][a]
1	320	244	15.0	31.15	38	-	-	-
2	275	262	21.0	4.96	36	33	28	26
3	310	318	5.9	2.52	38	34	27	24
4	335	320	6.2	4.69	38	-	-	-
5	325	367	8.0	11.44	38	39	30	26
6	330	338	10.1	2.37	38	-	-	-
7	345	252	10.2	36.90	40	-	-	-
8	355	573	31.2	38.05	100	65	51	42
9	355	504	16.2	29.56	100	-	-	-
10	415	490	13.2	15.31	106	-	-	-
11	420	-	-	-	106	-	-	-

[a]See footnote in Table 9.6.
[b]Shear strength of woven fabric-reinforced laminate obtained by ASTM D 5379.

Table 9.8: Comparison of strength values between unidirectional properties (UD), cross-ply symmetric laminate approximation (CP) and the woven fabric model results.

Sample type	V^o_{fiber} [%]	F_{xt} [MPa] calculated UD	CP	woven	F_{xt} [MPa] experimental woven	error [%] CP	woven
1	0.28	628	315	320	244	29.1	31.1
2	0.32	533	268	275	262	2.1	4.9
3	0.40	855	428	310	318	34.4	2.5
4	0.43	913	458	335	320	43.0	4.7
5	0.46	970	485	325	367	32.1	11.4
6	0.47	990	495	330	338	46.5	2.4
7	0.41	875	438	345	250	75.0	38.0
8	0.27	695	348	355	573	39.4	38.1
9	0.27	695	348	355	504	31.1	29.6
10	0.44	1110	555	415	490	13.3	15.3
11	0.44	1110	555	420	-	-	-

Table 9.9: Comparison of predicted and experimental values of $E[GPa]$ for vinyl ester matrix reinforced with E-glass continuous strand mat (CSM).

Fiber Volume Fraction	**Modulus E [GPa]**			
	Approximate	**Exact**	**Experimental**	
	Eq.(9.67)	Eq.(9.66)	Longitudinal	Transverse
0.16	8.57	7.66	8.58	7.04
0.25	11.63	10.22	10.81	9.94

Chapter 10

Beams

Most composite structures are designed as assemblies of beams, columns, plates, and shells. Beams are structural members that carry mainly bending loads and have one dimension (the length) much larger than the other two dimensions (width and height.) From a geometric point of view, beams and columns are one-dimensional elements, while plates and shells are two-dimensional. Columns carry mostly axial-compression loads. If a one-dimensional element carries both axial-compression as well as bending loads, it is called a *beam-column*. One-dimensional elements carrying tensile loads are called *rods* or *tendons*. If they carry mostly torque, they are called *shafts*, but beams and columns may also carry torque in addition to bending and axial loads. Each engineering discipline has particular names for beams that are used in particular applications. Civil engineers use girders in bridges, stringers in building construction, rafters and purlins in roof construction, and so on. Aircraft engineers use longitudinal stiffeners to reinforce aircraft fuselages, beam spar in aircraft wings, and so on. All these can be analyzed as beams as a first approximation.

Composite beams are thin-walled and composed of an assembly of flat panels (Figure 10.1). Most beams are prismatic but they can be tapered. They are produced by pultrusion, filament winding, hand lay-up, etc. To reduce cost and weight, the cross-sectional area of a composite beam should be as small as possible. The required bending stiffness is achieved by increasing the moment of inertia as much as possible. This is done by enlarging the dimensions of the cross-section while reducing the thickness of the walls. The typical example is the use of an I-beam as opposed to a rectangular solid beam. Furthermore, common processing methods used to produce beams, such as pultrusion, work best for relatively thin walls.

To further illustrate the reasons for thin-walled construction, consider a rectangular solid beam of width b and height h. For a constant cross-sectional area $A = bh$, the moment of inertia $I = Ah^2/12$ grows with the square of the height. For the same cross-section, a tall and thin beam will provide a larger bending stiffness (EI). Wide-flange I-beams or box-beams will perform even better. When glass fibers are used as a reinforcement, a high moment of inertia is necessary to compensate for the relatively low modulus of elasticity of the fibers (72.3 GPa) and of the resulting composite (about 36 GPa for a 50% fiber volume fraction and a polymer

Figure 10.1: Typical structural shapes produced by pultrusion.

matrix, compared to 207 GPa for steel).

For a given cross-sectional area A and given material strength F_x, the maximum moment that can be applied to a rectangular beam grows linearly with the height as $M = F_x Ah/6$. Therefore, a tall and thin beam will carry a higher load, which is limited only by lateral-torsional buckling under bending. When the geometry is fixed (and buckling is prevented) the maximum moment is only limited by the material strength according to $\sigma = F_x = Mc/I$, where c is the distance from the neutral axis to the outer surface. Since composites are very strong, very large and thin-walled sections can be built. Finally, all manufacturing processes favor thin-walled construction. In general, thin composites are less expensive and can be manufactured with better quality and uniformity.

Deflections, strength, and buckling are equally important aspects in the design of composite beams. Analysis and design for deflections and strength are described in this chapter. Section 10.1 deals with preliminary design, which is greatly simplified by using carpet plots. Thin-walled beams under bending and torsional effects are dealt with in Section 10.2. Buckling is considered briefly in Sections 10.1.3 and 10.1.4.

10.1 Preliminary Design

The deflection of a composite beam has two components, bending and shear

$$\delta = \delta_b + \delta_s$$

The bending deflection δ_b is controlled by the bending stiffness (EI) and the shear deflection δ_s by the shear stiffness (GA). Shear deformations are neglected for metallic beams because the shear modulus of metals is high ($G \approx E/2.5$), but shear deformations are important for composites because the shear modulus is low (about $E/10$ or less). The significance of the shear deflection δ_s with respect to the bending deflection varies with the span, the larger the span the lesser the influence of shear (compared to bending). The values of the bending (EI) and shear stiffness (GA) can be computed using the analysis presented in Section 10.2 or approximated using carpet plots.

Maximum deflections can be computed using the formulas in Figure 10.2 or other tables widely available in the literature [220]. Deflections at any point along the span can be computed by the dummy-load method. In this method [178], a fictitious load of unit value is applied to the beam at the location where the deflections is sought. Bending and shear diagrams are constructed for the real load $(M(x), V(x))$ and for the dummy load $(M^0(x), V^0(x))$ independently. The deflection δ, at the position where the dummy load is placed, is computed as

$$\delta = \int_0^L \left(\frac{M}{(EI)} M^0 + \frac{V}{(GA)} V^0 \right) dx \tag{10.1}$$

10.1.1 Design for Deflections

In preliminary design, an average modulus of elasticity can be obtained from the carpet plot of E_x (Figure 6.8). Then, the geometry of the section can be designed to achieve the required bending stiffness (see Example 6.6, p. 197). The carpet plots can be used to estimate the laminate modulus E_x of various laminates as a function of the amount of 0-degree, 90-degree, and $(\pm\theta)$ laminae in the laminate. Selecting laminates with high values of α (see Section 6.5) yields the highest modulus E_x but such laminates may have low shear modulus G_{xy}, which may result in unacceptable values of shear deflection. Also, a laminate with a high value of α will have low shear strength F_{xy}, which may be inadequate to carry the shear loads.

The shear deflection δ_s is usually computed after the geometry has been sized. If the shear deflection is excessive, the laminate or the geometry will have to be changed. The shear deflection is controlled by the shear stiffness (GA) of the section. For preliminary design, it is assumed that the flanges do not contribute to the shear stiffness. Therefore, A can be taken as the area of the webs and G_{xy} can be taken as the apparent shear modulus of the laminate, obtained from the carpet plot (Figure 6.10). Selecting a laminate with higher values of γ (in the carpet plots) yields higher values of G_{xy}.

The webs and flanges can be made with different laminates, provided such configurations can be manufactured. For example, a rectangular tube made by filament winding must have the same laminate construction on all the walls (webs and flanges) because of manufacturing constraints. If different laminates can be used, the designer will try to maximize E_x on the flanges and G_{xy} in the webs.

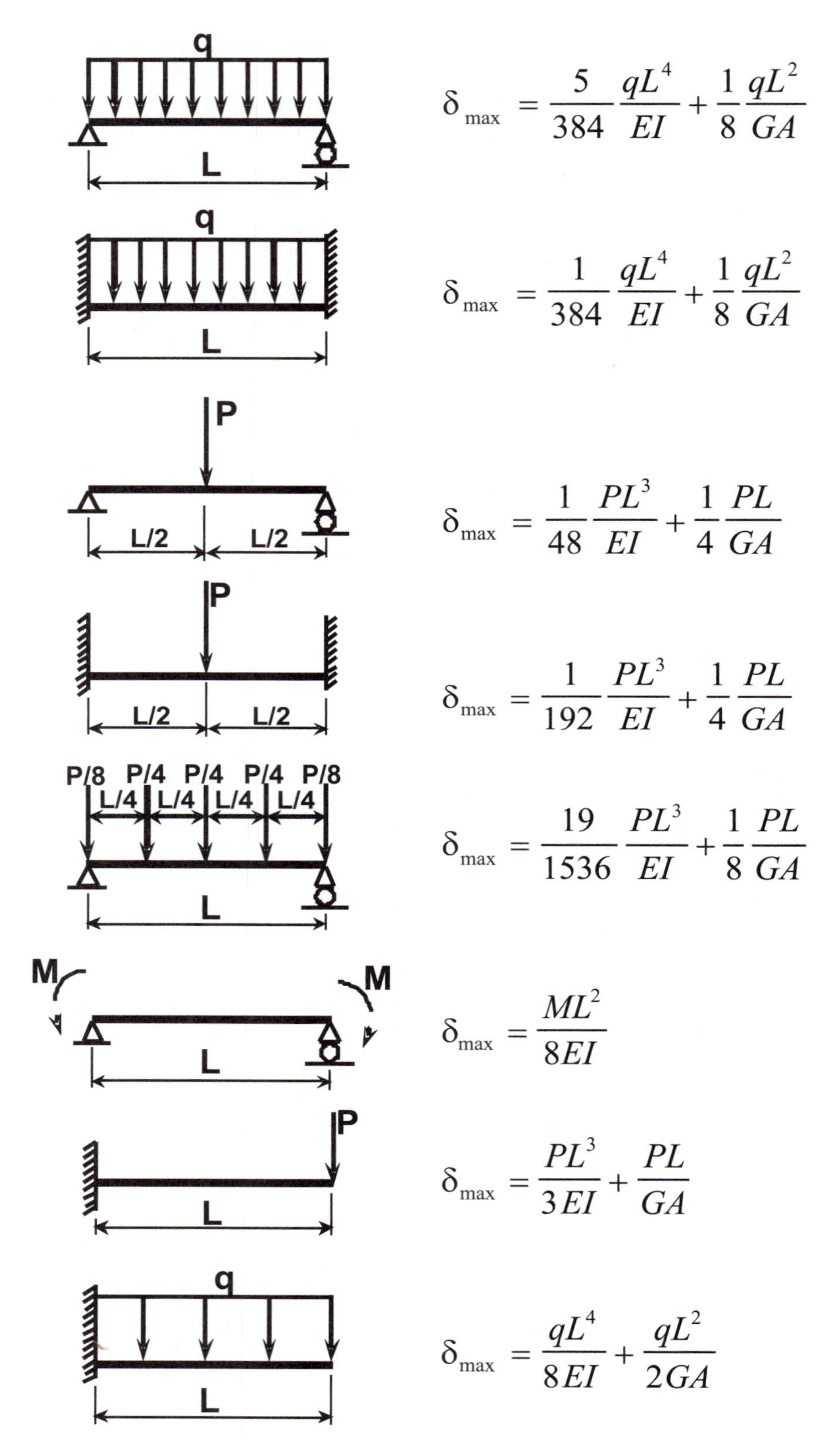

Figure 10.2: Formulas for maximum deflections of beams including bending and shear components.

10.1.2 Design for Strength

In preliminary design, it is assumed that the bending moment is carried primarily by the flanges and the shear force is carried by the webs. The bending stress can be approximated by the formula used for isotropic materials

$$\sigma = \frac{Mc}{I} \tag{10.2}$$

where M is the bending moment (from the moment diagram), I is the moment of inertia, and c is the distance from the neutral axis to the top or bottom of the section. For composite beams, both positive and negative values of stress must be checked (not just the largest absolute value) because the tensile and compressive strengths of the laminate are different. Defining the section modulus

$$Z = \frac{I}{c}$$

and replacing the stress σ in (10.2) by the factored strength value ϕF_{xt} or ϕF_{xc} (tensile or compressive), the required section modulus is

$$Z = \frac{\alpha\, M}{\phi\, F_x} \tag{10.3}$$

where α is the load factor (see Section 1.4.3.)

For most common cross-sections used (I-beams, box-beams, etc.), the state of stress in the flanges is mostly axial stress σ_x. Then, the strength F_x is either the tensile or compressive strength of the laminate (F_{xt} or F_{xc}) depending on the sign of the stress. The strength of a laminate can be estimated from carpet plots (Figures 7.12–7.15). On the other hand, for rectangular solid beams with the laminae perpendicular to the load direction, the axial stress σ_x changes sign through the thickness of the laminate and the flexural strength F_x^b of the laminate should be used instead.

The section modulus allows the designer to size the cross-section for bending strength. Using Jourawski's formula for isotropic materials [221, p. 361],[1] the shear stress in the web is

$$\tau = \frac{Q\, V}{I\, t} \tag{10.4}$$

where V is the shear force from the shear diagram, I is the moment of inertia of the beam, t is the thickness of the web, and Q is the first moment of area.[2] Since Q is maximum at the neutral axis, the maximum shear stress acts at the neutral axis. The shear stress τ must be resisted by the laminate in the webs, which have

[1] The first edition is used as a reference here because the presentation of torsion, unsymmetric bending, shear center, and thin-walled beams was abridged for the second edition [222].

[2] For an explanation of first moment of area, see the section on "shear stresses in beams" in any basic strength of materials textbook [223, p. 48] or advanced strength of materials textbooks [221, 224].

a laminate shear strength F_{xy} given in the carpet plot, Figure 7.15. Therefore, the design must satisfy

$$\phi\, F_{xy} > \frac{\alpha\, Q\ V}{I\ t} \tag{10.5}$$

If the shear strength of the laminate is exceeded, the thickness of the webs or the strength of the laminate should be increased. The shear strength of the laminate can be increased by selecting a laminate with higher γ in the carpet plots.

Example 10.1 *Verify that the box-beam (Figure 6.13) designed in Example 6.6 (p. 197) is sufficiently strong, for a reliability of 99.5%, assuming the load and strength variability are represented by* $C_\sigma = 0.2, C_F = 0.3$.

Solution to Example 10.1 *Example 6.6 (p. 197) concluded with a design having dimensions* $h = 188\ mm$, $b = 94\ mm$, $t = 4.7\ mm$ *and a laminate with* $\alpha = 0.8$ *and* $\gamma = 0.2$. *The moment of inertia is*

$$I \cong t\left[\frac{h^3}{6} + \frac{bh^2}{2}\right] = 13.0(10^6)\ mm^4 = 1,300\ cm^4$$

and the first moment of area with respect to the neutral axis (Figure 6.13) is

$$Q = tb\frac{h}{2} + 2t\frac{h}{2}\frac{h}{4} = 83,058.4\ mm^3$$

For a laminate with $\alpha = 0.8$, $\gamma = 0.2$, $\beta = 0$, *from Figure 7.12–7.15 the strength values at last ply failure are*

$$F_{xt} = 425\ MPa$$
$$F_{xc} = 305\ MPa$$
$$F_{xy} = 80\ MPa$$

Based on the bending moment at the root of the cantilever beam

$$M = PL = 1000(2) = 2000\ Nm$$

and shear force

$$V = P = 1000\ N$$

From (10.2) (see Figure 6.13), the average stress over the thickness of the flange is

$$\sigma_x = \frac{M}{I}\frac{h}{2} = \frac{2000}{13(10^{-6})}\left[\frac{188\ 10^{-3}}{2}\right] = 14.462\ MPa$$

giving a safety factor for bending

$$R_b = \frac{305}{14.462} = 21.1$$

At the neutral axis, using (10.4), the shear stress is

$$\tau = \frac{Q\,V}{I\,t} = \frac{83,058.4\,(10^{-9})(1000)}{13(10^{-6})\ [(2)4.7(10^{-3})]} = 0.680\ MPa$$

giving a safety factor for shear

$$R_s = \frac{80}{0.680} = 117.6$$

For a reliability of 99.5%, the standard variable is, from Table 1.1, $z = -2.5076$. *Then, using (1.12) and the load and strength COV,* $C_\sigma = 0.2, C_F = 0.3$, *requires a safety factor*

$$R = \frac{1 + \sqrt{1 - (1 + 2.5076^2 \times 0.3^2)(1 + 2.5076^2 \times 0.2^2)}}{1 + 2.5076^2 \times 0.3^2} = 4.17$$

Therefore, the box-beam is sufficiently strong. The design could be optimized to make better use of the material (see Exercise 10.12).

10.1.3 Design for Buckling

When a thin-walled beam is subject to bending, compressive loads are induced in some portions of the cross-section, usually in one of the flanges. Compressive loads may cause buckling of that portion of the cross-section. For example, Figure 11.4 may represent the buckling shape of the compression flange of a box-beam, and Figure 11.6 the same for an I-beam. The webs of the cross-section carry most of the shear and they may also buckle in a shear buckling mode.

The buckling resistance depends on the material properties of the laminate, the geometry, and the way various elements of the cross-section are supported. For example, the flange of a box-beam is supported on both edges by the webs. Assuming that the flange is simply supported at the two edges resting on the webs (Figure 11.4), the buckling load per unit width can be approximated by (Section 11.2.1)

$$N_x^{CR} = \frac{2\pi^2}{b^2}\left(\sqrt{D_{11}D_{22}} + D_{12} + 2D_{66}\right) \tag{10.6}$$

Assuming that the bending moment is carried fully by the two flanges of the box-beam, the compressive load (per unit width of flange) can be obtained by summation of forces and moments

$$N_x = \frac{M}{bh} \tag{10.7}$$

where h is the height of the beam and b is the width of both flanges. The compressive load computed with (10.7) should not exceed the limit load given by (10.6). Equation (10.6) provides a conservative estimate of the buckling load of the flange because it assumes a simple support at the connections of the flange to the webs. In actual sections, the flanges are more or less rigidly connected to the webs. Therefore, the rotation at the edges of the compression flange are somewhat restricted. In the limiting case of very thick webs, the edges of the flange are almost clamped (Section 11.2.2). In this case, the buckling load can be predicted using

$$N_x^{CR} = \frac{\pi^2}{b^2}\left(4.6\sqrt{D_{11}D_{22}} + 2.67D_{12} + 5.33D_{66}\right) \tag{10.8}$$

Since the compression flange is neither simply supported nor clamped, the actual buckling load of the flanges is between the values predicted by (10.6) and (10.8).

In the case of an I-beam, each flange can be modeled as two plates simply supported by the web, with the other edge free (Figure 11.6). Then, the buckling load of each half flange can be predicted with (Section 11.2.3)

$$N_x^{CR} = \frac{12D_{66}}{b^2} + \frac{\pi^2 D_{11}}{a^2} \tag{10.9}$$

where b is half of the flange width and a is the unsupported length of the flange or the region over which the compressive load N_x acts. Because the rotational constraint provided by the web, the actual flange buckling load is larger than the one predicted by (10.9). Buckling of the flange can be prevented by attaching the flange to a panel. This is typically done when the beam supports a floor system or a bridge deck in civil engineering construction.

The buckling formulas presented in this section only provide estimates of the actual buckling strength of a thin-walled beam. A more accurate and elaborate treatment falls outside the scope of this textbook, requiring numerical analysis [225, 226] or a combined analytical/experimental approach [227]. Lateral–torsional and Lateral–distortional buckling of beams are additional modes of failure that must be considered in design [228].

10.1.4 Column Behavior

A thin-walled column loaded axially in compression may fail in one of several modes of failure, which are described next.

Euler Buckling A slender column loaded axially in compression buckles with a lateral deflection resembling bending of a beam. This is called *Euler buckling* and it is described in introductory textbooks [178, p. 577]. For a slender, long column, buckling occurs for axial stress values much lower than the compressive strength of the material. As the column buckles with lateral deflection, the shape of the cross-section remains undeformed. The critical load (or buckling load) of a long column is a function of the bending stiffness of the section (EI), the length, and the type of support at the ends of the column

$$P_{CR} = \frac{\pi^2 (EI)}{L_e^2} \tag{10.10}$$

where L_e is the effective length defined in terms of the actual length L and the coefficient of restraint as

$$L_e = KL \tag{10.11}$$

The coefficient K can be computed theoretically or it can be adjusted to fit experimental data. Values are given in many handbooks and textbooks [223, p. 123], [229, p. 644]. Theoretical values as well as recommended values for conventional materials are given in Table 10.1.

Table 10.1: End-restraint coefficients for long column buckling.

End-restraint	$\mathbf{K}_{theory}$	$\mathbf{K}_{steel}$	$\mathbf{K}_{wood}$
pinned-pinned	1.0	1.0	1.0
clamped-clamped	0.5	0.65	0.65
pinned-clamped	0.7	0.8	0.8
clamped-free	2.0	2.1	2.4

The bending stiffness (EI) is the same used to predict the deflections of the beam. In preliminary design, (EI) can be computed using carpet plots, selecting a value of E_x, and designing the geometry to give the required moment of inertia I. A more accurate prediction can be obtained using methodology presented in Section 10.2. Euler buckling is usually prevented in practice by bracing. The column is braced to reduce the effective length, thus increasing the buckling load until the mode of failure changes to local buckling or material failure.

Local Buckling Thin-walled columns usually display buckling of the walls without the overall deflection of Euler buckling. In this case the shape of the cross-section changes while the axis of the column remains straight. While the walls can be analyzed individually for buckling, using the formulas in Section 11.2 (see also Section 11.3.2), the loads predicted in this way can have significant errors. The whole cross-section should be considered in the analysis to obtain accurate results, and the Finite Element Method is often used for this purpose [225].

Material Failure The compressive strength of the material is reached only when the walls of the cross-section are thick and narrow. In this case, the local buckling stress could be higher than the compressive strength of the material. Material failure can be predicted simply by comparing the applied compressive stress to the compressive strength of the laminate F_{xc} (Figure 7.14) or by performing a laminate stress analysis and failure evaluation as described in Chapter 7.

Mode Interaction The column can fail with a combination of the modes described above. For example, local and Euler modes interact for pultruded wide-flange I-beams of intermediate length [226, 227, 230]. In this case, an empirical interaction factor can be used to estimate the critical load [227].

$$\frac{P_{CR}}{P_L} = \frac{1+1/\lambda^2}{2c} - \sqrt{\left(\frac{1+1/\lambda^2}{2c}\right)^2 - \frac{1}{c\lambda^2}} \tag{10.12}$$

$$\lambda = KL\sqrt{\frac{P_L}{\pi^2(EI)}}$$

where P_l is the local buckling load, (EI) is the bending stiffness, c is the interaction coefficient, K is the coefficient of restraint, and λ^2 is the dimensionless buckling load.

10.2 Thin-Walled Beams

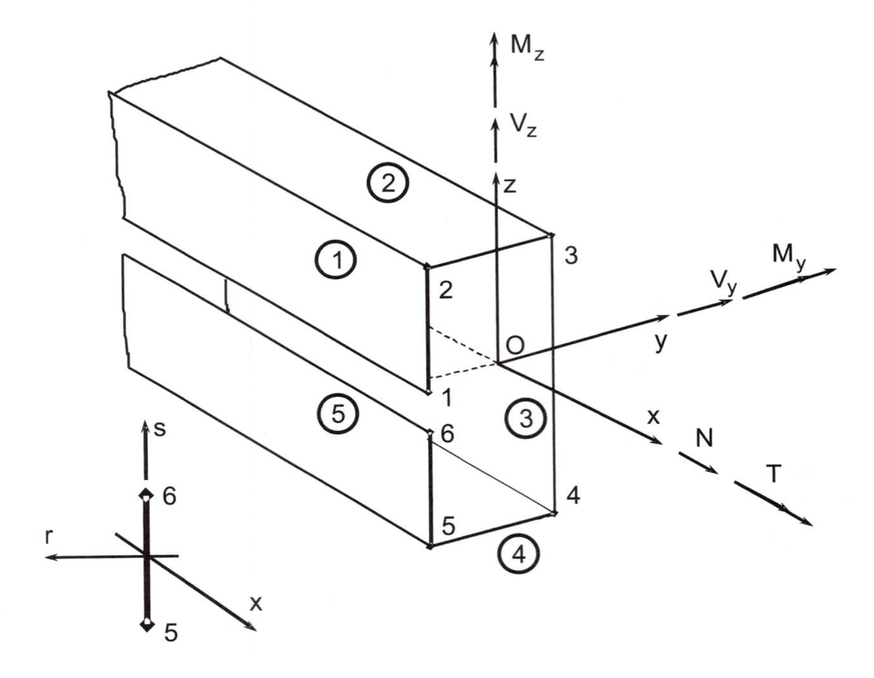

Figure 10.3: Definition of the global and local coordinate systems, and node and segment numbering.

The objective of this section is to present equations that can be used for design of open or closed sections of arbitrary shape, subjected to bending, torque, shear, and axial forces. The theory, first presented in [231], follows closely the classical thin-walled beam theory, as presented in classical textbooks [221, 224]. For the sake of brevity, many details of the classical theory are not extensively explained. The presentation in this section emphasizes the generalization of the classical theory to composite materials. For more detailed explanation of the classical theory, the reader is advised to consult the literature [221, 224].

The analysis presented in this section applies to most laminated beams where very dissimilar materials, or severely unsymmetrical, unbalanced lay-ups are excluded to avoid the undesirable effects that those configurations produce, including warping due to curing residual stresses, etc. Special applications, such as helicopter rotor blades [232] and swept forward aircraft wings [233], where strong coupling effects are desirable, are outside the scope of this section.

Practical composite beams are thin-walled (Figure 10.1), built with several walls. Each wall is usually a complex laminate, optimized to maximize the performance of the beam. Webs are likely to contain ±45 laminae to resist shear, and flanges usually contain more unidirectional reinforcement to maximize bending stiffness.

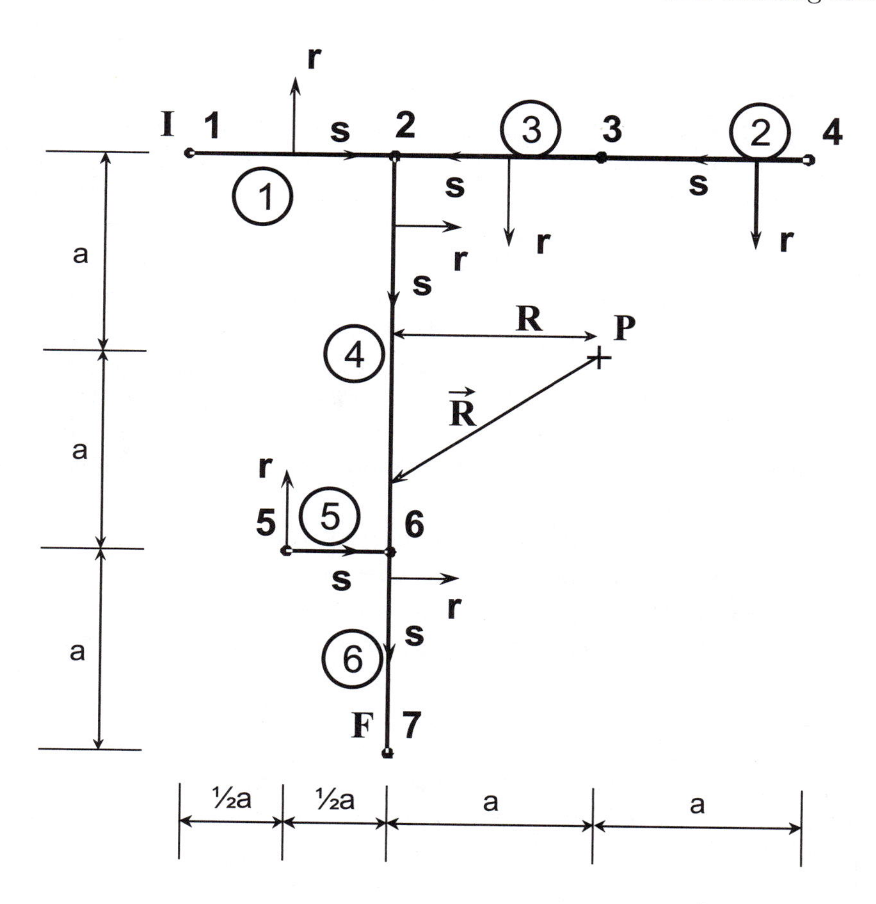

Figure 10.4: Definition of the local coordinate system, node and segment numbering for an open section.

The middle surface of the beam cross-section is represented by a polygonal line called the *contour* (Figure 10.3). The joints of the cross-section are modeled at the intersection of the middle surfaces of the walls. Each wall is described by one or more segments (Figure 10.4). Curved segments can be modeled as a collection of flat elements and concentrated axial-stiffness can be used to represent the contributions

Table 10.2: Contour definition for Figure 10.4. Used for the computation of contour integrals except $\omega(s)$.

Segment number	$\mathbf{n}_i$	$\mathbf{n}_f$	**Segments converging to $\mathbf{n}_i$**		
1	1	2	0	0	0
2	4	3	0	0	0
3	3	2	2	0	0
4	2	6	1	3	0
5	5	6	0	0	0
6	6	7	4	5	0

of attachments that are not modeled explicitly.

All segments are defined by a initial node n_i and a final node n_f. For a closed single-cell section, the segments and the nodes are numbered consecutively around the contour. The same applies for an open section without branches, such as a C-channel. An open section with branches, such as an I-beam, requires special consideration to assure that the static moment and the sectorial area are computed correctly. For the section shown in Figure 10.4, the definition of the contour used to compute the static moment is given in Table 10.2. The definition starts at a free edge. A segment is added only if the contributions of all segments converging to its initial node n_i are known. In Table 10.2, segment 1 is defined starting at a free edge with the local coordinate s pointing towards the branching point 2; segment 2 is defined before segment 3, and segment 4 cannot be added to the contour until all of the above are defined.

The definition of a segment must reference all the segments converging to its initial node. In Table 10.2, segment 4 is defined with $n_i = 2$ and the segments 1 and 3 are listed as converging at that point.

The coordinates of each node are given in terms of an arbitrary global coordinate system x, y, z. The principal axes of bending are aligned with the coordinate system x, η, ζ, rotated an angle ϑ about the x-axis (Figure 10.5). For each segment, the contour coordinate s is oriented from the initial node to the final node. The other two local coordinates are the global x-axis and the r-axis that is obtained as the cross-product of x times s, or $r = x \times s$, in such a way that the r-coordinate spans the thickness of the segment.

The fiber orientation of lamina k is given by the angle θ^k (Figure 10.6), positive counterclockwise around the r-axis and starting from the x-axis. The first lamina ($k = 1$) is located at the bottom of the laminate on the negative r-axis. Since the orientation of the r-axis depends on n_i and n_f, the order in which the laminae are specified and the angle θ^k depend on the segment definition.

Geometric properties such as moment of inertia, first moment of area, etc., are used in classical beam theory to define section properties such as bending stiffness (EI), etc. When dealing with beams made of a single homogeneous material, such as steel or aluminum, the geometric property can be separated from the material

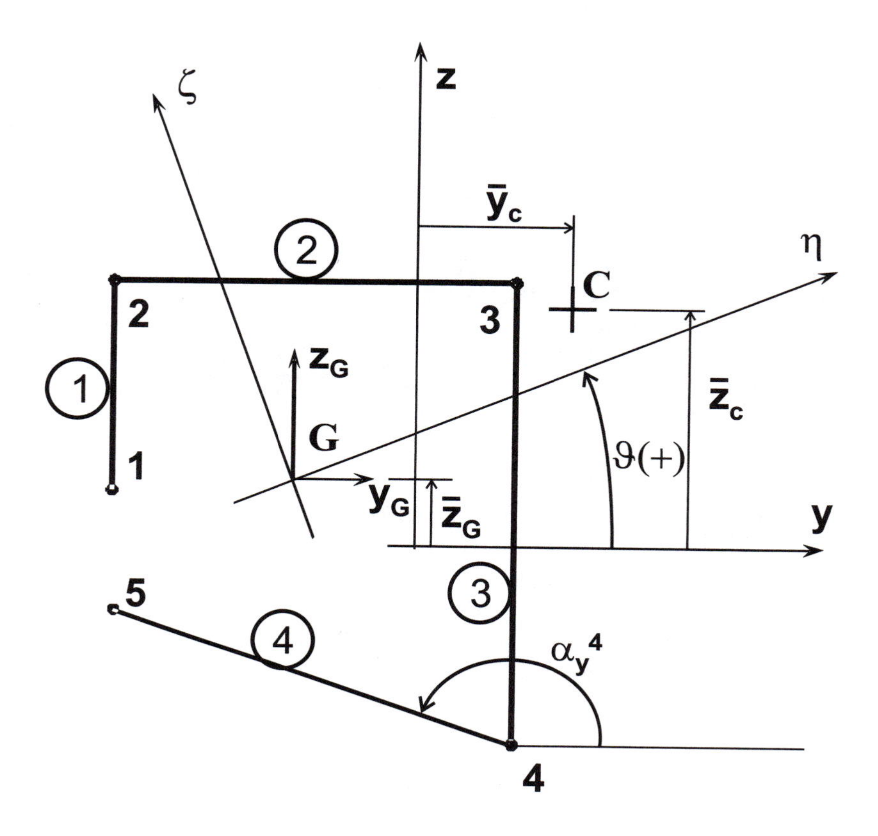

Figure 10.5: Definition of principal axes of bending for a general open section.

property because the later is constant through the cross-section of the beam. Since the material properties of composite beams vary from lamina to lamina and from segment to segment, it is not possible to separate the elastic properties (E and G) from the geometric properties (area, moment of inertia, etc.). Therefore, *mechanical* properties are defined in this section, that contain combined information about the geometry and the material.

All *mechanical* properties are defined by integrals over the area of the cross-section. The area integral is divided into an integral over the contour $\int_s ds$ and an integral over the thickness of the segment $\int_t dr$; the integral over the thickness of the segment $\int_t dr$ is solved first. Then, the contour integrals are reduced to algebraic summations over the number of segments.

Example 10.2 *To begin illustrating the concept of mechanical properties, consider a two-layer rectangular beam of width b and total thickness $2t$. The bottom layer has thickness t, density ρ_1, and modulus E_1. The top layer has $\rho_2 = 2\rho_1$ and $E_2 = 31E_1$. Compute (a) the geometric center of gravity, (b) the mass center of gravity, and (c) the mechanical center of gravity.*

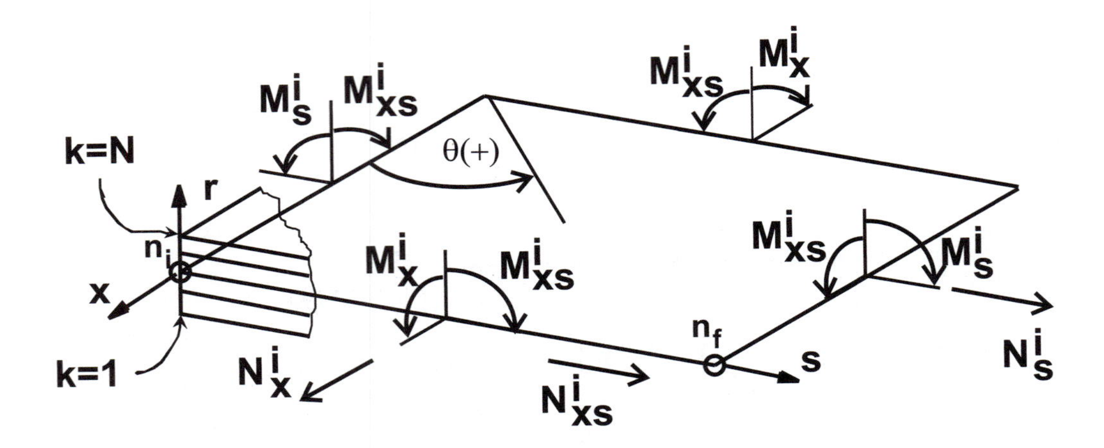

Figure 10.6: Definition of stress resultants in the i-th wall and convention for fiber orientation.

Solution to Example 10.2 *Taking the z-axis through the thickness of the beam, the geometric center of gravity of the cross-section is found as*

$$\overline{z}_G = \int_A z dA / \int_A dA$$

It can be shown that $\overline{z}_G$ lays in the interface between the two laminae regardless of the coordinate system used. Using a coordinate system with origin at the geometric center of gravity just found, the mass center of gravity is located at

$$\overline{z}_\rho = \int_A \rho z dA / \int_A \rho dA = t/6$$

That is, the mass center of gravity is located at $t/6$ into the top layer, which is twice as dense as the bottom layer. The mechanical center of gravity is located at

$$\overline{z}_M = \int_A E z dA / \int_A E dA = \frac{15}{32} t$$

That is, the mechanical center of gravity almost coincides with the geometric and mass center of gravity of the top layer, which is 31 times stiffer than the bottom layer.

10.2.1 Wall Constitutive Equations

The constitutive equations of a laminated plate (6.20) are used to model each flat wall (substituting s for y). Neglecting transverse shear deformations, the compliance equations are

$$\left\{\begin{array}{c}\epsilon_x^i\\ \epsilon_s^i\\ \gamma_{xs}^i\\ \kappa_x^i\\ \kappa_s^i\\ \kappa_{xs}^i\end{array}\right\} = \left[\begin{array}{cccccc}\alpha_{11} & \alpha_{12} & \alpha_{16} & \beta_{11} & \beta_{12} & \beta_{16}\\ & \alpha_{22} & \alpha_{26} & \beta_{12} & \beta_{22} & \beta_{26}\\ & & \alpha_{66} & \beta_{16} & \beta_{26} & \beta_{66}\\ & & & \delta_{11} & \delta_{12} & \delta_{16}\\ & & & & \delta_{22} & \delta_{26}\\ sym. & & & & & \delta_{66}\end{array}\right]\left\{\begin{array}{c}N_x^i\\ N_s^i = 0\\ N_{xs}^i\\ M_x^i\\ M_s^i = 0\\ M_{xs}^i\end{array}\right\} \quad (10.13)$$

where the classical assumption[3] $\sigma_s = 0$ has been replaced by $N_s^i = M_s^i = 0$, which is more appropriate for a laminated composite. In (10.13), N_x^i, N_s^i, and N_{xs}^i are the normal and shear forces per unit length along the boundary of the plate (Figure 10.6) with units [N/m], M_x^i, M_s^i, M_{xs}^i are the moments per unit length on the sides, with units [N]. The bending moments are positive when they produce a concave deformation looking from the negative r-axis. The midplane strains are ϵ_x^i, ϵ_s^i, and γ_{xs}^i, and the curvatures are κ_x^i, κ_s^i, and κ_{xs}^i.

The superscript i is used not only to indicate the segment number but also to differentiate the plate quantities N_x^i, N_s^i, N_{xs}^i, M_x^i, M_s^i, M_{xs}^i, ϵ_x^i, ϵ_s^i, γ_{xs}^i, κ_x^i, κ_s^i, and κ_{xs}^i (all of which vary along the contour) from the beam quantities N, V_η, V_ζ, T, M_η, M_ζ, ε_x, κ_η, κ_ζ, and β (Figure 10.3) to be defined later.

In order to arrive at practical equations, the off-axis laminae should be arranged in a balanced symmetric configuration. The laminate can be unsymmetrical as a result of specially orthotropic laminae (isotropic, unidirectional, random and balanced fabrics of any type including $\pm\theta$ fabrics) that are not symmetrically arranged with respect to the middle surface. Under these conditions $\alpha_{16} = \beta_{16} = 0$ and δ_{16} decreases rapidly in magnitude with increasing number of laminae (see Example 6.3 and 6.5, pp. 186, 191). If these conditions are not met, the theory still yields approximate results.

Then, using the beam theory assumptions $N_s^i = 0$, $M_s^i = 0$, and assuming uncoupling between normal and shearing effects ($\alpha_{16} \approx 0$, $\beta_{16} \approx 0$, $\delta_{16} \approx 0$), the compliance equations (10.13) can be reduced to

$$\left\{\begin{array}{c}\epsilon_x^i\\ \kappa_x^i\\ \gamma_{xs}^i\\ \kappa_{xs}^i\end{array}\right\} = \left[\begin{array}{cccc}\alpha_{11} & \beta_{11} & 0 & 0\\ \beta_{11} & \delta_{11} & 0 & 0\\ 0 & 0 & \alpha_{66} & \beta_{66}\\ 0 & 0 & \beta_{66} & \delta_{66}\end{array}\right]\left\{\begin{array}{c}N_x^i\\ M_x^i\\ N_{xs}^i\\ M_{xs}^i\end{array}\right\} \quad (10.14)$$

The deformations ϵ_s and κ_s can be computed from (10.13) but they are usually not needed. The reduced constitutive equation for the i-th segment is obtained by inverting (10.14) as

$$\left\{\begin{array}{c}N_x^i\\ M_x^i\\ N_{xs}^i\\ M_{xs}^i\end{array}\right\} = \left[\begin{array}{cccc}A_i & B_i & 0 & 0\\ B_i & D_i & 0 & 0\\ 0 & 0 & F_i & C_i\\ 0 & 0 & C_i & H_i\end{array}\right]\left\{\begin{array}{c}\epsilon_x^i\\ \kappa_x^i\\ \gamma_{xs}^i\\ \kappa_{xs}^i\end{array}\right\} \quad (10.15)$$

[3]In classical thin-walled theory, the undeformability of the contour implies $\epsilon_s = 0$ and $\kappa_s = 0$; $\sigma_s = 0$ is used in the constitutive equations to recover Hooke's law.

All the section properties needed to solve the general problem of bending and torsion are computed in terms of the segment stiffness A_i, B_i, D_i, F_i, C_i, and H_i. The segment stiffness A_i is the axial stiffness per unit length of the segment; B_i represents the coupling between bending curvature κ_x^i and extensional force per unit length N_x^i that appears when the laminate is not symmetric with respect to the midsurface of the segment; D_i is the bending stiffness of the segment under bending M_x^i; F_i is the in-plane shear stiffness under shear N_{xs}^i; H_i is the twisting stiffness under twisting moment M_{xs}^i; and C_i is the coupling between the twisting curvature κ_{xs}^i and the shear flow $q = N_{xs}^i$. For isotropic materials, they reduce to

$$\begin{aligned} A_i &= Et = EA/b \\ D_i &= Et^3/12 = EI/b \\ F_i &= Gt = GA/b \\ H_i &= Gt^3/12 = \tfrac{1}{4}GJ_R/b \end{aligned} \tag{10.16}$$

where E, G, t, b, and I are the modulus, shear modulus, thickness, width, moment of inertia, and $J_R = bt^3/3$.

10.2.2 Neutral Axis of Bending and Torsion

Assuming a state of deformation $\epsilon_x^i \neq 0$ with all the other strains and curvatures equal to zero, the first two of (10.15) reduce to

$$M_x^i = \frac{B_i}{A_i} N_x^i = e_b\, N_x^i \tag{10.17}$$

with the location of the neutral surface of bending for the segment i given by

$$e_b = \frac{B_i}{A_i} \tag{10.18}$$

According to (10.14), a force N_x^i acting at the middle surface of the segment ($r = 0$; axis s) produces a bending curvature $\kappa_x^i = \beta_{11} N_x^i$. An additional moment $M_x^i = e_b N_x^i$ is required at the middle surface to restore $\kappa_x^i = 0$. The combination of N_x^i and M_x^i at the middle surface are statically equivalent to a force N_x^i acting with eccentricity e_b. Therefore, e_b indicates the location r of the point where a force N_x^i only produces $\epsilon_x^i \neq 0$ with $\kappa_x^i = 0$, which is the neutral axis of bending.

The neutral axis of bending (Figure 10.7) is described by

$$\begin{aligned} y(s') &= y(s) - e_b \sin\alpha_y^i \\ z(s') &= z(s) + e_b \cos\alpha_y^i \end{aligned} \tag{10.19}$$

where α_y^i is the angle of the segment with respect to axis y.

The bending stiffness of the segment is computed with respect to the neutral axis of bending s' (Figure 10.7). Assuming that $N_x^i = 0$ and replacing the first equation in (10.15) into the second equation, the moment-curvature relationship becomes

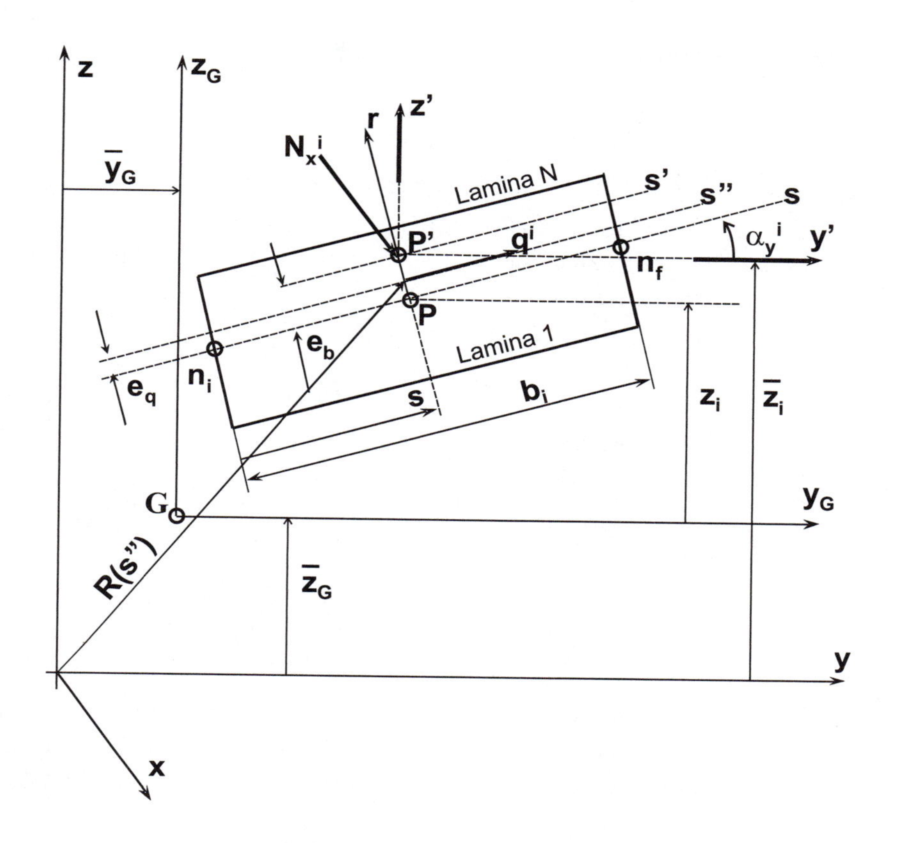

Figure 10.7: Cross-section of segment i showing the definition of the various variables.

$$M_x^i = \left(D_i - \frac{(B_i)^2}{A_i} \right) \kappa_x^i \tag{10.20}$$

The first two equations in (10.15) become uncoupled when N_x^i, M_x^i, ϵ_x^i and κ_x^i are defined with respect to s'-axis (neutral surface of bending in Figure 10.7)

$$\begin{Bmatrix} N_x^i \\ M_x^i \end{Bmatrix} = \begin{bmatrix} A_i & 0 \\ 0 & \overline{D}_i \end{bmatrix} \begin{Bmatrix} \epsilon_x^i \\ \kappa_x^i \end{Bmatrix} \tag{10.21}$$

where

$$\overline{D}_i = D_i - e_b^2 \, A_i \tag{10.22}$$

Next, assuming a state of deformation $\gamma_{xs}^i \neq 0$ and all other strain and curvatures equal to zero, the last two equations in (10.15) also become uncoupled and can be reduced to

$$M_{xs}^i = e_q \, N_{xs}^i \tag{10.23}$$

with the location of the neutral axis of torsion for the segment i given by

$$e_q = \frac{C_i}{F_i} \tag{10.24}$$

A shear flow $q = N^i_{xs}$ acting with eccentricity $r = e_q$ (axis s'') respect to the midsurface (contour s) in Figure 10.7, produces only shear strain and no twisting curvature κ^i_{xs}, regardless of the lack of symmetry of the laminate. The axis of torsion (axis s'') is described by (see Figure 10.7)

$$\begin{aligned} y(s'') &= y(s) - e_q \sin \alpha^i_y \\ z(s'') &= z(s) + e_q \cos \alpha^i_y \end{aligned} \tag{10.25}$$

The last two equations of (10.15) become uncoupled when N^i_{xs}, M^i_{xs}, γ^i_{xs},and κ^i_{xs} are defined with respect to s''-axis (the neutral surface of torsion in Figure 10.7), reducing to

$$\left\{ \begin{array}{c} N^i_{xs} \\ M^i_{xs} \end{array} \right\} = \left[\begin{array}{cc} F_i & 0 \\ 0 & \overline{H}_i \end{array} \right] \left\{ \begin{array}{c} \gamma^i_{xs} \\ \kappa^i_{xs} \end{array} \right\} \tag{10.26}$$

where $\overline{H}_i$ is the torsional stiffness of the segment, recomputed with respect to the axis s''. Assuming that $N^i_{xs} = 0$ and replacing the third equation in (10.15) into the fourth equation, the torsional stiffness per unit length of the segment is obtained as

$$\overline{H}_i = H_i - e_q^2 F_i \tag{10.27}$$

and the *torsional stiffness of the segment* is defined as

$$(GJ^i_R) = 4 \overline{H}_i b_i \tag{10.28}$$

where the factor 4 is explained in Section 10.2.6 (10.45) and b_i is the length of the i-th segment (Figure 10.7).

10.2.3 Axial Stiffness

To obtain the axial stiffness (EA) of the section, assume that a constant state of strain $\epsilon^i_x = \epsilon_x$ is applied with all other strains and curvatures equal to zero. Integrating the axial stress over the area of the cross-section, the axial force is

$$N = \int_s \int_t \sigma_x \, dr \, ds = \int_s N^i_x(s') \, ds \tag{10.29}$$

The integration over the thickness yields N^i_x, which is applied at the neutral surface of bending s'. The contour s' is chosen because only at s' the integral of σ_x results in axial force N^i_x and no bending moment. Then, using (10.21) results in

$$N = \int_s A_i \, \epsilon^i_x \, ds = \epsilon_x \int_s A_i \, ds = \epsilon_x \, (EA) \tag{10.30}$$

where (EA) represents the *axial stiffness* of the section, and $\int_s$ indicates an integration over the contour describing the cross-section. The contour integral reduces to adding the contribution of all the segments (Figure 10.3 through 10.5)

$$(EA) = \sum_{i=1}^{n} A_i\, b_i \tag{10.31}$$

where n is the number of segments describing the cross-section. Note that (EA) is not separated into E and A as it is done for beams made of a single isotropic material. Since all the section properties, such as (EA), result in algebraic expressions, they can be programmed into a general computer program for cross-sections of arbitrary shape.

10.2.4 Mechanical Center of Gravity

The mechanical center of gravity of the cross-section (point G in Figure 10.7) is the point of application of the axial force N, which is the resultant of the axial stresses caused by a constant state of strains ϵ_x. Equilibrium of moments with respect to the y-axis leads to

$$N\,\overline{z}_G = \int_s \int_t z(s')\,(\sigma_x\, dr\, ds) = \int_s z(s') N_x^i(s') ds \tag{10.32}$$

Note that the bending moment produced by the axial stress σ_x is computed applying N_x^i at the neutral surface of bending s', because s' is the only surface with respect to which σ_x does not produce a bending moment. From now on, all bending moments produced by σ_x will be computed in this way. Similarly, all the computations of torque produced by shear σ_{xs} are performed integrating the shear flow $q^i = N_{xs}^i$ along the neutral axis of torsion s''. Furthermore, note that the length of a segment is the same regardless of the contour used (s, s', or s'') to measure the quantities in the integral.

Next, using (10.21), (10.19), and (10.30) yield

$$\epsilon_x\,(EA)\;\overline{z}_G = \epsilon_x \int_s \left(z(s) + e_b \cos\alpha_y^i\right) A_i\, ds \tag{10.33}$$

Solving for $\overline{z}_G$ and repeating the same procedure for $\overline{y}_G$, the coordinates of the *mechanical center of gravity* are

$$\overline{z}_G = \frac{(EQ_y)}{(EA)}$$
$$\overline{y}_G = \frac{(EQ_z)}{(EA)} \tag{10.34}$$

where (EQ_y) and (EQ_z) are the *mechanical static moments* defined as

$$(EQ_y) = \int_s z\left(s'\right)\ A_i\, ds = \sum_{i=1}^{n} \overline{z}_i A_i b_i$$

$$(EQ_z) = \int_s y\left(s'\right)\ A_i\, ds = \sum_{i=1}^{n} \overline{y}_i A_i b_i \tag{10.35}$$

and $\overline{y}_i$, $\overline{z}_i$, are the coordinates of the point P′ (Figure 10.7) where $s = b_i/2$.

10.2.5 Bending Stiffness

To obtain the bending stiffness (EI) of the section, the product of the modulus of elasticity E times the moments of inertia I_{ss}, I_{rr}, and the product of inertia I_{sr} of classical beam theory are replaced by the *mechanical properties of each segment* defined as

$$\begin{aligned}
(EI^i_{s'}) &= \overline{D}_i\, b_i \\
(EI^i_{r'}) &= A_i\, b_i^3/12 \\
(EI^i_{r's'}) &= 0
\end{aligned} \tag{10.36}$$

where the *mechanical product of inertia* $(EI^i_{r's'})$ vanishes because the $s'r'$ are the principal axes of bending of the segment (Figure 10.7). The *mechanical moments of inertia* and the *mechanical product of inertia* of a segment with respect to axes $y'z'$ (Figure 10.7) are obtained by a rotation $-\alpha^i_y$ around x, as

$$\begin{aligned}
(EI^i_{y'}) &= (EI^i_{s'}) \cos^2 \alpha^i_y + (EI^i_{r'}) \sin^2 \alpha^i_y \\
(EI^i_{z'}) &= (EI^i_{s'}) \sin^2 \alpha^i_y + (EI^i_{r'}) \cos^2 \alpha^i_y \\
(EI^i_{y'z'}) &= \left[(EI^i_{r'}) - (EI^i_{s'})\right] \sin \alpha^i_y \cos \alpha^i_y
\end{aligned} \tag{10.37}$$

Using the parallel axis theorem and adding the contributions of all the segments, the *mechanical moments of inertia (bending stiffness)* and the *mechanical product of inertia* with respect to axes y_G, z_G are obtained as

$$\begin{aligned}
(EI_{y_G}) &= \sum_{i=1}^{n} \left[(EI^i_{y'}) + A_i\, b_i (z_i + e_b \cos \alpha^i_y)^2\right] \\
(EI_{z_G}) &= \sum_{i=1}^{n} \left[(EI^i_{z'}) + A_i\, b_i \left(y_i - e_b \sin \alpha^i_y\right)^2\right] \\
(EI_{y_G z_G}) &= \sum_{i=1}^{n} \left[(EI^i_{y'z'}) + A_i\, b_i (z_i + e_b \cos \alpha^i_y) \left(y_i - e_b \sin \alpha^i_y\right)\right]
\end{aligned} \tag{10.38}$$

where $(y_i - e_b \sin \alpha_y^i)$ and $(z_i + e_b \cos \alpha_y^i)$ are the coordinates of the center of the segment (point P′ in Figure 10.7) over the axis s' (located at $r = e_b$) and y_i, z_i are the coordinates of the center of segment i (point P in Figure 10.7) over the contour s (located at $r = 0$) with respect to the global system y_G, z_G. The system y_G, z_G has its origin at the *mechanical center of gravity* of the section and it is parallel to the global system y, z.

The rotation ϑ locating the principal axes of bending η, ζ, with respect to the axes y_G, z_G, is found by imposing the condition $(EI_{\eta\zeta}) = 0$, leading to

$$\tan 2\vartheta = \frac{2(EI_{y_G z_G})}{(EI_{z_G}) - (EI_{y_G})} \tag{10.39}$$

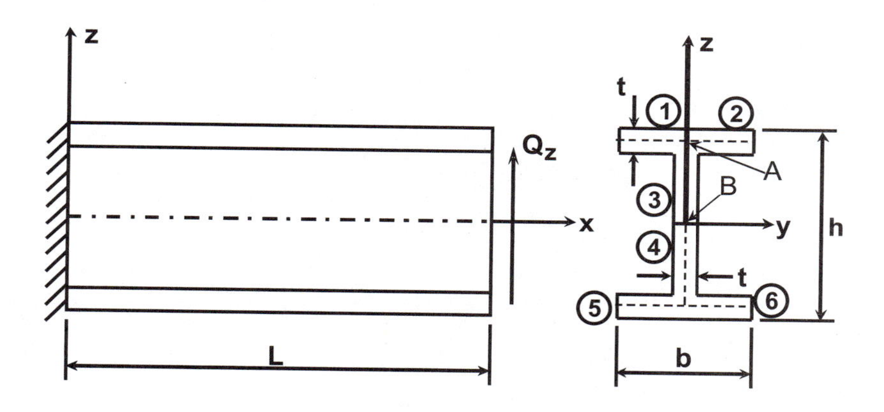

Figure 10.8: The cantilever I-beam of Example 8.4.

Finally, the maximum and minimum *bending stiffness* with respect to the principal axes of bending are

$$(EI_\eta)\,;\,(EI_\zeta) = \frac{(EI_{y_G}) + (EI_{z_G})}{2} \pm \sqrt{\left(\frac{(EI_{y_G}) - (EI_{z_G})}{2}\right)^2 + (EI_{y_G z_G})^2} \tag{10.40}$$

Example 10.3 *A cantilever beam of length $L = 54$ mm with rectangular cross-section of width $b = 9$ mm and thickness $t = 9$ mm is loaded by a tip load $P = -1000$ N perpendicular to the laminate. The laminate is a $[\pm 45/0]_T$. Each lamina, of thickness $t_k = 3$ mm, is made of Kevlar^{TM}–epoxy with properties $E_1 = 76$ GPa, $E_2 = 5.56$ GPa, $G_{12} = 2.3$ GPa ($G_{13} = G_{12}$), $G_{23} = 1.56$ GPa, and $\nu_{12} = 0.34$. Compute the deflection under the load.*

Solution to Example 10.3 *First, compute the compliance matrix (6.20) by inverting the stiffness matrix of the laminate (6.16)*[4]

[4]Most of the computations in this example can be done with CADEC [1] or similar laminate analysis software.

$$\begin{bmatrix} [\alpha] & [\beta] \\ [\beta] & [\delta] \end{bmatrix} = \begin{bmatrix} [A] & [B] \\ [B] & [D] \end{bmatrix}^{-1}$$

$$= \begin{bmatrix} .00823 & -.00559 & .000668 & -2.07 & 1.34 & 1.28 \\ -.00559 & .0140 & .00450 & 1.51 & .381 & 1.67 \\ .000668 & .00450 & .0142 & -0.0107 & 1.71 & 3.26 \\ -2.07 & 1.51 & -.0107 & 925 & -542 & -401 \\ 1.34 & .381 & 1.71 & -542 & 2320 & -455 \\ 1.28 & 1.67 & 3.26 & -401 & -455 & 3490 \end{bmatrix} 10^3$$

Next, write the reduced compliance matrix in (10.14) as

$$\begin{bmatrix} 0.00823 & -2.07 & 0 & 0 \\ -2.07 & 925 & 0 & 0 \\ 0 & 0 & 0.0142 & 3.26 \\ 0 & 0 & 3.26 & 3490 \end{bmatrix} 10^3$$

and invert it to obtain the coefficients in (10.15) as

$$\begin{array}{ll} A = 0.278\ GPa\ m, & F = 0.09\ GPa\ m \\ B = 6.22(10^{-4})\ GPa\ m^2, & C = -8.374(10^{-5})\ GPa\ m^2 \\ D = 2.473(10^{-6})\ GPa\ m^3, & H = 3.648(10^{-7})\ GPa\ m^3 \end{array}$$

The neutral axis of bending is located at

$$e_b = \frac{B}{A} = 2.238(10^{-3})\ m$$

The bending stiffness with respect to axes s', r' passing through the neutral axis of bending are (see (10.22) and (10.36)),

$$(EI_{s'}) = (D - e_b^2 A)b = 9.73(10^{-9})\ GPa\ m^4$$
$$(EI_{r'}) = Ab^3/12 = 16.89(10^{-9})\ GPa\ m^4$$

The shear strain γ_{xr} transverse to the wall has a negligible effect on the overall deflection of most thin-walled beams, and it was neglected at the onset of this section. However, it may be significant for a solid rectangular beam. In this case, the shear stiffness of the cross-section can be computed as

$$(GA) = \frac{b}{h_{44}} = \frac{0.009}{72.595} = 0.124(10^{-3})\ GPa\ m^2$$

Then, the deflection can be computed using Figure 10.2 as

$$\delta = \frac{PL^3}{3(EI_{s'})} + \frac{PL}{(GA)} = (5,395 + 435)P$$

with δ in meters and P in GN. Note that the shear deflection accounts for 7.5% of the total. For $P = 1000\ N$, $\delta = 0.006\ m$.

> **Example 10.4** *Compute the bending stiffness of a cantilever I-beam (Figure 10.8) of length $L = 30$ cm, subjected to a tip-shear force $V_z = 445$ N. The dimensions of the beam are: $h = 2.5$ cm, $b = h/2$, $t = h/16$. The material is Kevlar–epoxy with the following elastic constants: $E_1 = 76.0$ GPa, $E_2 = 5.56$ GPa, $G_{12} = 2.30$ GPa, $\nu_{12} = 0.34$. The lay-up sequence is $[\pm 30/0]_S$, six laminae ($n = 6$), with lamina thickness $t_k = t/6$.*

Solution to Example 10.4 *There are three walls: two identical flanges and a web. Since the same lay-up is specified for all walls, (10.14) needs to be evaluated only once. Using a standard laminate analysis software such as CADEC [1], the coefficients in (10.14) are found to be*

$$\alpha_{11} = 14.7\ (GPa\ m)^{-1}$$
$$\alpha_{66} = 58.5\ (GPa\ m)^{-1}$$
$$\beta_{11} = \beta_{66} = 0$$
$$\delta_{11} = 1.27(10^8)\ (GPa\ m^3)^{-1}$$
$$\delta_{66} = 2.50(10^8)\ (GPa\ m^3)^{-1}$$

for all walls ($i = 1...6$). Since the laminate is symmetric

$$A_i = 1/\alpha_{11} = 1/14.7 = 0.068\ GPa\ m = 68(10^6)\ N/m$$
$$B_i = C_i = 0$$
$$D_i = 1/\delta_{11} = 1/1.27(10^8) = 7.874\,(10^{-9})\ GPa\ m^3 = 7.874\ Nm$$
$$F_i = 1/\alpha_{66} = 1/58.5 = 0.017\ GPa\ m = 17(10^6)\ N/m$$
$$H_i = 1/\delta_{66} = 1/2.5(10^8) = 4.0(10^{-9})\ GPa\ m^3 = 4.0\ Nm$$
$$e_b = e_q = 0$$

For the top and bottom flanges, and setting $i = f$ to indicate the whole flange (segments 1 plus 2 or 5 plus 6 in Figure 10.8), (10.36) yields

$$(EI^f_{s'}) = D_f b = 7.874\,(10^{-9})0.0125 = 98.43(10^{-12})\ GPa\ m^4$$
$$(EI^f_{r'}) = A_f b^3/12 = 0.068\frac{(0.0125)^3}{12} = 11.07(10^{-9})\ GPa\ m^4$$

and since both flanges are parallel to the y-axis, $\alpha_y = 0$. Then, according to (10.37), the two bending stiffness of the flange are

$$(EI^f_{y'}) = (EI^f_{s'}) = 98.43(10^{-12})\ GPa\ m^4$$
$$(EI^f_{z'}) = (EI^f_{r'}) = 11.07(10^{-9})\ GPa\ m^4$$

For the web, since $\alpha_y = \pi/2$, and taking into account that b_3 is the width of the web between the middle surface of both flanges ($b_3 = h - t$), (10.36–10.37) lead to

$$(EI^w_{z'}) = (EI^w_{s'}) = D_w(h-t) = 7.813\,(10^{-9})0.02344 = 0.1845(10^{-9})$$
$$(EI^w_{y'}) = (EI^w_{r'}) = \frac{A_w}{12}\,(h-t)^3 = \frac{0.068}{12}(0.02344)^3 = 72.99(10^{-9})$$

Since the I-beam is doubly-symmetric (symmetry of geometry and material with respect to two perpendicular axes), the neutral axis as well as the mechanical center of gravity coincide with the geometrical centroid. Therefore, the mechanical moment of inertia (bending stiffness) is computed with (10.38) as

$$(EI_{y_G}) = 2\left[(EI^f_{y'}) + A_f b\left(\frac{h-t}{2}\right)^2\right] + EI^w_{y'} = 306.7(10^{-9})\;GPa\;m^4 = 306.7\,N\,m^2$$

which is the same as using the parallel axis theorem but replacing areas and moment of inertia by mechanical properties (EA) and (EI).

Example 10.5 *Compute the bending stiffness in principal axis $(EI_\eta), (EI_\zeta)$, for an angle section with equal flanges of thickness $t = 2\,mm$, width $b = 20\,mm$, and properties $E = 210,000\,MPa$, $\nu = 0.3$, resulting in calculated values of $G = 80,769\,MPa$, $A_i = 0.42 \times 10^6\,MPa\,mm$ $D_i = 0.14 \times 10^6\,MPa\,mm^3$, $F_i = 0.1615 \times 10^6\,MPa\,mm$, $H_i = 0.5385 \times 10^6\,MPa\,mm^3$. The coordinates of the nodes, in mm, are (see schematic on p. 374):*

node	*y*	*z*
1	*20*	*0*
2	*0*	*0*
3	*0*	*20*

The contour is defined clockwise (cw) starting at node #1,

segment	*node i*	*node f*	*segment contribution*
1	*1*	*2*	*0*
2	*2*	*3*	*1*

Solution to Example 10.5 *Below are the calculations in MATLAB®. First, define the section by specifying the coordinates of the nodes and the segment connectivity in the format of Table 10.2:*

```
YZ = [20 0;0 0; 0 20.001]; %nodal coordinates
conect = [1 2 0;2 3 1];    %segment connectivity and contributions
ti = [2 2]; %mm
Ei = [1 1]*210E3; %MPa
nui = [.3 .3];
```

Get the number of nodes, number of segments, from the data:

```
nn = length(YZ(:,1));      %# of nodes
ns = length(conect(:,1));  %# of segments
Gi = Ei/2./(1+nui);
```

Calculate the length, angle, and centroid of each segment from the data:

```
for i = 1:ns     %for each segment
    no = conect(i,1); %initial node for segment i
    nf = conect(i,2); %final node
    dY = YZ(nf,1)-YZ(no,1);
    dZ = YZ(nf,2)-YZ(no,2);
    bi(i) = ((dY)^2+(dZ)^2)^(1/2);  %length of segment
    alphai(i) = atan(dZ/dY);         %angle of segment
    yi_(i) = (YZ(no,1)+YZ(nf,1))/2; %centroid of each segment
    zi_(i) = (YZ(no,2)+YZ(nf,2))/2;
end
```

Calculate the segment properties:

```
Ai = Ei.*ti       %extension
Bi = [0 0]        %coupling
ebi = Bi./Ai      %neutral axis of bending for the segment
Di = Ei.*ti.^3/12   %bending
Di_ = Di-ebi.^2.*Ai %shifted to neutral axis of bending
Fi = Gi.*ti
Ci = [0 0]
Hi = Gi.*ti.^3/12
eqi = Ci./Fi
Hi_ = Hi-eqi.^2.*Ci %shifted to neutral axis of torsion
EIsi = Di_.*bi      %bending stiffness of segment
EIri = Ai.*bi.^3/12
EIsri = [0 0]       %mechanical product of inertia
```

Rotate segment stiffness to global coordinates:

```
m = cos(alphai)
n = sin(alphai)
EIyi = m.^2.*EIsi + n.^2.*EIri
EIzi = n.^2.*EIsi + m.^2.*EIri
EIyzi = (EIri-EIsi).*m.*n %1st ed.
```

Add all segments into the section:

```
EA  = sum(Ai.*bi)
GJR = 4*sum(Hi_.*bi)
EQy = sum(zi_.*Ai.*bi)
EQz = sum(yi_.*Ai.*bi)
```

Calculate the mechanical center of gravity (C.G.):

```
yG_ = EQz/EA
zG_ = EQy/EA
```

New coordinates, translated to the C.G.

```
yi = yi_-yG_
zi = zi_-zG_
```

Add the whole section:

```
EIyy = sum(EIyi+Ai.*bi.*(zi+ebi.*m).^2)
EIzz = sum(EIzi+Ai.*bi.*(yi-ebi.*n).^2)
EIyz = sum(EIyzi+Ai.*bi.*(zi+ebi.*m).*(yi-ebi.*n))
```

Find the principal axes and rotate to them:

```
theta = 1/2*atan(2*EIyz/(EIzz-EIyy)); %1st ed.
disp(['theta ',num2str(theta*180/pi)])
EInn = 1/2*(EIyy+EIzz) + sqrt((EIyy-EIzz)^2/4+EIyz^2)
EIqq = 1/2*(EIyy+EIzz) - sqrt((EIyy-EIzz)^2/4+EIyz^2)
```

When executed in MATLAB, the code above yields,

$$(EI_\eta) = 1,123 \times 10^6\, N\, m^2$$
$$(EI_\varsigma) = 287 \times 10^6\, N\, m^2$$

The MATLAB code is available in the Web site [1].

10.2.6 Torsional Stiffness

The torsional stiffness (GJ_R) is found by using energy balance between the work done by external torque T and the strain energy due to shear

$$\frac{1}{2} T \beta L = \frac{L}{2} \int_s \left(\gamma^i_{xs}\, q^i + M^i_{xs}\, \kappa^i_{xs} \right) ds \tag{10.41}$$

where β is the rate of twist and $q_i = N^i_{xs}$ is the shear flow.

Open Section

Since the shear flow of an open section under torsion is zero ($q_i = N^i_x = 0$), using (10.26) the energy balance (10.41) can be written as

$$T\beta = \kappa^2_{xs} \int_s \overline{H_i}\; ds \tag{10.42}$$

with the twisting curvature κ_{xs} constant along the contour and equal to twice the twist rate β

$$\kappa_{xs} = 2\, \frac{\partial^2 w}{\partial x\, \partial s} = 2\, \beta \tag{10.43}$$

where w is the transverse deflection of the laminate. Combining (10.43) and (10.42) results in

$$\frac{T}{\beta} = 4 \int_s \overline{H_i}\; ds \tag{10.44}$$

The *torsional stiffness* (GJ_R), which is equal to T/β, is obtained simply adding the contributions of all the segments

$$(GJ_R) = 4\sum_{i=1}^{n} \overline{H_i}\, b_i \tag{10.45}$$

For the isotropic case, (10.45) leads to the classical result $J_R = \frac{1}{3}\sum b_i\, t_i^3$.

Single-Cell Closed Section

For a closed section under torsion, the contribution of the twisting curvature to the strain energy in (10.41) is negligible compared to the contribution of the shear flow. Taking into account that the shear flow is constant along the contour and using (10.26), the energy balance (10.41) reduces to

$$T\beta = q^2 \oint_s \frac{ds}{F_i} = q^2 \sum_{i=1}^{n} \frac{b_i}{F_i} \tag{10.46}$$

Integrating the shear flow over the contour s'' yields an expression for the torque

$$T = \oint_s \left(q\, R(s'')\right) ds \ = q\,[2\Gamma_{s''}] \tag{10.47}$$

where $\Gamma_{s''}$ is the area enclosed by the contour s''. Note that the arm-length $R(s'')$ is measured at the neutral surface of torsion s''. The rate of twist β is determined dividing (10.46) by (10.47)

$$\beta = \frac{q}{2\Gamma_{s''}} \oint_s \frac{ds}{F_i} = \frac{q}{[2\Gamma_{s''}]} \sum_{i=1}^{n} \frac{b_i}{F_i} \tag{10.48}$$

The *torsional stiffness* is obtained dividing (10.47) by (10.48)

$$(GJ_R) = \frac{[2\Gamma_{s''}]^2}{\sum_{i=1}^{n} b_i/F_i} \tag{10.49}$$

The expression above can be improved to account for the nonuniform distribution of shear through the thickness of the laminate by adding the *torsional stiffness* of the open cell affected by a correction factor equal to 3/4 [231]

$$(GJ_R) = \frac{[2\Gamma_{s''}]^2}{\sum_{i=1}^{n} b_i/F_i} + \frac{3}{4}\left[4\sum_{i=1}^{n} \overline{H}_i\, b_i\right] \tag{10.50}$$

Example 10.6 *A rectangular tube is clamped at one end and subject to a torque* $T = 0.113\ Nm$ *at the free end. The dimensions of the tube at the midsurface are: height* $d = 25.273\ mm$*, width* $c = 51.562\ mm$*, and length* $L = 0.762\ m$*. The walls are made of carbon–epoxy cross-ply with 6 laminae, each* $0.127\ mm$ *thick, in a* $[0/90]_3$ *configuration, with a total thickness of* $0.762\ mm$*. Note that the laminate is not symmetric, but the tube is symmetric since the interior lamina has the same* $0°$ *angle in all the four walls. The elastic properties of each lamina are* $E_1 = 141.685\ GPa$*,* $E_2 = 9.784\ GPa$*,* $G_{12} = 5.994\ GPa$*, and* $\nu_{12} = 0.42$*. Compute the angle of twist at the free end.*

Solution to Example 10.6 *For each wall, the coefficients of the compliance matrix in (10.13) can be easily obtained using standard laminate analysis software. Then, the coefficients in (10.15) are obtained inverting (10.14) or directly as*

$$
\begin{aligned}
A_i &= \delta_{11}/\Delta_1 = 0.378/\Delta_1 = 58.328\ GPa\ mm \\
B_i &= -\beta_{11}/\Delta_1 = -0.0209/\Delta_1 = -3.225\ GPa\ mm^2 \\
D_i &= \alpha_{11}/\Delta_1 = 0.0183/\Delta_1 = 2.824\ GPa\ mm^3 \\
\Delta_1 &= \alpha_{11}\delta_{11} - \beta_{11}^2 = 0.0183(0.378) - 0.0209^2 = 0.006481 \\
F_i &= \delta_{66}/\Delta_2 = 4.52/\Delta_2 = 4.565\ GPa\ mm \\
C_i &= -\beta_{66}/\Delta_2 = 0 \\
H_i &= \alpha_{66}/\Delta_2 = 0.219/\Delta_2 = 0.221\ GPa\ mm^3 \\
\Delta_2 &= \alpha_{66}\delta_{66} - \beta_{66}^2 = 4.52(0.219) = 0.990
\end{aligned}
$$

Since the $[0/90]_3$ *laminate is not symmetric, the neutral surface of bending is located at a distance given by (10.18) as*

$$e_b = \frac{B_i}{A_i} = \frac{-3.225}{58.328} = -0.055\ mm$$

from the midsurface. The neutral surface of torsion is given by (10.24) as

$$e_q = \frac{C_i}{F_i} = 0$$

From (10.49), the torsional stiffness of the closed section is

$$(GJ_R) = \frac{T}{\beta} = \frac{[2\Gamma_{s"}]^2}{\sum_{i=1}^{n} b_i/F_i}$$

Since $e_q = 0$*, the contour s" through the neutral surface of torsion coincides with the contour s through the midsurface. Therefore*

$$\Gamma_{s"} = \Gamma_s = cd = 51.562(25.273) = 1303.1\ mm^2$$

Since all walls have the same laminate, $F_i = 4.565\ GPa\ mm$ *for all walls, and the torsional stiffness is*

$$(GJ_R) = \frac{(2(1303.1))^2}{2(25.273 + 51.562)/4.565} = 201,774.9\ GPa\ mm^4$$

The rate of twist is equal to the torque divided by the torsional stiffness (10.82)

$$\beta = \frac{T}{(GJ_R)} = \frac{0.113\ Nm}{201,774.9(10^9)(10^{-12})Nm^2} = 560(10^{-6})rad/m$$

from which the twist angle at $L = 0.762\ m$ *is*

$$\phi = \beta L = 560\,(10^{-6})(0.762) = 0.427(10^{-3})\ rad$$

which compares well with the experimental value of $0.420(10^{-3})$ *rad reported in [234, 235].*

10.2.7 Shear of Open Sections

As in the case of homogeneous materials, the shear flow q caused by shear forces V is obtained from equilibrium using Jourawski's formula [221, p. 361]

$$q = \frac{V\ Q}{I} \tag{10.51}$$

where Q, I, are the first and second moments of area, respectively. But in the case of inhomogeneous materials, the resultant of axial forces that causes no bending curvature, and the resultant of shear flow that causes no twisting curvature act on different local axis. Therefore, the notation $\eta(s')$, $\zeta(s')$, $r(s'')$, etc., is used to indicate on what local axis the global coordinate is measured. Then, using principal axes ζ, η, the shear flow q_η caused by a shear force V_η is given by adapting (10.51) for non-homogeneous material by replacing geometric properties by *mechanical properties* as follows

$$q_\eta(s) = -\frac{(EQ_\zeta(s))V_\eta}{k_\zeta(EI_\zeta)} \tag{10.52}$$

where the minus sign is used to assure that the integral of the shear flow is equal to the shear force when the sign convention of Figure 10.3 is used. The shear flow is positive when oriented from n_i to n_f, along the s-coordinate in Figure 10.6. The first moment of the area at one side of the point where shear is evaluated is replaced by the *mechanical static moment* $(EQ_\zeta(s))$. The moment of inertia is replaced by the *bending stiffness* (EI_ζ), defined in (10.40), and k_ζ is a correction factor explained later.

The mechanical static moment is obtained by rewriting (10.35) as

$$(EQ_\zeta(s)) = \int_0^s \eta(s')A_i ds = \int_0^s (\eta(s) - e_b \sin\alpha_\eta^i)\, A_i\, ds \tag{10.53}$$

where α_η^i is the angle of the segment with respect to the principal axis η. Note that the mechanical static moment $(EQ_\zeta(s))$ defined in (10.53) is variable along the contour, while the values defined in (10.35) are the total values for the beam cross-section. The integral in (10.53) must start at the free edge ($s = 0$) to satisfy the equilibrium condition $q(s) = 0$ at the free edge of the open section.

For the case of moderately thick walls, the accuracy of Jourawski's formula can be improved introducing a correction factor k_ζ, computed to restore equilibrium between the applied force V_η and the integral of the shear flow q_η over the contour [231]

$$V_\eta = \int_s (q_\eta(s)\,ds)\cos\alpha_\eta^i = -\frac{V_\eta}{k_\zeta(EI_\zeta)}\int_s (EQ_\zeta(s))\cos\alpha_\eta^i\,ds \tag{10.54}$$

which results in

$$k_\zeta = \frac{-1}{(EI_\zeta)}\int_s (EQ_\zeta(s))\cos\alpha_\eta^i\,ds \tag{10.55}$$

In classical textbooks, $k_\zeta = 1$, which is valid for thin walls. Similarly

$$q_\zeta(s) = -\frac{(EQ_\eta(s))V_\zeta}{k_\eta(EI_\eta)} \tag{10.56}$$

where the *mechanical static moment* is

$$(EQ_\eta(s)) = \int_0^s \zeta(s')A_i ds = \int_0^s \left(\zeta(s) + e_b\cos\alpha_\eta^i\right)A_i\,ds \tag{10.57}$$

and the correction factor is

$$k_\eta = \frac{-1}{(EI_\eta)}\int_s (EQ_\eta(s))\sin\alpha_\eta^i\,ds \tag{10.58}$$

For most sections having flat walls, the integrals in (10.53) and 10.57 can be reduced to algebraic summations, as illustrated in Example 10.7, p. 368.

Example 10.7 *Compute the mechanical static moment about the y-axis and the shear flow at points A:($z = (h-t)/2$) and B:($z = 0$) of the web in Figure 10.8, using the properties in Example 10.4, p. 361.*

Solution to Example 10.7 *Since the contour is modeled by the dashed line shown in Figure 10.8, only the flange contributes to the mechanical static moment at point A. The mechanical static moment of interest is given by (10.57). Taking ζ in the direction of z (Figure 10.3) the value of $\zeta = (h-t)/2$ is constant in the flange. Therefore, the mechanical static moment accumulated from the free edge to the intersection with the web (segment # 1 in Figure 10.8) is*

$$(EQ_\eta) = \int_0^{b/2}\frac{(h-t)}{2}A_1 ds = \frac{(h-t)}{2}A_1\frac{b}{2} = 4.9825(10^{-6})\ GPa\ m^3$$

Taking $k_\eta = 1$ in 10.56, and noting that $(EI_\eta) = (EI_{yG})$ from 10.56, the shear flow at point A in the flange is

$$q_\zeta = -\frac{(EQ_\eta(s=A))V_\zeta}{(EI_\zeta)} = -\frac{4.9825(10^{-6})(445)}{306.7(10^{-9})} = -7,229.25\ N/m$$

and the direction of the shear flow is from the center of the flange towards the free edges. In the web, immediately below point A, the static moment and shear flow from both segments #1 and #2 accumulate, resulting in

$$(EQ) = (2)4.9825(10^{-6}) = 9.965(10^{6})\ GPa\ m^3$$
$$q_\varsigma = -(2)7,229.25 = -14,458.5\ N/m$$

Since the section is doubly-symmetric, point B is located at the centroid, which coincides with the origin of the coordinate system The mechanical static moment increases from point A to B due to the contribution of the web, resulting at point B in

$$(EQ_\eta) = \frac{(h-t)}{2} A_1 b + \frac{(h-t)^2}{8} A_3 = 14.634(10^{-6})\ GPa\ m^3$$

Consequently, the shear flow at point B is

$$q_\varsigma = -\frac{14.634(10^{-6})(445)}{306.7(10^{-9})} = 21,232.9\ N/m$$

with the shear flow in the web having the same direction as the applied load.

In preliminary design codes and procedures, it is usual to compute the shear stress as the shear force divided by the area of the web, or the shear flow as the shear force divided by the depth of the beam. For this example, the approximated value of shear flow would be $q = -V/h = -445/0.025 = -17.8(10^3)\ N/m$, *thus underpredicting the shear flow at point B by 16%.*

Shear Stiffness

To obtain the shear stiffness (GA) of the section, consider an infinitesimal segment dx of the beam subject to shear V_η. The balance between the external work and the strain energy is

$$\frac{1}{2} V_\eta \gamma dx = \frac{1}{2} \frac{V_\eta^2}{(GA_\varsigma)} dx = \frac{1}{2} \int_s \left(M_{xs}^i\, \kappa_{xs}^i + q^i\, \gamma_{xs}^i\right) ds dx \tag{10.59}$$

Taking into account only the shear deformations ($\kappa_{xs}^i \cong 0$), and using the third of (10.15) and (10.52) yields

$$\gamma_{xs}^i = \frac{q^i}{F_i} = -\frac{V_\eta\,(EQ_\varsigma(s))}{k_\varsigma (EI_\varsigma) F_i} \tag{10.60}$$

Replacing into (10.59) with $\kappa_{xs}^i = 0$, the *shear stiffness* of the section is

$$(GA_\varsigma) = \frac{[k_\varsigma(EI_\varsigma)]^2}{\int_s [(EQ_\varsigma(s))]^2 \frac{ds}{F_i}} \tag{10.61}$$

Similarly

$$(GA_\eta) = \frac{[k_\eta(EI_\eta)]^2}{\int_s [(EQ_\eta(s))]^2 \frac{ds}{F_i}} \tag{10.62}$$

For sections having flat walls, the integrals in the denominator of (10.61) and 10.62 can be easily performed segment by segment.

Example 10.8 *Compute the shear stiffness of the section in Figure 10.8 using the results of Example 10.4 and 10.7 (pp. 361, 368).*

Solution to Example 10.8 *Begin the computation of mechanical static moment (10.57) on segment #1 at the free edge (with "s" pointing to the right in Figure 10.8). Take $\zeta = z$ with the origin at the geometric centroid and $c = (h - t)/2$. Then for segment #1*

$$(EQ_\eta(s)) = \int_0^s \zeta A_1 ds = cA_1 s$$

Next, the contribution of segment #1 to the denominator of (10.62) is

$$\Delta_1 = \int_0^{b/2} [cA_1 s]^2 \frac{ds}{F_1} = \frac{b^3 c^2 A_1^2}{24 F_1}$$

Next, compute the contribution of segment #2, also starting at the edge (with s pointing to the left in Figure 10.8). It can be shown that the contributions of segment #2 are identical to those of segment #1. The accumulated mechanical static moment at the top of the flange is $(EQ) = cA_1 b$. The contribution of segment #3 is computed starting at the top, with "s" pointing downward ($s = c - \zeta$)

$$(EQ_\eta(s)) = cA_1 s + \int_0^s (c - s)A_3 ds = cA_1 b + A_3 \left(cs - \frac{s^2}{2} \right)$$

and

$$\Delta_3 = \int_0^c [EQ_\eta(s)]^2 \frac{ds}{F_3} = \frac{c^3(2A_3^2c^2 + 10cA_1bA_3 + 15A_1^2b^2)}{15F_3}$$

At the centroid, the accumulated mechanic static moment is $(EQ) = cA_1 b + A_3 c^2/2$. For segment #4, taking "s" downward ($s = -\zeta$) the contributions are

$$(EQ_\eta(s)) = cA_1 b + A_3 \frac{c^2}{2} + \int_0^s (-s)A_3 ds = cA_1 b + A_3 \frac{c^2}{2} - A_4 \frac{s^2}{2}$$

with $A_4 = A_3$ and

$$\Delta_4 = \int_0^c [EQ_\eta(s)]^2 \frac{ds}{F_4} = D_3$$

with $F_4 = F_3$. The accumulated mechanical static moment at the bottom of the web is $(EQ) = cA_1 b$. For segment #5, start at the free edge to obtain

$$(EQ_\eta(s)) = \int_0^s (-c)A_5 ds = -cA_5 s$$

with $A_5 = A_1$, and

$$\Delta_5 = \int_0^{b/2} (EQ_\eta(s))^2 \frac{ds}{F_5} = \frac{b^3 c^2 A_1^2}{24 F_1}$$

with $F_5 = F_1$. Now, all the static moment must be accumulated at the initial point of segment #6. Taking "s" pointing toward the free edge results in

$$[EQ_\eta(s)] = cA_1\frac{b}{2} + \int_0^s (-c)A_6 ds = cA_1\frac{b}{2} - cA_6 s$$

with $A_6 = A_1$ and

$$\Delta_6 = \frac{b^3c^2A_1^2}{24F_1}$$

with $F_6 = F_1$. Finally, setting $A_3 = A_1$ and $F_3 = F_1$, the denominator of (10.62) is

$$\Delta = \frac{c^2A_1^2(5b^3 + 8c^3 + 40c^2b + 60cb^2)}{30F_1} = 0.248(10^{-9})\ GPa\ m^6$$

and the shear stiffness (10.62) is computed using $(EI_\eta) = (EI_{yG})$ from Example 10.4, p. 361, as

$$(GA_\eta) = \frac{[306.7(10^{-9})]^2}{0.248(10^{-9})} = 0.379(10^{-3})GPa\ m^2$$

As a comparison, compute an approximate value for shear stiffness as the laminate shear moduli G_{xy} times the area of the web A_w. For the laminate used in this example, using (6.35),

$$G_{xy} = \frac{1}{t\alpha_{66}} = \frac{A_{66}}{t} = \frac{0.0171}{0.025/16} = 10.944\ GPa$$

leading to

$$G_{xy}A_w = 10.944(0.025)\frac{0.025}{16} = 0.428(10^{-3})\ GPa\ m^2$$

which is 13% higher than the actual value.

Sectorial Properties

The *sectorial area* is defined as (see [221, p. 307])

$$\omega(s) = \int_0^s R(s'')\, ds \tag{10.63}$$

where $R(s'')$ is the projection of the radius $\vec{R}(s'')$ on the normal to ds (Figure 10.4). Although the integral in (10.63) can be computed using arbitrary points for the pole P and for the initial point I (where $\omega(s) = 0$, Figure 10.9), computation of the shear center (10.71) requires that the mechanical center of gravity ((10.34), Figure 10.4) be used as pole. The arm-length $R(s'')$ in (10.63) must be evaluated on the neutral axis of torsion s'' because $\omega(s)$ will be used to define twisting moment caused by shear flow.

Computation of sectorial area $\omega(s)$ may require to treat the segments in a different order of that used initially to define the contour, because the computations for a segment can start only if the value of $\omega(s)$ in either extreme is already known. Once the pole P and initial point I are chosen, the diagram of $\omega(s)$ can be constructed as in Figure 10.9, for the section shown in Figure 10.4. Then the *mechanical sectorial properties* are defined as:

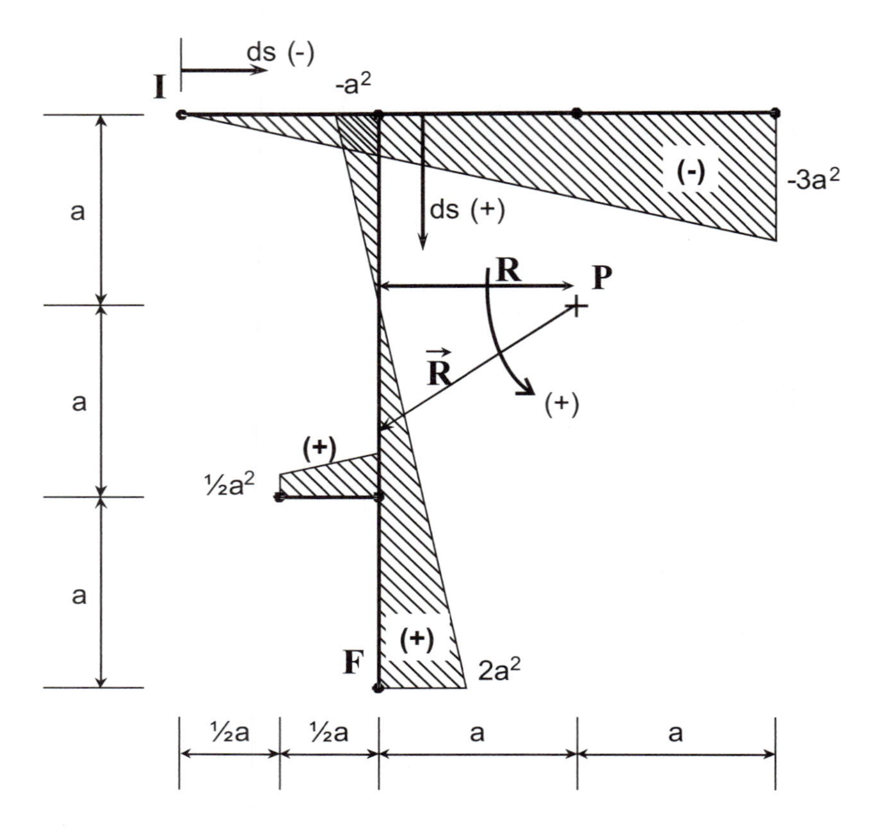

Figure 10.9: Sectorial area diagram.

Mechanical sectorial static moment

$$(E\omega) = \int_s \omega(s)\, A_i\, ds \tag{10.64}$$

Mechanical sectorial linear moments

$$(EQ\omega_\zeta) = \int_s (\eta(s) - e_b \sin\alpha_\eta^i)\, \omega(s)\, A_i\, ds \tag{10.65}$$

and

$$(EQ\omega_\eta) = \int_s (\zeta(s) + e_b \cos\alpha_\eta^i)\, \omega(s)\, A_i\, ds \tag{10.66}$$

Mechanical sectorial moment of inertia

$$(EI\omega) = \int_s [\omega(s)]^2 A_i\, ds \tag{10.67}$$

Mechanical Shear Center

The coordinates η_c, ζ_c, of the *mechanical shear center* are calculated by equilibrium of moments due to shear forces V_ζ, V_η, and its associated shear flows $q_\zeta(s), q_\eta(s)$. Taking moments with respect to the mechanical center of gravity (y_G, z_G, in Figure 10.5) results in

$$V_\zeta\,\eta_c = \int_I^F R(s'')\,q_\zeta(s)\,ds$$
$$-V_\eta\,\zeta_c = \int_I^F R(s'')\,q_\eta(s)\,ds \tag{10.68}$$

where I and F are the initial and final point (Figure 10.4), and the minus sign on the LHS of the second equation is necessary to satisfy the sign convention in Figure 10.5. Using (10.52) and (10.53) yields

$$\zeta_c = \frac{1}{k_\zeta(EI_\zeta)}\int_I^F R(s'')\left[\int_I^s \eta(s')\,A_i\,ds\right] ds \tag{10.69}$$

Recognizing that $[R(s'')\,ds]$ is the differential of sectorial area $d\omega(s)$ and that $[\eta(s')\,A_i\,ds]$ is the differential of mechanical static moment $d(EQ_\zeta(s))$, integration by parts is used to change the order of integration. Since the mechanical static moment is zero at the end points, the boundary terms drop, leading to

$$\zeta_c = -\frac{1}{k_\zeta(EI_\zeta)}\int_I^F \eta(s')\,A_i\left[\int_I^s d\omega\right] ds = -\frac{1}{k_\zeta(EI_\zeta)}\int_I^F \eta(s')\,A_i\,\omega(s)\,ds \tag{10.70}$$

and using (10.65)

$$\zeta_c = -\frac{(EQ\omega_\zeta)}{k_\zeta(EI_\zeta)}$$
$$\eta_c = \frac{(EQ\omega_\eta)}{k_\eta(EI_\eta)} \tag{10.71}$$

In (10.71), $(EQ\omega_\zeta)$ and $(EQ\omega_\eta)$ are computed by (10.65)–(10.66) using the mechanical center of gravity as the pole for $\omega(s)$ because the moment equilibrium is formulated with respect to y_G, z_G. If the *mechanical shear center* is used as the pole, then $\zeta_c = 0$ and $\eta_c = 0$, and both mechanical sectorial linear moments are zero.

Example 10.9 *Compute the location of the shear center in principal coordinates η_c, ζ_c, as well as in structural coordinates $\overline{y}_c, \overline{z}_c$, for the angle section of Example 10.5. The mechanical center of gravity is located at $\overline{y}_G = 5\,mm, \overline{z}_G = 5\,mm$, and the principal axes of bending are rotated an angle $\theta = 45^o$ with respect to the structural coordinate system y, z, resulting in $(EI_\eta) = 1,123 \times 10^6\,N\,m^2$, $(EI_\zeta) = 287 \times 10^6\,N\,m^2$. Take node #1 as initial point **I**, and remember to use the mechanical center of gravity (CG) as the pole **P**.*

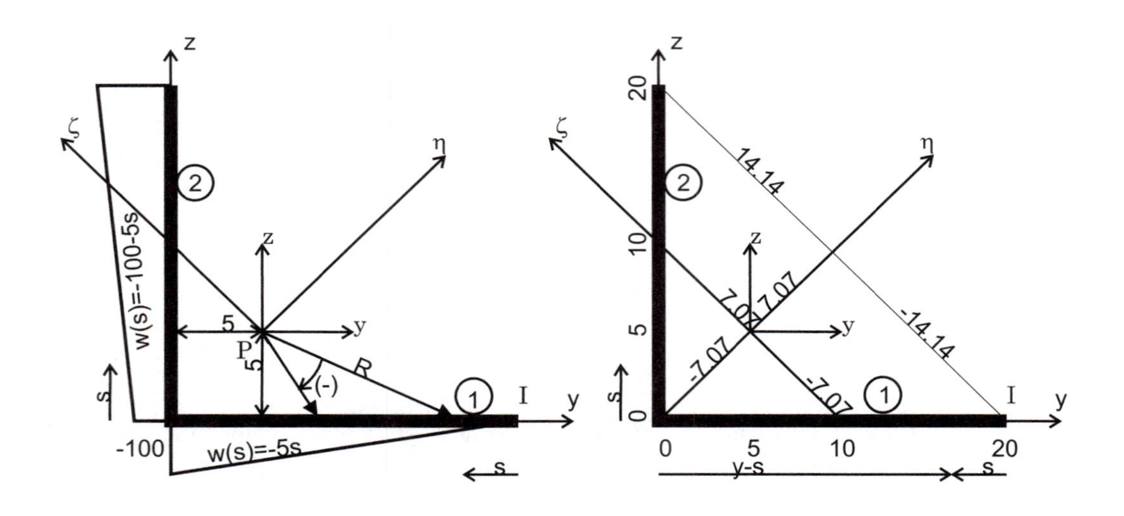

Figure 10.10: Schematic for Examples 10.5 and 10.9.

Solution to Example 10.9 *The mechanical center of gravity is at $\overline{y}_G = 5, \overline{z}_G = 5$. A system parallel to y, z, but translated to the C.G. is (see Figure 10.5),*

$$y_G = y - 5$$

$$z_G = z - 5$$

The principal axes η, ζ, are rotated by an angle $\theta = 45^o$.

Next, calculate the sectorial moment and the coordinates $\eta(s), \zeta(s)$ for each segment. The initial point is node #1.

Segment #1. *The s-coordinate comes left from I (Figure 10.10), and the segment is described by*

$$s = 20 - y = 15 - y$$

or

$$y_G = 15 - s$$

$$z_G = -5$$

Using (5.25),

$$\left\{ \begin{array}{c} \eta \\ \zeta \end{array} \right\} = \left[\begin{array}{cc} m & n \\ -n & m \end{array} \right] \left\{ \begin{array}{c} 15 - s \\ -5 \end{array} \right\} = \frac{\sqrt{2}}{2} \left\{ \begin{array}{c} 10 - s \\ s - 20 \end{array} \right\}$$

with

$$m = \cos(\theta) = \cos 45 = \frac{\sqrt{2}}{2}$$

$$n = \sin(\theta) = \sin 45 = \frac{\sqrt{2}}{2}$$

The pole must be the C.G. The normal from the pole to segment #1 is 5 units long. Starting with "s" at point I, the radius R sweeps clockwise (cw), that is negatively, so the sectorial area is negative,

$$\omega(s) = -5s$$

so that by the end of the segment it has a value of -100.

Segment #2. *The s-coordinate goes up from node #2, (Figure 10.10),*

$$s = z = z_G + 5$$

or

$$y_G = -5$$

$$z_G = s - 5$$

Using (5.25),

$$\left\{ \begin{array}{c} \eta \\ \zeta \end{array} \right\} = \left[\begin{array}{cc} m & n \\ -n & m \end{array} \right] \left\{ \begin{array}{c} -5 \\ s-5 \end{array} \right\} = \frac{\sqrt{2}}{2} \left\{ \begin{array}{c} s-10 \\ s \end{array} \right\}$$

The normal from the pole to segment #2 is 5 units long. Starting at node #2, the radius R sweeps cw, that is negatively, so the sectorial area is negative,

$$\omega(s) = -100 - 5s$$

so that by the end of the segment it has a value of -200.

Mechanical sectorial moments.

$$\begin{aligned}(EQ\omega_\zeta) &= \int_s \eta(s'')\omega(s)ds \\ &= -\frac{\sqrt{2}}{2}A_1 \int_0^{20} [(10-s)5s + (s-10)(100+5s)]ds = 0\end{aligned}$$

so,

$$\zeta_c = -\frac{(EQ\omega_\zeta)}{(EI_\zeta)} = 0$$

$$\begin{aligned}(EQ\omega_\eta) &= \int_s \zeta(s'')\omega(s)ds \\ &= -\frac{\sqrt{2}}{2}A_1 \int_0^{20} [(s-20)5s + s(100+5s)]ds = -5,600\sqrt{2} \times 10^6\end{aligned}$$

so,

$$\eta_c = \frac{(EQ\omega_\eta)}{(EI_\eta)} = \frac{-5,600\sqrt{2} \times 10^6}{1,123 \times 10^6} = -5\sqrt{2}$$

or

$$\overline{y}_c = 0$$
$$\overline{z}_c = 0$$

10.2.8 Shear of Single-Cell Closed Section

Jourawski's formula (10.52) can be used also to compute the shear flow caused by shear forces in closed sections. For the open section, the initial point ($s = 0$) for the computation of the *mechanical static moment* is set at the free edge so that $q(s = 0) = 0$. For the closed section, the initial point is set so that a shear force applied at the shear center does not induce twisting ($\beta = 0$). The initial point cannot be located by simple inspection of the contour but rather results from the analysis, as explained next.

First, consider the shear force V_η acting on the *mechanical shear center* which does not produces shear flow by torsion. Equation (10.52) yields

$$q_{1\eta}(s) = -\frac{V_\eta\,(EQ_\varsigma(s))}{k_\varsigma(EI_\varsigma)} \tag{10.72}$$

where an arbitrary initial point was used in (10.53) for the computation of the mechanical static moment. To change the initial point is equivalent to add a constant $q_{0\eta}$ such that

$$q_\eta(s) = q_{1\eta}(s) + q_{0\eta} \tag{10.73}$$

The rate of twist due to torsion is given by (10.48), but the shear flow must be inside the integral because the shear flow due to shear (10.73) is not constant along the contour (see [221, p. 371])

$$\beta = \frac{1}{2\Gamma_{s''}} \oint_s (q_{1\eta}(s) + q_{0\eta})\frac{ds}{F_i} \tag{10.74}$$

A shear load applied at the *mechanical shear center* should not induce twisting ($\beta = 0$). Therefore the constant $q_{0\eta}$ is

$$q_{0\eta} = -\oint_s q_{1\eta}(s)\,\frac{ds}{F_i} \Big/ \oint_s \frac{ds}{F_i} \tag{10.75}$$

Similarly, the shear flow caused by V_ς can be obtained as

$$q_\varsigma(s) = q_{1\varsigma}(s) + q_{0\varsigma} \tag{10.76}$$

where

$$q_{1\varsigma}(s) = -\frac{V_\varsigma\,(EQ_\eta(s))}{k_\eta(EI_\eta)} \tag{10.77}$$

and

$$q_{0\zeta} = -\oint_s q_{1\zeta}(s)\frac{ds}{F_i} \; / \; \oint_s \frac{ds}{F_i} \tag{10.78}$$

Similarly to the case of the open section, the coordinates ζ_c, η_c of the *mechanical shear center* are calculated by equilibrium of moments caused by the shear forces and shear flows with respect to the mechanical center of gravity, yielding

$$\begin{aligned}\zeta_c &= -\frac{1}{V_\eta}\oint_s \left(q_{1\eta}(s) + q_{0\eta}\right) R(s'')\, ds \\ \eta_c &= \frac{1}{V_\zeta}\oint_s \left(q_{1\zeta}(s) + q_{0\zeta}\right) R(s'')\, ds\end{aligned} \tag{10.79}$$

The derivation of the shear stiffness is similar to that of the open section. For the closed section, the contribution of twisting ($M^i_{xs}\kappa^i_{xs}$) in (10.59) is negligible compared to the contribution of the shear flow ($q^i\gamma^i_{xs}$), and using (10.26) with $q_i = N^i_{xs}$ yields

$$\begin{aligned}(GA_\zeta) &= V_\eta^2 / \oint_s \frac{[q_\eta(s)]^2}{F_i}\, ds \\ (GA_\eta) &= V_\zeta^2 / \oint_s \frac{[q_\zeta(s)]^2}{F_i}\, ds\end{aligned} \tag{10.80}$$

where $q_\eta(s)$ and $q_\zeta(s)$ are defined in (10.73)–(10.76).

The shear flow due to torsion is obtained from (10.47). For moderately thick walls, the expression can be improved recognizing that part of the moment is equilibrated by Saint Venant stresses, like in the open section, leading to the following expression [231]

$$q_T = \frac{T}{2\Gamma_{s''}}\left[1 - \frac{3}{4}\frac{(GJ_R)_{open}}{(GJ_R)_{closed}}\right] \tag{10.81}$$

where $(GJ_R)_{open}$ and $(GJ_R)_{closed}$ are the torsional stiffness of the open section (10.45) and closed section (10.50), respectively.

10.2.9 Beam Deformations

The deformations of the beam are now computed using the classical formulas referred to the principal axes. The axial strain ϵ_x, the two curvatures κ_η, κ_ζ, and the twist rate β are computed as

$$\epsilon_x = \frac{N}{(EA)}$$
$$\kappa_\eta = \frac{M_\eta}{(EI_\eta)}$$
$$\kappa_\zeta = \frac{M_\zeta}{(EI_\zeta)}$$
$$\beta = \frac{T}{(GJ_R)} \tag{10.82}$$

where the axial stiffness is given by (10.31), two bending stiffness are given by (10.40) and the torsional stiffness is given by either (10.45) or (10.50) for open and closed cross-sections, respectively.

If the beam is a component of a structural system (indeterminate or not), the *mechanical* properties (EA), (EI_η), (EI_ζ),(GA_η), (GA_ζ), and (GJ_R), can be easily transformed into *equivalent geometrical properties* (dividing by arbitrary *equivalent moduli* E and G). The nodal displacements of the structure and stress resultants at any section can be obtained using the *equivalent properties* as input for any structural analysis program, either a beam finite element analysis or a matrix structural analysis program. Later on, these stress resultants can be used in (10.82) to compute the beam deformations at any section of the structure.

10.2.10 Segment Deformations and Stresses

The axial strain in the midsurface at any point can be computed in terms of the beam deformations (10.82)

$$\epsilon_x^i(s) = \frac{N}{(EA)} + \zeta(s)\frac{M_\eta}{(EI_\eta)} - \eta(s)\frac{M_\zeta}{(EI_\zeta)} \tag{10.83}$$

where $\epsilon_x^i(s)$ varies along the contour because $\zeta(s)$ and $\eta(s)$ are the coordinates in principal axis of a point in the contour s.

The bending curvature, which is unique for each segment, can be computed in terms of the beam deformations (10.82)

$$\kappa_x^i = \kappa_\eta \cos\alpha_\eta^i + \kappa_\zeta \sin\alpha_\eta^i \tag{10.84}$$

where α_η^i is the angle between the orientation of the i-th segment and the principal axis η.

The twisting curvature, constant for all segments, is calculated from the twist rate β (see (10.43) and (10.48))

$$\kappa_{xs}^i = \left\{ \begin{array}{ll} -2T/(GJ_R) & \text{open section} \\ -T/(GJ_R) & \text{closed section} \end{array} \right\} \tag{10.85}$$

where the minus sign is introduced to account for the definition of the positive torque T in Figure 10.3 and the positive twisting moment M_{xs}^i in Figure 10.6.

The shear strain $\gamma_{xs}^i(s)$ depends on the total shear flow, from torsion and shear, acting in the local s'' axis

$$q(s) = q_T + q_\eta(s) + q_\zeta(s) \tag{10.86}$$

The constant shear flow due to torsion q_T is given by (10.81) for closed sections and $q_T = 0$ for open sections. The shear flow $q_\eta(s)$, $q_\zeta(s)$ due to shear forces V_η, V_ζ are given by (10.52), (10.56), for open sections and by (10.73), (10.76), for closed sections.

The shear strain $\gamma_{xs}^i(s)$ in the midsurface can be calculated from the third equation in (10.15)

$$\gamma_{xs}^i(s) = \frac{N_{xs}^i - C_i\,\kappa_{xs}^i}{F_i} = \frac{q(s)}{F_i} - e_q^i\,\kappa_{xs}^i \tag{10.87}$$

where the twisting curvature is computed according to (10.85).

The stress resultants are computed at each point by (10.15). The four stress resultants are complemented with $N_s^i = 0$ and $M_s^i = 0$, reordered, and introduced in a standard laminate analysis program to compute stresses and to predict failure using a failure criterion.

Since the strains and the corresponding stress resultants vary along the contour, the failure criterion should be evaluated at several points along the contour. If applied loads (M_η, M_ζ, T) vary along x, the computations should be repeated for various sections along the length of the beam as well.

Example 10.10 *Compute the stresses at the clamped end of the solid, rectangular beam in Example 10.3, p. 359.*

Solution to Example 10.10 *The tip load produces a uniform shear diagram along the beam ($V = V_z$) and a bending moment distribution with a maximum at the support*

$$V = V_z = -1000\ N$$
$$M_{\max} = -V_z\ L = 1000(0.054) = 54\ Nm$$

To use the standard procedure for computation of stresses in laminates (Section 6.2), the shear force and bending moment must be expressed on a per unit width basis

$$V_x = \frac{V}{b} = \frac{-1000}{0.009} = -111,111\ N/m$$
$$M_x = \frac{M_{\max}}{b} = \frac{54}{0.009} = 6,000\ N$$

Eight stress resultants are used in Section 6.2 for the computation of stresses, but for a beam with no axial load and no torsion $N_x = N_s = N_{xs} = M_s = M_{xs} = V_s = 0$. Therefore, the loading on the laminate becomes (substituting y for s)

$$N_x = 0$$
$$N_y = 0$$
$$N_{xy} = 0$$
$$M_x = 6,000 \ N$$
$$M_y = 0$$
$$M_{xy} = 0$$
$$V_x = -111,111 \ N/m$$
$$V_y = 0$$

Using (6.20), (6.23), and (6.24), or a laminate analysis software, the following values of stresses are obtained (values in GPa)

Lamina #	*Location*	σ_x	σ_y	σ_{xy}	σ_{yz}	σ_{xz}
3	Top	.953	-.00716	-.0249	-.000512	-.0168
3	Bottom	-.305	.0158	-.00838	-.000512	-.0168
2	Top	.0482	.0864	-.0736	.00208	-.0140
2	Bottom	-.289	-.129	.190	.00208	-.0140
1	Top	-.163	-.00301	-.0508	-.00334	-.0143
1	Bottom	-.244	.0370	-.0323	-.00334	-.0143

These values of stress can be transformed to material coordinates in each lamina (5.37) and used into a failure criterion to evaluate the strength of the beam.

Example 10.11 *Compute the deflection at the tip of the beam in Figure 10.8 using the results of Examples 10.4, 10.7, and 10.8 (pp. 361, 368, 370).*

Solution to Example 10.11 *Using Figure 10.2, the deflection is*

$$\delta = \delta_b + \delta_s = \frac{V_z L^3}{3(EI)} + \frac{V_z L}{(GA)} = \left[\frac{(0.3)^3}{3(306.7)} + \frac{0.3}{(379,000)}\right](445) = 13.411 \ mm$$

where the shear deformation accounts for 2.6% of the total.

Example 10.12 *Compute the First Ply Failure load of the beam in Example 10.4, 10.7 and 10.8 (pp. 361, 368, 370) according to the max. stress criterion. The strength values for Kevlar–epoxy are taken from Table 1.3–1.4:* $F_{1t} = 1380 \ MPa$, $F_{1c} = 586 \ MPa$, $F_{2t} = 34.5 \ MPa$, $F_{2c} = 138 \ MPa$, $F_6 = 44.1 \ MPa$. *Neglect any increase of strength represented by in-situ values (see Exercise 10.13).*

Solution to Example 10.12 *For the cantilever beam, the maximum shear and bending moment occur at the root, with values*

$$M = -V_z L = -445(0.30) = -133.5 \ Nm$$
$$V = V_z = 445 \ N$$

Therefore, the critical points that need to be checked are: the flange and the web at the web-flange intersection, both at top and bottom of the beam, and the web at the centroid. At the top, $z = (h-t)/2$, and using (10.83) with no axial force ($N=0$) yields

$$\epsilon_x^i = \frac{zM}{(EI)} = \frac{0.025(1-1/16)(0.5)(-133.5)}{306.7} = -0.005101$$

for both flange and web ($i = 1,2,3$) at point A in Figure 10.8. Since $k_{xs}^i = 0$ in the flange, (10.82) and 10.87 yield

$$\kappa_x^{(1)} = \frac{M}{(EI)} = \frac{-133.5}{306.7} = -0.435$$
$$\gamma_{xs}^{(1)} = \frac{q}{F_1} = \frac{-7229.25}{0.017(10^9)} = -425.25(10^{-6})$$

where $N_{xs}^{(1)} = q$ is the shear flow computed in Example 10.7 (p. 368) and the minus sign indicates that the direction of the shear flow in segment #1 is opposite to the s-coordinate in that segment. Using (10.15) and the values computed in Example 10.4, the forces and moments per unit length on the top flange are

$$\begin{Bmatrix} N_x^{(1)} \\ M_x^{(1)} \\ N_{xs}^{(1)} \\ M_{xs}^{(1)} \end{Bmatrix} = \begin{bmatrix} 68(10^6) & 0 & 0 & 0 \\ 0 & 7.874 & 0 & 0 \\ 0 & 0 & 0.017(10^9) & 0 \\ 0 & 0 & 0 & 4.0 \end{bmatrix} \begin{Bmatrix} -0.005101 \\ -0.435 \\ -425.25(10^{-6}) \\ 0.0 \end{Bmatrix}$$
$$= \begin{Bmatrix} -346,862 \\ -3.425 \\ -7229.25 \\ 0.0 \end{Bmatrix}$$

In summary, the following values are obtained for the stress resultants (substituting y for s above)

	Flange	*Flange*	*Web*	*Web*	*Web*
	$z = h/2$	$z = -h/2$	$z = h/2$	$z = -h/2$	$z = 0$
N_x [N/m]	-346,862	346,862	-346,862	346,862	0
N_y [N/m]	0	0	0	0	0
N_{xy} [N/m]	-7229.25	-7229.25	-14,458.5	-14,458.5	-21,232.9
M_x [N]	-3.425	-3.425	0	0	0
M_y [N]	0	0	0	0	0
M_{xy} [N]	0	0	0	0	0
R	1.36	1.79	1.40	1.88	14.60

Then, each column in the table is input as the loading to the corresponding laminate (flange or web) using a laminate analysis program such as CADEC [1] to evaluate strains, stresses, and strength. The strength values are taken from Table 1.3. Neglect any increase of strength related to in-situ effects (see Exercise 10.13). The strength ratio R at the five chosen locations were computed using FPF and the maximum stress criterion, with the results reported in the table above. The minimum value can be used to estimate the FPF load of the beam.

10.2.11 Restrained Warping of Open Sections

When free warping due to torsion is restrained, axial secondary forces appear. They can be computed [221, Section 8.11] as

$$N_x^i(s) = -\frac{d^2\phi}{dx^2}\,\overline{\omega}(s)\,A_i \tag{10.88}$$

where ϕ is the angle of twist and $\overline{\omega}(s)$ is the principal sectorial area (10.98). Equilibrium of forces requires a secondary shear flow

$$q_\omega(s) = \frac{d^3\phi}{dx^3}\int_0^s \overline{\omega}(s)\,A_i\,ds \tag{10.89}$$

The angle of twist ϕ is computed solving the general equation for torsion

$$\frac{d^4\phi}{dx^4} - k^2\frac{d^2\phi}{dx^2} = -k^2\frac{T'}{(GJ_R)} \tag{10.90}$$

where $T' = dT/dx$, $k^2 = (GJ_R)/(EI\overline{\omega})$, and $(EI\overline{\omega})$ is the *mechanical warping stiffness* defined as in [221, p. 320]. Particular solutions to (10.80) are given in [176, Table 22] for various loads and boundary conditions. The *mechanical warping stiffness* is computed as in (10.67) but using principal sectorial area

$$(EI\overline{\omega}) = \int_s [\overline{\omega}(s)]^2 A_i\,ds \tag{10.91}$$

When warping is restrained, the axial strain due to secondary forces

$$\epsilon_x^i(s) = -\frac{d^2\phi}{dx^2}\,\overline{\omega}(s) \tag{10.92}$$

should be added to the axial strain computed in (10.83). Also, the shear flow due to restrained warping $q_\omega(s)$ given by (10.89), should be added to the shear flow computed in (10.86).

Principal Sectorial Area Diagram

The bending moments M_ζ and M_η caused by the secondary axial forces must also vanish. Using (10.88), the conditions $M_\zeta = 0$ and $M_\eta = 0$ lead to

$$\int_s (\eta(s) - e_b \sin\alpha_\eta^i)\,\overline{\omega}(s)\,A_i\,ds = 0 \tag{10.93}$$

$$\int_s (\zeta(s) + e_b \cos\alpha_\eta^i)\,\overline{\omega}(s)\,A_i\,ds = 0 \tag{10.94}$$

These equations are satisfied using the *mechanical shear center* as a pole, regardless of the initial point. In the case of torsion T acting alone, the secondary axial forces integrated in the contour s' must vanish because no axial force N is applied

$$N = \int_s N_x^i(s)\, ds = 0 \tag{10.95}$$

and using (10.88) leads to

$$\int_s \overline{\omega}(s)\, A_i\, ds = 0 \tag{10.96}$$

To satisfy (10.96), the initial point is changed, which is equivalent to subtracting a constant. Using the *mechanical shear center* as the pole to satisfy (10.93) and (10.94), a sectorial area $\omega_1(s)$ is computed with an arbitrary initial point

$$\omega_1(s) = \int_I^s R(s'')\, ds \tag{10.97}$$

where the arm-length $R(s'')$ is measured on the axis s'' (Figure 10.9). Then, the *principal sectorial area* $\overline{\omega}(s)$ is defined as

$$\overline{\omega}(s) = \omega_1(s) - \omega_0 \tag{10.98}$$

where the constant ω_0 is obtained from (10.96) as

$$\omega_0 = \frac{1}{(EA)} \int_s \omega_1(s)\, A_i\, ds \tag{10.99}$$

Exercises

Exercise 10.1 *Using the simplified method described in Section 10.1, design an I-beam of length $L = 4$ m to be supported at the ends and loaded at the midspan by a load $P = 1000$ N. The maximum deflection should not exceed 1/800 of the span. Use the carpet plots for E-glass–polyester with $V_f = 0.5$. Use the same laminate configuration for the flanges and the web and use your judgement to select values of α, β, γ in the carpet plots. Let the height-to-width ratio be $h/b = 2$ and the width-to-thickness ratio be $b/t = 10$. Optimum values for these geometric ratios could be obtained by analyzing the buckling modes of the beam.*

Exercise 10.2 *Determine the strength ratio at first ply failure of the beam designed in Exercise 10.1. Consider tensile, compressive, and shear stress. Use the carpet plots.*

Exercise 10.3 *Using the simplified method described in Section 10.1, determine the maximum tip load P that can be applied to the beam of Example 6.6, p. 197, before the bottom flange buckles. Assume the flange to be simply supported on the webs, which gives a conservative answer.*

Exercise 10.4 *Determine the Euler buckling load of a column clamped at the base and free at the loaded end. The column has the same dimensions and material of the beam in Example 6.6, p. 197.*

Exercise 10.5 *Determine an approximate estimate for the load P that causes local buckling to a column with the dimensions and material of Example 6.6, p. 197. Assume that the load is uniformly distributed around the perimeter of the cross-section. (Hint: See Section 11.3.2)*

Exercise 10.6 *A simply supported rectangular beam (solid) of length $L = 100$ mm, width $b = 10$ mm is made of a $[\pm 45/0_2]_S$ glass–polyester laminate with each lamina 0.125 mm thick. A uniform load of intensity 1,000 N/m is applied at the edge and parallel to the laminate (the load is in the plane of the laminate and perpendicular to the axis of the beam). Compute the maximum deflection. You may use the properties in Table 1.3 or the carpet plots for glass–polyester.*

Exercise 10.7 *A simply supported rectangular beam (solid) of length $L = 100$ mm and width $b = 10$ mm is made of a $[\pm 45/0_2]_S$ carbon–epoxy (AS4–3501-6; assume $F_4 = F_5 = F_6$) laminate with properties given in Table 1.3, with each lamina 0.125 mm thick. A transverse load P is applied at the midspan, perpendicular to the laminate. Compute (a) the deflection as a function of P and (b) the maximum load that can be applied to the beam for first ply failure using the maximum stress criterion.*

Exercise 10.8 *Recompute the deflection of the beam in Example 6.6, p. 197, using the formulation of Section 10.2.*

Exercise 10.9 *Compute the torsional stiffness (GJ_R) of the beam designed in Example 6.6, p. 197.*

Exercise 10.10 *Compute the torsional stiffness (GJ_R) of the beam in Example 10.4, p. 361.*

Exercise 10.11 *Consider a box-beam with constant area A, and variable depth h, width b, and thickness t.*

(a) Derive an expression for the moment of inertia per unit area in terms of the ratio $r = h/b$.

(b) Derive an expression for the failure moment per unit area in terms of the ratio $r = h/b$, for a constant material strength F_x.

(c) Comment on the significance of your results with regard to thin-walled construction.

Exercise 10.12 *Redesign the box-beam of Example 6.6 and 10.1 (pp. 197, 344) to make better use of the material; i.e., changing the geometry of the cross-section in such a way that the serviceability (deflection) requirement is met without overdimensioning for strength as much as it is done in Example 10.1. Perform the new design for a reliability of 99.5% assuming the load and strength variability are represented by $C_\sigma = 0.2, C_F = 0.3$.*

Exercise 10.13 *Recalculate the strength ratio values in Example 10.12 (p. 380) taking into account in-situ values (see Section 7.2.1), and ply thickness $t_k = 0.2$ mm.*

Chapter 11

Plates and Stiffened Panels

A plate is a flat structural element having two dimensions much larger than the thickness. Essentially, the laminates described in Chapter 6 are plates. They can be loaded by edge loads (N_x, N_y, N_{xy}) or transverse loads; the later produce bending moments (M_x, M_y, M_{xy}). While bending moments may be transferred to the supports on the edges, the bending moments are usually unknown at the onset of the analysis. The design of the plates seldom begins with a set of specified bending moments M_x, M_y, M_{xy}. However, all the analysis presented in Chapter 6 assumed that these bending moments and in-plane loads where known. Therefore, it is necessary to determine these values as a function of the loading and boundary conditions before the stresses in the plate can be computed.

For beams, it is always possible to draw the bending moment diagram without information about the material or cross-sectional dimensions. For most shell problems, the in-plane stress resultants N_x, N_y, and N_{xy} can be obtained using only equilibrium equations; that is, without involving the shell thickness or material properties. This is not the case for plates, in which the moments M_x, M_y, and M_{xy} cannot be found from the applied loads without specifying the material and the thickness. For this reason, approximate methods, which are valid only for particular cases, are used in preliminary design. These provide a first approximation to the thickness and laminate architecture. Then, preliminary design is usually followed by a refined analysis using Finite Element Analysis or computer programs based on complex analytical solutions [236]. A compressive study of analytical and numerical solutions for composite plates is presented in [175]. Only simple preliminary design methods are presented in this chapter.

The design for bending is further complicated by the fact that the bending stiffness coefficients (D_{ij}) and the bending strength values are sensitive to the changes in the relative position of various laminae, namely the laminate stacking sequence (LSS), even when the number and type of laminae remains unchanged. This is in contrast to the in-plane behavior of laminates, which is, to a first approximation, insensitive to the laminate stacking sequence.

The in-plane stiffness coefficients (A_{ij}) are insensitive to changes in the relative position of various laminae (stacking sequence). Therefore, the laminate in-plane moduli E_x, E_y, and G_{xy} completely characterize the in-plane stiffness of balanced symmetric laminates regardless of the laminate stacking sequence. This is not the case for bending, where the orientation of the laminae next to the surface have more influence on the bending stiffness and strength than inner laminae. Laminate bending moduli E_x^b, E_y^b, and G_{xy}^b are defined in (6.36), but it must be noted that these values change if the laminate stacking sequence varies. If the laminate bending moduli are not available, the laminate in-plane moduli (E_x, E_y, G_{xy}) provide a rough approximation. Comparing Figure 6.8 with Figure 6.11, it is evident that the laminate in-plane moduli could lead to errors in bending and buckling computations. Therefore, bending and buckling computations should always be checked using the correct bending stiffness matrix $[D]$ computed for the actual laminate stacking sequence being used.

11.1 Plate Bending

Preliminary design of laminates for bending is based on carpet plots of Flexural Strength for Laminates. *Flexural strength* is defined as the value of stress at failure, on the plate surface, computed using the following formulas for isotropic materials [177, (3.11)],

$$F_x^b = \frac{6M_x}{t^2} \tag{11.1}$$

$$F_y^b = \frac{6M_y}{t^2} \tag{11.2}$$

$$F_{xy}^b = \frac{6M_{xy}}{t^2} \tag{11.3}$$

where M_x, M_y, and M_{xy} are the values of the bending moments that cause failure of the actual laminate. Flexural strength values are sometimes reported by material suppliers, based on experimental data for a limited set of laminate configurations such as unidirectional, quasi-isotropic, etc. Experimental values of flexural strength can be determined following ASTM D 790.

The flexural strength values may be predicted using one of the failure criteria described in Chapter 7. The values may be based on first ply failure (FPF) or last ply failure (LPF), depending of the objectives of the design. Failure of the actual laminate is predicted by the procedures described in Chapter 7, by applying only one bending moment at a time. For example, applying a unit moment $M_x = 1$, the strength ratio for last ply failure is the value of M_x that causes ultimate failure of the laminate. Then, (11.1) is used to determine the flexural strength F_x^b. Carpets plots of flexural strength for E-glass–polyester (see Table 1.3) are shown in Figure 11.1–11.3. Since the flexural strength is a function of the stacking sequence, the plots were obtained by interspersing the 0, 90, and ± 45 laminae as much as possible. The value F_y^b can be read from the F_x^b plot by interchanging the values of α and β.

Having defined the flexural strength, the laminate thickness can be estimated in

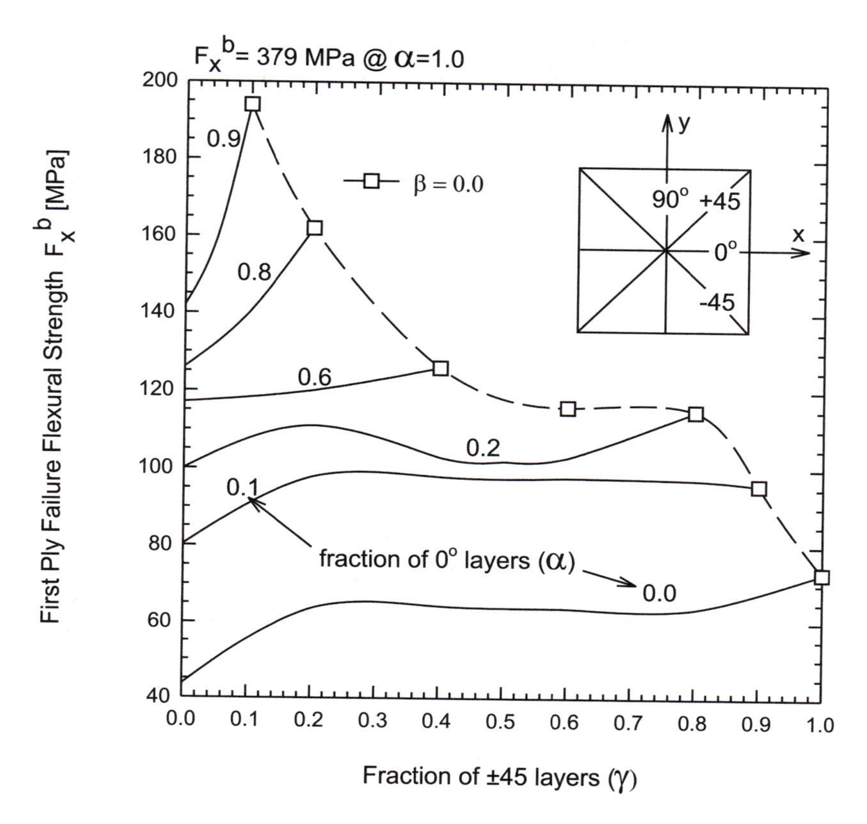

Figure 11.1: Carpet plot of FPF flexural strength F_x^b (loading: M_x).

preliminary design using (11.1), (11.2), and (11.3), for known values of the applied moments (see Example 11.1).

11.2 Plate Buckling

Buckling is the loss of stability of the structure, or part of the structure, when subjected to compressive stresses. Buckling results in large deflections and reduction of stiffness of the structure with respect to the unbuckled condition and it may lead to structural collapse. Buckling of composites is a complex topic that is outside the scope of this book. However, a few particular cases can be solved exactly [175, 177] or simple approximate equations can be proposed [237].

The solutions presented in this section are for symmetric, specially orthotropic laminates with $[B] = [0]$ and $D_{16} = D_{26} = 0$. For balanced symmetric laminates with nonzero values of D_{16} and D_{26}, the equations may predict erroneous results. The magnitude of the error depends on the magnitude of D_{16} and D_{26}. An effort should be made to minimize these coupling coefficients by placing the off-axis laminae in pairs ($\pm\theta$) and locate them as close as possible to the middle surface. The use of balanced, biaxial woven or stitched mats helps reduce the coupling effects

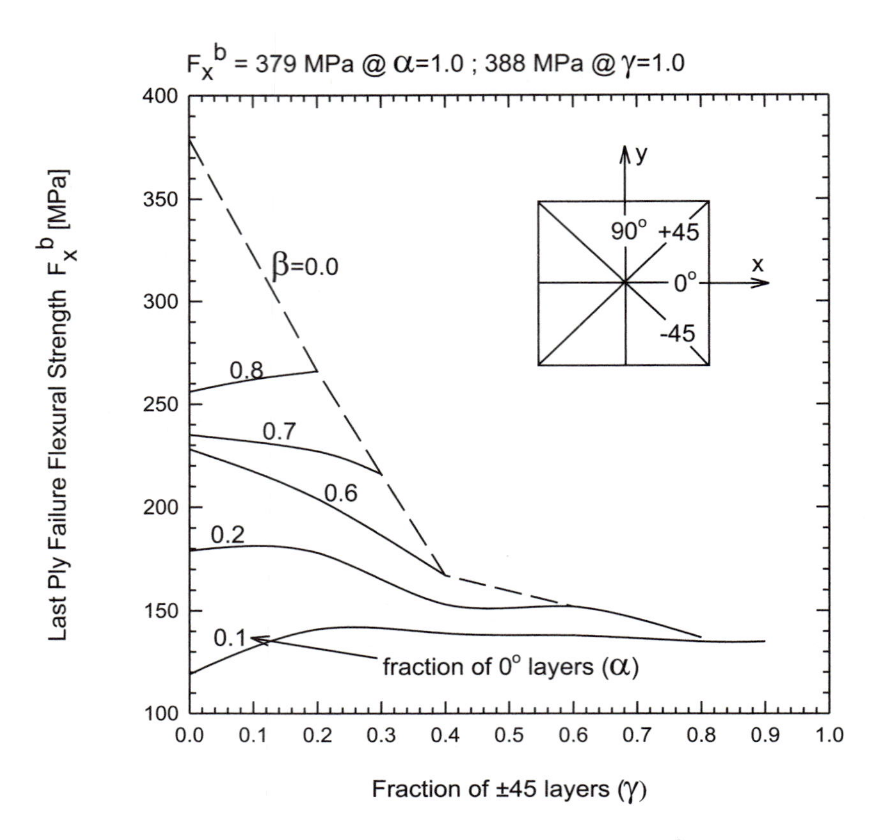

Figure 11.2: Carpet plot of LPF flexural strength F_x^b (loading: M_x).

(see Example 6.5). Transverse shear deformation effects, which are important for buckling, were neglected in the derivation of the equations in this section. Therefore, these equations are not conservative unless both dimensions of the plate (a and b) are much larger than the thickness ($a, b > 20t$).

11.2.1 All Edges Simply Supported

In the case of a rectangular plate, simply supported around the boundary, and subject to one edge load (Figure 11.4), the buckling load per unit length is [175]

$$N_x^{CR} = \frac{\pi^2 D_{22}}{b^2} \left[m^2 \frac{D_{11}}{D_{22}} \left(\frac{b}{a}\right)^2 + 2\frac{D_{12} + 2D_{66}}{D_{22}} + \frac{1}{m^2} \left(\frac{a}{b}\right)^2 \right] \tag{11.4}$$

where m is the number of waves of the buckled shape along the loading direction (e.g., $m = 2$ in Figure 11.4). The number of waves is $n = 1$ for this type of loading and boundary conditions. The integer m can be computed as the nearest integer to the real number R_m

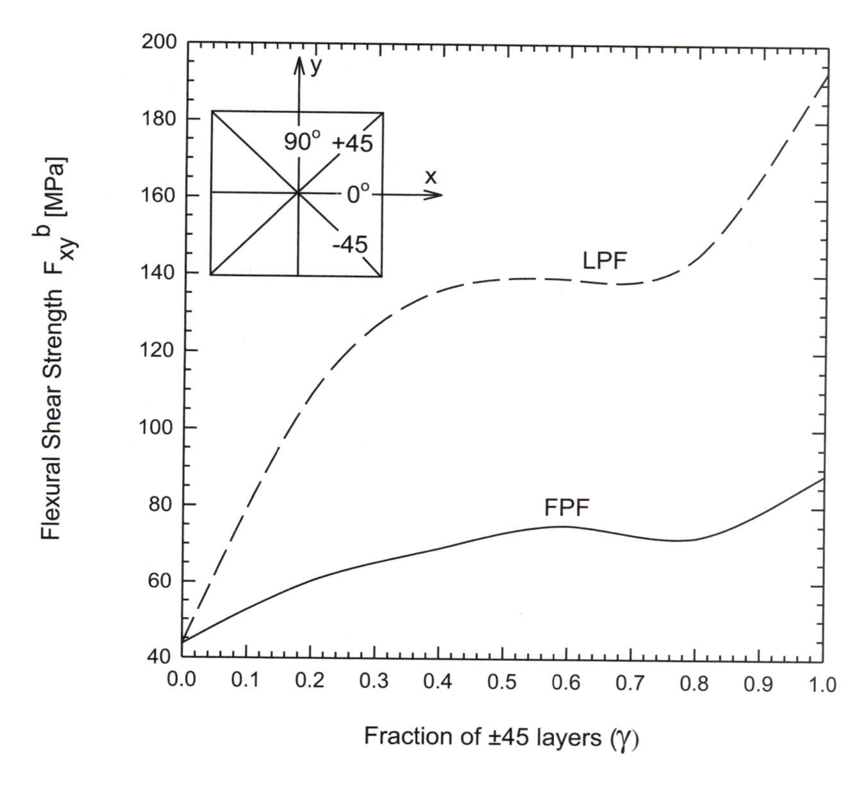

Figure 11.3: Carpet plot of FPF and LPF flexural shear strength F_{xy}^b (loading: M_{xy}).

$$R_m = \left(\frac{a}{b}\right)\left(\frac{D_{22}}{D_{11}}\right)^{\frac{1}{4}} \tag{11.5}$$

If the plate is very long ($a >> b$)

$$N_x^{CR} = \frac{2\pi^2}{b^2}\left(\sqrt{D_{11}D_{22}} + D_{12} + 2D_{66}\right) \tag{11.6}$$

Equation 11.6 also provides a good approximation when the loaded edges are clamped, where clamped means that the rotation and transverse deflection are restricted but the in-plane displacements are free [237].

In preliminary design, the coefficients of the bending stiffness matrix $[D]$ can be computed using laminate bending moduli given in carpet plots (Figure 6.10–6.12, see Section 6.5). Using (6.45) and dividing by the laminate thickness, (11.4) and (11.5) become

$$\sigma_x^{CR} = \frac{\pi^2 E_y^b}{12\Delta}\left(\frac{t}{b}\right)^2\left[m^2\frac{E_x^b}{E_y^b}\left(\frac{b}{a}\right)^2 + 2\frac{E_y^b\nu_{xy}^b + 2G_{xy}^b\Delta}{E_y^b} + \frac{1}{m^2}\left(\frac{a}{b}\right)^2\right] \tag{11.7}$$

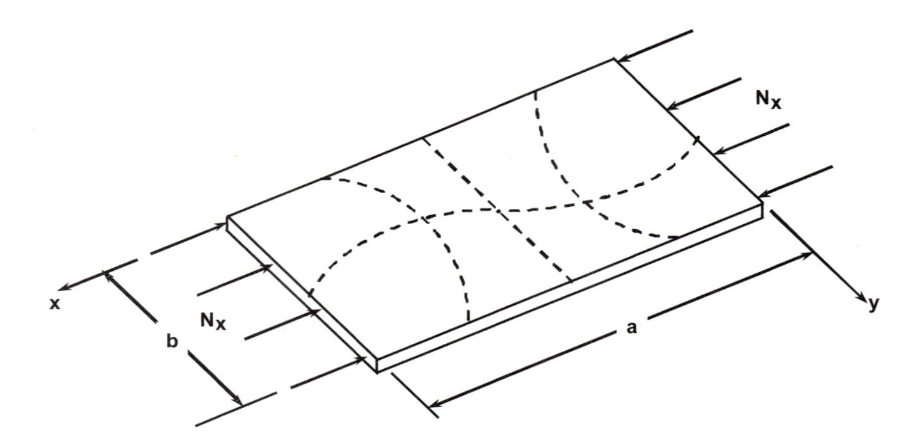

Figure 11.4: Buckling of rectangular plate with all edges simply supported (mode $m = 2$).

and

$$R_m = \left(\frac{a}{b}\right)\left(\frac{E_y^b}{E_x^b}\right)^{\frac{1}{4}} \tag{11.8}$$

where

$$\Delta = 1 - \nu_{xy}^b \nu_{yx}^b \quad ; \nu_{yx}^b = \frac{E_y^b}{E_x^b}\nu_{xy}^b \tag{11.9}$$

In terms of laminate moduli and dividing by the laminate thickness, (11.6) can be written as

$$\sigma_x^{CR} = \frac{N_x^{CR}}{t} = \left(\frac{t}{b}\right)^2 \frac{\pi^2}{6\Delta}\left(\sqrt{E_x^b E_y^b} + E_y^b \nu_{xy}^b + 2G_{xy}^b \Delta\right) \tag{11.10}$$

A plot of dimensionless buckling stress $(\sigma_x^{CR} 12\Delta b^2/\pi^2 E_y^b t^2)$ as a function of the length/width ratio a/b is shown in Figure 11.5 using the material properties in Figure 6.10–6.12, for $\alpha = \beta = 0.2$, $\gamma = 0.6$ (see Section 6.5). Note how the number of modes m increases with the length of the plate. The locii of minimum values is given by (11.10), which is often used as a conservative estimate for design. Note that the buckling strength in (11.10) grows with the square of the thickness, thus providing motivation for the use of sandwich plates. Also, it is evident that the ratio (t/b) plays a crucial role in the buckling strength of the plate. Therefore, the simplest way to increase the buckling load is to reduce the effective width of the plate by intermediate supports. This is accomplished by adding longitudinal stringers to support the plate at intermediate points through the width of the plate (Section 11.3).

Usually, the geometric parameters, such as $(t/b)^2$ in (11.10), play a more significant role than the material parameters (α, β, γ) and the laminate stacking sequence.

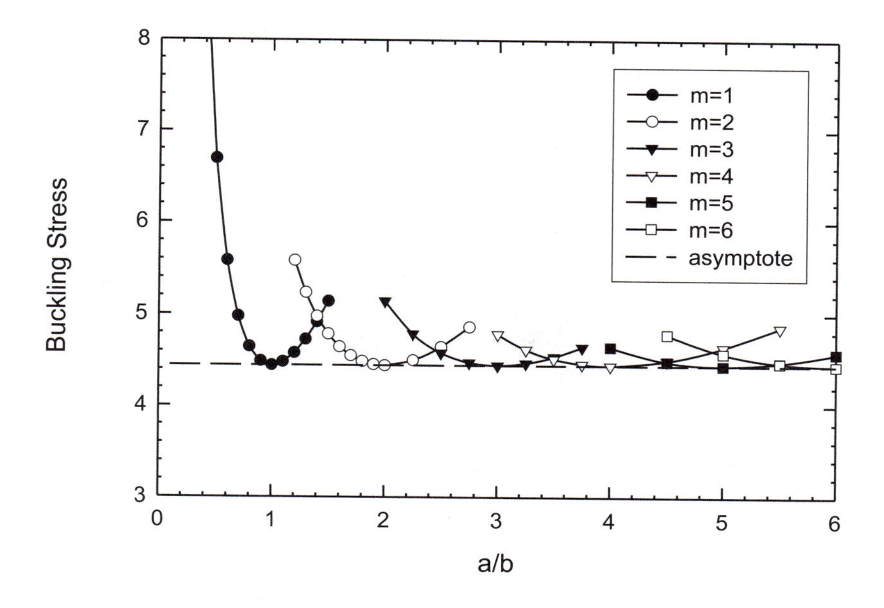

Figure 11.5: Dimensionless buckling stress v.s. a/b.

This means that the geometry dominates the buckling response. However, different material systems (carbon vs. E-glass) drastically affect the stiffness of the material and, as a consequence, the buckling strength.

For a given material system, empirical equations based on experimental data or parametric studies have been proposed for preliminary design of compression members (see Section 11.3.2). The experimental data confirms that the fact that, for a given material system, the geometric parameters dominate the buckling behavior.

11.2.2 All Sides Clamped

For a long plate ($a/b > 4$) with all sides clamped and loaded in the x-direction only, the buckling load per unit width can be estimated by [237]

$$N_x^{CR} = \frac{\pi^2}{b^2}\left(4.6\sqrt{D_{11}D_{22}} + 2.67D_{12} + 5.33D_{66}\right) \tag{11.11}$$

or, in terms of laminate bending moduli (see Section 6.5), the buckling stress is

$$\sigma_x^{CR} = \frac{N_x^{CR}}{t} = \left(\frac{t}{b}\right)^2 \frac{\pi^2}{12\Delta}\left(4.6\sqrt{E_x^b E_y^b} + 2.67E_y^b\nu_{xy}^b + 5.33G_{xy}^b\Delta\right) \tag{11.12}$$

where Δ is given by (11.9).

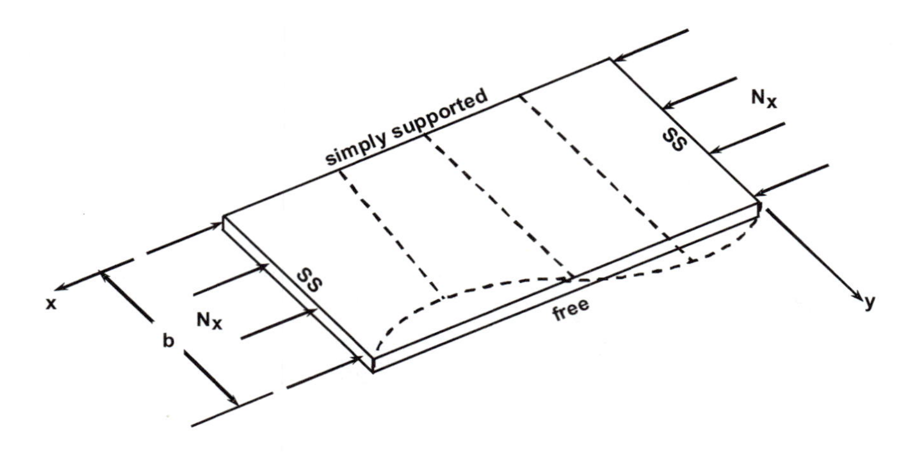

Figure 11.6: Buckling of a rectangular plate with an unloaded edge free.

11.2.3 One Free Edge

A long plate of length a and with b ($a/b > 4$) with three edges simply supported, one unloaded edge free, and loaded along length (Figure 11.6), has a buckling load per unit width given by [237]

$$N_x^{CR} = \frac{12D_{66}}{b^2} + \frac{\pi^2 D_{11}}{a^2} \tag{11.13}$$

or in terms of the apparent bending moduli (see Section 6.5), the buckling stress is

$$\sigma_x^{CR} = \frac{N_x^{CR}}{t} = \left(\frac{t}{b}\right)^2 \left(G_{xy}^b + \pi^2 \left(\frac{b}{a}\right)^2 \frac{E_x^b}{12\Delta}\right) \tag{11.14}$$

where Δ is given by (11.9).

11.2.4 Biaxial Loading

A rectangular plate ($1 \leq a/b < \infty$) with all edges simply supported and subject to loads N_x and $N_y = kN_x$, with k constant, has a buckling load [175] given by the minimum with respect to m and n of

$$N_x^{CR} = \frac{\pi^2}{b^2} \frac{D_{11}m^4c^4 + 2Hm^2n^2c^2 + D_{22}n^4}{m^2c^2 + kn^2} \tag{11.15}$$

with

$$\begin{aligned} c &= b/a \\ H &= D_{12} + 2D_{66} \\ k &= \frac{N_y}{N_x} = \frac{N_y^{CR}}{N_x^{CR}} \end{aligned} \tag{11.16}$$

The minimum value of (11.15) is found by trying all the combinations of integer numbers m and n, that represent the number of half-waves in the x and y directions, respectively.

11.2.5 Fixed Unloaded Edges

Consider a rectangular plate ($1 \leq a/b < \infty$) loaded in the x-direction only. If the loaded edges are simply supported and the unloaded sides are clamped, the buckling load is obtained by minimizing the following equation with respect to m

$$N_x^{CR} = \frac{\pi^2}{b^2}\left[D_{11}m^2c^2 + 2.67D_{12} + 5.33\frac{D_{22}}{c^2} + D_{66} + \frac{1}{m^2}\right] \qquad (11.17)$$

where m is the number of half-waves in the x-direction and $c = b/a$ [237].

11.3 Stiffened Panels

Stiffened panels consist of a flat or curved laminate reinforced by a grid of longitudinal and transverse stiffeners (Figure 11.7). The stiffeners provide increased buckling resistance under in-plane load and increased bending stiffness under transverse and bending loads.

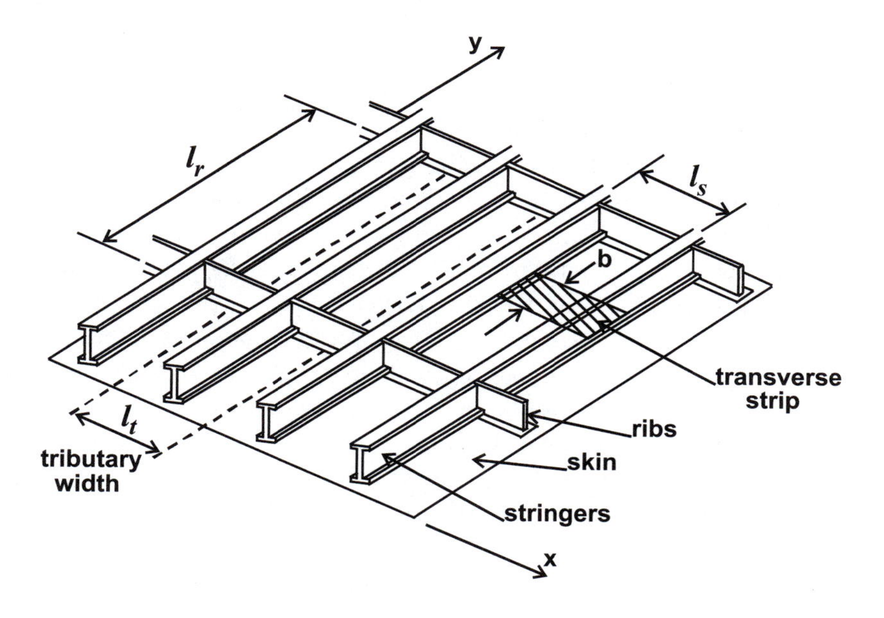

Figure 11.7: Typical reinforced panel.

The main components of stiffened panels have particular names depending of the application. A portion of an aircraft skin is shown in Figure 11.7. The ribs and stringers reinforce and support the skin. In the aircraft fuselage the reinforcements along the length of the fuselage are also called *stringers* but the transverse reinforcements are called *frames*. In shipbuilding the reinforcements along the hull are called *longitudinals* and the transverse reinforcements are called *bulkheads*. In bridge construction, the skin is called a *deck*, which is supported by stringers and cross-beams. In floor systems, the floor deck is supported by longitudinal joists and transverse beams. In roof construction, the roof panels are supported by purlings and rafters. The differences between ribs, frames, bulkheads, etc., are related to function, stiffness, etc. The interested reader should consult specialized literature such as [238–241]. For the purpose of this section, it is sufficient to realize that the preliminary design of each component can be done separately. The design approach depends on whether the main load is in-plane or bending. Under in-plane load, the main consideration is buckling. Under bending, deflections and strength are the main considerations, but lateral-torsional buckling of the stringers may also be a concern. Comprehensive treatment of the design of stiffened panels falls outside the scope of this book. Therefore, only the basic aspects of the design will be covered with emphasis on the particular aspects of the use of composite materials.

11.3.1 Stiffened Panels under Bending Loads

As discussed earlier, rigorous analysis of bending of plates (stiffened or not) is complicated. The design can be simplified by noting that usually the stringer spacing is much smaller than the spacing of ribs (frames, bulkheads, etc.)

Skin Design

The design of the skin is based on the assumption that the stringer spacing is much smaller than the rib or frame spacing (typically 6/20 for an aircraft fuselage). Under this condition, the bending moment in the skin in the transverse direction (transverse to the stringers) is much larger than the bending moment in the skin in the longitudinal direction. Therefore, the skin can be designed isolating a narrow transverse strip (Figure 11.7) and treating it as a beam of width b, thickness t, and length equal to the stringer spacing l_s. The support conditions at the end of the beam depend on the torsional stiffness of the stringers. If the stringers are very stiff, they do not twist, and the skin can be analyzed as a clamped beam. If the stringers are not stiff in torsion, the skin can be analyzed as a continuous beam. Assuming a uniformly distributed pressure p_0 on the skin, the load on the beam is $q_0 = p_0 b$. The maximum moment and deflection of the skin are, for the clamped case

$$M = -q_0 l_s^2/12 \text{ (at supports)} \tag{11.18}$$

$$M = q_0 l_s^2/24 \text{ (at mid-span)} \tag{11.19}$$

$$\delta = -\frac{q_0 l_s^4}{384(EI)} \tag{11.20}$$

and for the continuous beam case [179, p. 282]

$$M \cong \pm q_o l_s^2/10 \tag{11.21}$$

where (EI) is the bending stiffness of the strip. The moment of inertia of the strip is

$$I = \frac{b\,t^3}{12} \tag{11.22}$$

If the design is deflection controlled, (11.20) and (11.22) allow for sizing of the thickness after selecting a laminate and a value of E_x, which can be obtained from Figure 6.11. In this case, the design method is illustrated in Section 6.5.1.

If the design is strength controlled, (11.18) or (11.21) along with (11.2) allow for the sizing of the thickness of the skin, after selecting a laminate and a value of flexural strength F_x^b in Figures 11.1 or 11.2. Appropriate load and resistance factors should be used in the design (see Section 1.4.3.)

Since the bending moment M_y in the skin has been neglected, the design method leads to the selection of laminates reinforced primarily in the x-direction ($\alpha \to 1$). Such a selection would be incorrect since significant values of bending moment M_y are likely to appear near the ribs. Also, significant values of twisting moment M_{xy} and shear force N_{xy} are likely to appear near the rib-stringer intersection. Therefore, the designer is advised to avoid predominantly unidirectional laminates. It must be noted that this simplified design method has been applied mostly with quasi-isotropic laminates, woven-cloth reinforced laminates, or continuous or chopped strand mat reinforcements.

Stringer Design

Consistent with the skin design, the design of the stringer can be done using the method of tributary area. The load taken by the stringer is the load applied on the skin over the area halfway through the two adjacent stiffeners (Figure 11.7), which is called the *tributary area.* The stringer is then designed as a beam with a uniformly distributed load $q_0 = p_0 l_t$. If the stringer spacing is uniform, then the tributary width l_t is equal to the stringer spacing (or pitch l_s). The support conditions depend on the stiffness of the cross-members. Since frames and bulkheads are usually quite stiff, the stringer is modeled as a clamped beam, with the maximum moment and deflection given by (11.18)–(11.20).

The stringer shape is selected as a compromise among fabrication, cost, torsional and bending stiffness, and other considerations specific to the application at hand [238,241]. Various stiffener geometries are illustrated in Figure 11.8. Once the shape is selected, the stiffener is designed as a beam to provide adequate strength and to satisfy any deflection constrains. The stiffener works as a beam with the primary stresses oriented along the length of the stiffener. Therefore, the apparent properties used are the in-plane properties: E_x and F_x for bending, G_{xy} and F_{xy} for shear.

The contribution of the skin to the stiffness and strength in the direction of the stringer is usually accounted by adding a portion of the skin to the stringer.

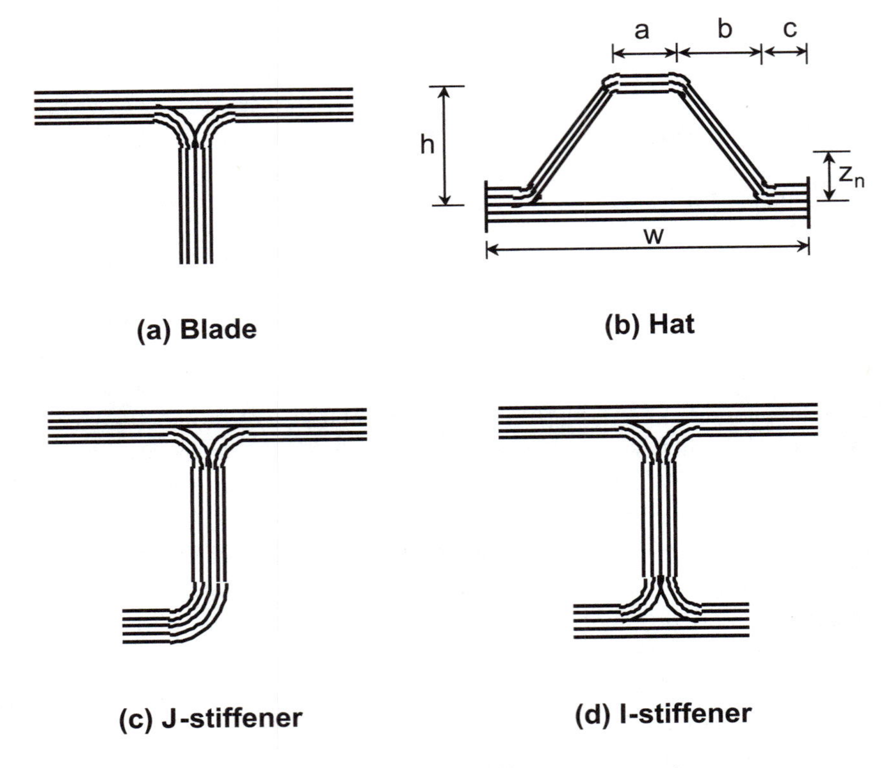

Figure 11.8: Typical stiffener geometries.

However, not all of the tributary width contributes fully to the stringer because of nonuniform shear distribution (shear lag), less than perfect skin-stringer connection (loss of composite action), or skin buckling. Therefore, only a portion (or effective width) of the skin is added to the stringer.

Example 11.1 *Perform a preliminary design of a portion of the hull for a planing power boat with the following specifications: waterline length: $L_w = 5.8$ m, cruising speed: $v = 17$ m/s.*

Solution to Example 11.1 *There are two main forces acting on the hull, hydrostatic water pressure and wave slamming [242]. Since wave slamming forces are much larger for this design, hydrostatic forces can be ignored. The impact pressure due to wave slamming can be estimated by empirical formula described in [241]*

$$p = \frac{Av^2 + Bv\sqrt{L_w} + CL_w}{D}$$

where p is the pressure in psi, v is the speed in m.p.h., L_w is the waterline length in feet, $D = 144.55$ is a constant involving the density of water and dimension conversion factors; A, B, C are empirical constants given in [241] as $A = 2.151$, $B = 1.267$, $C = 16.884$. For this example, the empirical formula yields $p = 173.5\ kPa$.

The pressure p is the maximum pressure occurring over a small portion of the skin. Reduction factors to take into account the location of the panel being designed and other pertinent factors are given in [241]. The maximum pressure is used in this example assuming that the panel being designed is at the most critical location in the hull. A relatively low load factor $\alpha = 1.5$ is suggested because the pressure used correspond to an extreme operating condition [241].

The boat hull is composed of a skin internally reinforced with longitudinal stiffeners, which in turn are supported at the bulkheads. Since the stiffener spacing or pitch l_s is much smaller than the bulkhead spacing l_r, the procedure described in Section 11.3.1 can be used.

The skin is designed analyzing a narrow strip of width $b = 1.0\ cm$ as shown in Figure 11.7, which is assumed to work as a continuous beam supported on the stiffeners (also called longitudinals). As a first trial, the following dimensions are assumed:

bulkhead spacing: $l_r = 1.25\ m$

stiffener spacing : $l_s = 0.15\ m$

From (11.21), the maximum bending moment of the beam (strip) due to a distributed load $q_0 = p_0 b$ *is*

$$M = \frac{(173,500)(0.01)(0.15)^2}{10} = 3.9\ Nm$$

Choosing a cross-ply configuration, which is similar to the woven material used in boat construction, take $\alpha = \beta = 0.5$, $\gamma = 0$. Then, interpolate in Figure 11.2–11.3 to get the static flexural strength of the dry composite

$$F_x^b = 180 + \frac{(228 - 180)}{(0.6 - 0.2)}(0.5 - 0.2) = 216\ MPa$$
$$F_{xy}^b = 44\ MPa$$

In this example, the wet strength values are estimated using wet strength retention factor[1] $\phi = 0.8$*, resulting in*

$$\phi F_x^b = 0.8 \times 216 = 172.8\ MPa$$

Any such reduction, including additional knock-down factors for the material properties, should be accounted at this point. Next, noting that M_x in (11.1) is the moment per unit length ($M_x = M/b$), the required thickness of the skin is

$$t_{sk} = \sqrt{\frac{\alpha\, 6\, M/b}{\phi F_x^b}} = \sqrt{\frac{1.5 \times 6 \times 3.9/0.01}{172.8 \times 10^6}} = 4.5\ mm$$

The longitudinal stiffeners are designed as beams that support a tributary area of the skin. Therefore, the load per unit length on the stiffeners is

$$q = p_0 l_s = 173.5(0.15) = 26\ kPa$$

[1]Strength of the laminate after immersion in water. May be estimated using the wet strength retention factor which is a percentage of the static dry strength.

Assuming the stiffeners are clamped at the bulkheads, use (11.18) to compute the bending moment in the beam at the supports, using a bulkhead spacing $l_r = 1.25$ m

$$M = \frac{26\,(10^3)\,(1.25)^2}{12} = 3,385\ Nm$$

Unlike the skin, the lips and the hat portion of the stiffener work in tension or compression, not in bending, even though the stiffener as a whole is in bending. Therefore, the strength values to be used are the tensile and compressive strength of the laminate under in-plane loads F_{xt} and F_{xc}, not the flexural strength values. Using Figures 7.12–7.15, and a wet strength retention factor $\phi = 0.8$, the following values are obtained for an E-glass–polyester laminate with $\alpha = \beta = 0.5$ (cross-ply)

$$\phi\, F_{xt} = 0.8 \times 465 = 372\, MPa$$
$$\phi\, F_{xc} = 0.8 \times 185 = 148\, MPa$$
$$\phi\, F_{xy} = 0.8 \times 40 = 32\, MPa$$

Note that the compressive strength is lower, as expected, and its value is close to the flexural strength used before. The hat portion of the stiffener is in compression at the supports, where the moment is maximum. Therefore, the required section modulus (10.3) is

$$Z_{req} = \frac{\alpha\, M}{\phi\, F_{xc}} = \frac{1.5(3,385)}{148\,(10^6)} = 34.28\, cm^3$$

Once the required Z is known, the geometry and dimensions of the stiffener are chosen to provide an adequate value of Z. The design method is restricted to the case of having the same laminate properties in the skin and the stiffener. This is usually sufficient in preliminary design but a more refined analysis is possible analyzing the skin-stiffener combination as a thin-walled beam (Section 10.2). In this example, different thicknesses of the skin (t_{sk}) and the walls of the stiffener (t_s) are handled through the ratio t_s/t_{sk}.

Hat stiffeners (Figure 11.8) are selected in this example because of their high torsional stiffness and excellent buckling resistance. They are filled with foam to prevent local buckling of the stiffener walls and to facilitate the fabrication process (molding).

The section modulus is a purely geometric property $Z = I/c$, where I is the moment of inertia and "c" is the distance from the neutral axis to the tension or compression flanges of the stiffener. For a hat stiffener as shown in Figure 11.8, simple strength of materials computations can be used to obtain the location of the neutral axis with respect to the base of the stiffener

$$z_n = \frac{h\left[b + \sqrt{h^2 + c^2}\right]}{2a + w_e t_{sk}/t_s + b + 2\sqrt{h^2 + c^2}}$$

and the moment of inertia with respect to the neutral axis

$$I = \left\{ b\,(h - z_n)^2 + \left(2a + w_e \frac{t_{sk}}{t_s}\right) z_n^2 + 2\sqrt{h^2 + c^2}\left[\frac{h^2}{12} + \left(\frac{h}{2} - z_n\right)^2\right]\right\} t_s$$

To achieve the required section modulus, and as a first trial, choose the following dimensions for the stiffener (see Figure 11.8):

$$
\begin{aligned}
b &= 38.1\ mm \\
w &= 2b = 76.2\ mm \\
a &= b/2 = 19.05\ mm \\
c &= (w - b - 2a)/2 = 0
\end{aligned}
$$

The effective width of the skin contributing to the stiffener is set to $w_e = w = 76.2\ mm$. The thickness of the walls of the stiffener are set equal to 1.2 times the thickness of the skin: $t_s = 1.2(4.5) = 5.4\ mm$. Therefore,

$$
\begin{aligned}
z_n &= 29.8\ mm \\
Z &= 29.9\ cm^3
\end{aligned}
$$

which is not sufficient. Increasing the thickness of the walls to $t_s = 1.4\ t_{sk} = 6.3\ mm$ results in

$$Z = 34.5\ cm^3$$

which barely exceeds the required value of $z = 34.28\ cm^3$. The stiffener can now be checked for compressive failure at the midspan and shear of the webs.

A stiffener with constant thickness and the same laminate configuration in all the walls is likely to be less than optimum. Different thicknesses for various walls can be incorporated easily in this example by deriving a new expression for z_n and I. Different laminate stacking sequences and materials for various walls can be incorporated if the stiffener is analyzed using the procedure described in Section 10.2. The advantage of the proposed design methodology is that it allows for a fast design of many different portions of the hull. More accurate analysis, including finite element modeling, can be performed only after a preliminary design of the geometry and materials has been formulated. At a preliminary design stage, changes in stiffener geometry, bulkhead and stiffener spacing, lay-up, and material can be easily evaluated using simple formulas. For example, various lay-ups can be investigated by just selecting different values of α, β, γ to be used in reading the carpet plots.

11.3.2 Stiffened Panel under In-plane Loads

Many modes of failure can occur when a stiffened panel is subjected to compressive edge loads (N_x, N_y) or shear load (N_{xy}) but the design is usually controlled by buckling. Several modes of buckling can occur. Overall or global buckling occurs when the whole panel buckles with a mode (or shape) that deforms the entire panel. Skin buckling occurs when only the skin buckles but the stiffeners retain their undeformed shape. In this case, the panel may still be able to carry the applied loads.

Torsional buckling of the stiffeners occurs mainly in open-section stiffeners (blade, J, I, etc., Figure 11.8). This mode of buckling seriously debilitates the panel and may lead to global buckling and collapse. Close-section stiffeners, such as the hat stiffener in Figure 11.8, have high torsional stiffness that not only prevents torsional buckling of the stiffeners but also contributes to the overall stability of the panel.

Finally, local buckling of the stiffener occurs when the walls of the stiffener (flange, web) buckle locally. Local buckling of the stiffener debilitates the panel and may lead to structural collapse. Open-section stiffeners are more susceptible to local buckling, mainly on the walls with free edges, but close-section stiffeners may also buckle locally. A filler (foam or honeycomb) may be used to prevent local buckling of closed-section stiffeners, but the possibility of water retention in the filler must be considered.

Many practical considerations affect the design of a stiffened panel, including fabrication cost, ease of splicing, etc. [238–240]. Therefore, no attempt is made in this section to present a comprehensive review of panel design. Instead, a brief description of the simplest design methodology available is presented to illustrate how composite materials fit into the classical panel design methods. Better predictions of buckling loads and postbuckling behavior can be achieved using design-optimization programs based on semi-analytical methods [243–246]. These programs represent a compromise between limitations of closed-form solutions presented here and the complexity and computational expense of nonlinear finite element analysis. These software packages can be used to analyze in detail a particular panel or directly to design a panel. When a new situation arises, a preliminary design of a stiffened panel can be done, using simple equations, to reveal the main features of various design alternatives such as stiffener spacing, etc. Then, a detailed analysis using a software package will reveal a better approximation to the true load carrying capacity of the panel and these results can be used to redefine the design. Ultimately, prototype testing is likely to be used to verify the performance of the panel.

Strength of Stiffeners and Skin

The compressive edge load is applied to the panel over its entire cross-section. The load is distributed to the skin and the longitudinal stiffeners proportionally to their respective axial stiffness (EA). The load carried by each component can be computed taking into account that the in-plane strain is the same in the stiffener and the skin. Therefore, the in-plane strain ϵ_x under edge load N_x applied at the neutral surface of the panel is approximated, neglecting Poisson effects, as

$$\epsilon_x = \frac{N_x l_s}{(EA)} \tag{11.23}$$

where l_s is the stiffener pitch and (EA) is the axial stiffness of a portion of the panel containing a stiffener and a portion of the skin of width l_s. The value of (EA) can be computed treating the portion of the panel as a thin-walled beam, using the tools presented in Chapter 10. In terms of laminate moduli E_x^{sk} of skin and E_x^s of stiffener, the axial stiffness is

$$(EA) = E_x^{sk} A_{sk} + E_x^s A_s \tag{11.24}$$

where $A_{sk} = t_{sk} l_s$ and A_s are the areas of skin and stiffener, respectively.

As long as no buckling occurs, the area is designed to carry the load using in-plane strength F_{xc} of the laminate (Figure 7.14). In terms of strain, the material will not fail in compression as long as the predicted strain ϵ_x is less than the compressive strain to failure of the laminate ($\epsilon_{xc} = F_{xc}/E_x$, see Section 7.3.2). In terms of stress, since the strain ϵ_x is the same in the skin and the stiffener, the load on each element is

$$\begin{aligned} P_s &= E_x^s A_s \epsilon_x \\ N_x^{sk} &= E_x^{sk} t_{sk} \epsilon_x \end{aligned} \tag{11.25}$$

where P_s is the load in the stiffener, N_x^{sk} is the load per unit width (N/m) of skin, E_x^{sk} and E_x^s are the laminate moduli of skin and of stiffener, respectively, A_s and t_{sk} are the area of the stiffener and thickness of the skin, respectively. With this information, the material compressive strength of the skin and stiffener can be easily checked. For the skin, N_x^{sk} can be used in a laminate analysis program to evaluate the laminate using a failure criteria. Otherwise, if the laminate in-plane strength F_{xc} is available from a carpet plot, then the material compressive strength is not reached as long as $N_x^{sk}/t_{sk} < F_{xc}$. The stiffener can be analyzed in a similar way. However, most practical designs will not fail due to material strength but rather due to buckling.

Buckling of Stiffened Panels

The critical buckling stress for a simply supported plate of width b increases with the square of the ratio t/b, as shown in (11.10). Therefore, a practical method to increase the buckling strength of a laminate is to reduce the width b by dividing the plate into narrower segments. This can be done effectively by adding longitudinal stiffeners (called *stringers*) to provide support for the narrower plates. The longitudinal stiffeners carry part of the axial load and they can buckle as columns. Their effective length can be reduced by adding transverse stiffeners (ribs, frames, or bulkheads) thus increasing the column buckling load.

Under in-plane load N_x, the whole panel can buckle as a column if the sides of the panel are not supported. The buckling load of the panel is given by (10.10), namely

$$N_x^{CR} = \frac{1}{l_s} \frac{\pi^2 (EI)}{L_e^2} \tag{11.26}$$

where L_e is the effective length, defined in (10.11) and (EI) is the bending stiffness (with respect to neutral surface of bending) of a representative portion of the panel consisting of one stiffener and a portion of the skin of width equal to the stiffener spacing l_s. If both, the skin and the stiffeners are made with the same material configuration (same values of α, β, γ), then the panel has a unique value of laminate in-plane moduli E_x. Therefore, the buckling load per unit width is

$$N_x^{CR} = \frac{1}{l_s}\frac{\pi^2 E_x I}{L_e^2} \tag{11.27}$$

where I is the moment of inertia of the portion of the panel (of width l_s) with respect to the neutral surface of bending and L_e is the effective length of the column that depends on the actual length and the boundary conditions (10.11). For a stiffened panel with arbitrary material configuration, the bending stiffness (EI) in (11.26) can be computed following the procedure described in Section 10.2 for thin-walled beams. Then, the critical strain at which Euler buckling occurs is computed simply as

$$\epsilon_x^{CR} = \frac{N_x^{CR}}{(EA)} \tag{11.28}$$

where (EA) is the axial stiffness (11.24) of the portion of the panel being analyzed.

If the panel is supported on all sides, it works like a plate. The overall buckling load of the panel can be estimated by the equations in Section 11.2 provided the bending stiffness coefficients are replaced by equivalent values for the panel. These equivalent plate stiffnesses are obtained by averaging the contribution of the stiffeners onto the plate. The averaging technique is described in [247], within the context of isotropic materials.

In addition to overall buckling (column or plate buckling), individual plate elements in the skin or the walls of the stiffeners may buckle as plates (long plates). The buckling loads of these plate elements can be computed using the equations in Section 11.2. Portions of the skin limited by longitudinal and transverse stiffeners can be analyzed as simply supported plates (Section 11.2.3) and the local buckling load of stiffener walls can be computed with (11.6) or (11.13) depending on the edge supports (see also Section 10.1.3).

The methodology consists of dividing the panel and the stiffeners into a collection of narrow flat plates and analyzing each plate individually. For example, the cross-section shown in Figure 10.3 has been divided into segments which represent flat plates to be analyzed separately. The buckling load N_x^{CR} of plates that are supported at both edges (e.g., segments 2, 3, and 4) is approximated with (11.6). Equation 11.13 is used for segments with a free edge such as segments 1 and 5. Finally, the portion of the skin between two stiffeners is supported on both edges; thus analyzed with (11.6). The portion of the skin bonded to the stiffener is taken as part of the stiffener.

Once the buckling load N_x^{CR} for each segment is known, the buckling strain for each segment is computed simply as

$$\epsilon_x^{CR} = \frac{N_x^{CR}}{tE_x} \tag{11.29}$$

where t and E_x are the thickness and laminate modulus of the element being considered. The minimum value of ϵ_x^{CR} obtained with (11.28) and (11.29) indicates incipient buckling. The corresponding load per unit width of panel is simply

$$N_x^{CR} = \frac{1}{l_s}(EA)\epsilon_x^{CR} \quad (11.30)$$

where (EA) is the axial stiffness of the portion of the panel of width l_s. Both, the stiffener and the skin may carry additional loads after buckling, in what is called *postbuckling.*

Postbuckling and Crippling

Laminated composite panels are often able to sustain loads higher than the buckling loads. The postbuckling range extends from the initial buckling load up to the load that causes failure of the material or collapse of the structure. During postbuckling, the deformations (mainly the transverse deflections) grow significantly and the buckling shape or mode grows in magnitude. Ultimately, the large strains induced cause material failure that leads to collapse. The collapse load is called *crippling* strength. Accurate prediction of crippling strength values requires numerical methods capable of following the postbuckling path and also a good model of material degradation [248]. Preliminary design accounting for crippling strength is usually based on empirical equations [237, 249].

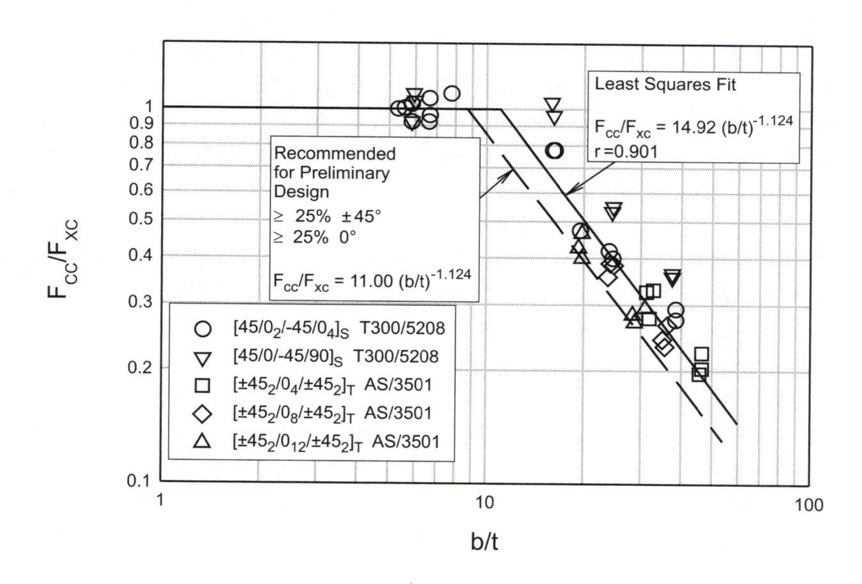

Figure 11.9: Normalized crippling data for the no-edge-free case [250].

Experimental crippling data for carbon–epoxy narrow plates simulating individual walls of a stiffener is shown in Figures 11.9 and 11.10 from [237]. For the no-edge-free case, the two unloaded edges were simply supported. For the one-edge-free case, one of the unloaded edges was free and the other simply supported. In

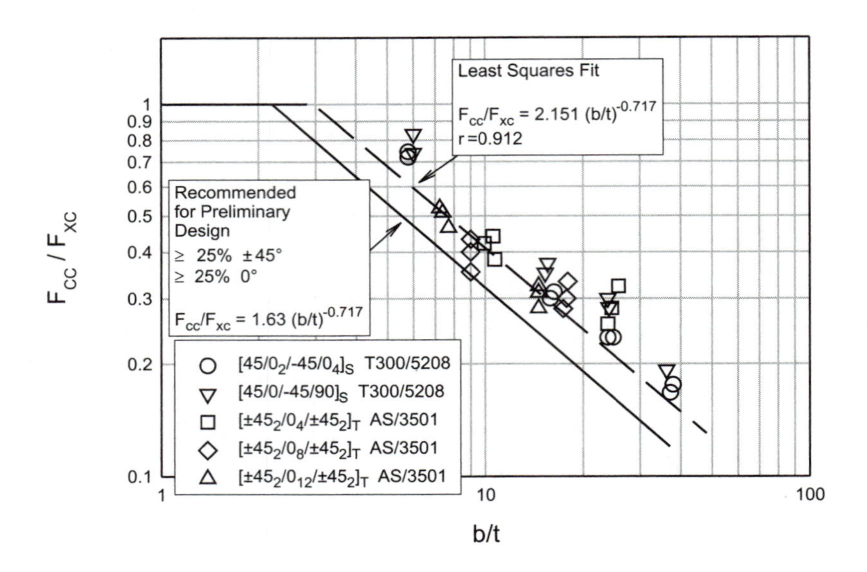

Figure 11.10: Normalized crippling data for the one-edge-free case [250].

both cases, the loaded edges were fixed. The data is normalized by the laminate compressive strength (F_{xc}) to produce crippling curves valid for different laminates. This normalization technique works for laminates with more than 25% of ± 45 laminae and more than 25% of 0-degree laminae, but does not work with all ± 45, all 0-degree, or all 0/90-degree laminates. A lower bound to the data, shown in the figures, is obtained by drawing a line parallel to the least square fit of the data in such a way that all data points lay above this line.

The crippling strength F_{cc} of each wall (or segment) of the stiffener is obtained from the graph, based on the width/thickness ratio (b/t) of the wall, its boundary conditions (one-edge-free or no-edge-free), and the compressive strength of the laminate (F_{xc}). Figures 11.9–11.10 suggest that, for the class of laminates considered, the geometric parameter b/t dominates the crippling behavior, in comparison to the laminate construction. This does not imply that laminate construction does not affect the crippling strength F_{cc}, but that the effect of laminate construction can be accounted for by the laminate compressive strength F_{xc}.

A similar empirical procedure is presented in [251, 252], where the crippling strain is computed as

$$\epsilon_{cc} = \alpha \left(\frac{\epsilon_{xc}}{\epsilon_x^{CR}} \right)^{\beta} \epsilon_x^{CR} \tag{11.31}$$

where α and β are empirical constants listed in Table 11.1, ϵ_{xc} is the strain to failure in compression of the plate element ($\epsilon_{xc} = F_{xc}/E_x$) and ϵ_x^{CR} is the buckling strain computed in Section 11.3.2

Table 11.1: Empirical constants for determination of crippling strain.

Edge Support	α	β
no-edge-free	0.56867	0.47567
one-edge-free	0.44980	0.72715

Once the crippling strain ϵ_{cc} or crippling strength F_{cc} of each wall is computed, it remains to compute the crippling load N_x^{CC} of the panel. According to [252], two cases must be considered. If the wall with lowest crippling strain is normal to the panel, such as a blade stiffener, collapse of the panel occurs. In this case, the load taken by the stiffener is

$$P_s = E_x^s A_s \epsilon_{cc} \tag{11.32}$$

If the wall with lowest crippling strain ϵ_{cc1} is parallel to the panel, such as the top of a hat stiffener, the crippled plate element will continue to take load until a second wall cripples with strain ϵ_{cc2}. Then, the load taken by the stiffener is

$$P_s = E_x^s b_1 t_1 (\epsilon_{cc1} - \epsilon_{cc2}) + \epsilon_{cc2} E_x^s A_s \tag{11.33}$$

where b_1, t_1, and ϵ_{cc1} are the width, thickness, and crippling strain of the wall that cripples first, and ϵ_{cc2} is the crippling strain of the second wall; E_x^s and A_s are the modulus and area of the stiffener, respectively.

If the skin is very thin, it may buckle long before the stiffeners buckle. In this case, the skin may be neglected altogether or a reduced "effective width" may be used to represent the reduced contribution of the skin. A well-known metal design formula was used in [252] to estimate the effective width of skin as

$$w_e = 1.9 t_{sk} (\epsilon_x)^{-1/2} \tag{11.34}$$

Then, the load taken by a portion of the skin of width l_s is

$$P_{sk} = 1.9 t_{sk}^2 E_x^{sk} (\epsilon_{cc})^{1/2} \tag{11.35}$$

where ϵ_{cc} is the crippling strain of the stiffener used in (11.32) or equal to ϵ_{cc2} of (11.33) depending on which situation controls.

It must be noted that most composite stiffened panels fail by separation of the skin from the stiffeners, unless the stiffeners are stitched, riveted, or otherwise properly attached to the skin. An empirical formula is presented in [252] to predict the failure load due to skin separation. The formula is based on the assumption that the skin will separate from the stiffener when the panel strain reaches the crippling strain of the skin. The skin crippling strain ϵ_{ss} was computed with (11.31) for the case of one-edge-free. Then, the failure load due to skin separation is

$$N_x^{ss} = \epsilon_{ss} \left(\frac{E_x^s A_s}{l_s} + E_x^{sk} t_{sk} \right) \tag{11.36}$$

The methodology outlined was used in [252] to predict the failure load of a number of stiffeners and stiffened panels for which experimental failure loads and failure strains (ϵ_F) are available. The ratio of experimental to analytical loads are depicted in Figure 11.11 for isolated stiffener crippling and in Figure 11.12 for compression panels, demonstrating the accuracy of this method for the examples considered. In both figures, ϵ_F is the experimental strain at failure and ϵ_{xc} is the intrinsic compressive strain to failure of the laminate (without buckling effects).

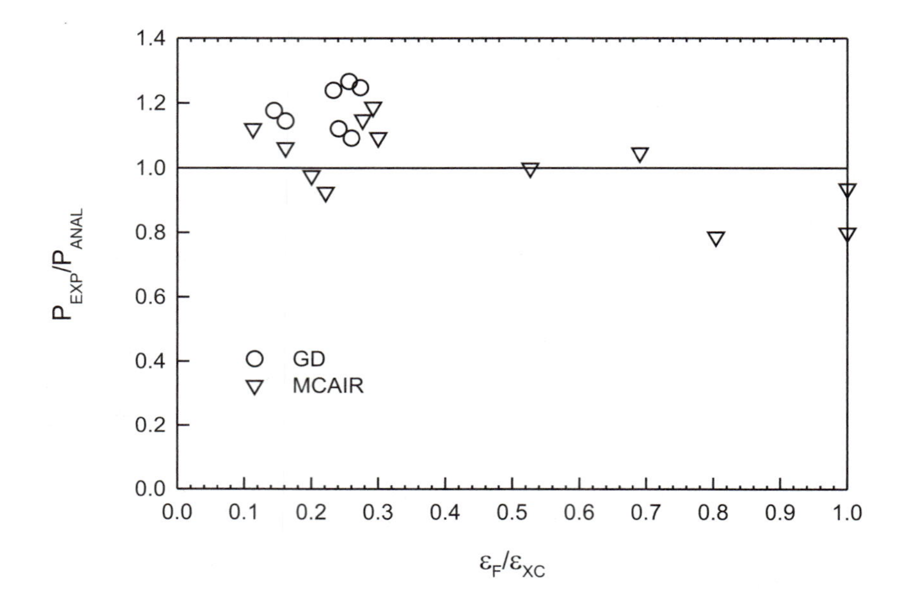

Figure 11.11: Comparison of analytical and experimental data for stiffener crippling [253].

Exercises

Exercise 11.1 *Select a material, laminate stacking sequence, and thickness, to carry a uniformly distributed load $p = 172$ kPa. The plate is 15 cm wide and 2 m long, simply supported around the boundary. You may base your design on the analysis of a plate strip 15 cm long and 1 cm wide, which can be analyzed as a beam.*

Exercise 11.2 *Compute the critical edge load N_x^{CR} for a plate simply supported around the boundary, with length $a = 2$ m (along x), width $b = 1$ m, $t = 1$ cm. The laminate is made of E-glass–polyester with $V_f = 0.5$ with a symmetric configuration defined by $\alpha = \beta = 0.2$, $\gamma = 0.6$. Compare your answer with Figure 11.5.*

Exercise 11.3 *Repeat computations of Exercise 11.2 for a plate with $a = 5$ m, $b = 1$ m. You may be able to use the simpler equation for long plates. Compare your answer with Figure 11.5.*

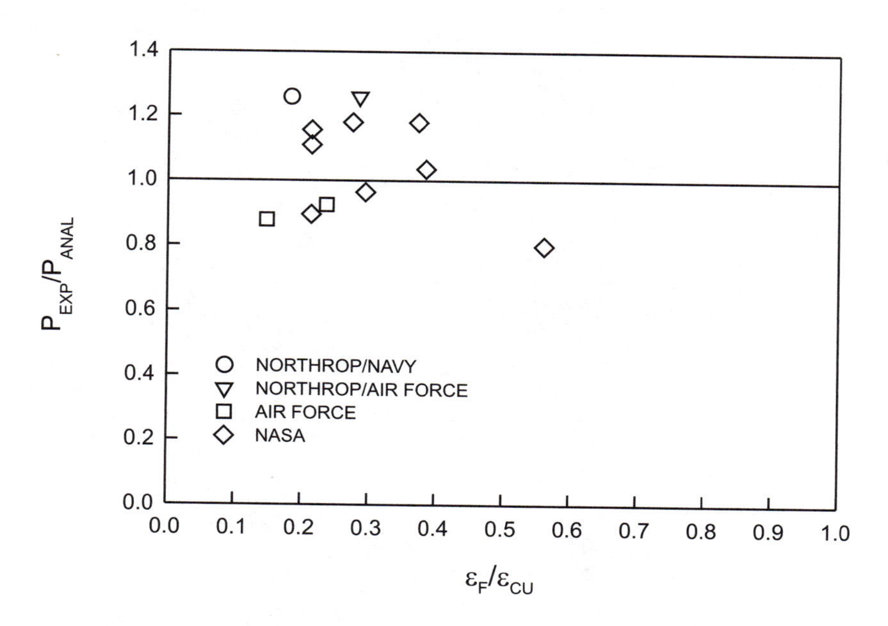

Figure 11.12: Comparison of analytical and experimental data for stiffened panels with edge load [253].

Exercise 11.4 *Repeat the computations of Exercise 11.3 for various values of γ between $\gamma = 0$ and $\gamma = 1$, keeping $\alpha = \beta = (1-\gamma)/2$. Plot and discuss the significance of your results. Interpret your findings in terms of the coefficients D_{ij}.*

Exercise 11.5 *Compute the critical edge load $N_x^{CR} = N_y^{CR}$ for the plate of Exercise 11.2 when it is loaded equally on all sides.*

Exercise 11.6 *Design a panel of length $a = 2$ m (along x), width $b = 1.5$ m, stiffened along the length with 5 blade stiffeners spaced at 25 cm. The panel is simply supported all around and it must carry a uniform load $p = 248$ kPa transverse to the panel. Design the panel (skin and stiffeners) with a load factor of 2.5 and a resistance factor 0.4 on the first ply failure strength. Use the carpet plots for E-glass–polyester. Assume a blade stiffener with a width-to-depth ratio $w/h = 1$ (see Figure 11.8, not to scale), and a depth-to-thickness ratio $h/t = 5$.*

Chapter 12

Shells

Shells are thin, curved structures such as pressure vessels, pipes, domes, caps, etc. While plates and beams resist transverse loads through bending, shells react to transverse loads mainly through membrane forces N_x, N_y, N_{xy}. Bending moments M_x, M_y, M_{xy} are usually quite small over most of the shell except around the points of application of concentrated loads and around the supports. Knowledge of the membrane forces is sufficient in most cases for preliminary design of shells, and often no other computations are required. If precise information about stresses near support or load points is required, a finite element analysis can be done at a later stage in the design process. However, the major decisions about materials, fiber architecture, and geometry are likely to be done at an early stage using membrane analysis.

The membrane forces can be obtained independently of the material properties, by using only the geometry and loads on the shell. Therefore, the membrane forces can be obtained from handbooks [176, Table 28] and textbooks [177, Ch. 10] that are not necessarily specialized in composites. Once the membrane forces N_x, N_y, N_{xy} are known, the design of the composite follows the guidelines presented earlier in this book. A small element of the shell is analyzed as a flat laminate to compute lamina stresses and strains and to evaluate a failure criterion. For the laminate analysis, the curvature of the shell can be ignored if the ratio of thickness to curvature t/r is less than 1/20. This is a good approximation in most cases because practical shells may have $r/t < 1/1000$.

In this chapter, membrane theory of shells is summarized and some useful results are presented. The interested reader is referred to [177] for additional details about shell theory. The results of membrane theory will then be used to illustrate the design of composite shells by developing specific examples.

Shells are more efficient than plates in carrying loads because the membrane stresses are uniform through the thickness, while the bending stresses in a plate only reach their maximum value at the surface (Figure 6.4). If the structure is designed so that the maximum stress is equal to the strength of the material, all of the thickness carries the maximum stress under membrane loading while only the surface carries the maximum stress under bending. Because of their high efficiency, shells tend to be very thin. When loaded in compression, shells are likely to fail because of buckling rather than material failure. Design procedures based on material failure as well as deformations is presented in this chapter. Buckling analysis of compression loaded shells can be done using finite element analysis.

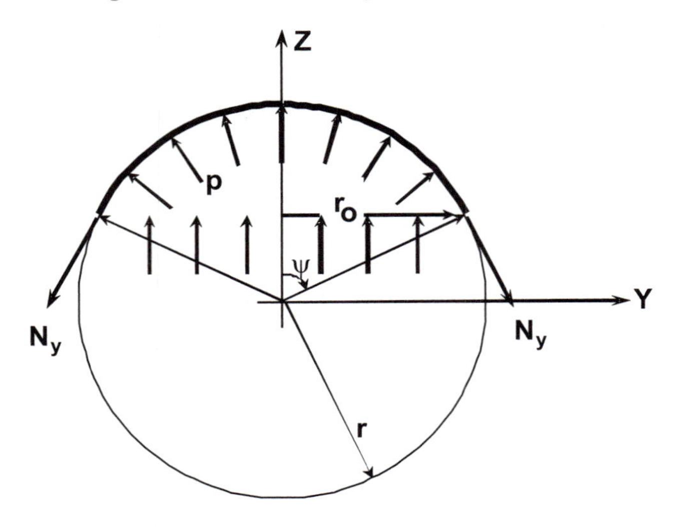

Figure 12.1: Membrane hoop force in a cylindrical pipe.

Example 12.1 *To illustrate the relative importance of the membrane stresses versus the bending stresses, consider a cylindrical pipe of radius r and thickness t, subjected to a uniform internal pressure p, with the axis of the cylinder along the x-axis. Compute the ratio of membrane to bending stress.*

Solution to Example 12.1 *As stated before, the equations of equilibrium are all that is needed to find the membrane forces and stresses. Summation of vertical forces (Figure 12.1) over a unit length of the shell along the x-axis, gives*

$$2N_y \sin\psi + 2r_0 p = 0$$

where the vertical component of the pressure acting on the curved surface has the same effect as a pressure acting on the projected area $2r_0$. Then, the hoop force per unit length is

$$N_y = p\frac{r_0}{\sin\psi} = pr$$

The hoop membrane force per unit length N_y is the same around the circumference; it depends only on the pressure and the radius, and it is independent of the material used for the pipe. If the material is homogeneous, the hoop membrane stress is

$$\sigma_y = \frac{N_y}{t} = \frac{pr}{t}$$

If the pipe has expansion joints that allow for free longitudinal expansion, the longitudinal stress is zero. In this case the hoop strain is

$$\epsilon = \frac{\sigma_y}{E} = \frac{pr}{Et}$$

Because of this strain, the perimeter stretches $2\pi r\epsilon$ *to a new perimeter and radius* $2\pi r' = 2\pi(r + r\epsilon)$. *Therefore, the new radius is* $r' = r(1+\epsilon)$. *The curvature of a shell is the inverse of the radius. The change in curvature is*

$$\chi = \frac{1}{r'} - \frac{1}{r} = \frac{1}{r}\left(\frac{1}{1+\epsilon} - 1\right) = -\frac{\epsilon}{r}\left(\frac{1}{1+\epsilon}\right)$$

$$\chi \approx -\frac{\epsilon}{r}(1 - \epsilon + \epsilon^2 - ...) \approx -\frac{\epsilon}{r}$$

$$\chi \approx -\frac{p}{Et}$$

The bending moment induced by the change in curvature is

$$M = D_{11}\chi = \frac{Et^3}{12(1-\nu^2)}\chi = -\frac{pt^2}{12(1-\nu^2)}$$

The maximum bending stress on the surface of the shell is

$$\sigma_b = \frac{6M}{t^2} = \pm\frac{p}{2(1-\nu^2)}$$

The ratio of membrane to bending stress is

$$\frac{\sigma}{\sigma_b} = \frac{r}{t}2(1-\nu^2)$$

For practical shells $r/t > 20$, *which indicates that the membrane stresses are at least 40 times larger than the bending stresses.*

12.1 Shells of Revolution

The geometry of a shell of revolution is generated by a line of arbitrary shape revolving around an axis of revolution (Figure 12.2). Examples include spheres, cylinders, cones, toroids, hemispherical and geodesic caps for cylindrical pressure vessels, elliptical shells, etc. Because of the relative simplicity of fabrication, shells of revolution are the most common type of shells in use. Most composite shells of revolution can be produced with filament winding (either polar or helical). Larger shells, such as domes, are designed as shells of revolution because of aesthetical reasons and construction convenience.

The shell is described by a surface located halfway through the thickness of the actual structure. The shell geometry is described by a meridian, which generates the shell surface by revolving around the shell axis. The intersection of the shell with any plane perpendicular to the shell axis is a circle, which is called a *parallel*. The position of a point on the shell is described by r_0, φ, and ψ (phi and psi). The shell has two radii of curvature, r_1 and r_2. Both are normal to the surface of the shell. Each parallel has a unique radius r_0, but r_0 is not normal to the shell. The radius r_2, which is normal to the shell, is constant along a parallel, and it is related to r_0 and φ by

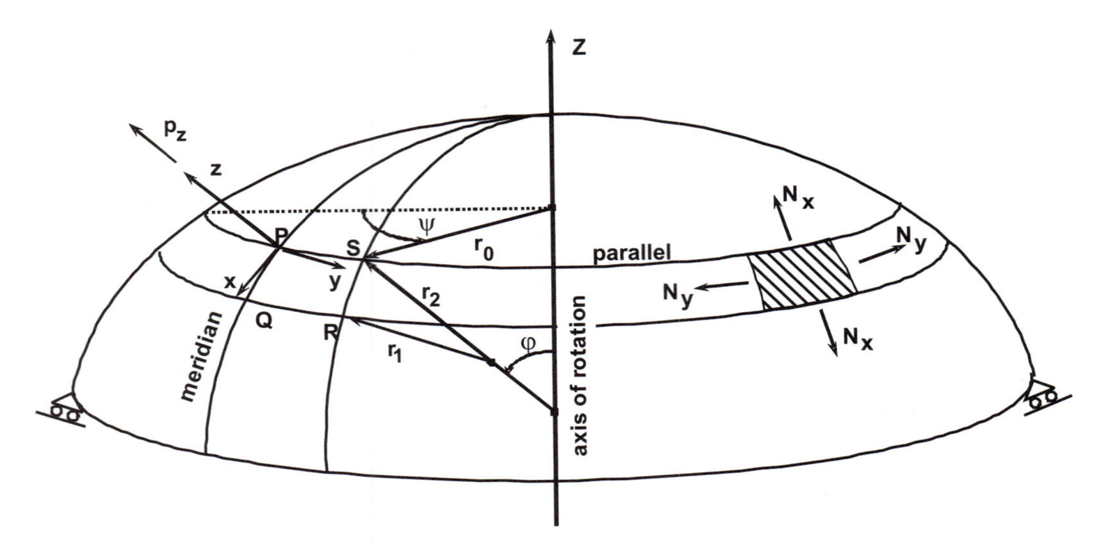

Figure 12.2: Geometry and coordinates for shells of revolution.

$$r_0 = r_2 \sin\varphi \tag{12.1}$$

The radius r_2 has its origin on the shell axis. The meridional radius r_1 has its origin at some point along r_2 and it is also normal to the shell surface. The shape of the meridian describes the entire shell. It is usually necessary to derive expressions for both radii of curvature $(r_1(\varphi), r_2(\varphi))$ as functions of φ to analyze the shell. The process is illustrated in the examples.

12.1.1 Symmetric Loading

If the load has also symmetry of revolution with respect to the shell axis, no shear forces develop, but only membrane forces N_x and N_y. A small element of the shell (PQRS in Figure 12.2) is used to illustrate the membrane stress resultants (forces per unit length) at any point on the shell. The meridional force (per unit length) N_x points along the meridian, and the hoop force N_y points along the parallel.

A local coordinate system (x, y, z), different from the global system (X, Y, Z), is used to describe the pressure distribution on the shell (Figure 12.2). The local x-axis points along the meridian, in the direction of positive angle φ, which is zero along the shell axis. The local y-axis points along the parallel, in the direction of positive angle ψ counterclockwise looking from the positive global Z-axis. Finally, the local z-axis is normal to the surface of the shell and positive in the outward direction $z = x \times y$. A pressure distribution on the shell has, in general, a normal component p_z and two shear components p_x and p_y (positive in the direction of the axes). When the material, geometry, and pressure distribution are all symmetric around the shell axis, the problem is said to be axial-symmetric, and the computation of $N_x(\varphi)$ and $N_y(\varphi)$ is quite simple. The meridional force $N_x(\varphi)$ and hoop force $N_y(\varphi)$ can be found simply by considering two equations of equilibrium. The derivation of these equations is presented in [177]. One of the two equations can be obtained from equilibrium of forces along the local z-coordinate, resulting in

$$\frac{N_x}{r_1} + \frac{N_y}{r_2} = p_z \tag{12.2}$$

The second equation can be obtained considering summation of forces acting on a portion of the shell limited by an angle φ (Figure 12.2)

$$N_x = \frac{F}{2\pi r_o \sin\varphi} \tag{12.3}$$

where F is the resultant of all forces applied to the portion of the shell above the parallel described by the angle φ (see Example 12.2). Because of symmetry, the resultant force F has only a component along the shell axis Z.

Equations 12.2 and 12.3 are used to determine $N_x(\varphi)$ and $N_y(\varphi)$ at all points on the shell. For membrane theory to be valid, the supports of the shell must provide only a reaction N_x, free of any bending constraint. Concentrated loads and constrained boundaries create bending stresses that are not accounted for in membrane theory.

Example 12.2 *Design a spherical shell to hold a volume $V = 1.0\ m^3$ of gas with an internal pressure $p = 1.0\ MPa$. The shell is to be constructed with E-glass–polyester composite with $V_f = 0.5$. A high density polyethylene liner is used on the inside to prevent gas diffusion through the composite shell. Assume that the liner does not contribute to the stiffness or strength of the shell (nonstructural liner). Use a load factor $\alpha = 1.25$ on the load and a resistance factor $\phi = 0.5$ on the FPF strength values.*

Solution to Example 12.2 *The volume of a sphere of radius a is $V = \frac{4}{3}\pi a^3$, from which the radius is*

$$a = \left(\frac{3}{4}\frac{V}{\pi}\right)^{1/3} = 0.62\ m$$

For a spherical shell, both radii of curvature are equal to the radius of the sphere

$$r_1 = r_2 = a$$

Then, (12.2) and (12.3), simplify to

$$N_x + N_y = p_z a$$
$$N_x = \frac{F}{2\pi a \sin^2\varphi}$$

When the sphere is subjected to an internal pressure, $p_z = p$, F can be obtained considering the pressure acting on one half of the sphere ($\varphi = \pi/2$). The radial pressure acting on the *one-half of the sphere produces the same resultant as the pressure acting on the projected area πa^2, so $F = \pi a^2 p$. Therefore,*

$$N_x = \frac{\pi a^2 p}{2\pi a \sin^2 \phi} = \frac{pa}{2} = 0.31(10^6)\ N/m$$
$$N_y = pa - \frac{pa}{2} = \frac{pa}{2} = 0.31(10^6)\ N/m$$

Since the membrane force is the same in any direction on a spherical shell ($N_x(\varphi) = N_y(\varphi) = N$), and the available carpet plots are restricted to $[0_\alpha/90_\beta/(\pm 45)_\gamma]_S$ combinations, it is reasonable to choose the quasi-isotropic laminate $[0/90/\pm 45]_S$, or $\alpha = \beta = 0.25, \gamma = 0.5$. From Figure 6.8, 6.9, 7.12, and 7.13, the properties are

$$E_x = E_y = E = 19,000\ MPa$$
$$\nu_{xy} = \nu = 0.34$$
$$FPF : F_{xt} = F_{yt} = 130\ MPa$$
$$LPF : F_{xt} = F_{yt} = 340\ MPa$$

The required thickness to avoid first ply failure is obtained from the condition

$$\phi\, F_{xt} > \frac{\alpha\, N}{t}$$

Then, the required thickness is

$$t_{FPF} = \frac{\alpha\, N}{\phi\, F_{xt}} = \frac{1.25 \times 0.31}{0.5 \times 130} \approx 6\, mm$$

The strain in any direction is given by (6.34)

$$\epsilon_x = \frac{N_x}{E_x\ t} - \nu_{xy} \frac{N_x}{E_x\ t}$$

which, simplified for quasi-isotropic materials becomes

$$\epsilon = \frac{N}{Et}(1-\nu) = \frac{pa}{2Et}(1-\nu) = \frac{0.31(1-0.34)}{19,000(0.006)} = 1,795(10^{-6})$$

The perimeter of the sphere (at a meridian) increases by $2\pi a\epsilon$ from its original value of $2\pi a$. Therefore, the new radius is $a + a\epsilon$, and the radial expansion of the sphere is

$$\delta = \frac{pa^2}{2Et}(1-\nu) = 1,795(10^{-6})\ 0.62 = 1.1\ mm$$

Example 12.3 *Design a circular cylindrical tank with closed ends subjected to an internal pressure $p = 1.0 MPa$. The dimensions are : length $L = 2.0$ m and diameter $2a = 1.0$ m. Use a load factor $\alpha = 1.15$ on the load and a resistance factor $\phi = 0.5$ on the FPF strength values.*

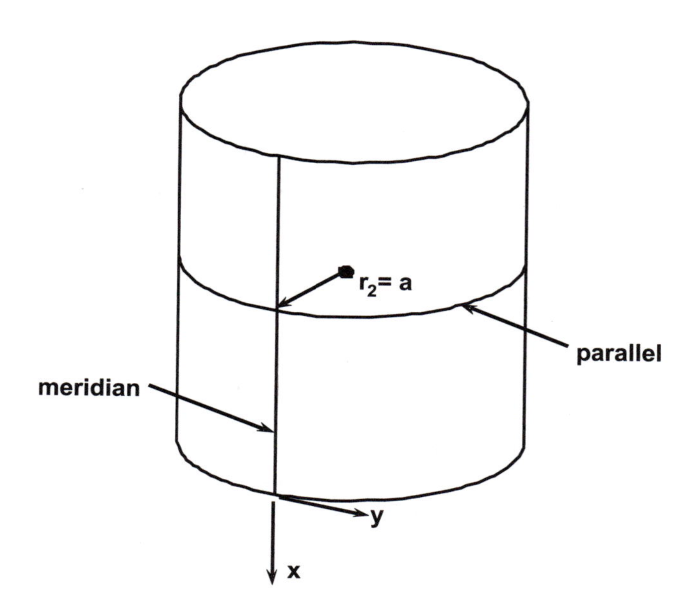

Figure 12.3: Circular cylindrical shell.

Solution to Example 12.3 *For a circular cylindrical shell, the hoop radius of curvature is $r_2 = a$ (Figure 12.3). The meridians, or generators of the shell surface are straight lines. Therefore, the meridional radius of curvature is $r_1 = \infty$. For this geometry, the hoop force is given by (12.2) with $p_z = p$, that is*

$$N_y = pa = 0.5(10^6)\ N/m$$

The longitudinal or meridional force N_x for the case of closed ends and internal pressure is, from (12.3)

$$N_x = \frac{\pi a^2 p}{2\pi a} = \frac{pa}{2} = 0.25(10^6)\ N/m$$

Since $N_y = 2N_x$, a laminate with $\alpha = 0.3$, $\beta = 0.6$, $\gamma = 0.1$ is selected (see Section 6.5 and 7.4). Assuming that the y-axis of the laminate coincides with the hoop direction, the laminate selected has twice as much fibers in the hoop direction than in the longitudinal direction. A 10% thickness of ±45 reinforcement is included to provide additional integrity to the structure. For this laminate, from Figure 6.8–6.10, 7.12–7.15, the properties are (values in MPa)[1]

	FPF	*LPF*
$E_x = 19,750$	$F_{xt} = 85$	$F_{xt} = 320$
$E_y = 27,500$	$F_{yt} = 170$	$F_{yt} = 580$
$\nu_{xy} = 0.15$		
$G_{xy} = 4,100$		

Then, the required thickness should satisfy the following conditions:

[1]The values of E_y and F_{yt} are read from the graph for E_x and F_{xt} interchanging the values of α and β.

$$\phi\, F_{yt} > \frac{\alpha\, N_y}{t} \rightarrow t = \frac{1.15 \times 0.5}{0.5 \times 170} \approx 6.8\ mm$$
$$\phi\, F_{xt} > \frac{\alpha\, N_x}{t} \rightarrow t = \frac{1.15 \times 0.25}{0.5 \times 85} \approx 6.8\ mm$$

resulting in $t = 6.8\ mm$. Note that the tensile strength values have been used because internal pressure produces tensile stress.

The design of the ends can be done also using membrane theory unless the ends are flat, in which case they work as plates, not shells. If the transition from the cylindrical portion to the tank-ends is not smooth, bending stresses will be generated. In any case, the final design can be reanalyzed using a finite element analysis program to evaluate any stress concentration that may arise at areas of transition or discontinuity. Useful recommendations regarding design of tanks and pressure vessels can be found in the specialized literature [180], [254].

To compute the radial expansion, first use (6.34) to compute the hoop strain

$$\epsilon_y = -\frac{\nu_{xy}}{E_x}\frac{N_x}{t} + \frac{1}{E_y}\frac{N_y}{t} = \frac{pa}{2E_y t}\left(2 - \nu_{xy}\frac{E_y}{E_x}\right)$$
$$= \frac{0.25\left(2 - 0.15\frac{27{,}500}{19{,}750}\right)}{27{,}500(0.0068)} = 2{,}395\, 10^{-6}$$

The elongation of the perimeter is $2\pi a \epsilon_\theta$ and the increase in radius is

$$\delta = a\epsilon_y = \frac{pa^2}{2E_y t}\left(2 - \nu_{xy}\frac{E_y}{E_x}\right) = 0.5(2{,}395)(10^{-6}) = 1.2\ mm$$

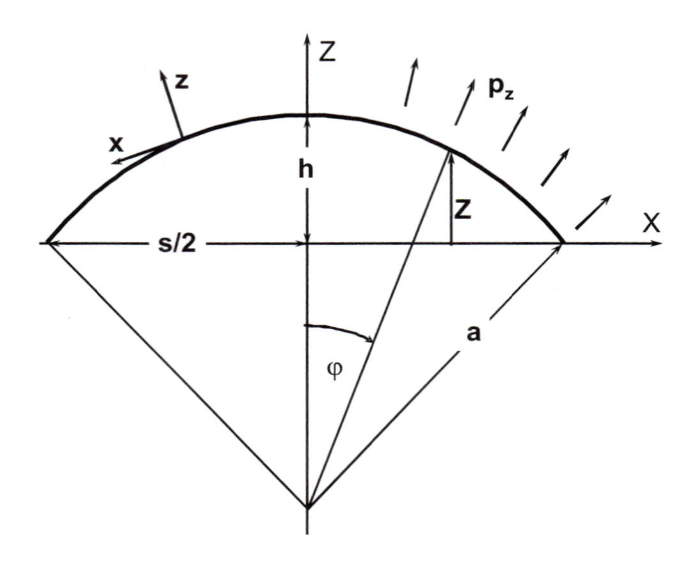

Figure 12.4: Spherical dome.

Example 12.4 ***Design of a dome.*** ***Part I.*** *Compute the stress resultants on the dome. The dome intended to cover communication equipment. A composite material is selected to avoid interference with the signals being transmitted. Housing antennas and other electronics under the roof saves on equipment costs because the electronics does not have to be built for outdoor exposure. The dome must withstand its own weight and the weight of fixtures hanging from it, snow load, and eventual live load of workers performing maintenance. The dome, in the shape of a spherical cap, has span $s = 24$ m and height $h = 2.4$ m (Figure 12.4).*

Solution to Example 12.4 *The radius of the sphere is given by*

$$a = \frac{h^2 + (s/2)^2}{2h} = 31.2 \ m$$

and the half-center angle spanned by the dome is

$$\varphi_{\max} = \arctan\left(\frac{s/2}{a-h}\right) = 22.62°$$

A dead load $p_D = 765 \ N/m^2$ (16 psf) is assumed, which can be corrected after the material and thickness are selected. The live load is chosen to be $p_L = 955 \ N/m^2$ (20 psf) [255]. Note that this load assumes that the weight of a worker is distributed over a large area. The snow load is assumed to be $p_S = 1915 \ N/m^2$ (40 psf). The dead load (self-weight) of the shell is equal to the surface area of the shell time the weight per unit area P_D. Considering a strip of the shell spanned by an angle $d\varphi$ (Figure 12.2) results in

$$\begin{aligned} \textit{radious of strip} &= r_0 = \sin\varphi \\ \textit{width of strip} &= a d\varphi \\ \textit{surface area of strip} &= (2\pi a \sin\varphi) a d\varphi \end{aligned}$$

The live and snow load per unit area $(p_L + p_S)$ act on the plan area, which is the projection of the surface of the shell onto an horizontal plane. In this case

$$\begin{aligned} \textit{projected width of the strip} &= a d\varphi \cos\varphi \\ \textit{plan area of the strip} &= (2\pi a \sin\varphi)(a d\varphi \cos\varphi) \end{aligned}$$

The total vertical load acting on a portion of the shell spanning an angle φ is, by integration

$$F = F_{L+S} + F_D = -\int_0^{\varphi} p_D 2\pi a^2 \sin\varphi' d\varphi' - \int_0^{\varphi} (p_L + p_S) 2\pi a^2 \sin\varphi' \cos\varphi' d\varphi'$$

where φ' is a dummy variable used for the integration. After integration

$$F = -2\pi a^2 \left[p_D(1 - \cos\varphi) + (p_L + p_S)\frac{1}{2}\sin^2\varphi \right]$$

Using (12.3) and taking into account that $r_0 = a \sin\varphi$, the meridional force (per unit length) is

$$N_x = \frac{F}{2\pi a \sin^2 \varphi}$$

which introduced in (12.2) with $r_1 = r_2 = a$ *yields the hoop force (per unit length)*

$$N_y = -(p_L + p_S + p_D) a \cos\varphi - N_x$$

where $\cos\varphi$ *has been used to obtain the pressure* p_Z *normal to the shell from the vertically applied pressure* $p_L + p_S + p_D = 3635\ N/m^2$.

Table 12.1: Loading conditions.

φ[o]	N_x [N/m]	N_y [N/m]
0.001	-56706.0	-56706.0
2.262	-56710.7	-56613.0
4.524	-56724.6	-56334.0
6.786	-56747.9	-55869.5
9.048	-56780.7	-55220.1
11.310	-56823.0	-54386.6
13.572	-56875.0	-53370.1
15.834	-56936.8	-52171.9
18.096	-57008.6	-50793.7
20.358	-57090.7	-49237.2
22.620	-57183.4	-47504.5

The resulting forces on the laminate are shown in Table 12.1. Note that no shear load is present because of symmetry of the geometry, material, and pressure distribution. The meridional force N_x *increases slightly with increasing values of* φ*. The hoop force* N_y *decreases with* φ*. If using noninteracting failure criteria, such as maximum stress, the stress at the edge of the dome control the design. But, if using interacting criteria, the stress at* $\varphi = 0$ *may control the design.*

Example 12.5 ***Design of a dome. Part II.*** *Using the results of Example 12.4, determine the thickness required to resist buckling. Assume the stiffness of the material has no variability and it is adequately estimated by using carpet plots. However, if the structure is imperfection sensitive, small imperfections may yield large variations in buckling load. For this reason, a relatively small resistance factor is specified,* $\phi = 0.5$*. The variability of the loads is accounted for by specifying a load factor* $\alpha = 1.6$*. A more refined partition of load factors that takes into account the diverse nature of the loads, i.e, dead, live, and snow loads, is presented in Chapter 13.*

Solution to Example 12.5 *As a first approximation, an empirical formula for buckling of isotropic spherical shells is used here. For a spherical cap of radius* $a = 31.2\,m$*, spanning a half-center angle* $\varphi = 22.62°$*, and subject to a uniform pressure* q *normal to the surface, the buckling pressure* q_{CR} *can be estimated as [176, Table 35.22]*

$$q_{CR} = 0.3E\left(\frac{t}{a}\right)^2[1-0.00875(\varphi_{\max}-20°)]\left(1-0.000175\frac{a}{t}\right)$$

where E is the modulus of the isotropic material and $\varphi_{\max} = 22.62°$ is the angle in degrees at the edge of the cap.

For a quasi-isotropic laminate, $\alpha = \beta = 0.25$, $\gamma = 0.5$, and selecting glass–polyester, the carpet plot of stiffness (Figure 6.8) yields $E_x = E_y = 19.0\ GPa$.

Since the dome is very shallow, the normal pressure q is close to the vertical pressure p_z. The later is known from Example 12.4, as $p_z = p_L + p_S + p_D$. Then, the factored vertical resistance ϕq has to exceed the factored applied pressure αp_z as follows

$$\phi \times q > \alpha \times (p_L + p_S + p_D) = 1.6 \times 3,635 = 5,816\,N/m^2$$

or

$$q = \frac{5,816}{0.5} = 11,632\,N/m^2$$

and solving for the thickness t, using MATLAB®,

```
t = fzero(@(x) 0.3*E*(x/a)^2*(1-.00875*(phimax-20))*(1-.000175*a/x)-q, 0.1);
```

yields $t = 47.9\,mm$.

Example 12.6 ***Design of a dome. Part III.*** *Check the design of Example 12.5 for strength. The variability of strength is accounted for by a resistance factor $\phi = 0.75$. The variability of the load is the same as in Example 12.5, i.e., $\alpha = 1.6$.*

Solution to Example 12.6 *The choice of fiber orientations for the laminate largely depends on the manufacturing process. A quasi-isotropic laminate was selected in Example 12.5, although it may not be the optimum for structural efficiency or ease of fabrication (see Exercises 12.5–12.8). For a quasi-isotropic laminate, $\alpha = \beta = 0.25$, $\gamma = 0.5$, and selecting glass–polyester, the carpet plot of LPF, Figures 7.13 and 7.14 yield $F_{xt} = F_{yt} = 340\,MPa$ and from [1], $F_{xc} = F_{yc} = 165\,MPa$. Recall that transverse properties can be estimated from the plots of longitudinal properties by interchanging the values of α and β.*

At the edge of the dome, $N_x = -57,090.7\,N/m$, $N_y = -47,504.5\,N/m$. Taking into account that the dome works in compression, and using (7.58) to check the design for strength yields

$$\phi\,F_x > \alpha\,N_x/t$$
$$0.75 \times 165 \times 10^6 > 1.6 \times 57,090.7/0.0479$$
$$123.75 \times 10^6 > 1.907 \times 10^6$$

and

$$\phi F_y > \alpha N_y/t$$
$$0.75 \times 165 \times 10^6 > 1.6 \times 47,504.5/0.0479$$
$$123.75 \times 10^6 > 1.59 \times 10^6$$

The shell is so much overdesigned for strength that we don't bother checking at $\varphi = 0$. A sandwich shell would be more efficient. See Example 12.7.

Example 12.7 ***Design of a dome. Part IV.*** *Design a sandwich shell to be more efficient than the design in Examples 12.4–12.6.*

Solution to Example 12.7 *First, find the thickness required to satisfy the strength requirement $\phi F_{xc} > \alpha N_x/t$, so*

$$t = \frac{1.6 \times 57,090}{0.5 \times 165 \times 10^6} = 1.11\, mm$$

Since a commonly available ply thickness is 0.25 mm, and a quasi-isotropic face requires eight plies, we select for the cade of the sandwich, $t_f = t/2 = 8 \times 0.25 = 2\, mm$.

Next, select a core material and core thickness. A polyurethane foam weighs only $51\, kg/m^3$, has a modulus of only $E = 20\, MPa$, $\nu = 0.23$, and $G = E/2/(1+\nu) = 8.13\, MPa$. The buckling requirement in Example 12.4 was $t = 2t_f = 47.9\, mm$. The mechanical moment of inertia (10.2.5) of the solid shell in Example 12.4 is

$$(EI)_c = \frac{E\, h^3}{12} = \frac{19 \times 10^9 \times \left(4.79 \times 10^{-3}\right)^3}{12} = 174 \times 10^3\, Nm$$

For the sandwich, there are two faces, each at a distance $t_c/2$ from the neutral surface, which gives a mechanical moment of inertia

$$(EI)_s = 2 \times E \times t_f \times (t_c/2)^2$$

Taking into account that the sandwich has to sustain α-times the buckling load, and the composite might only carry ϕ-times its share of the burden,

$$\phi\,(EI) = \alpha \times 2 \times E \times t_f \times (t_c/2)^2$$

from which the required thickness of the core is

$$t_c = 2\sqrt{\frac{(EI)_c}{2Et_f}} = 2\sqrt{\frac{1.6 \times 174 \times 10^3}{0.5 \times 2 \times 19 \times 10^9 \times 0.002}} = 0.171\, m \approx 175\, mm$$

So, the laminate has two faces, each t_f thick, separated by a core of thickness t_c, for a total thickness $t = t_c + 2t_f$. By modeling the sandwich in such a way, equation (6.36) can be used to estimate the laminate bending moduli of the sandwich laminate. Note that bending moduli are used because the buckling problem is controlled by bending effects. Assigning properties $E_x = 37,900\, MPa$, $E_y = 11,300\, MPa$, $G_{xy} = 3,300\, MPa$, $\nu_{xy} = 0.3$ to the *faces (from Table 1.3), the following values are obtained using laminate analysis software such as CADEC [1] or (6.36),*

$$E_x^b = E_y^b = 1,268\,MPa$$
$$G_{xy}^b = 473.4\,MPa$$
$$\nu_{xy}^b = 0.338\,MPa$$
$$r_M^b = 0.022$$

Using $E = 1,268\,MPa$, and $t = 179\,mm$, the empirical buckling formula in Example 12.5 yields $q_{CR} = 11,861\,N/m^2$ which exceeds the required $\alpha p_z/\phi = 11,632\,N/m^2$. Therefore, the sandwich shell resists both buckling and material failure.

While the empirical formula involves many approximations, it serves to illustrate the point. In-plane loads are carried by the thin composite faces and buckling can be addressed by using a core. For increased accuracy, buckling can be checked using a standard finite element program, entering the laminate moduli E_x^b, E_y^b, ν_{xy}^b, G_{xy}^b as the properties of an equivalent orthotropic material. The finite element model would also facilitate the analysis of concentrated loads and other local effects that must be considered in the final design. Since the forces have been considered independently without taking into account possible interaction between two or more forces acting at a given point, the design should be checked by the procedure explained in Chapter 7.

12.2 Cylindrical Shells with General Loading

The surface of a cylindrical shell is formed by a straight line, called the *generator,* moving parallel to its initial position and describing a curve in the shape of the shell cross-section, also called the *contour* (Figure 12.5). If the contour is a circle, the shell is a circular cylindrical shell, as depicted in Figure 12.3. The equations presented in this section are valid for any cross-section without negative curvature; that is the local radius of curvature r_2 should always points outwards (Figure 12.5).

For general loading, three membrane forces per unit length act at any point on the shell, as depicted in Figure 12.5. The three differential equations of equilibrium are derived in [177], namely

$$
\begin{aligned}
N_y &= p_z r \\
\frac{\partial N_{xy}}{\partial x} + \frac{1}{r}\frac{\partial N_y}{\partial \psi} &= -p_y \\
\frac{\partial N_x}{\partial x} + \frac{1}{r}\frac{\partial N_{xy}}{\partial \psi} &= -p_x
\end{aligned}
\qquad (12.4)
$$

where p_x, p_y, and p_z are the components of the surface pressure in the direction of the local coordinates x, y, and $z = x \times y$ (Figure 12.5). These equations can be solved one at a time. The first of (12.4) is used to obtain $N_y(x, \psi)$ as a function of the position on the shell. Once N_y is known, the second of (12.4) is solved for

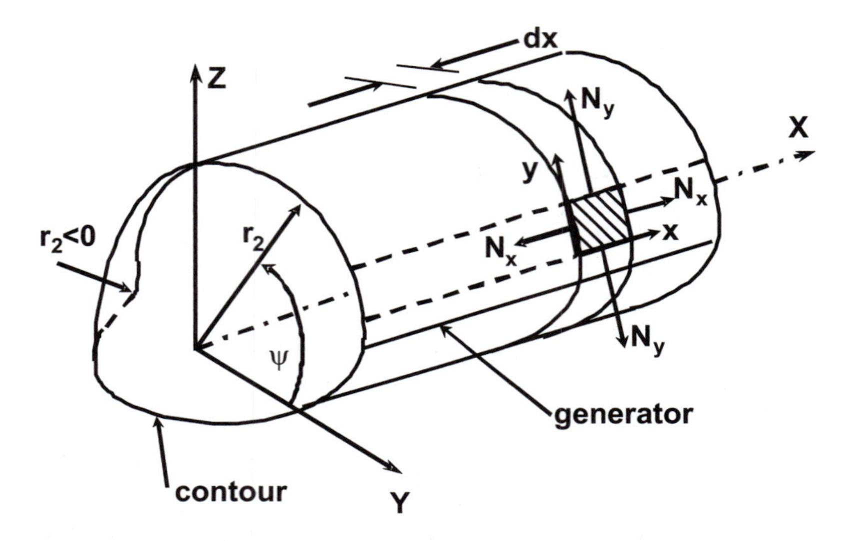

Figure 12.5: Cylindrical shell with general cross-section.

$N_{xy}(x, \psi)$ by integrating with respect to x. Once N_y and N_{xy} are known, N_x is obtained by integrating the third of (12.4). The complete solution is given in [177], namely

$$\begin{aligned} N_y &= p_z r \\ N_{xy} &= -\int \left(p_y + \frac{1}{r}\frac{\partial N_y}{\partial \psi} \right) dx + f_1(\psi) \\ N_x &= -\int \left(p_x + \frac{1}{r}\frac{\partial N_{xy}}{\partial \psi} \right) dx + f_2(\psi) \end{aligned} \tag{12.5}$$

where $f_1(\psi)$ and $f_2(\psi)$ are arbitrary functions of ψ to be determined using the boundary conditions or symmetry conditions.

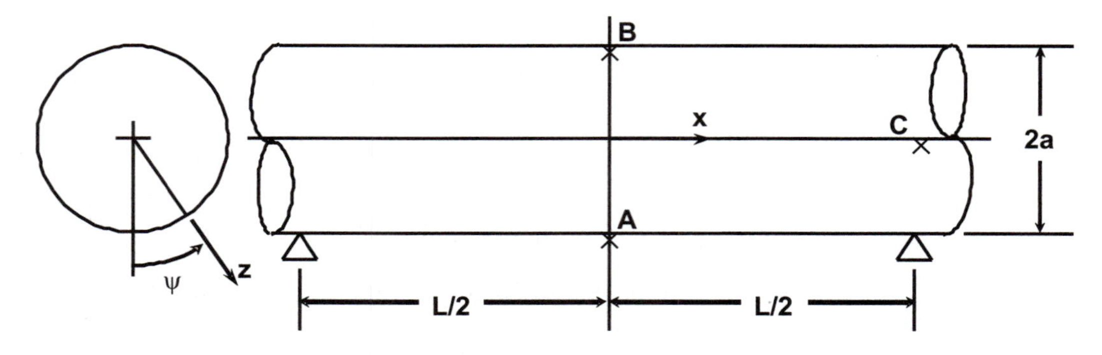

Figure 12.6: Simply supported pipe.

Example 12.8 *Design a composite cylindrical pipe of radius $a = 1.0$ m full of water. Assume the pipe is simply supported every 4.0 m and it is free to expand along the length (Figure 12.6). Use a load factor $\alpha = 2$ (to account for dynamic effects during filling) and a resistance factor $\phi = 0.25$ (to account for very long-term loading, i.e., creep rupture) on the FPF strength values.*

Solution to Example 12.8 *The hydrostatic pressure is equal to the specific weight γ times the depth measured from the top, and it acts radially on the wall of the pipe*

$$p_z = \rho\ g\ a(1 + \cos\psi) = \gamma\ a(1 + \cos\psi)$$

where ρ is the density, $g = 9.81\ m/s^2$ is the acceleration of gravity and $\gamma = \rho g$ is the specific weight, which for water is 9810 N/m^3. Using (12.5)

$$\begin{aligned}
N_y &= \gamma\ a^2(1 + \cos\psi) \\
N_{xy} &= \int \gamma\ a \sin\psi\ dx + f_1(\psi) = \gamma\ a\ x \sin\psi + f_1(\psi) \\
N_x &= -\int \gamma\ x \cos\psi\ dx - \frac{1}{a}\int \frac{df_1}{d\psi} dx + f_2(\psi) \\
&= -\frac{\gamma x^2}{2}\cos\psi - \frac{x}{a}\frac{df_1}{d\psi} + f_2(\psi)
\end{aligned}$$

Since the pipe is free to expand at the supports,[2] *$N_x = 0$ at $x = \pm\ L/2$, and using the equation for N_x at both ends yields*

$$\begin{aligned}
0 &= -\frac{\gamma L^2}{8}\cos\psi - \frac{L}{a}\frac{df_1}{d\psi} + f_2(\psi) \\
0 &= -\frac{\gamma L^2}{8}\cos\psi + \frac{L}{a}\frac{df_1}{d\psi} + f_2(\psi)
\end{aligned}$$

which solved for $f_1(\psi)$ and $f_2(\psi)$ yields

$$\begin{aligned}
f_2(\psi) &= \frac{\gamma L^2}{8}\cos\psi \\
\frac{df_1}{d\psi} &= 0
\end{aligned}$$

The last equation means that f_1 is constant. Replacing f_1 by a constant C in the solution for N_{xy} and evaluating at the midspan ($x = 0$) results in $N_{xy} = C$. The integral of N_{xy} around the contour is the torque $T = 2\pi a C$. Since there is no torque applied, the constant C must be zero. Then, the forces are

[2] Since the pipe is a continuous beam, the assumption of simple supports leads to a conservative design because the maximum bending moment is larger for a simply supported beam.

$$N_y = \gamma\ a^2 (1 + \cos\psi)$$
$$N_{xy} = \gamma\ a\ x \sin\psi$$
$$N_x = \frac{\gamma}{8}(L^2 - 4x^2)\cos\psi$$

The hoop force N_y is maximum at the bottom of the pipe ($\psi = 0$) and constant along the length. The axial force N_x is maximum at the midspan ($x = 0$), tensile at the bottom ($\psi = 0$) and compressive at the top ($\psi = \pi$).

The shear force is maximum at the supports ($x = L/2$) for $\psi = \pi/2$. Therefore, there are three critical points for the design, labeled A, B, and C in Figure 12.6.

At point A ($x = 0$, $\psi = 0$) the forces are: $N_y = 2\gamma a^2$, $N_x = \gamma L^2/8$, $N_{xy} = 0$.

At point B ($x = 0$, $\psi = \pi$) the forces are: $N_y = 0$, $N_x = -\gamma L^2/8$, $N_{xy} = 0$.

At point C ($x = L/2$, $\psi = \pi/2$) the forces are: $N_y = \gamma a^2$, $N_x = 0$, $N_{xy} = \gamma a L/2$.

The numerical values are given in the table below[3] (in N/m)

	A	B	C
N_x	19,620	-19,620	0
N_y	19,620	0	9,810
N_{xy}	0	0	19,620

Considering all the critical points, the laminate needs about the same strength in all directions. Therefore, a quasi-isotropic laminate is selected ($\alpha = \beta = 0.25$, $\gamma = 0.5$). The FPF strength values are obtained from carpet plots (available on the Web site [1]), taking into account that the hoop direction on the pipe corresponds to the y-direction on the material

$$F_{xt} = F_{yt} = 137.5\,MPa$$
$$F_{xc} = F_{yc} = 130\,MPa$$
$$F_{xy} = 95\,MPa$$

where $F_{xt} = F_{yt}$ and $F_{xc} = F_{yc}$ because $\alpha = \beta$. The required thickness is the maximum of those obtained using (7.58), using the largest compressive, tensile, and shear forces

$$t = \frac{\alpha\ N_x}{\phi\ F_{xc}} = \frac{2\ (-19,620)}{0.25\ (-130 \times 10^6)} = 1.20\ mm$$
$$t = \frac{\alpha\ N_y}{\phi\ F_{xt}} = \frac{2\ (19,620)}{0.25\ (137.5 \times 10^6)} = 1.14\ mm$$
$$t = \frac{\alpha\ N_{xy}}{\phi\ F_{xy}} = \frac{2\ (19,620)}{0.25\ (95 \times 10^6)} = 1.65\ mm$$

Therefore, the required thickness is $t = 1.65\,mm$ and shear controls the design. Since the forces have been considered independently without taking into account possible interaction between two or more forces acting at a given point, the design should be checked by the procedure explained in Chapter 7. See Example 12.9.

[3] Beam theory could be used to obtain values of N_x and N_{xy} but not of N_y, see Example 12.11.

Example 12.9 *Verify the design in Example 12.8 using laminate analysis software [1]. Use material properties for E-glass–isophthalic-polyester from Table 1.3 and ply thickness* $t_k = 0.25\,mm$*. Neglect any increase in strength due to in-situ effects.*

Solution to Example 12.9 *The laminate used in Example 12.8 is a* $[\pm 45/0/90]_S$*. Lamina properties are taken from Table 1.3. The loads given in Example 12.8 must be increased by the load factor* $\alpha = 2.0$*, given in given in Example 12.8. The strength values taken from Table 1.3 must be decreased by the load factor* $\alpha = 0.25$ *given in Example 12.8.*

Property	*Value [MPa]*	$\phi = 0.25$
E_1	*37900*	-
E_2	*11300*	-
G_{12}	*3300*	-
ν_{12}	*0.3*	-
F_{1t}	*903*	*225*
F_{2t}	*40*	*10*
F_{1c}	*357*	*89*
F_{2c}	*68*	*17*
F_6	*40*	*10*

The load must be increased by a load factor and the strength reduced by a resistance factor, so a minimum strength ratio $R > 1$ *is sought. Using CADEC [1] and the maximum stress criterion yields* $R = 0.971$ *at point A,* $R = 1.580$ *at point B, and* $R = 0.725$ *at point C (first ply failure). This design, with a laminate thickness* $t = 8 \times 0.25 = 2.0\,mm$ *does not meet the requirements. The design in Example 12.8 did work with* $t = 1.65\,mm$ *because the carpet plots account for in-situ effects while the analysis in this example did not (see Exercise 12.9). Since the shell is very thin,* ($a/t = 1000/2.0 = 500$)*, the design should be checked for buckling (see Example 12.10).*

Example 12.10 *Verify the design in Examples 12.8–12.9 for buckling. While the resistance factor for strength in Example 12.8 was prescribed at* $\phi = 0.25$*, the resistance factor for buckling is usually different. In this example, assume* $\phi = 0.75$ *as reduction factor for the computed buckling strength* M_{CR}*, to take into account uncertainties in the parameters entering in the computation of buckling strength.*

Solution to Example 12.10 *Since a quasi-isotropic material was selected in Examples 12.8–12.9, buckling load can be estimated using formulas for isotropic materials [176, Table 35.16]. For a thin-walled circular tube with bending moments* M *applied at the ends, the critical moment is*

$$M_{CR} = K \frac{E}{1-\nu^2} a t^2$$

where K *is a constant obtained from theory or experiments. Using* $K = 0.72$ *(minimum value from experiments on metal tubes) and the moduli from the carpet plots (Figure 6.8, 6.9)*

$$E = E_x = E_y = 19\ GPa$$
$$\nu = \nu_{xy} = 0.34$$

the critical moment is

$$M_{CR} = 0.72\frac{19(10^9)}{(1-0.34^2)}\left[1.45\left(10^{-3}\right)\right]^2 = 32,522\,Nm$$

The pipe is subject to a uniform distributed load due to the weight of the water

$$q = \frac{\gamma\pi a^2 L}{L} = 9810\ \pi = 30,819\,N/m$$

which for a simply supported beam gives a maximum moment

$$M = \frac{qL^2}{8} = 61,638\,Nm$$

Now, according to (), the resistance must exceed the load

$$\phi\ M_{CR} > \alpha\ M$$

However,

$$0.75 \times 32,522 < 2.0 \times 30,819$$
$$24,391.5 < 61,638\,Nm$$

Therefore, the shell cannot resist the load because of buckling. Since the critical buckling moment M_{CR} grows with the square of the thickness and only linearly with the modulus E, it is convenient to use more thickness and a material with a lower E. If cost is a controlling factor in the design, a lower fiber volume fraction and higher filler content in the resin would be an option since the cost of the filler is negligible. If weight is the controlling factor, a sandwich construction with foam or honeycomb core could be beneficial. Of course, a shell stiffened with longitudinal stiffeners and/or hoop ribs could be used, but the design of such a structure is more complex (see Section 11.3).

Example 12.11 *Using beam-bending equations, estimate the required thickness for a pipe of radius $a = 1.0\,m$ to be filled with water. The pipe is simply supported every $4.0\,m$ and it is free to expand. Use the same material of Example 12.8 on p. 423. Use a load factor $\alpha = 2$ (to account for dynamic effects during filling) and a resistance factor $\phi = 0.25$ (creep rupture) on the FPF strength values.*

Solution to Example 12.11 *The volume of water in the pipe is*

$$V = \pi a^2 L = 12.566\ m^3$$

Since the density of water is $\gamma = 1000\ kg/m^3 = 9810\ N/m^3$, the weight of water is

$$W = \gamma V = 123,272\ N$$

resulting in a distributed load of

$$q = W/L = 30,819\ N/m$$

The moment of inertia of the cross-section is

$$I = \pi a^3 t = \pi\ t$$

and the maximum stress is

$$\sigma = \frac{Ma}{I}$$

The maximum bending moment for a continuous-beam (simply supported every 4.0 m) is [179, p. 282]

$$M \cong \frac{qL^2}{10} = \frac{30,819(4^2)}{10} = 49,310\ Nm$$

Assuming that tensile strength controls the design, $F_{xt} = 130\,MPa$ *from Example 12.8, and*

$$\phi\,F_{xt} > \alpha\,\frac{M\ a}{I}$$

Therefore, the required moment of inertia is

$$I = \frac{\alpha\,M\ a}{\phi\,F_{xt}} = \frac{2.0 \times 49,310 \times 1}{0.25 \times 77 \times 10^6} = 0.003034\ m^4$$

and using the expression for the moment of inertia, the required thickness is

$$t = \frac{I}{\pi\,a^3} \approx 1.0\,mm$$

For a continuous-beam, the maximum shear force is $V \cong 0.6qL$. *This can be used along with (10.5) to check the shear strength of the pipe. The solution presented does not account for the hoop stress generated by the internal pressure (see Example 12.8). More importantly, buckling should be considered for such a thin shell.*

Exercises

Exercise 12.1 *Design a spherical pressure vessel with a radius* $a = 0.5\ m$ *and internal pressure* $p = 10\,MPa$ *with variability* $C_p = 0.15$ *for a LPF reliability of 99.5%. Use one or more layers of M1500 laminate with laminate strength and moduli given in Table 1.5, assuming the COV for strength is 0.35. Note that M1500 is a continuous strand mat fabric, thus quasi-isotropic.*

Exercise 12.2 *Optimize the design of Exercise 12.1 for minimum weight by selecting a laminate configuration among those available in Table 1.5. As in Exercise 12.1, one or more layers can be used. Note that only M1500, CM1808, and XM2408 have the same longitudinal and transverse properties (*$E_y = E_x$, $F_y = F_x$*). The transverse properties are not available in Table 1.5 for TVM3408 and UM1608. Specify the number of layers to be used based on the thickness per layer in Table 1.5.*

Exercise 12.3 *Optimize the cylindrical tank of Example 12.3 on p. 414 for minimum weight by changing the laminate configuration but not the material. Use the carpet plots E-glass–polyester. Note that approximate values of* E_y *and* F_y *can be obtained from the plots for* E_x *and* F_x *by interchanging the values of* α *and* β.

Exercise 12.4 *Redesign the cylindrical tank of Example 12.3 on p. 414 using one or more layers of M1500 laminate with properties given in Table 1.5. Since FPF strength is required in Example 12.3, but Table 1.5 lists only LPF, include an additional resistance factor* $\phi = 0.5$, *in addition to* ϕ *in Example 12.3. Specify the number of layers to be used.*

Exercise 12.5 *Optimize the design of Exercise 12.4 for minimum weight by using a CM1808 from Table 1.5 laminate. Since CM1808 has the same amount of* $0°$ *and* $90°$ *fibers, the transverse strength* F_y *is the same as the longitudinal strength* F_x. *Note that* $0°$*-orientation in Table 1.5 is the direction of the* $0°$*-layers (if any) in the laminate.*

Exercise 12.6 *Redesign the laminate in Examples 12.4-12.5 on p. 417 by using a sandwich shell. Determine the thickness of the core material and the faces. The core material has a density of* $50\ kg/m^3$, $E = 20\ MPa$, $\nu = 0.23$. *The faces are built with one or more layers of M1500 with properties given in Table 1.5. Specify the number of layers to be used.*

Exercise 12.7 *Redesign the dome in Example 12.4 on p. 417 for the case of added internal pressure* $p = 2\,kPa$. *Use the carpet plots for E-glass–polyester.*

Exercise 12.8 *Optimize the laminate in Exercise 12.6 for minimum weight. Consider alternative laminate configurations among those available in Table 1.5. Specify the number of layers and the thickness of the foam to be used.*

Exercise 12.9 *Recompute Example 12.9 on p. 425 taking into account in-situ effects and a lamina thickness* $t_k = 0.25\,mm$. *Take into account that the laminate thickness must be a multiple of the lamina thickness.*

Chapter 13

Strengthening of Reinforced Concrete

E. J. Barbero,[1] F. Greco,[2] and P. Lonetti[2]

Fiber-reinforced composite (FRP) materials are used to rehabilitate and retrofit steel-reinforced concrete (RC) members such as beams, columns, and so on. This technique is applied also for masonry, steel, and timber structures. A review of the behavior of RC constituent materials, i.e., concrete and steel, is reported first. The reader is assumed familiar with composite materials, which are described in detail in previous chapters. The FRP strengthening methodology presented here is consistent with American Concrete Institute (ACI) guidelines [256]. The main hypotheses and assumptions for the design of RC members are reported in accordance to ACI guidelines [257].

Reinforced concrete (RC) structural members, such as beams, columns, and slabs, are made of concrete reinforced by steel bars. The concrete mixture contains cement, water, and aggregates. Since concrete has high resistance in compression but low in tension, longitudinal reinforcing bars and stirrups are incorporated to increase the tensile strength of the RC member (Figure 13.1) [258, Chapters 2,3]. Note that in customary civil engineering terminology, the word *reinforced* is used to denote the steel reinforcement. Therefore, the word *strengthened* is used to denote the addition of fiber-reinforced composites (FRP).[3]

Strengthening of RC members by FRP is accomplished by external bonding to the concrete faces (Figure 13.2). Typically, the composite system consists of thin strips (less than 2 mm thick) of various widths. Carbon and glass fiber-reinforced epoxy laminates are the most common products utilized in practice. FRP is bonded

[1]West Virginia University, USA.

[2]Università della Calabria, Italy.

[3]In civil engineering construction, the acronym FRP (fiber-reinforced plastics) is used to denote *fiber-reinforced composites*, due to historical reasons.

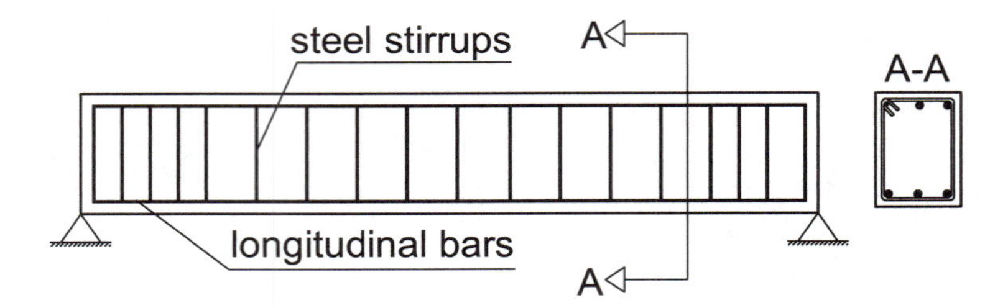

Figure 13.1: Reinforced concrete beam displaying the steel reinforcement (bars and stirrups).

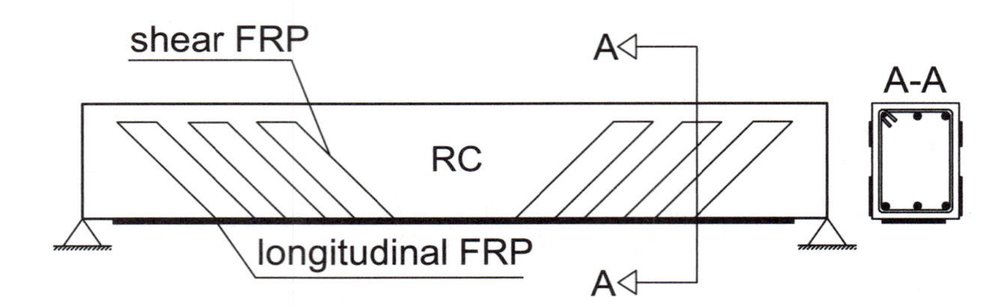

Figure 13.2: Strengthened concrete beam displaying the FRP used for longitudinal (flexural) and shear strengthening.

to concrete faces by means of compatible adhesives to guarantee shear and peel bonding between dissimilar materials [259, Chapter 2].

FRP strengthening is utilized to increase the load carrying capacity of RC members in several ways (Figure 13.3). The bending strength of beams is improved by bonding FRP plates to the tensile face of the section. The shear strength of beams is increased by the application of inclined or vertical FRP strips. Also, FRP strips are used in RC columns to reduce the lateral deformability of the section, improving the strength of the concrete material by confinement.

FRP systems used for strengthening of RC members are classified as follows:

1. Wet lay-up systems are based on unidirectional or multidirectional fiber sheets impregnated on-site with a resin.

2. Preimpregnated (prepreg) systems are based on unidirectional or multidirectional fiber sheets or fabrics impregnated with a resin. The composites is applied directly to the concrete face, and then cured on-site.

3. Precured systems are available in a wide variety of shapes and dimensions. Adhesive, primer, and putty are required for on-site application.

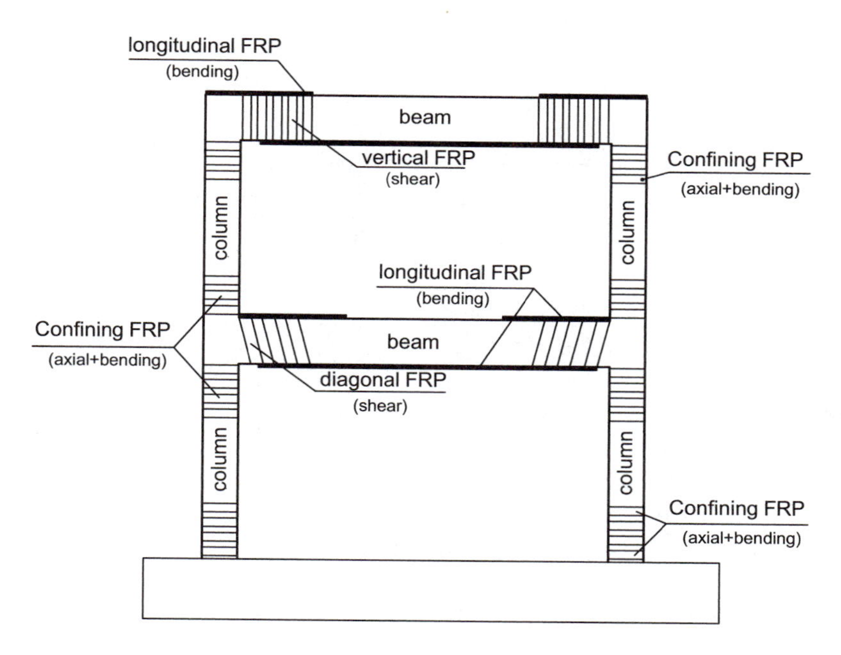

Figure 13.3: Strengthening possibilities using FRP on beams and columns.

13.1 Strengthening Design

The analysis and design of the FRP strengthening is performed in the framework of *limit states design* (1.4.3), which is a probabilistic design methodology used to ensure that the structure is able to resist load combinations reproducing the *ultimate limit states* and the *serviceability limit states*, defined as follows:

1. *Ultimate limit states*: the structure is subjected to the most severe loading combinations. The design should anticipate damage of the structural elements, avoiding any partial or complete collapse mechanisms of the structure.

2. *Serviceability limit states*: the structure is subjected to loading combinations reproducing the in-service conditions. The performance of structure is checked to avoid deformations, vibrations, and cracking that would impair the usability of the structure.

The probabilistic nature of external loads, material strengths and construction tolerances are considered. The *nominal capacity (strength)* F of the structure, or of parts of the structure, depends on the strength characteristics of the constituent materials, which are obtained by standardized experimental methods. To take into account uncertainties in the material properties and the construction tolerances of RC elements (beams, columns, slabs, etc.), a *strength reduction factor*[4] ϕ is

[4]See Section 1.4.3 for an introduction to reliability based design. The "strength reduction factor" of [257] is also called "resistance factor" in the literature.

introduced to penalize the *nominal capacity* of the RC member [257], down to a *factored capacity* ϕF (see Table 13.1).

Table 13.1: Strength reduction factor ϕ.

Structural element	ϕ
Beam or slab: Bending or flexure	0.65 - 0.9
Columns with ties	0.65
Columns with spirals	0.75
Beam shear and torsion	0.75

The *required capacity (strength)* U of the structure or of a part of the structure is expressed in terms of factored loads and moments, including axial, shear and bending stress resultants. Due to the variability of external loads, *load factors* (see Section 1.4.3) are introduced in the loading combinations. The following elementary loading conditions must be taken into account in the design procedure:

1. D, self-weigh and permanent loads of the structure.

2. L, live loads of accidental nature [257, p. 116], such as pedestrian, lateral fluid pressure (F_d), snow (S_w), rain (R_n), wind (W_d), earthquake (E_q), creep or shrinkage and temperature (T_p), soil loads (H_h), and roof live loads (L_r).

ACI guidelines recommend load factors for each loading combination taking into account the uncertainties of the load intensities. For example, dead loads present less uncertainty than live loads and thus a lower factor is utilized. Recommended values of the loading combinations for the ultimate limit states are given by

$$
\begin{aligned}
U &= 1.4\,(D + F_d) \\
U &= 1.2\,(D + F_d + T_p) + 1.6\,(L + H_h) + 0.5\,(L_r \text{ or } S_w \text{ or } R_n) \\
U &= 1.2D + 1.6\,(L_r \text{ or } S_w \text{ or } R_n) + (1.0L_r \text{ or } 0.8W_d) \\
U &= 1.2D + 1.6W_d + 1.0L + 0.5\,(L_r \text{ or } S_w \text{ or } R_n) \\
U &= 1.2D + 1.0E_q + 1.0L + 0.2S_w \\
U &= 0.9D + 1.6W_d + 1.6H_h \\
U &= 0.9D + 1.0E_q + 1.6H_h
\end{aligned}
\tag{13.1}
$$

All seven equations must be evaluated to find the required capacity U. Each structural element of the structure is characterized by an internal strength, known as factored capacity ϕF that must be larger than the required capacity U, as follows

$$
\phi F \geq U = \sum_i \alpha_i Q_i \tag{13.2}
$$

where α_i, Q_i, are partial load factors and applied load pairs reproducing the loading combinations presented in (13.1).

Next, the serviceability limit states must be checked. ACI guidelines provide limitations on stress, crack opening, deformations under serviceability conditions, and so on. The performance of the structure is evaluated with loading combinations that reproduce in-service loads. The loading combinations for serviceability limit states are similar to these for the ultimate limit states (13.1) but taking all the partial load factors equal to unity, i.e., $\alpha_i = 1$.

To avoid unexpected failures in which FRP should be damaged by fire, vandalism, or construction defects, ACI restricts the amount of FRP that can be used. Finally, the ratio between unstrengthened and strengthened capacity of the structure is limited to avoid excessive stress levels in the FRP reinforcement. ACI guidelines require that the unstrengthened system must be able to resist at least the following loading combination

$$\phi F \geq 1.1 S_{DL} + 0.75 S_{LL} \tag{13.3}$$

where S_{DL} and S_{LL} are the stress resultants that equilibrate dead (DL) and live (LL) loads, respectively.

In the analysis of beams and columns, stress resultants are zero- and first-order weighted integrals of stress over the section of the beam, where the weight factors are either 1 or z, where z is the coordinate on the plane of loading and normal to the axis of the member, measured with respect to the neutral axis (see Section 6.1.3). Stress resultants equilibrate external loads through axial load, shear load, and bending moments.

If the FRP were to be damaged, structural collapse mechanisms are prevented by limiting the *maximum allowable strengthening*. The maximum is obtained by combining required capacities for strengthened and unstrengthened sections given by (13.1) and (13.3) as

$$\%^{\max} = \frac{(S)^{str}_{req.} - (S)^{unstr}_{req.}}{(S)^{unstr}_{req.}} \times 100 \tag{13.4}$$

For example, if only dead loads D and live loads L are applied, the strengthening system is designed to resist a loading combination given by (13.1) as

$$U = 1.2\,(S_{DL}) + 1.6\,(S_{LL}) \tag{13.5}$$

and (13.4) yields

$$\%^{\max} = \frac{0.1 + 0.85 S_{LL}/S_{DL}}{1.1 + 0.75 S_{LL}/S_{DL}} \times 100 \tag{13.6}$$

13.2 Materials

RC structures strengthened by FRP are based on a combination of concrete, steel, and FRP. This section is devoted to describe the mechanical behavior of concrete and reinforcing steel [258], whereas the behavior of the FRP is described in previous chapters.

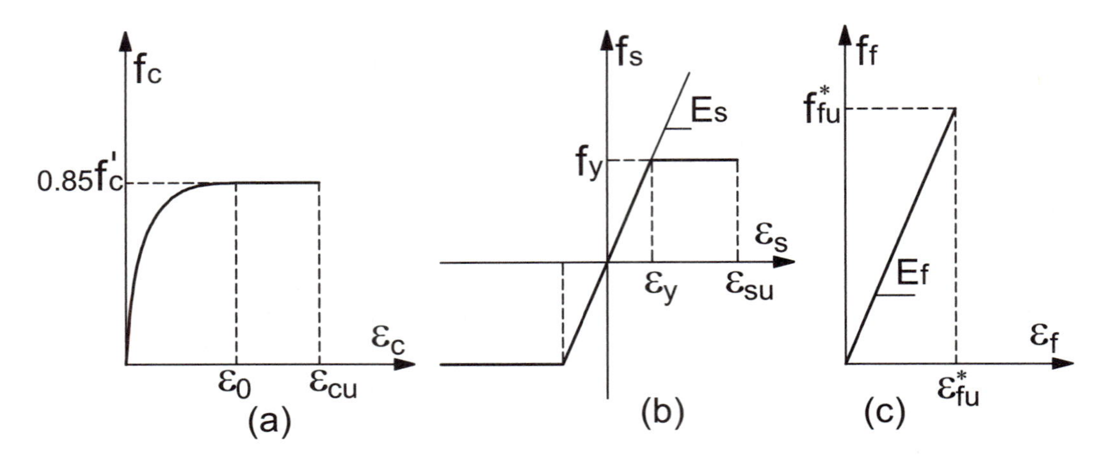

Figure 13.4: Design stress-strain curves for (a) concrete, (b) steel, and (c) FRP.

13.2.1 Concrete

The compressive strength of concrete f_c' is 10 to 20 times higher than its tensile strength. As a result, the tensile strength of concrete is considered negligible. The behavior is linear as long as the stress is less than 40% of the strength of concrete. Subsequently, nonlinearities appear at about 70% when concrete develops cracks perpendicular to the loading direction. Thereafter, complete failure is observed. The initial modulus of elasticity for normal-weight concrete is defined conventionally as the slope of the line connecting the origin and the point of the experimental curve at a stress level equal to $0.45f_c'$ [257]. ACI guidelines [257, Sect. 8.5] are often used to estimate the modulus of elasticity as follows

$$E_c = 4700\sqrt{f_c'}\,[MPa] \tag{13.7}$$

Moreover, ACI guidelines [257, Sect. 10.2.6] allow the designer to use a compressive stress-strain relationship based on rectangular, trapezoidal, or parabolic models in agreement with results of compression tests. Several models are proposed in the literature to reproduce the experimental data. For example, the following stress-strain relationship for concrete material is often used [260] (Figure 13.4a)

$$\begin{aligned} f_c &= 0.85f_c'\left[2\left(\varepsilon_c/\varepsilon_0\right) - \left(\varepsilon_c/\varepsilon_0\right)^2\right] \text{ for } \varepsilon_c \leq \varepsilon_0 \\ f_c &= 0.85f_c' \text{ for } \varepsilon_0 < \varepsilon_c \leq \varepsilon_{cu} \end{aligned} \tag{13.8}$$

where $\varepsilon_0 = 2(0.85f_c'/E_c)$ and $\varepsilon_{cu} = 0.003$.

13.2.2 Steel Reinforcement

The behavior of steel is assumed identical in tension and compression, and approximated by elasto-perfectly-plastic behavior (Figure 13.4b)

$$f_s = E_s \varepsilon_s \qquad \text{for } \varepsilon_s < \varepsilon_y$$
$$f_s = f_y \qquad \text{for } \varepsilon_s \geq \varepsilon_y \tag{13.9}$$

where E_s, ε_y, f_y, are the modulus (200 kN/mm^2), the strain, and the yield stress, respectively.

13.2.3 FRP

FRP is assumed linear elastic until failure (Figure 13.4c). The behavior is described by the FRP *tensile strength* f_{fu}^*, *ultimate strain* ε_{fu}^*, and *modulus* E_f (from product literature, see also [1]). These material properties are usually reported by the vendor of FRP. Also, they can be measured by standard tests [256] and/or predicted using the methodology presented in Chapter 4. ACI guidelines suggest a reduction factor C_E to account for the effect of environmental conditions on the installed FRP reinforcement (Table 13.2). The *allowable tensile strength* f_{fu} and *allowable strain* ε_{fu} of FRP are defined by

$$f_{fu} = C_E f_{fu}^*, \quad \varepsilon_{fu} = C_E \varepsilon_{fu}^* \tag{13.10}$$

in terms of the FRP tensile strength f_{fu}^* and ultimate strain ε_{fu}^*.

Table 13.2: FRP reduction factor C_E due to environmental effects.

Exposure conditions	**Fiber type**	**Environmental reduction factor**
Interior exposure	carbon	0.95
	glass	0.75
	aramid	0.85
Exterior exposure	carbon	0.85
(bridges, piers, and	glass	0.65
enclosed parking garages)	aramid	0.75
Aggressive environment	carbon	0.85
(chemical plants and	glass	0.50
wastewater treatment plants)	aramid	0.70

13.3 Flexural Strengthening of RC Beams

The *moment capacity* of a reinforced concrete beam can be increased by introducing FRP laminates on the tensile surface of the member. The composite behaves analogously to steel reinforcement, increasing the tensile stress resultant and thus the moment capacity of the beam.

13.3.1 Unstrengthened Behavior

The initial behavior of RC beams (without FRP) is practically linear. Flexural cracks are formed at the outermost surface in tension where the tensile strength of the concrete is reached and a reduction of bending stiffness is observed. After tensile concrete cracking, the tensile force is transferred from the concrete to the reinforcing steel bars resulting in a second linear portion. After the yield stress in the steel reinforcement is reached, the stiffness depends on the ability of the structure to redistribute the loads onto other sections that are not affected by concrete cracking. For instance, in the case of a simply supported beam, redistribution is quite limited because a collapse configuration is likely once the midspan reaches the moment capacity. When yielding of the steel reinforcement precedes concrete crushing, high dissipation of energy in plastic work is possible before collapse. This behavior is known as *ductile*, because the beam exhibits extensive crack patterns in the concrete and substantial deflections before collapse. When the RC beam has a large amount of steel, the beam fails by concrete crushing and a *brittle* failure mechanism is observed.

13.3.2 Strengthened Behavior

When the RC beam is strengthened by FRP, the initial stiffness is practically unchanged because only thin FRP strips are used, which do not significantly increase the stiffness of the structure. In the postcracking range, the concrete is cracked and the traction force is shared by the steel and the FRP reinforcement. Since the FRP is farthest from the neutral axis of the beam, its contribution is most relevant. Once the steel yields, only the FRP is able to carry additional load. Collapse occurs at higher loads but lower deflections than for the unstrengthened system. Consequently, even though the ultimate load is increased, the ductility of the beam is reduced.

RC strengthened beams are affected by different flexural damage modes, which are reported schematically in Figure 13.5. From a safety point of view, the most desirable damage mode is yielding of the steel with the FRP still attached before concrete compressive failure is reached. However, there are a number of other possible damage mechanisms that may interact, producing coupled damage modes [259]. For example, strengthened beams may develop *delamination* or *debonding* between the concrete and the FRP, which may advance in the FRP, in the concrete, or in the adhesive, causing catastrophic failure damage mechanisms [261–263]. In order to prevent these failure mechanisms, ACI guidelines limit the maximum strain level in the FRP, namely ε_{fd}, which is defined as a function of the number of the layers n and the total thickness t_f of FRP (in mm)

$$\varepsilon_{fd} = 0.41\sqrt{\frac{f'_c}{nE_f t_f}} \leq 0.9\varepsilon_{fu} \tag{13.11}$$

Therefore, the effective strain level at failure in the FRP is given by

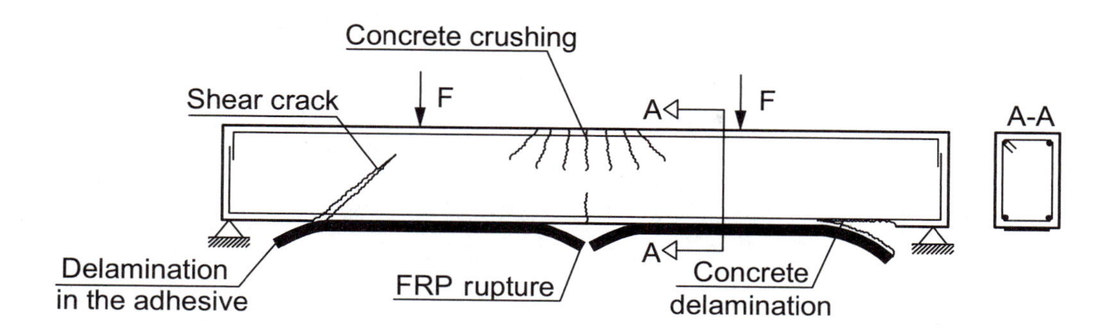

Figure 13.5: Failure modes of the RC strengthened beam.

$$\varepsilon_{fe} = \min\left(\varepsilon_{fd}, 0.9\varepsilon_{fu}\right) \tag{13.12}$$

where, due to the elastic behavior of the FRP, the stress in the FRP satisfies

$$f_{fe} \leq E_f \varepsilon_{fe} \tag{13.13}$$

13.3.3 Analysis

According to limit states design, the following inequality must be verified

$$\phi M_n \geq M_u \tag{13.14}$$

where M_n is the *nominal moment capacity* of the member (13.32), (13.47), or (13.57), M_u is the *required moment capacity* calculated by (13.1), and ϕ is the resistance factor defined by the following expression in terms of the tensile strain ε_s and yield strain ε_y in the steel reinforcement

$$\phi = \begin{cases} 0.9 \text{ for } \varepsilon_s \geq 0.005 \\ 0.65 + 0.25\left(\varepsilon_s - \varepsilon_y\right)/(0.005 - \varepsilon_y) \text{ for } \varepsilon_y \leq \varepsilon_s < 0.005 \\ 0.65 \text{ for } \varepsilon_s < \varepsilon_y \end{cases} \tag{13.15}$$

Note that the strength reduction factor ϕ penalizes mainly those failure modes for which the steel reinforcement is not yielded and the corresponding damage modes are dominated by concrete crushing. The brittle damage mechanisms with concrete crushing are called *compression-controlled.* When failure is reached in the steel reinforcement with strains greater than 0.005, ductile behavior of the section is expected and the behavior is said to be *tension-controlled.*

According to Figure 13.6, the top and bottom surfaces of the beam are subjected to compressive and tensile strains, respectively. The undeformed position on the section is known as the neutral axis. The prediction of the bending capacity of the section is based on the following hypotheses:

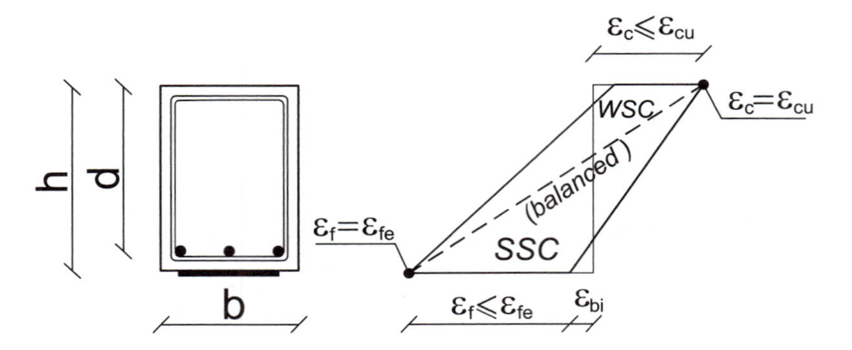

Figure 13.6: Strong strengthened configuration (SSC) and weak strengthened configuration (WSC).

1. Nonlinear constitutive relationships are used for concrete and steel reinforcement as described in Section 13.3.

2. FRP behavior is linear elastic until the effective stress f_{fe} and effective strain ε_{fe}, given by (13.11) and (13.13), are reached.

3. The section remains planar and perpendicular to the neutral axis and thus the strain distribution through the section is linear.

4. Concrete, steel reinforcement, and FRP are assumed perfectly bonded.

5. The tensile resistance of the concrete is negligible.

In order to satisfy (13.14) the *factored nominal moment capacity* ϕM_n needs to be calculated. The FRP is attached to the *original*, reinforced but unstrengthened structure, in which an *initial* strain distribution is present due to dead loads (self-weight). Therefore, the original configuration should be analyzed by calculating the stress-strain states at the soffit[5] of the beam. Calculation of the initial strain ε_{bi} is reported in the following sections, leading to (13.71).

In order to evaluate the moment capacity of the strengthened member, it is necessary to identify the failure mode of the section. By virtue of the limitations imposed on stress and strain in concrete, steel reinforcement, and FRP, one the following failure modes should be identified:

1. FRP failure on the tension side for $\varepsilon_f = \varepsilon_{fe}$, before or after steel reinforcement yield, i.e., $\varepsilon_s < \varepsilon_y$ or $\varepsilon_s \geq \varepsilon_y$, respectively. This failure mechanism is defined as *weak strengthening configuration* (WSC, Figure 13.6).

2. Concrete crushing on the compression side for $\varepsilon_c = \varepsilon_{cu} = 0.003$, before or after steel reinforcement yield, i.e., $\varepsilon_s < \varepsilon_y$ or $\varepsilon_s \geq \varepsilon_y$, respectively. This failure mechanism is known as *strong strengthening configuration* (SSC, Figure 13.6).

[5]The surface of the beam subjected to tensile stress where FRP is attached, usually the bottom or underside.

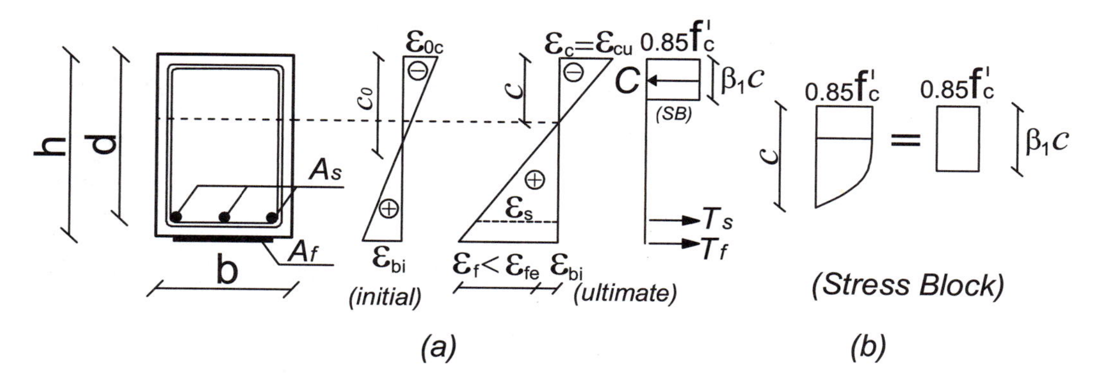

Figure 13.7: Strong strengthening configuration: (a) strain, stress and stress resultants (b) stress-block equivalence of the compressive stress resultant.

The moment capacity of the section is determined at the onset of the analysis by assuming a failure mode (e.g., WSC with steel yielded) and subsequently checking if the conditions for such failure mode are verified (i.e., $\varepsilon_s > \varepsilon_y$, $\varepsilon_c < 0.003$, $\varepsilon_f = \varepsilon_{fe}$).

If the assumed strain distribution does not match the calculated one, another failure mechanism should be considered. The failure modes associated to SSC and WSC are shown in Figure 13.6. The line between both domains, known as *FRP balanced configuration*, corresponds to the simultaneous failure of FRP and concrete. This failure mechanism is important as a reference for preliminary design purposes, because it corresponds to the full utilization of the section materials (i.e., the failure happens in the concrete and in the FRP simultaneously). The FRP-balanced configuration is defined as *inelastic* or *elastic* if the steel is yielded or not, respectively.

13.3.4 Strong Strengthening Configuration (SSC)

Strong strengthening configurations (SSC) refer to ultimate limit states for which the failure is concrete crushing. The concrete strain reaches $\varepsilon_{cu} = 0.003$ while the strain in the FRP is less than its effective strain, i.e., $\varepsilon_f < \varepsilon_{fe}$ (Figure 13.7). Since the strain distribution on the section is assumed to be linear, the following relationships among the strain on the top surface of concrete, in the FRP, and in the steel reinforcement apply

$$\varepsilon_c = \varepsilon_{cu} = 0.003 \qquad \text{(fiber compression)} \qquad (13.16)$$

$$\varepsilon_f = \varepsilon_{cu}\frac{h-c}{c} - \varepsilon_{bi} < \varepsilon_{fe} \qquad \text{(FRP)} \qquad (13.17)$$

$$\varepsilon_s = \varepsilon_{cu}\frac{d-c}{c} \qquad \text{(tensile steel reinforcement)} \qquad (13.18)$$

where ε_{bi} is the initial strain and c is the depth of the neutral axis measured from the top surface. The variable c is determined by balancing the compressive and tensile forces along the longitudinal direction as follows

$$C - T_s - T_f = 0 \tag{13.19}$$

with

$$C = \int_0^c \sigma_c(y)\, bdy \qquad \text{(concrete)} \tag{13.20}$$

$$T_f = E_f A_f \varepsilon_f = E_f A_f \left(\varepsilon_{cu}\frac{h-c}{c} - \varepsilon_{bi}\right) \qquad \text{(FRP)} \tag{13.21}$$

$$T_s = A_s f_y \text{ if } \varepsilon_s = \varepsilon_{cu}\frac{d-c}{c} \geq \varepsilon_y \qquad \text{(steel-yielded)} \tag{13.22}$$

$$T_s = A_s E_s \varepsilon_s \text{ if } \varepsilon_s = \varepsilon_{cu}\frac{d-c}{c} < \varepsilon_y \qquad \text{(steel-not yielded)} \tag{13.23}$$

In order to calculate the compressive force in the concrete, the integral (13.20) must be evaluated. For this purpose, the stress-strain relationship reported in (13.8) should be utilized. However, a simplified stress distribution, known in the literature as Whitney's stress-block, is frequently utilized in practice[6]. It assumes a constant distribution of stress $0.85\, f_c'$, spanning over the depth $\beta_1 c$, where β_1 is a parameter that relates the model (13.8) and stress-block distributions. The following expression provides an explicit evaluation of the compressive stress resultant

$$C = \int_0^c \sigma_c(y)\, bdy = 0.85 f_c' \beta_1 cb \tag{13.24}$$

with

$$\beta_1 = \begin{cases} 0.85 & \text{if} \quad f_c' \leq 27.57 \text{ MPa} \\ 0.85 - 0.05(f_c' - 27.57)/6.9 & \text{if} \quad 27.57 < f_c' \leq 55.15 \text{ MPa} \\ 0.65 & \text{if} \quad f_c' > 55.15 \text{ MPa} \end{cases} \tag{13.25}$$

Assuming that the tensile steel reinforcement is yielded, i.e., $\varepsilon_s \geq \varepsilon_y$, the position of the neutral axis is determined by substituting (13.24–13.25) and (13.21–13.22) into (13.19)

$$0.85 f_c' \beta_1 cb - E_f A_f \left(\varepsilon_{cu}\frac{h-c}{c} - \varepsilon_{bi}\right) - A_s f_y = 0 \tag{13.26}$$

which can be written as

[6]The stress-block analysis gives accurate results mainly when the compressive strain at the top surface is close to the strain to failure of concrete.

$$
\begin{aligned}
& A_1 c^2 + A_2 c + A_3 = 0 \\
& A_1 = 0.85 f_c' \beta_1 b \\
& A_2 = E_f A_f \left(\varepsilon_{cu} + \varepsilon_{bi}\right) - A_s f_y \\
& A_3 = -h E_f A_f \varepsilon_{cu}
\end{aligned} \tag{13.27}
$$

Solving (13.27) for c, the strains in the FRP and the steel reinforcement given by (13.21) and (13.22) must satisfy the following conditions

$$
\varepsilon_f < \varepsilon_{fe}, \quad \varepsilon_s \geq \varepsilon_y \tag{13.28}
$$

In the case of steel reinforcement not yielded, i.e., $\varepsilon_s < \varepsilon_y$, (13.23) holds and the position of the neutral axis is determined by solving the following equation

$$
0.85 f_c' \beta_1 bc - E_f A_f \left(\varepsilon_{cu} \frac{h-c}{c} - \varepsilon_{bi} \right) - A_s E_s \varepsilon_{cu} \frac{d-c}{c} = 0 \tag{13.29}
$$

Making some simplifications, (13.29) can be written as

$$
\begin{aligned}
& A_1 c^2 + A_2 c + A_3 = 0 \\
& A_1 = 0.85 f_c' \beta_1 b \\
& A_2 = E_f A_f \left(\varepsilon_{cu} + \varepsilon_{bi}\right) + A_s E_s \varepsilon_{cu} \\
& A_3 = -h E_f A_f \varepsilon_{cu} - d A_s E_s \varepsilon_{cu}
\end{aligned} \tag{13.30}
$$

Solving for c, the FRP and steel reinforcement strains given by (13.21) and (13.23) must satisfy the following conditions

$$
\varepsilon_f < \varepsilon_{fe}, \varepsilon_s < \varepsilon_y \tag{13.31}
$$

The nominal moment capacity M_n is obtained from moment equilibrium with respect to the point of application of the concrete stress resultant, as follows

$$
M_n = T_s \left(d - \frac{\beta_1 c}{2} \right) + \psi_f A_f E_f \epsilon_f \left(h - \frac{\beta_1 c}{2} \right) \tag{13.32}
$$

where ψ_f is a reduction factor, with a recommended value of 0.85 according to ACI guidelines, and T_s is given by (13.22) or (13.23) for yielded or not yielded steel reinforcement, respectively. Finally, the factored moment capacity is equal to ϕM_n, with ϕ given by (13.15).

13.3.5 Weak Strengthening Configuration (WSC)

For WSC, failure is reached in the FRP first. The strain distribution in the FRP is fixed to the effective value, i.e., $\varepsilon_f = \varepsilon_{fe}$, whereas the strain in the concrete is always below the ultimate, i.e., $\varepsilon_c < \varepsilon_{cu}$ (Figure 13.8). In view of the linear distribution of strain, the following relationships hold

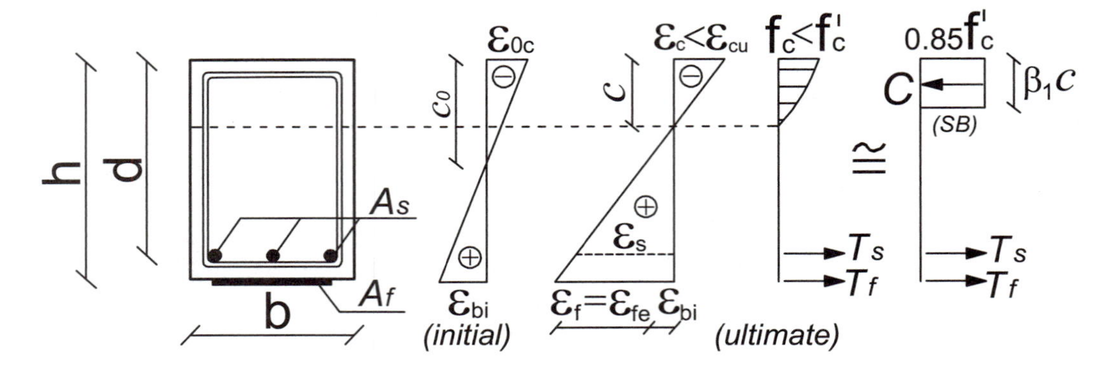

Figure 13.8: Weak strengthening configuration: strain, stress, and stress resultants in the strengthened system.

$$\varepsilon_c = (\varepsilon_{fe} + \varepsilon_{bi})\,\frac{c}{h-c} < \varepsilon_{cu} \qquad \text{(Compression fiber)} \qquad (13.33)$$

$$\varepsilon_f = \varepsilon_{fe} \qquad \text{(FRP)} \qquad (13.34)$$

$$\varepsilon_s = (\varepsilon_{fe} + \varepsilon_{bi})\,\frac{d-c}{h-c} \qquad \text{(Tensile steel)} \qquad (13.35)$$

where ε_{bi} is the initial strain. The depth c of the neutral axis is determined from the balance between compressive and tensile forces along the longitudinal direction

$$C - T_s - T_f = 0 \qquad (13.36)$$

with

$$C = \int_0^c \sigma_c\,(y)\,bdy \qquad \text{(Concrete)} \qquad (13.37)$$

$$T_f = A_f f_{fe} = A_f E_f \varepsilon_{fe} \qquad \text{(FRP)} \qquad (13.38)$$

$$T_s = A_s f_y \qquad \text{(steel-yielded)} \qquad (13.39)$$

$$T_s = A_s f_s = A_s E_s\,(\varepsilon_{fe} + \varepsilon_{bi})\,\frac{d-c}{h-c} \qquad \text{(steel-not yielded)} \qquad (13.40)$$

Since in this case the concrete strain is less than the ultimate, i.e., $\varepsilon_c < \varepsilon_{cu}$, Whitney's stress-block might not be accurate enough. However, the stress-block can be used to estimate the depth c of the neutral axis. In the case of steel reinforcement yielded, (13.39) is verified and the depth c of the neutral axis is determined by (13.36) as

$$0.85 f'_c \beta_1 bc - A_s f_y - A_f f_{fe} = 0 \qquad (13.41)$$

which gives

$$c = \frac{A_s f_y + A_f f_{fe}}{0.85 f'_c \beta_1 b} \tag{13.42}$$

For this value of c, the following conditions concerning the strains in steel and concrete must be checked

$$\varepsilon_c = (\varepsilon_{bi} + \varepsilon_{fe}) \frac{c}{h - c} < \varepsilon_{cu} \tag{13.43}$$

$$\varepsilon_s = (\varepsilon_{bi} + \varepsilon_{fe}) \frac{d - c}{h - c} \geq \varepsilon_y \tag{13.44}$$

In the case of steel reinforcement not yielded (13.40) holds and the depth c of the neutral axis is obtained by solving (13.36), rewritten as

$$0.85 f'_c \beta_1 bc - A_s E_s(\varepsilon_{fe} + \varepsilon_{bi}) \frac{d - c}{h - c} - A_f f_{fe} = 0 \tag{13.45}$$

Equation (13.45) can be rewritten as

$$\begin{aligned} & B_1 c^2 + B_2 c + B_3 = 0 \\ & B_1 = 0.85 f'_c \beta_1 b \\ & B_2 = -\left[E_s A_s \left(\varepsilon_{fe} + \varepsilon_{bi}\right) + E_f A_f \varepsilon_{fe} + 0.85 f'_c \beta_1 bh\right] \\ & B_3 = h E_f A_f \varepsilon_{fe} + d A_s E_s \left(\varepsilon_{fe} + \varepsilon_{bi}\right) \end{aligned} \tag{13.46}$$

Once the position of the neutral axis is determined by solving (13.46), the nominal moment capacity M_n is obtained from the moment equilibrium equation with respect to the point of application of the concrete stress resultant, as follows

$$M_n = T_s \left(d - \frac{\beta_1 c}{2}\right) + \psi_f A_f f_{fe} \left(h - \frac{\beta_1 c}{2}\right) \tag{13.47}$$

with T_s equal to (13.39) or (13.40) in the case of yielded or not yielded steel reinforcement respectively, and ψ_f is a reduction factor, with a recommended value of 0.85 from ACI guidelines. Finally, the factored moment capacity is equal to ϕM_n, with ϕ given by (13.15).

13.3.6 Balanced Strengthening Configuration (BSC)

The balanced configuration refers to a strain distribution for which simultaneous failure of concrete and FRP is predicted. This *limit case* is frequently analyzed for design purposes because it corresponds to the maximum utilization of the strength of both materials, but it is not used in practice because the WSC is safer, introducing ductility into the structure. Also BSC cannot be achieved in practice because the FRP area given by the vendors typically does not match exactly the value required for a balanced configuration. Since the strain distribution on the section is known, the position of the neutral axis c_b is evaluated as a function of concrete and FRP strain values as

$$c_b = \frac{\varepsilon_{cu} h}{\varepsilon_{cu} + (\varepsilon_{fe} + \varepsilon_{bi})} \tag{13.48}$$

where the strain in the steel reinforcement is determined as

$$\varepsilon_s = \frac{\varepsilon_{cu}}{c_b}(d - c_b) \tag{13.49}$$

Once the position of the neutral axis is known, equilibrium can be utilized to determine the FRP area as

$$C - T_s - T_f = 0 \tag{13.50}$$

where

$$C = 0.85 f'_c \beta_1 c_b b, \quad \text{(compressive, concrete)} \tag{13.51}$$

$$T_f = E_f A_{fb} \varepsilon_{fe}, \quad \text{(tensile, FRP)} \tag{13.52}$$

$$T_s = A_s f_y \text{ if } \varepsilon_s = \frac{\varepsilon_{cu}}{c_b}(d - c_b) \geq \varepsilon_y \quad \text{(tensile after steel yields)} \tag{13.53}$$

$$T_s = A_s E_s \varepsilon_s \text{ if } \varepsilon_s = \frac{\varepsilon_{cu}}{c_b}(d - c_b) < \varepsilon_y \quad \text{(tensile before steel yields)} \tag{13.54}$$

Substituting, (13.51–13.53) or (13.51–13.52), (13.54) for the case of steel reinforcement yielded or not-yielded, respectively, an estimate of the FRP area is obtained as

$$A_{fb} = \frac{1}{E_f \varepsilon_{fe}} \left(0.85 f'_c \beta_1 b c_b - A_s f_y\right) \quad \text{(steel yielded)} \tag{13.55}$$

$$A_{fb} = \frac{1}{E_f \varepsilon_{fe}} \left(0.85 f'_c \beta_1 b c_b - A_s E_s \frac{\varepsilon_{cu}}{c_b}(d - c_b)\right) \quad \text{(not yielded)} \tag{13.56}$$

Finally, the nominal moment capacity is obtained by moment equilibrium with respect to the point of application of the concrete stress resultant, as follows

$$M_{nb} = A_s f_s \left(d - \frac{\beta_1 c_b}{2}\right) + \psi_f A_{fb} f_{fe} \left(h - \frac{\beta_1 c_b}{2}\right) \tag{13.57}$$

where the factored moment capacity is equal to ϕM_{nb}, with ϕ given by (13.15). The area of FRP A_{fb} predicted by (13.55–13.56) needs to be rounded upward to the nearest available section available in the market. For $A_f < A_{fb}$, the member is in the range of WSC failure mode and failure occurs in the FRP. The distance to the neutral axis is less than for the balanced configuration, i.e., $c < c_b$; thus, the moment capacity is less than for the balanced configuration. For $A_f > A_{fb}$, the member is in the range of SSC failure mode and failure occurs in the concrete. The distance to the neutral axis is larger than for the balanced configuration, i.e., $c > c_b$, thus a larger value of the moment capacity is expected.

FRP reinforcement works well for beams with small amount of the steel reinforcement (underreinforced sections), in which the failure in the unstrengthened beam involves yielding of the steel bars. In this case, the FRP can be considered an addition to the steel reinforcement; the capacity of the member is improved as soon as the FRP is introduced. On the contrary, for beams reinforced with too much steel (overreinforced sections), the addition of FRP does not increase the moment capacity of the member, because the introduction of FRP does not influence significantly the failure mechanism of the section, which is based on the failure of the concrete.

13.3.7 Serviceability Limit States

The serviceability limit states check the performance of the structure under the service loading conditions to seek durability, efficiency, and appropriate working conditions. In the framework of FRP strengthening of RC beams, specific guidelines are provided to control the maximum allowable stresses in the FRP, the concrete, and the reinforcement steel bars.

First, the stress levels in the steel bars $f_{s,s}$ is limited to be less than 80% of the yield stress

$$f_{s,s} \leq 0.8 f_y \tag{13.58}$$

Second, the stress level in the concrete $f_{c,s}$ is limited to be less than 45% of the compressive strength

$$f_{c,s} \leq 0.45 f'_c \tag{13.59}$$

These restrictions are introduced to avoid excessive deformations under in-service loads and undesirable deflections that can limit the use and the efficiency of the structure. Similar limitations on the stresses are introduced to prevent creep rupture and fatigue of the FRP system, as follows

$$f_{f,s} \leq \begin{array}{l} 0.20 f_{fu} \text{ (glass FRP)} \\ 0.30 f_{fu} \text{ (aramid FRP)} \\ 0.55 f_{fu} \text{ (carbon FRP)} \end{array} \tag{13.60}$$

The following calculation can be applied for each loading stage associated to the unstrengthened and strengthened configurations. Typically, under service loads, it is assumed that the structure behaves as linear elastic, which is a good approximation if the actual load does not damage the structure. Therefore, instead of the nonlinear stress-strain relationship for concrete (13.8) and steel reinforcement (13.9), linear behavior should be used for both materials. The following assumptions are used for the analysis of the member under serviceability limit states:

1. The behavior of concrete, steel, and FRP are linear elastic.

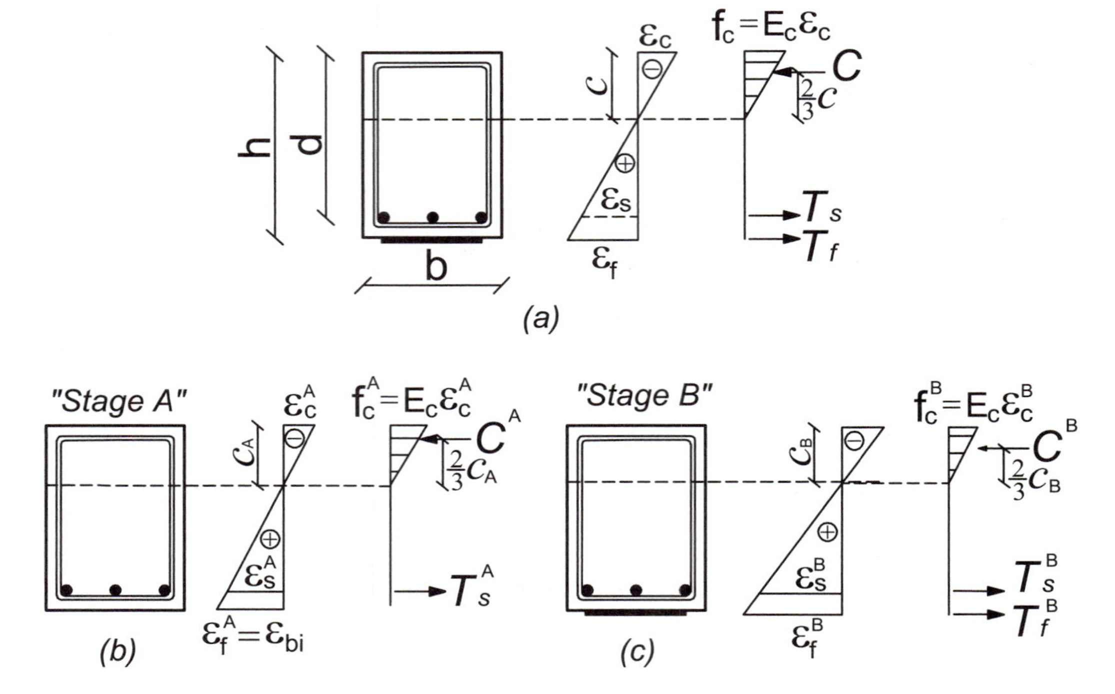

Figure 13.9: Serviceability limit state: (a) strain and stress distribution, (b) Stage A (original structure), (c) FRP strengthened structure.

2. The section remains planar and perpendicular to the deformed neutral axis.
3. The concrete, steel, and FRP are perfectly bonded.
4. The tensile resistance of the concrete is negligible.

First, the general case in which FRP, steel reinforcement and concrete are present is discussed. The stress distribution on the section requires the evaluation of the neutral axis position (Figure 13.9a). This is achieved by means of the equilibrium equation along the length of the beam

$$C - T_s - T_f = 0 \tag{13.61}$$

where

$$C = E_c \varepsilon_c cb/2; \quad T_s = E_s \varepsilon_s A_s; \quad T_f = E_f \varepsilon_f A_f \tag{13.62}$$

Due to linearity of the strain in the section, the following relationships hold

$$\varepsilon_f = \frac{(h-c)}{c}\varepsilon_c; \quad \varepsilon_s = \frac{(d-c)}{c}\varepsilon_c. \tag{13.63}$$

Substituting (13.63) into (13.62) and then into (13.61), the equilibrium equation is expressed as

$$E_c bc^2/2 - E_s A_s (d-c) - E_f A_f (h-c) = 0 \tag{13.64}$$

Equation (13.64) can be rewritten so that the position of the neutral axis (positive root) is obtained by solving the following quadratic equation

$$\begin{aligned} &C_1 c^2 + C_2 c + C_3 = 0 \\ &C_1 = E_c b/2 \\ &C_2 = E_s A_s + E_f A_f \\ &C_3 = -E_s A_s d - E_f A_f h \end{aligned} \tag{13.65}$$

The stress distribution in the steel reinforcement, in the FRP, and in the farthermost surface of the concrete are obtained as

$$\begin{aligned} &f_c = E_c M\ c/(EI)_{cr} \\ &f_s = E_s M\,(d-c)/(EI)_{cr} \\ &f_f = E_f M\,(h-c)/(EI)_{cr} \end{aligned} \tag{13.66}$$

where M is the required moment capacity corresponding to the serviceability loading combinations and $(EI)_{cr}$ is the bending stiffness of the cracked section given by the following formula

$$(EI)_{cr} = E_c bc^3/3 + E_s A_s (d-c)^2 + E_f A_f (h-c)^2 \tag{13.67}$$

The service stress in the steel reinforcing bar is calculated by taking into account two different stages. The first stage (stage A) refers to the original configuration of the structure before the FRP is applied. Only dead loads and a reduced amount of live loads are applied at this stage. Also, limitations on the stress given by (13.58–13.59) apply. Once the FRP is installed, the member works as a combined strengthened system (stage B); thus, the restriction given by (13.60) should be considered, taking into account the loading combinations given by the serviceability limit states, i.e., (13.1) not factored. Serviceability solutions for stage A and stage B are derived using (13.61–13.66) for $A_f = 0$ and $A_f \neq 0$, respectively.

Stage A: Initial configuration

In stage A (Figure 13.9b), the initial position of the neutral axis c_A is determined by solving (13.65) specialized for $A_f = 0$, as follows

$$\begin{aligned} &C_1 c_A^2 + C_2 c_A + C_3 = 0 \\ &C_1 = E_c b/2 \\ &C_2 = E_s A_s \\ &C_3 = -E_s A_s d \end{aligned} \tag{13.68}$$

The stresses in the concrete and the steel are given by substituting (13.67) with $A_f = 0$ into Eq (13.66) as follows

$$f_c^A = E_c \frac{M_A c_A}{E_c b c_A^3/3 + E_s A_s (d - c_A)^2} \tag{13.69}$$

$$f_s^A = E_s \frac{M_A (d - c_A)}{E_c b c_A^3/3 + E_s A_s (d - c_A)^2} \tag{13.70}$$

where M_A represents the external moment corresponding to stage A, which is frequently taken as a function of dead loads only, i.e., $M_A = M_{DL}$. It is also possible to calculate the initial strain ε_{bi}, (typically due to dead load only), at the surface where the FRP will be placed, as

$$\varepsilon_{bi} = \frac{M_{DL} (h - c_A)}{(EI)_{cr}} \tag{13.71}$$

Stage B: Strengthened configuration

Once the beam is strengthened, the stress distribution is calculated with reference to the loading combinations of the serviceability limit states, which reproduce the in-service loads. In stage B, the member is strengthened; thus, the neutral axis c_B is determined by (13.64), which takes into account the contribution of the FRP. The additional stresses in the concrete and the steel reinforcement are evaluated substituting (13.67) into the second of (13.66), with $M = M_B - M_A$ representing the supplementary moment due to the serviceability loading combinations added to stage A, as follows

$$f_c^B = E_c \frac{(M_B - M_A) c_B}{E_c b c_B^3/3 + E_s A_s (d - c_B)^2 + E_f A_f (h - c_B)^2} \tag{13.72}$$

$$f_s^B = E_s \frac{(M_B - M_A)(d - c_B)}{E_c b c_B^3/3 + E_s A_s (d - c_B)^2 + E_f A_f (h - c_B)^2} \tag{13.73}$$

Finally, the total stresses in the concrete and the steel reinforcement are determined by adding (13.69–13.70) and (13.72–13.73) as

$$f_c = f_c^A + f_c^B \tag{13.74}$$

$$f_c = E_c \frac{M_A c_A}{E_c b c_A^3/3 + E_s A_s (d - c_A)^2} + E_c \frac{(M_B - M_A) c_B}{E_c b c_B^3/3 + E_s A_s (d - c_B)^2 + E_f A_f (h - c_B)^2}$$

$$f_s = f_s^A + f_s^B \tag{13.75}$$

$$f_s = E_s \frac{M_A (d - c_A)}{E_c b c_A^3/3 + E_s A_s (d - c_A)^2} + E_s \frac{(M_B - M_A)(d - c_B)}{E_c b c_B^3/3 + E_s A_s (d - c_B)^2 + E_f A_f (h - c_B)^2}$$

The FRP is active in stage B only (Figure 13.9c). The corresponding stress can be computed by solving for f_f in the following relationship

$$\frac{\varepsilon_s}{(d - c_B)} = \frac{f_s}{E_s (d - c_B)} = \frac{f_f + E_f \varepsilon_{bi}}{E_f (h - c_B)} \tag{13.76}$$

that leads to a final value of the stress

$$f_f = f_s \left(\frac{E_f}{E_s} \right) \frac{h - c_B}{d - c_B} - \varepsilon_{bi} E_f \tag{13.77}$$

Example 13.1 *Determine the moment capacity of a RC beam reinforced on the tension side with three #5 (d=16 mm) grade 60 bars, $f'_c = 27.57\ MPa$, and $f_y = 413.68\ MPa$. The dimensions of the cross-section are: width b = 300 mm, height h = 500 mm, and the concrete cover thickness is $\delta = h - d = 30$ mm, where "d" is the distance between the tension reinforcing bar (rebar) and the topmost (compression) fiber (surface) of the beam. The initial deformation of the original, unstrengthened structure is $\varepsilon_{bi} = 0.0060$. The beam is subjected to interior environmental exposure, and strengthened with carbon FRP with the following properties: $f^*_{fu} = 2800\ MPa$, $\varepsilon^*_{fu} = 0.017$, $b_f = 200$ mm, $t_f = 1.0$ mm, $n = 1$.*

Solution to Example 13.1 *Preliminary design values for the FRP are found using (13.10) for $C_E = 0.95$ (interior environmental exposure, Table 13.2)*

$$f_{fu} = C_E f^*_{fu} = 0.95 \times 2800 = 2660\ MPa$$
$$\varepsilon_{fu} = C_E \varepsilon^*_{fu} = 0.95 \times 0.017 = 0.0161$$

The debonding strain ε_{fd}, effective strain ε_{fe}, and stress f_{fe} are determined by using (13.11) and (13.12) as follows

$$\varepsilon_{fd} = 0.41 \sqrt{\frac{f'_c}{nE_f t_f}} = 0.41 \sqrt{\frac{27.57}{1.0 \times 2800/0.017 \times 1.0}} = 0.0053 \leq 0.0144 = 0.9\ \varepsilon_{fu}$$
$$\varepsilon_{fe} = \min(\varepsilon_{fd}, 0.9\ \varepsilon_{fu}) = \min(0.0053, 0.0144) = 0.0053 = \varepsilon_{fd}$$
$$f_{fe} = E_f \varepsilon_{fe} = 873.68\ MPa$$

Strong Strengthening Configuration (SSC) *The position of the neutral axis is determined by assuming a trial strain distribution corresponding to SSC with steel reinforcement yielded (13.26) as follows*

$$A_1 c^2 + A_2 c + A_3 = 0 \rightarrow c \simeq 87.1\ mm$$

where

$$A_1 = 0.85 f'_c \beta_1 b = 0.85 \times 27.57 \times 0.85 \times 300 = 5975.79\ N/mm$$
$$A_2 = E_f A_f (\varepsilon_{cu} + \varepsilon_{bi}) - A_s f_y$$
$$A_2 = 164705 \times 200 \times (0.003 + 0.006) - 603 \times 413.68 = 47018\ N$$
$$A_3 = -hE_f A_f \varepsilon_{cu} = -500 \times 164705 \times 200 \times 0.003 = -4.941 \times 10^7\ N\ mm$$

Next, the assumed failure mode must be verified. Using (13.16–13.18), the strain distribution must be consistent with SSC, in which the FRP is in the elastic range, as follows

$$\begin{aligned} \varepsilon_f &= \varepsilon_{cu}(h-c)/c - \varepsilon_{bi} = 0.003(500-87.1)/87.1 - 0.006 \\ \varepsilon_f &= 0.0082 < \ 0.0053 = \varepsilon_{fe} && [\textit{not verified}] \\ \varepsilon_s &= \varepsilon_{cu}(d-c)/c = 0.003(470-87.1)/87.1 \\ \varepsilon_s &= 0.0131 > 0.002 = \varepsilon_y && [\textit{verified}] \end{aligned}$$

Since one of the conditions is not verified, i.e., the FRP is not in the elastic range, a new trial failure mode must be considered.

Weak Strengthening Configuration (WSC) *The WSC strain distribution refers to the failure of the FRP with steel yielded. The neutral axis is determined using (13.42), which uses the stress-block*

$$c = \frac{A_s f_y + A_f E_f \varepsilon_{fe}}{(0.85 f_c' \beta_1 b)} = \frac{603 \times 413.68 + 200 \times (164705 \times 0.0053)}{0.85 \times 27.57 \times 0.85 \times 300} \simeq 71.0 \ mm$$

Once the position of the neutral axis is determined, the inequalities (13.43–13.44) should be verified

$$\begin{aligned} \varepsilon_c &= (\varepsilon_{fe} + \varepsilon_{bi})\, c/(h-c) \\ &= (0.0053 + 0.006)\, 71/(500-71) = 0.0018 \leq 0.003 = \varepsilon_{cu} && [\textit{verified}] \\ \varepsilon_s &= (\varepsilon_{fe} + \varepsilon_{bi})\,(d-c)/(h-c) \\ &= (0.0053 + 0.006)\,(470-71)/(500-71) = 0.0105 \geq 0.002 = \varepsilon_y && [\textit{verified}] \end{aligned}$$

Finally, the nominal moment capacity is obtained by (13.47) with T_s given by (13.39) as follows

$$\begin{aligned} M_n &= A_s f_y\,(d - \beta_1 c/2) + \psi_f A_f f_{fe}\,(h - \beta_1 c/2) \\ M_n &= 603 \times 413.68 \times (470 - 0.85 \times 71/2) \\ &\quad + 0.85 \times 200 \times 873.68 \times (500 - 0.85 \times 71/2) \\ M_n &\simeq 1.795 \times 10^8 \ Nmm \end{aligned}$$

where, utilizing (13.15) the factored moment capacity is

$$(\phi M_n) = \left(0.9 \times 1.795 \times 10^8\right) = 1.615 \times 10^8 \ N\,mm$$

In the previous calculation, the stress-block method may not be accurate because the strain at the top surface of the concrete is lower than the strain to failure of concrete, i.e., $\varepsilon_c = 0.0018 < 0.003$. The exact solution should consider the actual stress distribution (13.8). The actual position of the neutral axis can be calculated by using (13.36) as

$$C - T_s - T_f = 0 \tag{13.78}$$

where the compressive resultant "C" is obtained by substituting (13.8) into (13.37), $T_s = A_s f_y = 603 \times 413.68 = 2.49 \times 10^5\ N$ and $T_f = A_f f_{fe} = 200 \times 873.68 = 1.74 \times 10^5\ N$. The position of the neutral axis "c" is the root of (13.78), c = 86.2 mm. Finally, the nominal moment capacity is determined by using (13.47), as $M_n = 1.83 \times 10^8\ Nmm$, which is in good agreement with the value obtained by the use of the stress-block method.

13.3.8 Summary Design Procedure: Bending

Input data: required moment capacity M_u, area of steel reinforcement A_s, geometry of the section (b, h), concrete, steel, and FRP material properties $f'_c, f_y, f^*_{fu}, \varepsilon^*_{fu}$.

Output data: factored moment capacity ϕM_n, and FRP area A_f.

Step 1 Obtain the values of the original design loads; that is, calculate the initial moment capacity of the original RC member, namely M_0, by specializing the analysis presented in previous sections for $A_f = 0$. Calculate the maximum allowable increment of moment capacity in the original structure attainable with FRP reinforcement by using (13.4).

Step 2 Determine the initial strain at the soffit of the beam and the initial stress distribution before strengthening, that is, before the FRP is applied. Typically, only dead loads are considered at this stage. The initial strain is calculated by using (13.71) for cracked section.

Step 3 Select a strengthening system among those available in the market. The properties, stress, and strain in the FRP are evaluated by (13.10–13.13) taking into account environmental reduction factors dictated by the type of exposure of the installation (Table 13.2).

Step 4 Calculate an initial estimate of the approximate area of FRP, as a function of the additional moment ΔM carried by the FRP, given by the difference between the required moment capacity M_u and the initial moment capacity of the unstrengthened section ϕM_0

$$A_f^{req} = \frac{M_u - \phi M_0}{\phi \psi_f \left(E_f \varepsilon_{fd}\right) \left(h - c_0/2\right)} \tag{13.79}$$

where c_0 is the neutral axis of the unstrengthened section calculated in Step 2.

Step 5 Determine the actual area of the FRP system and the corresponding mode of failure (SSC or WSC). Configure the FRP area in terms of number of layers, width, and layer thickness using FRP products available in the marketplace.

Step 6 Determine the capacity of the strengthened system. If (13.14) is not verified, increase the FRP area and iterate until the factored moment capacity ϕM_n is greater than the required moment capacity M_u.

Step 7 Check the serviceability requirements.

Example 13.2 *A simply supported RC beam has a span of* 6300 *mm and a rectangular section* 300 × 600 *mm with three #6 bars (d=19 mm), with* $f'_c = 34.5\ MPa$ *for concrete and* $f_y = 413.68\ MPa$ *for steel. The original loads, before strengthening, are* $w_{DL} = 13.5\ N/mm$ *(Dead) and* $w_{LL} = 10.5\ N/mm$ *(Live). Due to a change of use of the structure, a 60% increase of live loads is expected. A carbon FRP with material properties* $f^*_{fu} = 2200\ MPa, \varepsilon^*_{fu} = 0.011, t_f = 0.333\ mm$*, is chosen to strengthen the structure with an internal exposure condition. The concrete cover is 40 mm.*

Solution to Example 13.2

Step 1 *Determine the required moment capacity, the initial moment capacity and the maximum allowable reinforcement.*
The required moment M_u *for a simply supported beam is determined as a function of the beam geometry and the factored dead and live loads using loading combinations from (13.1)*

$$M_u = 1/8\,[1.2\,(w_{DL}) + 1.6\,(1.5 \times w_{LL})]\,L^2$$
$$M_u = 1/8\,[1.2\,(13.5) + 1.6\,(1.6 \times 10.5)]\,6300^2$$
$$M_u = 2.13 \times 10^8\ N\ mm\ (midspan)$$

Next, calculate the moment capacity of the original section. The neutral axis c_0*, and the initial moment capacity* M_0 *of the unstrengthened section member are given by (13.42), (13.47), specialized for* $A_f = 0$ *and by using (13.25) for* β_1*, that is*

$$c_0 = \frac{A_s f_y}{(0.85 f'_c \beta_1 b)} = \frac{850 \times 413.68}{0.85 \times 34.5 \times 0.80 \times 300} \simeq 50\ mm$$
$$M_0 = A_s f_y \cdot (d - \frac{c_0 \cdot \beta_1}{2}) = 850 \times 413.68 \cdot (560 - \frac{50 \times 0.80}{2}) = 1.89 \times 10^8\ N\ mm$$

Since the factored moment capacity $\phi M_0 = 0.9 \times 1.89 \times 10^8 = 1.70 \times 10^8\ Nmm$ *is lower than the required moment capacity* $M_u = 2.13 \times 10^8\ N\ mm$*, FRP strengthening is needed.*

Now, (13.3) should be checked

$$\phi F = \phi M_0 = 1.70 \times 10^8 \geq 1/8\,[1.1\,(w_{DL}) + 0.75\,(1.6 \times w_{LL})]\,L^2$$
$$\phi F = 1.36 \times 10^8\ Nmm\ [verified]$$

The maximum allowable reinforcement is given by (13.6) as

$$\%^{\max} = \frac{(S)^{Streng.}_{req.} - (S)^{Unstreng.}_{req.}}{(S)^{Unstreng.}_{req.}} = \frac{0.1 + 0.85 S_{LL}/S_{DL}}{1.1 + 0.75 S_{LL}/S_{DL}} \cdot 100$$
$$\%^{\max} = \frac{0.1 + 0.85 \times 1.25}{1.1 + 0.75 \times 1.25} \cdot 100 = 57\%$$

and assuming, conservatively, that $(M_u)_{req}^{unstr} \approx \phi M_o$, *the actual reinforcement ratio is*

$$\% = \frac{(M_u)_{req.}^{Streng.} - (M_u)_{req.}^{Unstreng.}}{(M_u)_{req.}^{Unstreng.}} = \frac{2.13 - 1.70}{1.70} \cdot 100 = 25\% < 57\% \text{ [verified]}$$

Step 2 *Determine the state of strain at the soffit of the beam.*
In order to calculate the initial strain at the soffit of the section, the position of the neutral axis is estimated by (13.68) as

$$C_1 c_A^2 + C_2 c_A + C_3 = 0 \rightarrow c_A = 132\ mm$$

with

$$C_1 = E_c b/2 = 27606 \times 300/2 = 4.14 \times 10^6\ N/mm$$
$$C_2 = E_s A_s = 200000 \times 850 = 1.70 \times 10^8\ N$$
$$C_3 = -E_s A_s d = -200000 \times 850 \times 560 = -9.52 \times 10^{10}\ N\ mm$$

The strain at the soffit is determined by means of Eq. (13.71) as follows

$$\varepsilon_{bi} = \frac{M_{DL}(h - c_A)}{(EI)_{cr}} = \frac{6.69 \times 10^7 \times (600 - 132)}{3.749 \times 10^{13}} \simeq 0.00083$$

where the flexural stiffness is calculated by using (13.67) specialized for $A_f = 0$ *and the bending moment refers to a loading condition with only dead loads applied*

$$(EI)_{cr} = E_c b c_A^3/3 + E_s A_s (d - c_A)^2 = 3.749 \times 10^{13}\ Nmm^2$$
$$M_{DL} = 1/8 w_{DL} L^2 = (1/8)13.5 \times 6300^2 \cong 6.69 \times 10^7\ Nmm$$

Step 3 *Choose the strengthened system.*
Preliminary design quantities regarding the FRP are determined by means of (13.10) with $C_E = 0.95$ *(Table 13.2) and* f_{fu}^* *from product literature (see also [1]),*

$$f_{fu} = C_E f_{fu}^* = 0.95 \times 2200 = 2090\ MPa$$
$$\varepsilon_{fu} = C_E \varepsilon_{fu}^* = 0.95 \times 0.011 = 0.0104$$
$$E_f = f_{fu}/\varepsilon_{fu} = 2.00 \times 10^5\ MPa$$

Step 4 *Design the FRP reinforcement.*
As a trial, the effective strain in the FRP is calculated for one layer $(n = 1)$

$$\varepsilon_{fd} = 0.41\sqrt{\frac{f_c'}{nE_f t_f}} = 0.41\sqrt{\frac{34.5}{1.0 \times 2.00 \times 10^5 \times 0.333}} = 0.0093 \leq 0.0094 = 0.9\varepsilon_{fu}$$
$$\varepsilon_{fe} = \min(\varepsilon_{fd}, 0.9\varepsilon_{fu}) = \min(0.0093, 0.0094) = 0.0093 = \varepsilon_{fd}$$

The FRP is dimensioned by means of (13.79), as

$$A_f^{req} = \Delta M / \left[\phi\psi_f \left(E_f\varepsilon_{fe}\right)\left(h - c_0/2\right)\right]$$
$$A_f^{req} = (2.13 - 1.70) \times 10^8 / \left[0.9 \times 0.85 \times 2.00 \times 10^5 \times 0.0093 \times (600 - 132/2)\right] \simeq 57\ mm^2$$

An estimate of the FRP area is determined using the balanced strengthening configuration (BSC). Assuming an FRP reinforcement with one layer, the neutral axis and the strain in the steel reinforcement are determined by (13.48–13.49) as

$$c_b = \frac{\varepsilon_{cu} h}{\varepsilon_{cu} + (\varepsilon_{fe} + \varepsilon_{bi})} = \frac{0.003 \times 600}{0.003 + (0.0093 + 0.00083)} = 137\ mm$$
$$\varepsilon_s = \frac{\varepsilon_{cu}}{c_b}(d - c_b) = \frac{0.003}{137}(560 - 137) = 0.0093 > \varepsilon_y = 0.002$$

Since the steel reinforcement has yielded, the required area of FRP is determined by (13.55)

$$A_{fb} = \frac{1}{E_f\varepsilon_{fe}}\left(0.85 f_c' \beta_1 b c_b - A_s f_y\right)$$
$$A_{fb} = \frac{(0.85 \times 34.5 \times 0.80 \times 300 \times 137 - 850 \times 413.68)}{(2.00 \times 10^5 \times 0.0093)} = 330\ mm^2$$

Step 5 *Configure the FRP. The actual FRP configuration is chosen among the geometries available in the market and consistently with the predicted value in the dimensioning procedure. One such geometry can be obtained is the following*

$$n = 1, t_f = 0.333\ mm, b_f = 200\ mm \rightarrow \left(A_f = 66.6\ mm^2\right)$$

Step 6 *Calculate the moment capacity.*
Since the actual area A_f is lower than the required area A_{fb}, WSC is obtained. The position of the neutral axis is found by assuming a strain distribution based on WSC with steel reinforcement yielded. The trial neutral axis is found using (13.42) as

$$c = \frac{A_s f_y + A_f E_f \varepsilon_{fe}}{(0.85 f_c' \beta_1 b)} = \frac{850 \times 413.68 + 66.6 \times \left(2.00 \times 10^5 \times 0.0093\right)}{0.85 \times 34.5 \times 0.80 \times 300} \simeq 67\ mm$$

Subsequently, the assumed failure mode must be verified. Using (13.33–13.35), the strains on the section should satisfy the following

$$\varepsilon_c = (\varepsilon_{fd} + \varepsilon_{bi})\frac{c}{h - c} = (0.0093 + 0.00083)\frac{67}{600 - 67} = 0.00127 \leq 0.003 = \varepsilon_{cu}\ \textit{[verified]}$$
$$\varepsilon_s = (\varepsilon_{fd} + \varepsilon_{bi})\frac{d - c}{h - c} = (0.0093 + 0.00083)\frac{560 - 67}{600 - 67} = 0.00937 \geq 0.002 = \varepsilon_y\ \textit{[verified]}$$

The nominal moment capacity is obtained by (13.45) with T_s given by (13.39) as follows

$$M_n = A_s f_y \left(d - \beta_1 c/2\right) + \psi_f A_f f_{fe} \left(h - \beta_1 c/2\right)$$
$$M_n = 850 \times 413.68 \times (560 - 0.80 \times 67/2) + 0.85 \times 66.6 \times 1866 \times (600 - 0.80 \times 67/2)$$
$$M_n = 2.47 \times 10^8\ N\ mm$$

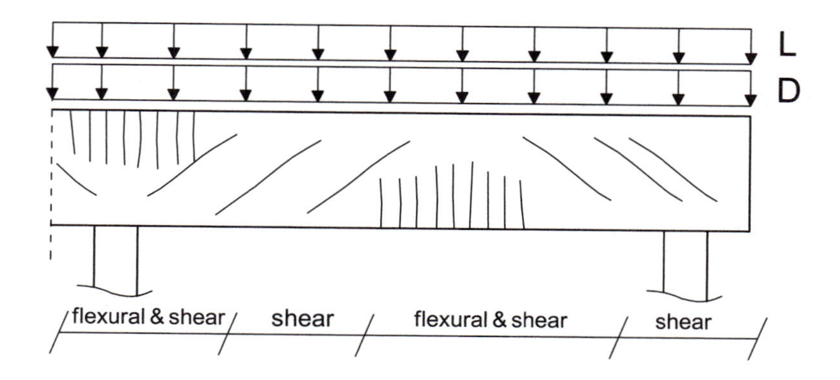

Figure 13.10: Crack distribution on a continuously supported beam subject to gravity and external loads.

Finally, the factored moment capacity is given by

$$\phi M_n = 0.9 \times 2.47 \times 10^8 = 2.22 \times 10^8 \ N\,mm > M_u = 2.13 \times 10^8 \ N\,mm$$

Step 7 *Check the serviceability requirements.*
According to (13.58,13.59) and the third of (13.60), the following limits on the applied stress level should be taken into account

$$\begin{aligned} f_{s,s} &\leq 0.8 f_y && \textit{(steel reinforcement)} \\ f_{c,s} &\leq 0.45 f_c' && \textit{(concrete)} \\ f_{f,s} &\leq 0.55 f_{fu} && \textit{(carbon FRP)} \end{aligned}$$

The moments in the serviceability condition are determined from

$$\begin{aligned} M_A &= 1/8\,[w_{DL}]\,L^2 = 1/8\,[13.5]\,6300^2 \cong 6.69 \times 10^7 Nmm && \textit{(Dead loads)} \\ M_B &= 1/8\,[w_{DL} + w_{LL}]\,L^2 \\ &= 1/8\,[13.5 + 16.8]\,6300^2 \cong 1.50 \times 10^8 \ Nmm && \textit{(Dead + Live loads)} \end{aligned}$$

The stress levels in the concrete and the steel are evaluated by determining the position of the neutral axis (stage A), which has been previously determined in Step 2, i.e., $c_A = 132$ *mm. The corresponding stress level is determined by (13.69–13.70) as*

$$\begin{aligned} f_c^A &= E_c \frac{M_A c_A}{E_c b c_A^3/3 + E_s A_s (d - c_A)^2} \\ f_c^A &= 27606 \frac{6.69 \times 10^7 \times 132}{27606 \times 300 \times 132^3/3 + 2.0 \times 10^5 \times 850 \times (560 - 132)^2} = 6.5 \ MPa \\ f_s^A &= E_s \frac{M_A (d - c_A)}{E_c b c_A^3/3 + E_s A_s (d - c_A)^2} \\ f_s^A &= 2.0 \times 10^5 \frac{6.69 \times 10^7 \times (560 - 132)}{27606 \times 300 \times 132^3/3 + 2.0 \times 10^5 \times 850 \times (560 - 132)^2} \cong 153 \ MPa \end{aligned}$$

When the FRP is installed, the serviceability loading condition including dead and live loads should be checked (stage B). In this configuration, the position of the neutral axis is determined by (13.65) as

$$C_1 c_B^2 + C_2 c_B + C_3 = 0 \rightarrow c_B = 137 \ mm \ (stage \ B)$$

with

$$\begin{aligned} C_1 &= \frac{E_c b}{2} = 4.14 \times 10^6 \ N/mm, \\ C_2 &= E_s A_s + E_f A_f = 1.83 \times 10^8 \ N, \\ C_3 &= -E_s A_s d - E_f A_f h = -1.03 \times 10^{11} Nmm \end{aligned}$$

The additional stress in the steel is determined by (13.72–13.73) as

$$\begin{aligned} f_c^B &= E_c \frac{(M_B - M_A) c_B}{E_c b c_B^3/3 + E_s A_s (d - c_B)^2 + E_f A_f (h - c_B)^2} \\ f_c^B &= 27606 \frac{(1.50 \times 10^8 - 6.91 \times 10^7) \times 137}{4.03 \times 10^{13}} \cong 7.8 \ MPa \\ f_s^B &= E_s \frac{(M_B - M_A)(d - c_B)}{E_c b c_B^3/3 + E_s A_s (d - c_B)^2 + E_f A_f (h - c_B)^2} \\ f_s^B &= 2.0 \times 10^5 \frac{(1.50 \times 10^8 - 6.91 \times 10^7) \times (560 - 137)}{4.03 \times 10^{13}} \cong 175 \ MPa \end{aligned}$$

and the total stresses obtained by adding the contributions arising from stage A and stage B as

$$\begin{aligned} f_{s,s} &= f_s^A + f_s^B = 328 \ MPa < 0.8 f_y = 330 \ MPa \\ f_{c,s} &= f_c^A + f_c^B = 14.30 \ MPa < 0.45 f_c = 15.52 \ MPa \end{aligned}$$

The corresponding stress level in the FRP is determined by (13.77) as

$$\begin{aligned} f_{f,s} &= f_{s,s} \left(\frac{E_f}{E_s} \right) \frac{h - c_B}{d - c_B} - \varepsilon_{bi} E_f \\ f_{f,s} &= 328 \times \left(\frac{2.0 \times 10^5}{2.0 \times 10^5} \right) \frac{600 - 137}{560 - 137} - 0.00083 \times 2.00 \times 10^5 \\ f_{f,s} &= 193 < 0.55 f_{fu} = 1149 \ MPa \end{aligned}$$

13.4 Shear Strengthening

RC beams under shear load are affected by several damage mechanisms. A schematic representation of possible crack distributions is shown in Figure 13.10. Vertical cracks affect the beam where flexural stresses prevail, diagonal cracks appear where shear stresses dominate, and mixed flexural-shear cracks are observed in the transition zones.

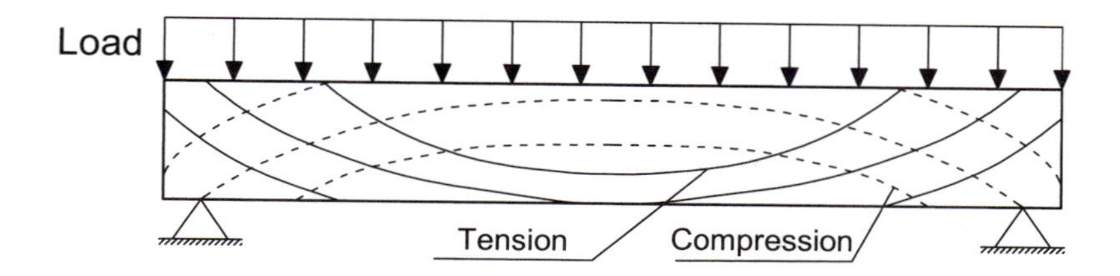

Figure 13.11: Crack pattern for a simply supported beam.

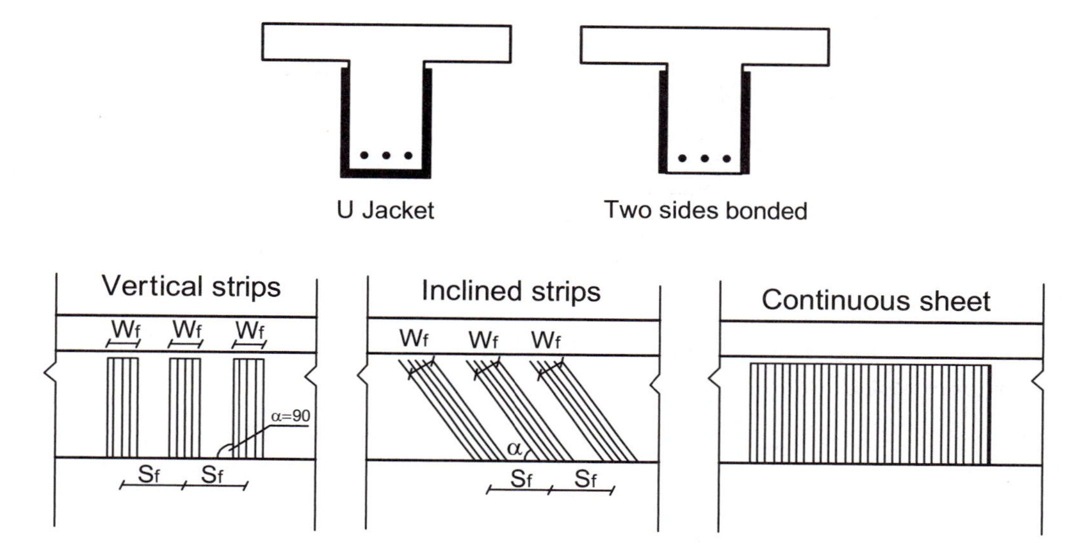

Figure 13.12: Possible configurations for shear strengthening with FRP.

Crack patterns correlate with the principal stress trajectories shown in Figure 13.11. Due to the low tensile strength of concrete, cracks are formed normal to the tensile principal stress directions. In RC beams, steel stirrups are placed perpendicularly to the tensile stress directions to prevent concrete cracking in shear.

The RC structure can be further strengthened by the use of FRP laminates and fabrics, which are applied externally to the concrete faces to increase the shear capacity of the member. Typical reinforcement configurations for continuous and discontinuous arrangements are reported in Figure 13.12. Complete wrapping of all sides of columns is customary, but typically beams are wrapped only where needed.

In order to prevent beams from reaching critical failure modes, such as delamination and debonding, the maximum strain in the FRP is limited according to the following expressions [256]

$$\varepsilon_{fe} = 0.004 \leq 0.75\varepsilon_{fu} \qquad \text{(Four-sided)} \qquad (13.80)$$

$$\varepsilon_{fe} = \kappa_v \varepsilon_{fu} \leq 0.004 \qquad \text{(Three- and two-sided)} \qquad (13.81)$$

where κ_v is the *bond reduction factor*. This quantity is expressed as a function of

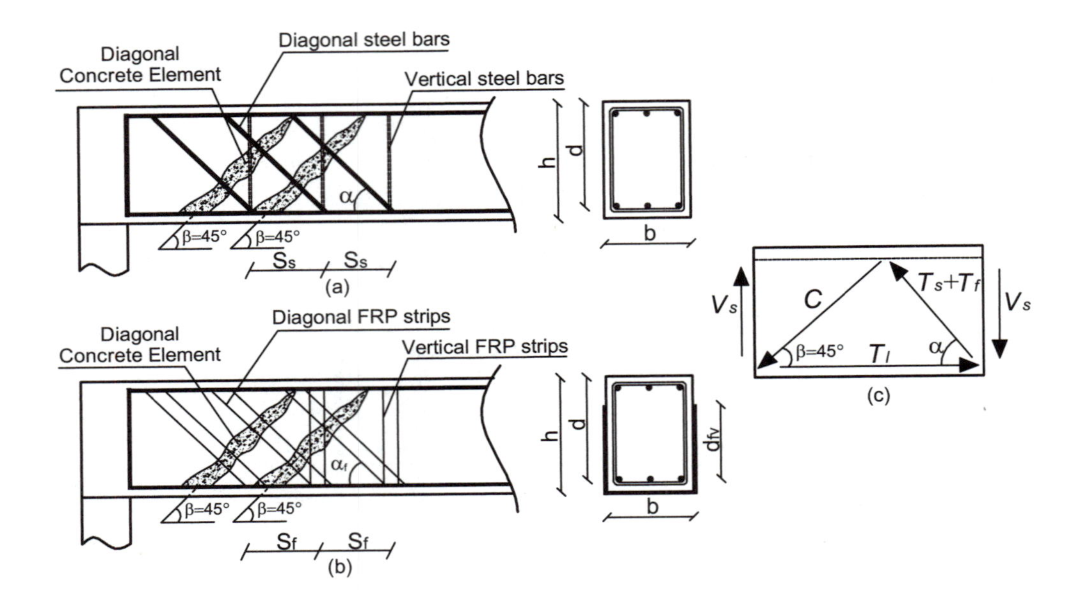

Figure 13.13: Possible configurations for shear strengthening: (a) steel, (b) FRP, (c) truss analogy.

the concrete strength f'_c (in MPa), the wrapping scheme, and the *active length of the bonding zone* L_e (in mm) as follows

$$\kappa_v = \frac{k_1 k_2 L_e}{11,900\varepsilon_{fu}} \leq 0.75 \tag{13.82}$$

with

$$L_e = \frac{23,300}{(n_f t_f E_f)^{0.58}} \qquad k_1 = \left(\frac{f'_c}{27}\right)^{2/3} \tag{13.83}$$

$$k_2 = \frac{d_{fv} - L_e}{d_{fv}} \qquad \text{(three-sided)} \tag{13.84}$$

$$k_2 = \frac{d_{fv} - 2L_e}{d_{fv}} \qquad \text{(two-sided)} \tag{13.85}$$

where d_{fv} is the depth of FRP shear reinforcement, as shown in Figure 13.13b.

The ultimate *shear capacity* is predicted by adding the resistance provided by the concrete C, the steel S, and FRP F. According to ACI guidelines, the shear strength of concrete is determined by

$$V_c = \lambda \frac{bd}{6}\sqrt{f'_c} \qquad \text{(shear and flexure)} \tag{13.86}$$

$$V_c = \lambda \left(1 + \frac{N_u}{14A_g}\right)\frac{bd}{6}\sqrt{f'_c} \qquad \text{(axial compression)} \tag{13.87}$$

where λ is 1 for normal weight concrete and 0.75 for light weight concrete, N_u is the factored axial force occurring in the loading combination given by (13.1), A_g is gross area of concrete, and N_u/A_g is the axial compression.

The failure mechanism involved in the FRP and the steel is based on truss analogy (Figure 13.13). The crack patterns reduce the beam to diagonal concrete elements under compressive loads, which are almost parallel to the isostatic lines of compression. These elements are connected to longitudinal and vertical steel and FRP elements, which are subjected to tensile forces. The ultimate shear capacity due to the steel V_s and FRP V_f are expressed as a function of number of intersecting FRP strips n_f and steel bars n_s (Figure 13.13), as

$$V_s = f_y A_{sv} n_s \sin\alpha \quad ; \quad V_f = f_{fe} A_{fv} n_f \sin\alpha_f \tag{13.88}$$

where $f_{fe} = E_f \varepsilon_{fe}$ with ε_{fe} given by (13.80–13.81); α, α_f are the inclination angle of the steel and FRP, respectively, and A_{sv}, A_{fv} are the areas of the steel and FRP reinforcements, respectively. The number of the steel and FRP elements can be expressed as a function of the spacing and depth of the steel (s_s, d) and the FRP (s_f, d_f), as

$$n_s = \frac{d}{s_s}\left[1 + \cot\alpha\right] \quad ; \quad n_f = \frac{d_f}{s_f}\left[1 + \cot\alpha_f\right] \tag{13.89}$$

Substituting the first of (13.89) into (13.88), the nominal shear capacity of inclined steel elements is found as

$$V_s = f_y A_{sv} \frac{d}{s_s}\left[\sin\alpha + \cos\alpha\right] \tag{13.90}$$

Similarity, substituting the second of (13.89) into (13.88), the nominal shear capacity of FRP based on intermittent strips of width w_f and spacing s_f, i.e., $A_{fv} = 2n_f t_f w_f$, is given by

$$V_f = f_{fe}\left(2n_f t_f w_f\right)\frac{d_{fv}}{s_f}\left[\sin\alpha_f + \cos\alpha_f\right] \tag{13.91}$$

In the case of continuous FRP reinforcement, the contribution of the FRP to the nominal shear capacity can be obtained from (13.91) by substituting $w_f = s_f \sin(\alpha_f)$ for w_f, leading to

$$V_f = f_{fe} 2n_f t_f d_{fv} \sin\alpha_f \left[\sin\alpha_f + \cos\alpha_f\right] \tag{13.92}$$

Usually, steel and FRP elements are placed vertically. Then, (13.90–13.92) with $\alpha = \alpha_f = 90°$ reduce to

$$V_s = f_y A_{sv} d/s_s \tag{13.93}$$

$$V_f = \left(2n_f t_f w_f\right) f_{fe} d_{fv}/s_f \quad \text{(discontinuous)} \tag{13.94}$$

$$V_f = \left(2n_f t_f\right) f_{fe} d_{fv} \quad \text{(continuous)}$$

Discontinuous FRP reinforcement is preferred because moisture can dry up while continuous reinforcement may trap moist between the FRP and concrete, which may precipitate debonding and delamination [259]. The nominal shear capacity is derived as the sum of the contributions of the FRP, concrete, and steel reinforcements

$$V_n = V_c + V_s + \psi_f V_f \tag{13.95}$$

where ψ_f is a reduction factor (0.95 for four-sided wrapped, 0.85 for two- and three-sided wrapped). ACI guidelines further limit the total shear reinforcement as follows

$$V_s + V_f \leq 0.66\sqrt{f_c'}bd \tag{13.96}$$

where the factored shear capacity ϕV_n with $\phi = 0.75$ (Table 13.1), must verify the following

$$\phi V_n \geq V_u \tag{13.97}$$

where V_u is the *required shear capacity* calculated in terms of the loading combination given by (13.1).

13.4.1 Summary Design Procedure: Shear

Input data: required shear capacity V_u; area of inclined or vertical shear steel reinforcement A_{sv}; concrete geometry b, h; concrete, steel and FRP material properties f_c', f_y, f_{fu}^*, ε_{fu}^*.

Output data: factored shear capacity ϕV_n; and shear FRP section area A_{fv}.

Step 1 Determine the initial shear capacity of the unstrengthened RC member and the required shear capacity for the strengthened member. The initial shear capacity is determined by setting the contribution of the FRP to zero in (13.95). The maximum allowable shear strengthening is calculated using (13.6).

Step 2 Calculate the number of the layers and the geometry of the shear reinforcement. An initial estimate of the number of layers can be obtained by (13.94) and (13.95) with the equality $V_u = \phi V_n$

$$n_f = \frac{1}{\psi_f}\left[\frac{V_u}{\phi} - V_c - V_s\right]\frac{1}{2f_{fe}t_f d_{fv}\sin\alpha_f\left(\sin\alpha_f + \cos\alpha_f\right)}$$

where f_{fe} is defined below (13.88), not to be confused with f_{fe} given by (13.13). Next calculate the factored shear capacity for the calculated number of layers n_f (integer value greater than the calculated one) and check (13.96) and (13.97). If it is not verified, increase the number of the layers.

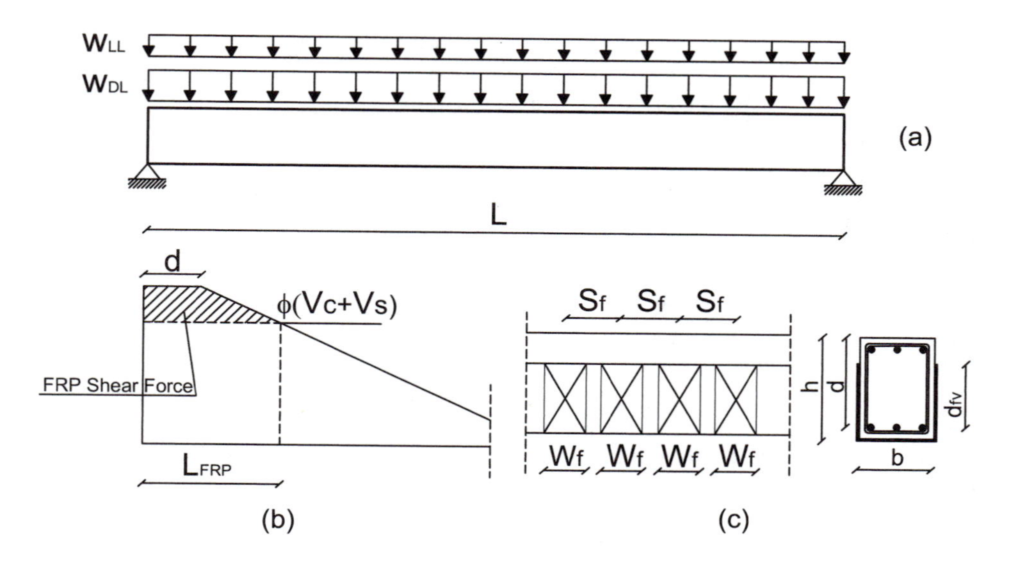

Figure 13.14: (a) Loading configuration, (b) Shear diagram, (c) Discontinuous FRP reinforcement.

Example 13.3 *Consider a simply supported RC beam with a span of* 6.3 *m and a rectangular section with* $b = 300$ *mm,* $h = 500$ *mm,* $d = 460$ *mm, located inside a building. The material properties of the RC structure are* $f_c' = 20.68$ *MPa for concrete and* $f_y = 275.79$ *MPa for steel. The shear steel reinforcement is made of 2-leg #3 (d = 9.5 mm) vertical steel bars with spacing* $s_s = 200$ *mm. The loads on the original structure are* $w_{DL} = 23$ *N/mm (Dead) and* $w_{LL} = 12$ *N/mm (Live). Due to a change of the use of the structure, a 50% increase of the live loads is expected. A carbon FRP with material properties* $f_{fu}^* = 3700$ *MPa,* $\varepsilon_{fu}^* = 0.016$, $t_f = 0.200$ *mm is selected to strengthen the RC structure.*

Solution to Example 13.3 *The following preliminary design parameters for concrete and steel reinforcement are calculated as:*
Concrete: $f_c' = 20.68\ MPa$, $A_c = 300 \times 500 = 1500\ mm^2$
Transverse steel reinforcement: 2-leg #3 rebar (d = 9.5 mm), $A_{sv} = 2\pi d^2/4 = 141.7\ mm^2$, $f_y = 275.79\ MPa$, $s_s = 200\ mm$.
Preliminary design quantities regarding the FRP are determined by means of (13.10) for $C_E = 0.95$ *(interior environmental exposure, see Table 13.2) as follows*

$$f_{fu} = C_E f_{fu}^* = 0.95 \times 3700 = 3515\ MPa$$
$$\varepsilon_{fu} = C_E \varepsilon_{fu}^* = 0.95 \times 0.016 = 0.0152$$

Step 1 *Determine the required shear capacity and the maximum allowable strengthening. The required shear capacity in this case occurs at the support (Figure 13.14), and it is determined as follows*

$$V_u = 1/2\left[1.2\left(w_{DL}\right) + 1.6\left(1.5 \times w_{LL}\right)\right] L$$
$$V_u = 1/2\left[1.2\left(23\right) + 1.6\left(18\right)\right] \times 6300 = 1.776 \times 10^5 \ N \ (support)$$

The factored shear capacity of the original structure is expressed as a function of the strength of concrete f'_c and steel reinforcement f_y only, by means of (13.86) and (13.93) as follows

$$V_c = \frac{1}{6} bd\sqrt{f'_c} = 1.046 \times 10^5 \ N$$
$$V_s = \frac{f_y A_{sv} d}{s_s} = 0.898 \times 10^5 \ N$$
$$\phi\left(V_c + V_s\right) = 0.75 \times \left(1.046 \times 10^5 + 0.898 \times 10^5\right) = 1.458 \times 10^5 \ N$$

In Figure 13.14, the shear diagram is compared to the resistance of the original structure, showing that the additional capacity is required only within a distance from the support equal to $L_{FRP} = 1$ m. The actual reinforcement $\%^{actual}$ given by (13.6) should be lower than the maximum allowable (13.4) $\%^{max}$, i.e.,

$$\%^{actual} = \frac{\left(V_u\right) - \phi\left(V_c + V_s\right)}{\phi\left(V_c + V_s\right)} \times 100 = 21.8\%$$
$$\%^{max} = \frac{0.1 + 0.85 w_{LL}/w_{DL}}{1.1 + 0.75 w_{LL}/w_{DL}} \times 100 = 45.3\%$$

Step 2 *Determine the number of the layers and the geometry of the reinforcement.*
It is important here to specify the typology of the shear strengthening. A three-sided wrapping ($\psi_f = 0.85$) with $d_{fv} = 300$ mm is considered. The required shear capacity of the FRP is

$$V_f^R = \frac{1}{\psi_f}\left[\frac{V_u}{\phi} - V_c - V_s\right] \simeq 0.499 \times 10^5 \ N (required)$$

Since $\varepsilon_{fe} = \min\left[0.004, 0.75\varepsilon_{fu}\right] = 0.004$, the number of layers is calculated with $\alpha = 90°$ and $d_{fv} = 300$ mm, as follows

$$n_f = \frac{1}{\psi_f}\left[\frac{V_u}{\phi} - V_c - V_s\right] \frac{1}{2 f_{fe} t_f d_{fv} \sin\alpha\left(\sin\alpha + \cos\alpha\right)}$$

With $n_f = 1$, the shear capacity for three-sided continuous wrapping is determined as follows

$$L_e = \frac{23300}{\left(n_f t_f E_f\right)^{0.58}} = \frac{23300}{\left(1 \times 0.2 \times 3700/0.016\right)^{0.58}} = 45.87 \ mm$$
$$k_1 = \left(\frac{f'_c}{27}\right)^{2/3} = \left(\frac{20.68}{27}\right)^{2/3} = 0.837$$
$$k_2 = \frac{d_{fv} - L_e}{d_{fv}} = 0.847$$
$$\kappa_v = \frac{k_1 k_2 L_e}{11900\varepsilon_{fu}} = 0.179 \leq 0.75$$
$$\varepsilon_{fe} = \kappa_v \varepsilon_{fu} = 0.0027 \leq 0.004$$
$$V_f = 2 n_f t_f E_f \varepsilon_{fe} d_{fv} = 0.759 \times 10^5 \ N > V_f^R$$

Since the shear capacity provided by the continuous wrapping is much greater than the required shear capacity, an intermittent sheet is preferred. Assuming the width to be $w_f = 200$ *mm, the maximum allowable spacing is*

$$s_f^{\max} = \frac{2n_f t_f w_f d_{fv} f_{fe}}{V_f^R} \approx 304 \; mm$$

Using $w_f = 200$ *mm and* $s_f = 300$ *mm, the shear capacity is*

$$V_f = (2n_f t_f w_f) \frac{d_{fv} f_{fe}}{s_f} = 0.506 \times 10^5 \; N > V_f^R$$

Step 3 *Check limitations on maximum allowable shear reinforcement.*
According to (13.96) the following limitations for discontinuous shear strengthening must be checked

$$V_s + V_f = 1.404 \times 10^5 \, N \leq 0.66\sqrt{f_c'}bd = 4.14 \times 10^5 \; N \quad (continuous)$$
$$V_s + V_f = 1.658 \times 10^5 \, N \leq 0.66\sqrt{f_c'}bd = 4.14 \times 10^5 \; N \quad (discontinuous)$$

13.5 Beam-column

A beam-column is subject mainly to axial load but may experience some bending due to lateral loads, eccentricity of the axial load, geometric imperfections, etc. RC columns are mainly subjected to axial and bending loads. They are utilized primarily to support vertical loads, horizontal wind forces, and seismic loads. Typical geometries of RC columns with rectangular or circular sections are shown in Figure 13.15. Similarly to RC beams, columns are reinforced by vertical and transverse steel bars.

FRP is used to increase the load capacity of RC columns subjected to pure axial load or combined axial and bending loads. FRP strengthening consists of wrapping the external face of the column using sheets or strips (Figure 13.16). Therefore, the effect of FRP strengthening is to confine the concrete, thus increasing the confined compressive strength of concrete.

A typical stress strain curve for confined and unconfined concrete is shown in Figure 13.17. The initial behavior is similar for both, until the unconfined strength of concrete $f_c = f_c'$ is reached. As soon as concrete material is affected by distributed cracks, the influence of FRP strengthening becomes noticeable and the mechanical behavior is improved. The postcracking stress-strain relationship of confined concrete shows linear behavior [264–266] and its slope depends mainly on the stiffness of the FRP material.

Since the FRP provides confinement only after a significant distribution of microcracks occurs in the concrete, a similarity between confined concrete and a liquid confined in a pipe is noted. Concrete behaves similarly to a liquid, since it is cracked and the FRP confining pressure on the perimeter are similar to those produced by a thin-walled pipe. The confinement effect of FRP provides an improvement in the stress-strain relationship only after substantial microcracks affect the concrete material. The concrete stress-strain relationship in the elastic range is not influenced

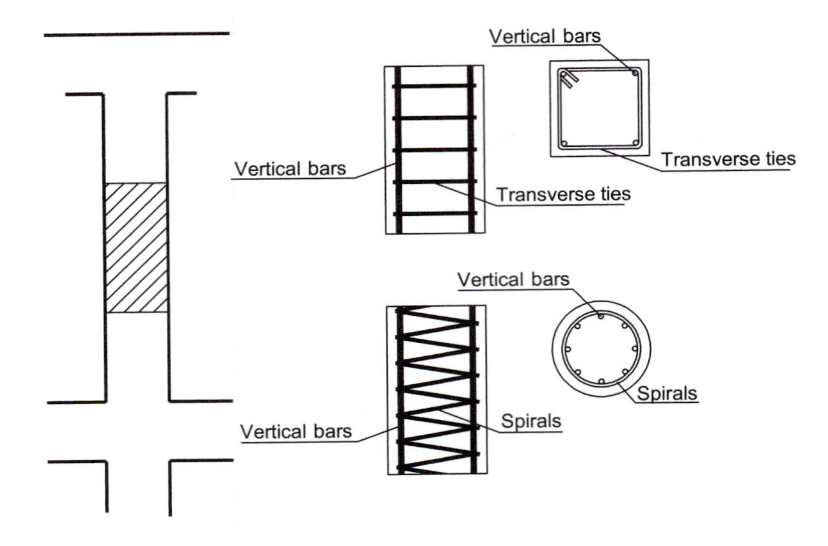

Figure 13.15: Type of columns: Vertical bars, transverse ties, and spirals steel reinforcement.

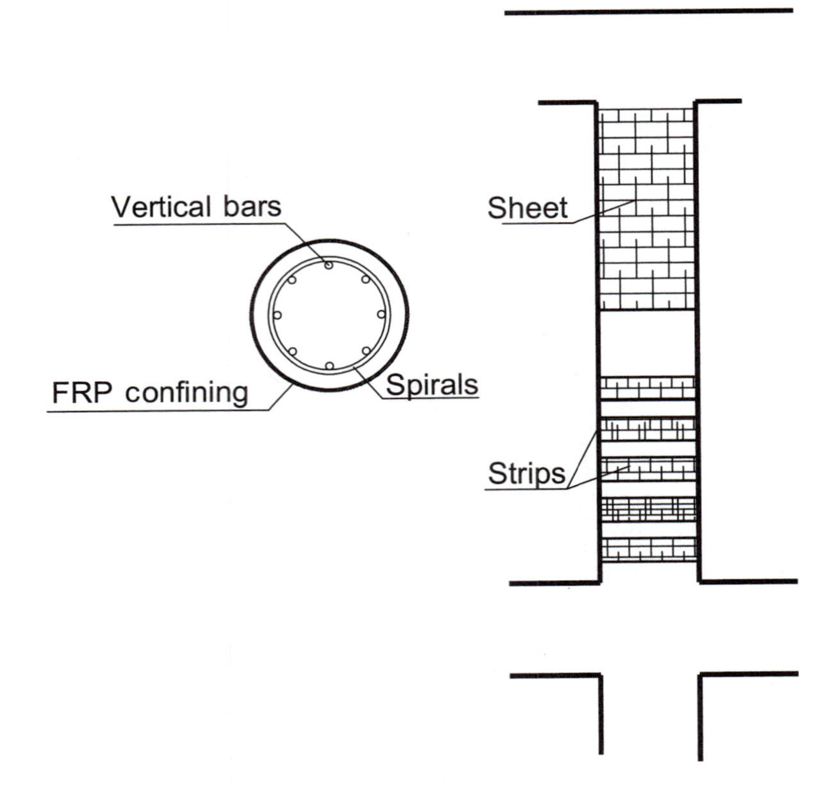

Figure 13.16: FRP reinforcement of the column.

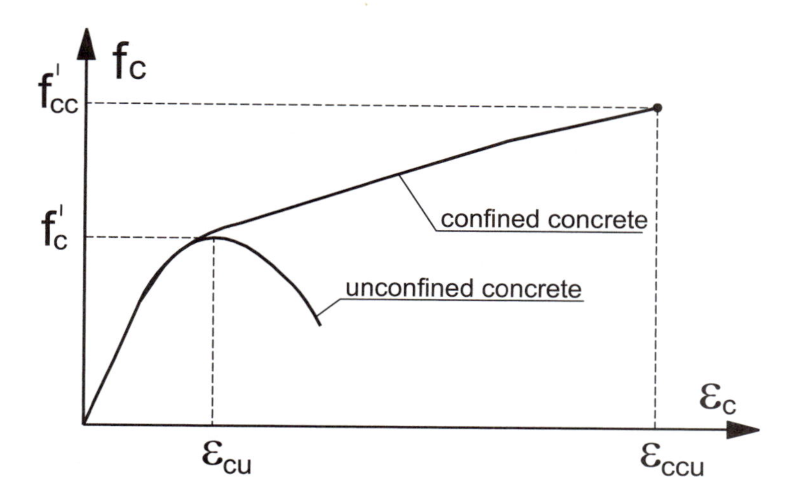

Figure 13.17: Idealized stress–strain behavior of FRP reinforcement for columns: confined and unconfined behavior.

by the FRP system; it increases the concrete resistance in the postcracking behavior only. As a result, the FRP system is considered a passive strengthening system.

13.5.1 Column: Pure Axial Compression

First, the evaluation of the confinement effects provided by the FRP should be identified. The ACI guidelines recommend calculating the confined compressive strength of the concrete f'_{cc} as a function of the maximum confining pressure f_l and the initial concrete strength f'_c provided by the FRP, as

$$f'_{cc} = f'_c + 3.3\psi_f \kappa_a f_l \tag{13.98}$$

where ψ_f is a reduction factor (equal to 0.95), κ_a is an efficiency factor that accounts for the geometry of the section and it is defined as

$$\kappa_a = 1 \qquad \text{(circular cross-section)} \tag{13.99}$$

$$\kappa_a = \frac{A_e}{A_c}\left(\frac{b^2}{h^2}\right) \qquad \text{(rectangular cross-section)} \tag{13.100}$$

where A_c is the concrete area of the section in compression and A_e is the sectional area of the effectively confined cross-section defined as follows

$$\frac{A_e}{A_c} = \frac{1 - \left[(b/h)\left(h - 2r_c\right)^2 + (h/b)\left(b - 2r_c\right)^2\right]/3A_g - \rho_g}{1 - \rho_g} \tag{13.101}$$

with $\rho_g = A_{st}/A_c$ is the longitudinal steel reinforcement ratio and r_c is the radius of the FRP corners (typically 25 mm). Experimental data support the use of (13.98) for FRP confinement as well [265, 267]. To evaluate the confining pressure, ACI guidelines propose the following relationship

$$f_l = \frac{2E_f n t_f \varepsilon_{fe}}{D} \text{ with } \varepsilon_{fe} \leq \kappa_\varepsilon \varepsilon_{fu} \tag{13.102}$$

where D is the diameter of circular section, or for noncircular columns, the diameter of an equivalent circular section external to the original one [256, Figure 12.3], [267, Figure 1, p. 1152].

The effective strain ε_{fe} is expressed as a function of a strain efficiency factor κ_ε to account for premature failure of the FRP system, with an experimentally adjusted value of 0.55. ACI guidelines introduce limitations on the minimum level of confinement ratio f_l/f'_c and on the maximum compressive strain in the confined concrete ε_{ccu} to prevent excessive concrete cracking by means of the following inequalities

$$f_l/f'_c \geq 0.08 \tag{13.103}$$

$$\varepsilon_{ccu} \leq 0.01 \text{ with } \varepsilon_{ccu} = \varepsilon'_c \left[1.5 + 12\kappa_b \frac{f_l}{f'_c}\left(\frac{\varepsilon_{fe}}{\varepsilon'_c}\right)^{0.45}\right] \tag{13.104}$$

where $\varepsilon'_c = 0.002$ is the strain of unconfined concrete corresponding to f'_c. Moreover, κ_b is an efficiency factor that accounts for the geometry of the section and it is defined as

$$\kappa_b = 1 \qquad \text{(circular cross-section)} \tag{13.105}$$

$$\kappa_b = \frac{A_e}{A_c}\left(\frac{h}{b}\right)^{0.5} \qquad \text{(rectangular cross-section)} \tag{13.106}$$

with A_e/A_c defined by (13.101). For noncircular sections, experimental observations have shown that the confinement effect provided by FRP is quite limited, especially for members with side aspect ratios h/b greater than 2.0 or face dimensions b or h greater than 900 mm [256].

The design of the strengthening for axially loaded beams is based on assumptions similar to those used for flexural strengthening, as follows

1. A constant strain distribution across the thickness of the column, limited by the concrete ($\varepsilon_c = \varepsilon_{ccu} \leq 0.01$) and FRP deformations ($\varepsilon_f = \varepsilon_{fe} \leq 0.55\varepsilon_{fu}$).

2. Perfect adhesion between concrete, steel, and FRP.

3. Negligible tensile resistance of the concrete.

The *nominal axial capacity* P_n of a nonslender, nonprestressed column strengthened by FRP system is determined taking into account the strength of confined concrete and steel, by means of the following additive formula (Figure 13.18)

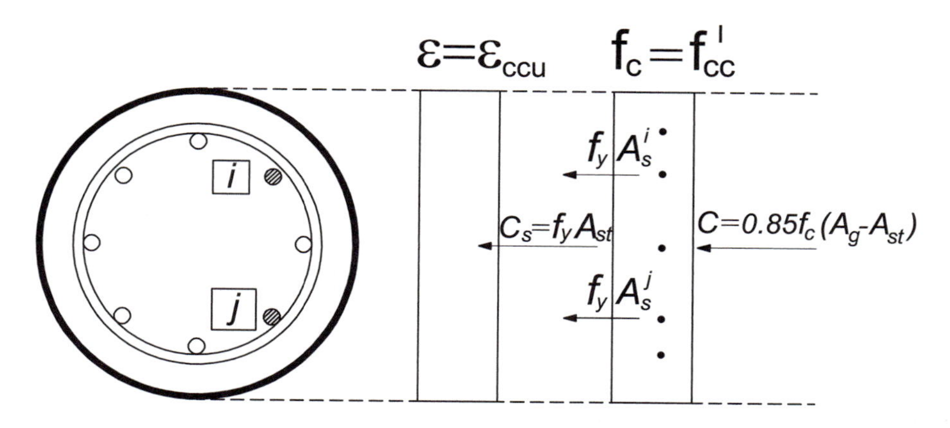

Figure 13.18: Column geometry, strain and stress distribution.

$$P_n = 0.85 f'_{cc} \left(A_g - A_{st}\right) + f_y A_{st} \tag{13.107}$$

where A_g is the gross area of the concrete and A_{st} is the steel area of the longitudinal reinforcement.

To take into account unknown eccentricities of the applied loads, the nominal axial capacity is penalized by introducing a factor ϕ_{ecc} (equal to 0.85 for spiral steel, 0.80 for tie steel), in addition to the customary strength reduction factor ϕ, as follows

$$\phi P_n = \phi \; \phi_{ecc} \left[0.85 f'_{cc} \left(A_g - A_{st}\right) + f_y A_{st}\right] \tag{13.108}$$

where ϕ is equal to 0.65 for tie steel, 0.75 for spiral steel reinforcement.

In order to prevent an excessive crack pattern under service loads, limitations to the allowable stress in concrete f_c and steel f_s are introduced accordingly to the ACI guidelines as follows

$$f_c \leq 0.65 f'_c; \quad f_s \leq 0.60 f_y \tag{13.109}$$

The stress distribution is evaluated taking into account the serviceability loading conditions. Due to assumed perfect adhesion between concrete and steel bars, the applied strain can be expressed by the following relationship

$$\varepsilon = \frac{P}{E_c \left(A_g - A_{st}\right) + E_s A_{st}} = \frac{f_c}{E_c} = \frac{f_s}{E_s} \tag{13.110}$$

where P is the applied load given by the serviceability loading conditions. Therefore, the applied stresses in concrete and steel reinforcement are calculated by means of (13.110) and limitations according to (13.109) should be checked

$$f_c = E_c \frac{P}{E_c\left(A_g - A_{st}\right) + E_s A_{st}} \leq 0.65 f'_c \qquad \text{(concrete)} \tag{13.111}$$

$$f_s = E_s \frac{P}{E_c\left(A_g - A_{st}\right) + E_s A_{st}} \leq 0.60 f_y \qquad \text{(steel reinforcement)} \tag{13.112}$$

The design of the FRP strengthening further requires the evaluation of the number of the layers n and their thickness t_f.

13.5.2 Summary Design Procedure: Column

Input data: required axial force capacity P_u, total area of longitudinal steel reinforcement A_{st}, concrete geometry b, h; concrete, steel and FRP material properties f'_c, f_y, f^*_{fu}, ε^*_{fu}.

Output data: factored nominal axial force ϕP_n, and FRP thickness t_f and number of layers n.

Step 1 Determine the required load capacity by the loading combination (13.1) and the factored axial strength of unconfined column by (13.108) with $f'_{cc} = f'_c$. Check the minimum and maximum allowable strengthening by (13.3) and (13.6).

Step 2 Calculate the confining pressure and the number of the layers. The confined compressive strength of concrete is determined by solving (13.108) with $\phi P_n = P_u$ as follows

$$f'_{cc} = \frac{1}{0.85\left(A_g - A_{st}\right)}\left(\frac{P_u}{\phi\, \phi_{ecc}} - f_y A_{st}\right) \tag{13.113}$$

Step 3 Evaluate κ_a from (13.99) or (13.100) utilizing (13.101) and calculate the confining pressure by solving (13.98) for f_l as

$$f_l = \frac{f'_{cc} - f'_c}{\psi_f 3.3 \kappa_a} \tag{13.114}$$

Step 4 Determine the number of the layers solving for n from (13.102)

$$n = \frac{D f_l}{2 t_f E_f \varepsilon_{fe}} \tag{13.115}$$

Step 5 For the chosen FRP dimension, recalculate the factored axial strength by using (13.108) and compare it with the required axial force.

Step 6 Check the service level of stresses as function of the serviceability loading combinations by means of (13.111) and (13.112).

Example 13.4 *Consider a nonslender, spirally wound column of diameter* 450 *mm, reinforced with 8 #4 bars* ($d = 13$ *mm) and spiral shear reinforcements, subjected to exterior exposure. The materials of the unstrengthened RC column are* $f'_c = 27.57$ *MPa for concrete and* $f_y = 275$ *MPa for steel reinforcement. The loads on the original structure are* $P_{DL} = 9.5 \times 10^5\ N$ *(Dead) and* $P_{LL} = 8.5 \times 10^5\ N$ *(Live). Due to a change on the use of the structure, a 50% increase of live loads is expected. Design the FRP reinforcement using the following material properties* $f^*_{fu} = 3300\ MPa, \varepsilon^*_{fu} = 0.016, t_f = 0.25\ mm.$

Solution to Example 13.4

Step 1 *Determine the required load capacity and the maximum allowable strengthening. According to (13.1) the required load capacity of the strengthened system for the new loading is*

$$P_u = 1.2\,(P_{DL}) + 1.6\,(1.5P_{LL}) = 1.2\left(9.5 \times 10^5\right) + 1.6\left(1.5 \times 8.5 \times 10^5\right) = 3.18 \times 10^6\ N$$

where the minimum load capacity of the unstrengthened is given by (13.3) as follows

$$(P_u)_{US} = 1.1\,(P_{DL}) + 0.75\,(1.5P_{LL}) = 1.1\left(9.5 \times 10^5\right) + 0.75\left(1.5 \times 8.5 \times 10^5\right) = 2.01 \times 10^6\ N$$

The unstrengthened factored load capacity, i.e., $(\phi P_n)_{US}$, *is determined by using (13.108) for the unconfined column with* $f'_{cc} = f'_c$, $\phi_{ecc} = 0.85$ *(spiral steel), and* $\phi = 0.75$ *(spiral steel) as follows*

$$\begin{aligned}
(\phi P_n)_{US} &= \phi\phi_{ecc}\left[0.85 f'_c\,(A_g - A_{st}) + f_y A_{st}\right] \\
(\phi P_n)_{US} &= 0.75 \times 0.85\left[0.85 \times 27.57 \times \left(1.589 \times 10^5 - 1061\right) + 275 \times 1061\right] \\
(\phi P_n)_{US} &= 2.54 \times 10^6\ N > 2.01 \times 10^6\ N
\end{aligned}$$

that satisfies

$$(\phi P_n)_{US} > (P_u)_{US}$$

Further limitations on FRP strengthening are imposed by (13.6) as follows

$$\begin{aligned}
\%^{max} &= \frac{0.1 + 0.85(1.5P_{LL})/P_{DL}}{1.1 + 0.75(1.5P_{LL})/P_{DL}} \times 100 \simeq 59\% \\
\%^{actual} &= \frac{(P_u) - (\phi P_n)_{US}}{(\phi P_n)_{US}} \times 100 = 25\%
\end{aligned}$$

Step 2 *Calculate the confining pressure and the number of the layers.*
Preliminary design quantities regarding the FRP are determined by means of (13.10) for $C_E = 0.85$ *(exterior environmental exposure, see Table 13.2) as follows*

$$f_{fu} = C_E f^*_{fu} = 0.85 \times 3300 = 2805\ MPa$$
$$\varepsilon_{fu} = C_E \varepsilon^*_{fu} = 0.85 \times 0.016 = 0.0136$$
$$E_f = f_{fu}/\varepsilon_{fu} = 2.062 \times 10^5\ MPa$$

An estimate of the compressive strength of the confined concrete is calculated by using (13.107) as

$$f'_{cc} = \frac{1}{0.85\,(A_g - A_{st})}\left(\frac{P_u}{\phi_{ecc}\phi} - f_y A_{st}\right) =$$
$$= \frac{1}{0.85 \times (1.589 \times 10^5 - 1061)}\left(\frac{3.18 \times 10^6}{0.85 \times 0.75} - 275 \times 1061\right) = 35.0\ MPa$$

where the confining pressure f_l *is estimated by (13.114) with* $\psi_f = 0.95$ *and* $\kappa_a = 1$ *because the section is circular as*

$$f_l = \frac{f'_{cc} - f'_c}{3.3\psi_f \kappa_a} = \frac{35.0 - 27.57}{3.3 \times 0.95} = 2.37\ MPa$$

The number of the layers is found by using (13.115) as

$$n = \frac{D f_l}{2 t_f E_f \varepsilon_{fe}} = \frac{450 \times 2.37}{2 \times 0.25 \times 2.062 \times 10^5 \times 0.00748} = 1.38 \approx 2$$

where $\varepsilon_{fe} = 0.55 \times 0.0136 = 0.00748$.

Step 3 *Recalculate the factored axial strength capacity.*
For $n = 2$, *the confining pressure is derived by means of (13.102) and*

$$f_l = \frac{2 E_f n t_f \varepsilon_{fe}}{D} = \frac{2 \times 2.062 \times 10^5 \times 2 \times 0.25 \times 0.00748}{450} = 3.428\ MPa$$

where the concrete strength is determined by using (13.98) as

$$f'_{cc} = f'_c + 3.3\psi_f \kappa_a f_l = 27.57 + 3.3 \times 0.95 \times 1 \times 3.428 = 38.32\ MPa$$

Finally, using (13.108) the factored-load capacity is given as

$$(\phi P_n) = \phi\phi_{ecc}\left[0.85 f'_{cc}\,(A_g - A_{st}) + f_y A_{st}\right]$$
$$(\phi P_n) = 0.75 \times 0.85 \left[0.85 \times 38.32 \times \left(1.589 \times 10^5 - 1061\right) + 275 \times 1061\right]$$
$$(\phi P_n) = 3.46 \times 10^6\ N > 3.18 \times 10^6\ N$$

Step 4 *Check limitations and the service level of stresses.*
Limitations given by (13.103), (13.104)

$$f_l/f_c' = \frac{3.428}{27.57} = 0.124 \geq 0.08$$

$$\varepsilon_{ccu} = \varepsilon_c' \left[1.5 + 12\kappa_b \frac{f_l}{f_c'} \left(\frac{\varepsilon_{fe}}{\varepsilon_c'}\right)^{0.45}\right]$$

$$\varepsilon_{ccu} = 0.002 \left[1.5 + 12 \times 0.124 \left(\frac{0.00748}{0.002}\right)^{0.45}\right]$$

$$\varepsilon_{ccu} = 0.0084 < 0.01$$

The required strength P_s for the serviceability loading condition is

$$P_s = P_{DL} + P_{LL} = 2.225 \times 10^6 \ N$$

The stress levels for concrete and steel reinforcement are given by (13.111) and (13.112)

$$\begin{aligned} f_c &= (E_c P_s)/(E_c (A_g - A_{st}) + E_s A_{st}) \\ &= (24,678 \times 2.225 \times 10^6)/(24,678 \times \left(1.589 \times 10^5 - 1061\right) + 2.0 \times 10^5 \times 1061) \\ &= 13.37 MPa \leq 0.65 f_c' \\ f_s &= (E_s P_s)/(E_c (A_g - A_{st}) + E_s A_{st}) \\ &= (2.0 \times 10^5 \times 2.225 \times 10^6)/(24,678 \times \left(1.589 \times 10^5 - 1061\right) + 2.0 \times 10^5 \times 1061) \\ &= 108.34 MPa \leq 0.60 f_y \end{aligned}$$

13.5.3 Beam-column: Combined Axial Compression and Bending

Eccentric axially loaded columns are characterized by combined axial forces and bending moments. The strength of the member is determined as a combination of previous cases. FRP reinforcement is used to confine the concrete thus increasing its compressive and flexural strength. The presence of hoop FRP wrapping enhances the shear strength of the column, preventing buckling of the steel bars. In order to analyze the stress-strain behavior for confined FRP reinforcement, ACI guidelines propose the following relationship [265, 267]

$$f_c = \begin{cases} E_c \varepsilon_c - (E_c - E_2)^2 / (4 f_c') \varepsilon_c^2 & 0 \leq \varepsilon_c \leq \varepsilon_t' \\ f_c' + E_2 \varepsilon_c & \varepsilon_t' \leq \varepsilon_c \leq \varepsilon_{ccu} \end{cases} \tag{13.116}$$

with

$$E_2 = \frac{f_{cc}' - f_c'}{\varepsilon_{ccu}}; \quad \varepsilon_t' = \frac{2 f_c'}{E_c - E_2}$$

The evaluation of the strength of the section is based on the following assumptions:

1. The section remains planar and perpendicular to the neutral axis.

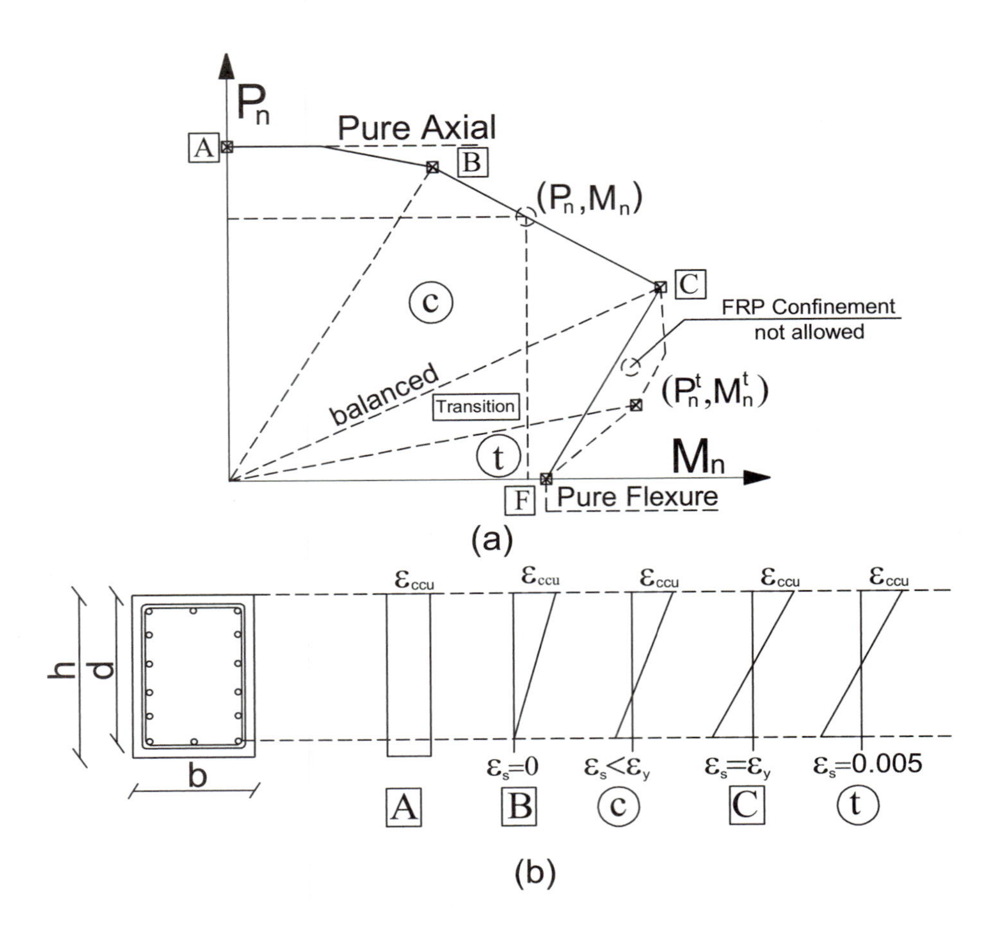

Figure 13.19: (a) Typical load-moment interaction diagram, (b) Strain distributions for the main failure modes.

2. The tensile resistance of concrete is negligible.

3. Concrete, steel reinforcement, and FRP are assumed to be perfectly bonded.

The confining effect can be analyzed by means of the following additional assumptions:

4. Compressive strength of concrete due to the presence of hoop wrapping is improved and it is taken equal to f'_{cc}, obtained by (13.98).

5. Due to confining, the maximum concrete strain ε_{ccu} is estimated according to (13.104) with limitation on the maximum value $\varepsilon_{ccu} \leq 0.01$.

6. To ensure shear integrity of the confined concrete due to FRP reinforcement, the effective strain in the FRP should be limited to $\varepsilon_{fe} = 0.004 \leq \kappa_{\varepsilon}\varepsilon_{fu}$.

The nominal capacity of the member is expressed as a function of the stress resultants, i.e., the axial force and bending moment P_n, M_n. Interaction diagrams

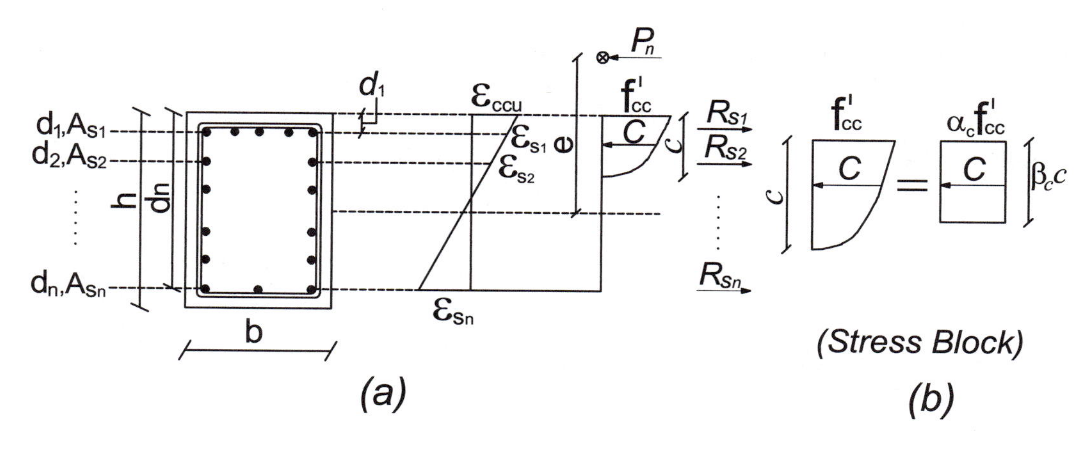

Figure 13.20: (a) Strain, stress, and resultants forces. (b) Stress-block equivalence for confined section.

reproduce all possible failure modes of the column (Figure 13.19). ACI guidelines limit the reinforcement effect provided by the FRP to those states in which the concrete fails reaching the ultimate compressive strain, i.e., $\varepsilon_c = \varepsilon_{ccu}$. Therefore, the interaction diagram of the strengthened section differs with respect to the unconfined section only above the balanced configuration line in the P-M diagram (Figure 13.19). Moreover, to facilitate the construction of the interaction diagrams, ACI guidelines suggest to evaluate bilinear curves passing through the following points:

- pure compression at uniform strain ε_{ccu} (point A).
- strain distribution with extreme compressive fiber equal to ε_{ccu} and zero strain at the steel reinforcement close to the tensile fiber (point B).
- balanced failure configuration with compressive strain of concrete equal to ε_{ccu} and tensile strain of steel reinforcement equal to yielding (point C).

The interaction diagram is developed taking into account the equilibrium of the axial force and bending moment resulting from a given strain and stress distribution (Figure 13.20). The equilibrium equations for internal forces are written as

$$P_n = C - \sum_{i=1}^{n_s} R_{si} \tag{13.117}$$

$$M_n = P_n e = C\left(\frac{h}{2} - c_g\right) + \sum_{i=}^{n_s} R_{si}\left(d_i - \frac{h}{2}\right) \tag{13.118}$$

where n_s is total number of steel bars, c_g is the distance from the extreme compression fiber of the compressive stress resultant, d_i is the distance of the i-th steel bar from the compressive concrete fiber, R_{si} is the tensile force of the i-th steel bar, and e is the eccentricity of the stress resultant with respect to the centroid of the geometric section. Since the assumed linear distribution of strain, the strains of the i-th steel bars can be expressed as a function of the concrete strain at the top fiber of the section as

$$\varepsilon_{si} = \frac{\varepsilon_c}{c}\,(d_i - c)\;\; i = 1, ..., n \tag{13.119}$$

In Figure 13.20, the compression stress resultant in concrete C for a confined beam-column and its position $c_g = \beta_c\, c/2$ (Figure 13.20) with respect to the compression fiber ε_{ccu} are determined by the integration of stress-strain relationships (13.116)

$$C = \int_0^c f_c\,(y)dy = \alpha_c f'_{cc}\beta_c\; c\; b \quad \text{(confined FRP)} \tag{13.120}$$

Alternatively, they can be determined as a function of the ratio f'_{cc}/f'_c and the failure compressive concrete strain ε_{ccu} by utilizing confined stress-block coefficients α_c, β_c, which are obtained by substituting a constant distribution of compressive stresses equal to $\alpha_c\, f'_{cc}$, spanning over the depth $\beta_c c$, for the actual distribution of the compressive stresses given by (13.116), as follows

$$\int_0^c f_c\,(y)\; b\; dy - \beta_c\,(f'_{cc}\alpha_c)\; b\; c = 0 \tag{13.121}$$

$$\int_0^c f_c\,(y)\, b\,(c-y)dy - \beta_c\,(f'_{cc}\alpha_c)\; b\; c\left(c - \frac{\beta_c c}{2}\right) = 0 \tag{13.122}$$

The confined stress-block parameters α_c, β_c, are available in [1] in tabular and graphic form for various values of compressive strength of concrete f'_c. For example, curves of α_c, β_c vs. ε'_{cc} for $f'_c = 27.57$ MPa are shown in Figure 13.21. In the plot, values of the ratio $r = f'_{cc}/f'_c$ are shown incrementally from $r = 1$ to $r = 3$ on increments of 0.2.

For an unconfined section, the compressive resultant and its centroid measured with respect to the compressive fiber are determined by classical definition of Whitney's stress-block formula as

$$C = \int_0^c f_c\,(y)bdy = 0.85 f'_c b\beta_1 c \quad \text{(unconfined FRP)} \tag{13.123}$$

$$\tag{13.124}$$

with $c_g = \beta_1\, c/2$ and β_1 expressed by (13.25).

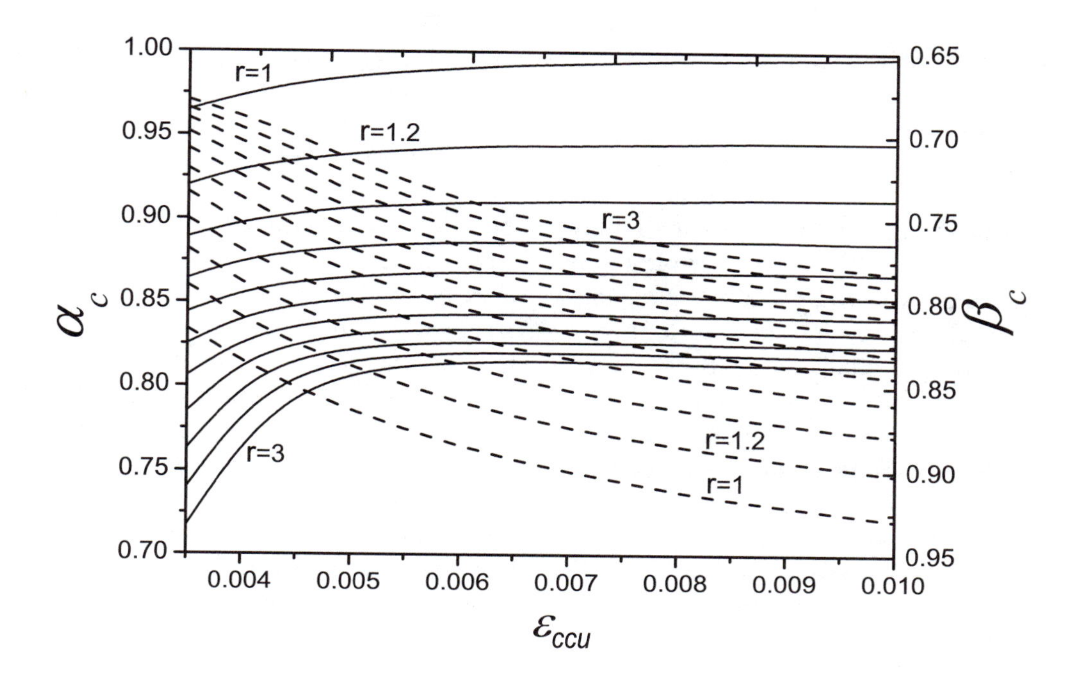

Figure 13.21: Confined stress-block parameters α_c (solid lines) and β_c (broken lines) for concrete with $f'_c = 27.57$ MPa, parametrically for $r = f'_{cc}/f'_c$ on 0.2 increments. See [1] for other values of f'_c.

The internal forces in the steel bars, for the cases of yielded and not yielded bars, are given by the following expressions

$$R_{si} = f_{si} A_{si} \text{ with } f_{si} = \begin{cases} f_y & \text{if} \quad \varepsilon_{si} > \varepsilon_y \\ -f_y & \text{if} \quad \varepsilon_{si} < -\varepsilon_y \\ E_s \varepsilon_{si} & \text{if} \quad -\varepsilon_y < \varepsilon_{si} < \varepsilon_y \end{cases} \tag{13.125}$$

Development of interaction diagrams requires an iterative calculation, with the pair P_n, M_n, evaluated by (13.117) and (13.118) for various strain distributions at failure. For example, the point in the interaction diagram corresponding to the steel-balanced configuration presents a strain distribution reaching simultaneously the yield-strain ($\varepsilon_{sn} = \varepsilon_y$) and strain to failure ($\varepsilon_c = \varepsilon_{ccu}$) in steel reinforcement and concrete, respectively. Therefore, the position of the neutral axis as well as the deformations of the compressive steel reinforcement are determined by a linear distribution of strains as

$$c = \frac{\varepsilon_{ccu}}{\varepsilon_{ccu} + \varepsilon_y} d_n \tag{13.126}$$

$$\varepsilon_{si} = \frac{\varepsilon_{ccu}}{c} (d_i - c) \;\; i = 1, ..., n_s \tag{13.127}$$

The compressive and tensile stress resultants are derived by using (13.120) and (13.125). The strength point of the section in the interaction diagram P_b, M_b, is determined from the equilibrium equations (13.117) and (13.118). The pair P_b, M_b, or P_b, e_b, corresponds to a point in the interaction diagram with failure mechanism based on the balanced strengthening configuration (BSC). For lower eccentricities $e \leq e_b$, the concrete crushes before yield of steel reinforcement, i.e., $\varepsilon_s \leq \varepsilon_y$ or $\varepsilon_s \leq 0.002$ for the case of grade 60 steel.

The failure modes involving failure in the concrete with the steel not yielded are known in the literature as compression-controlled. In this case, a brittle mode of failure is expected without significant warning before collapse. On the other hand, when the strain on the tensile steel reinforcement is greater or equal than 0.005, the mode of failure is tension-controlled. In this case, the behavior of the member is ductile and the corresponding failure mode results in high deflections and cracks with ample warning before failure. Between tension- and compression-controlled cases, there is a transition zone, in which combined ductile and brittle behaviors are possible (Figure 13.19). In the transition and tension-controlled zones, the failure mode is based on the yielding of the steel reinforcement with the concrete below its ultimate value. Consequently, the FRP is not effective, since it only increases the strength of the concrete due to confinement effect and thus, unconfined and confined strengths are virtually identical.

Alternately, interaction diagrams can be developed by assuming a value of the eccentricity $e = M_n/P_n$, then writing the equilibrium equations to determine the neutral axis position c and the load and moment capacities P_n, M_n. This procedure is based on the assumption that the failure mode is known priori. If the capacity and the position of the neutral axis are sought for a given value of the eccentricity, the following parameters are unknown

- the neutral axis, c
- the strain or stress level in i-th steel bars, ε_{si}, f_{si}
- the load P_n for the given capacity eccentricity, e

Then, the interaction diagram can be developed by means of the following iterative procedure

1. Select a trial value of the neutral axis position c. Typical values should be chosen in the range between tension- and compression-controlled cases for unstrengthened sections, i.e.,
   (a) compression-controlled: $c/d = \varepsilon/(0.002 + \varepsilon) = 0.6$
   (b) tension-controlled: $c/d = \varepsilon/(0.005 + \varepsilon) = 0.375$
2. Calculate the stress resultants in the tension and compression steel reinforcement by (13.120) and (13.125).
3. Determine the load and moment capacities by (13.117–13.118).

4. Check if the calculated eccentricity $e^* = M_n/P_n$ is equal to the assumed one

$$ERR = \frac{e - e^*}{e^*} \leq \text{tolerance}$$

5. If the tolerance condition is satisfied (typically 10^{-3}), the pair P_n, M_n represents the capacity of the member. Otherwise, a new value of the neutral axis position is chosen in Step 1.

13.5.4 Summary Verification Procedure: Beam-column

Input data: required axial force and moment capacity P_u, M_u; geometry and area of the longitudinal steel reinforcement d_i, A_{si}; FRP thickness and number of layers t_f, n; concrete geometry b, h; and concrete, steel, and FRP material properties $f'_c, f_y, f^*_{fu}, \varepsilon^*_{fu}$.

Output data: factored nominal axial and moment capacity $\phi(P_n, M_n)$.

Step 1 Determine preliminary design parameters, namely ρ_g, A_e/A_c, ε_{fe}, κ_a, f_l, f'_{cc}, and ε_{ccu}; read the confined stress-block parameters α_c, β_c, form tables or graphs [1] such as Figure 13.21; then, check the following inequalities

$$f_l/f'_c > 0.08$$

and

$$\varepsilon_{ccu} < 0.01$$

Step 2 Assume the failure mode of the section by adopting the strain ultimate values for concrete and steel bars. Calculate the neutral axis position c and the strain and stress in the steel bars

$$c = \frac{\varepsilon_c}{(\varepsilon_c + \varepsilon_{sn})} d$$

$$\varepsilon_{si} = \frac{\varepsilon_c}{c}(d_i - c)$$

$$f_{si} = \begin{cases} f_y \text{ if } \varepsilon_{si} \geqslant \varepsilon_y \text{ and } \varepsilon_{si} \leqslant -\varepsilon_y \\ E_s \varepsilon_{si} \text{ if - } \varepsilon_y < \varepsilon_{si} < \varepsilon_y \end{cases}$$

Step 3 Calculate the compressive resultants in concrete by (13.120) or (13.123) for confined or unconfined section, respectively and the tensile resultants in the steel reinforcement by (13.125).

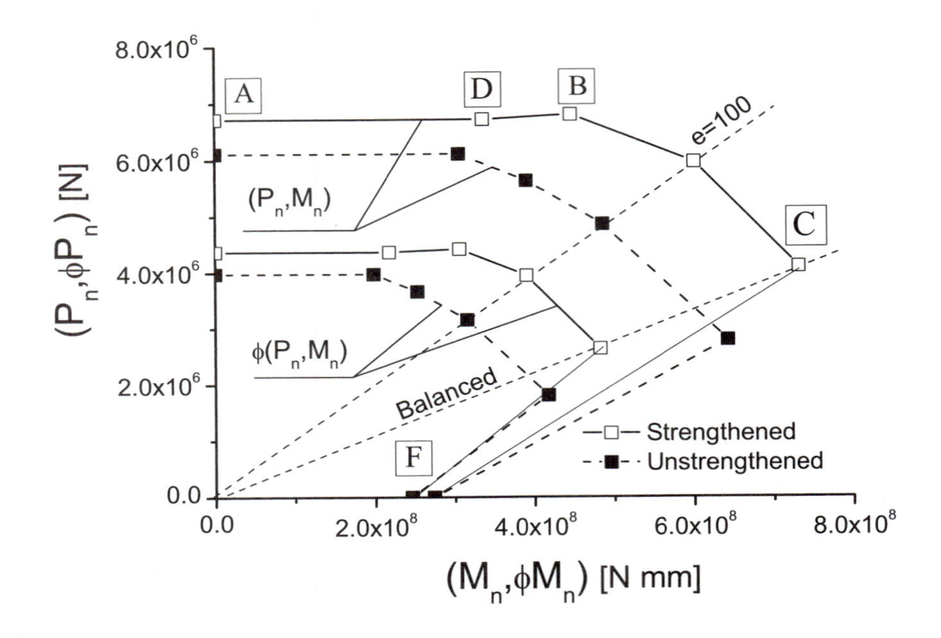

Figure 13.22: Interaction diagram for unstrengthened and strengthened sections.

Step 4 Determine load and moment P_n, M_n, by (13.117), (13.118) and the corresponding factored value $\phi\,(P_n, M_n)$ by calculating ϕ from (13.15).

Example 13.5 *A nonslender column with internal exposure condition and the cross-section reported in Figure 13.20 (b = 500, d_1 = 40 mm, d_2 = 185 mm, d_3 = 315 mm, d_4 = 460 mm) is reinforced with #7 bars (d = 22 mm). The steel reinforcement area are: $A_{s1} = A_{s4} = 1520$ mm^2, $A_{s2} = A_{s3} = 760$ mm^2. The RC properties are $f'_c = 27.57$ MPa, $f_y = 413.68$ MPa. Calculate the interaction diagram for the following cases:*

1. *Unstrengthened member*
2. *Strengthened member with three layers ($n = 3$) of carbon–epoxy FRP with properties $f^*_{fu} = 3792$ MPa, $\varepsilon^*_{fu} = 0.0166$, $t_f = 0.333$ mm, $E_f = f^*_{fu}/\varepsilon^*_{fu} = 2.275 \times 10^5$ MPa*

Solution to Example 13.5 *The interaction diagram is constructed by interpolating P, M, values corresponding to the following points (Figure 13.22):*

Pure axial point (A,D): strain distribution with external compressive fiber equal to ε_{ccu} and zero strain at the steel reinforcement close to the tensile fiber (B), balanced configuration (C), eccentricity equal to 100 (E), pure flexure (F).

For the analysis of the strengthened member, the following preliminary design parameters are calculated:

$$\rho_g = \frac{A_{s1} + A_{s2} + A_{s3} + A_{s4}}{b \times h} = \frac{4560}{250000} = 0.01824$$

$$\frac{A_e}{A_c} = \frac{1 - \left[(b/h)(h - 2r_c)^2 + (h/b)(b - 2r_c)^2 \right] / 3A_g - \rho_g}{1 - \rho_g} = 0.449$$

$$\varepsilon_{fe} = \min(0.004, \kappa_\varepsilon \varepsilon_{fu}) = \min(0.004, 0.55 \times 0.017 \times 0.95) = 0.004$$

$$\kappa_a = \frac{A_e}{A_c}\left(\frac{b^2}{h^2}\right) = 0.449, \quad \kappa_b = \frac{A_e}{A_c}\left(\frac{h}{b}\right)^{0.5} = 0.449$$

$$f_l = 2E_f n t_f \varepsilon_{fe}/D$$

$$= 2 \times 2.275 \times 10^5 \times 3 \times 0.333 \times 0.004/\sqrt{500^2 + 500^2} = 2.57 \ MPa$$

$$\text{with } f_l/f'_c = 2.57/27.57 = 0.093 > 0.08$$

$$f'_{cc} = f'_c + \psi_f 3.3 \kappa_a f_l = 31.2 \ MPa$$

$$\varepsilon_{ccu} = \varepsilon'_c \left[1.5 + 12\kappa_b \frac{f_l}{f'_c} \left(\frac{\varepsilon_{fe}}{\varepsilon'_c} \right)^{0.45} \right] = 0.0043 < 0.01$$

For $f'_{cc}/f_{c'} = 1.13$ *and* $\varepsilon_{ccu} = 0.0043$, *interpolating in Figure 13.21, the confined stress-block parameters are* $\alpha_c = 0.947$, *and* $\beta_c = 0.829$.

Step 1. Axial loading condition

Strengthened case (Point A) The nominal capacity for confined concrete is calculated by (13.108) with $\phi_{ecc} = 0.80$ *for columns with tie-shear reinforcement*

$$P_n = 0.80 \left[0.85 f'_{cc} (A_g - A_{st}) + f_y A_{st} \right]$$

$$P_n = 0.80 \left[0.85 \times 31.2 \times \left(2.5 \times 10^5 - 4{,}560 \right) + 413.68 \times 4{,}560 \right] = 6.71 \times 10^6 \ N$$

and the factored load capacity is given by (13.108)

$$\phi P_{n\,\max} = 0.65 \times 6.71 \times 10^6 = 4.36 \times 10^6 \ N$$

Unstrengthened case The nominal capacity is determined by using (13.108) taking the confined concrete strength equal to the unconfined one, i.e., $f'_{cc} = f'_c$

$$P_n = 0.80 \left[0.85 f'_c (A_g - A_{st}) + f_y A_{st} \right]$$

$$P_n = 0.80 \left[0.85 \times 27.57 \times \left(2.5 \times 10^5 - 4{,}560 \right) + 413.86 \times 4{,}560 \right] = 6.11 \times 10^6 \ N$$

and the factored load capacity is given by (13.108) with ϕ *obtained from Table 13.1.*

$$\phi P_{n\,\max} = 0.65 \times 6.11 \times 10^6 = 3.97 \times 10^6 \ N$$

In order to take into account load eccentricities due to imperfection in the construction or in the loads estimation, ACI guidelines introduce a minimum eccentricity $e = 0.1h$ *where* h *is the depth or width of the member, in this case* $h = 500 \ mm$ *(Figure 13.20). The corresponding points for the strengthened and unstrengthened cases in the interaction diagram are defined as follows.*

Strengthened case (Point D) The nominal capacity is:

$$(P_n, M_n) = \left(6.71 \times 10^6\ N, 6.71 \times 10^6 \times 0.1 \times 500\ N\ mm\right)$$

where the factored capacity is

$$\phi\,(P_n, M_n) = 0.65 \times \left(6.71 \times 10^6, 3.35 \times 10^8\right) = \left(4.36 \times 10^6\,N, 2.18 \times 10^8\ N\ mm\right)$$

Unstrengthened case The nominal capacity is:

$$(P_n, M_n) = \left(6.11 \times 10^6\ N, 6.11 \times 10^6 \times 0.1 \times 500\ N\right)$$

where the factored capacity is

$$\phi\,(P_n, M_n) = 0.65 \times \left(6.11 \times 10^6\ N, 3.05 \times 10^8\right) = \left(3.97 \times 10^6\,N, 1.98 \times 10^8\ N\ mm\right)$$

Step 2. Point B *The strain distribution on the extreme compressive fiber is ε_{ccu} and the strain at the steel reinforcement close to the tensile fiber is zero.*
Unstrengthened case: The concrete strain is assumed equal to its strain to failure $\varepsilon_c = \varepsilon_{cu} = 0.003$, while the strain in the reinforced steel close to the tensile fiber is zero, i.e., reaches yielding, i.e., $\varepsilon_{s4} = 0$. Therefore, the position of the neutral axis is given by the linearity distribution and the compressive strain of the steel reinforcement is

$$\begin{aligned}
c &= \frac{\varepsilon_{cu}}{(\varepsilon_{cu} + 0)} d = 460\ mm \\
\varepsilon_{s1} &= \frac{\varepsilon_{cu}}{c}(d_1 - c) = \frac{0.003}{460}(40 - 460) = -0.00274, && f_{s1} = -413.68\ N/mm^2 \\
\varepsilon_{s2} &= \frac{\varepsilon_{cu}}{c}(d_2 - c) = \frac{0.003}{460}(185 - 460) = -0.00179, && f_{s2} = -359\ N/mm^2 \\
\varepsilon_{s3} &= \frac{\varepsilon_{cu}}{c}(d_3 - c) = \frac{0.003}{460}(315 - 460) = -0.000946, && f_{s3} = -189\ N/mm^2 \\
\varepsilon_{s4} &= 0, && f_{s4} = 0
\end{aligned}$$

The stress resultants in the concrete and steel reinforcement are

$$\begin{aligned}
C &= 0.85 f_c' b \beta c = 0.85 \times 27.57 \times 500 \times 0.85 \times 460 = 4.58 \times 10^6\,N \\
R_{s1} &= f_{s1} A_{s1} = -413.68 \times 1520 = -6.28 \times 10^5\,N \\
R_{s2} &= f_{s2} A_{s2} = -359 \times 760 = -2.72 \times 10^5\,N \\
R_{s3} &= f_{s3} A_{s3} = -189 \times 760 = -1.43 \times 10^5\,N \\
R_{s4} &= f_{s4} A_{s4} = 0 \times 1520 = 0
\end{aligned}$$

The load and moment capacities are

$$P_n = C - \sum_{i=}^{n_s} R_{si} = 5.62 \times 10^6\ N$$

$$M_n = P_n e = C\left(\frac{h}{2} - \beta_1 c/2\right) + \sum_{i=}^{n_s} R_s\left(d_i - \frac{h}{2}\right) = 3.90 \times 10^8\ N\ mm$$

with

$$e = M_n/P_n = 69 \ mm$$

Finally, the factored capacity is

$$\phi(P_n, M_n) = 0.65 \times \left(5.62 \times 10^6, 3.90 \times 10^8\right) = \left(3.65 \times 10^6 \ N, 2.53 \times 10^8 \ N \ mm\right)$$

Strengthened case *Due to the presence of the FRP, the compressive strain is equal to the confined one, i.e.,* $\varepsilon_c = \varepsilon_{ccu} = 0.0043$. *The neutral axis and the strains in the steel reinforcement bars are equal to*

$$c = \frac{\varepsilon_{ccu}}{(\varepsilon_{ccu} + 0)} d = \frac{0.0043}{(0.0043 + 0)} 460 = 460 \ mm$$

$$\varepsilon_{s1} = \frac{\varepsilon_{cu}}{c}(d_1 - c) = \frac{0.0043}{460}(40 - 460) = -0.0040, \qquad f_{s1} = -413.68 \ N/mm^2$$

$$\varepsilon_{s2} = \frac{\varepsilon_{cu}}{c}(d_2 - c) = \frac{0.0043}{460}(185 - 460) = -0.0026, \qquad f_{s2} = -413.68 \ N/mm^2$$

$$\varepsilon_{s3} = \frac{\varepsilon_{cu}}{c}(d_2 - c) = \frac{0.0043}{460}(315 - 460) = -0.00136, \qquad f_{s3} = -271 \ N/mm^2$$

$$\varepsilon_{s4} = 0, \qquad f_{s4} = 0$$

The resultants in the concrete and steel reinforcement are

$$C = \alpha_c f'_{cc} b \beta_c c = 0.947 \times 31.20 \times 500 \times 0.829 \times 460 = 5.63 \times 10^6 \ N$$

$$R_{s1} = f_{s1} A_{s1} = -413.68 \times 1520 = -6.28 \times 10^5 \ N$$

$$R_{s2} = f_{s2} A_{s2} = -413.68 \times 760 = -3.14 \times 10^5 \ N$$

$$R_{s3} = f_{s3} A_{s3} = -275 \times 760 = -2.06 \times 10^5 \ N$$

$$R_{s4} = f_{s4} A_{s4} = 0 \times 1520 = 0$$

The load and moment capacities are

$$P_n = C - \sum_{i=}^{n_s} R_{si} = 6.78 \times 10^6 \ N$$

$$M_n = C\left(\frac{h}{2} - \beta_c c/2\right) + \sum_{i=}^{n_s} R_s \left(d_i - \frac{h}{2}\right) = 4.73 \times 10^8 \ N \ mm$$

with

$$e = M_n/P_n = 70 \ mm$$

Finally, the factored capacity is

$$\phi(P_n, M_n) = 0.65 \times \left(6.78 \times 10^6, 4.73 \times 10^8\right) = \left(4.41 \times 10^6 \ N, 3.00 \times 10^8 \ N \ mm\right)$$

Step 3. FRP balanced strain configuration (Point C) *The assumed strain distribution produces yield of the steel and crushing of concrete simultaneously.*

Unstrengthened case: The concrete strain is assumed equal to its strain to failure $\varepsilon_c = \varepsilon_{cu} = 0.003$, *while the strain in the reinforced steel reaches yielding, i.e.,* $\varepsilon_y = f_y/E_s = 0.00207$. *Therefore, the position of the neutral axis is given by (13.126) for* $\varepsilon_{ccu} = \varepsilon_c$ *and the compressive steel reinforcement strain is derived by using (13.127) as follows*

$$
\begin{aligned}
c &= \frac{\varepsilon_{cu}}{(\varepsilon_{cu} + \varepsilon_y)} d = \frac{0.003}{(0.003 + 0.00207)} 460 = 272 \ mm \\
\varepsilon_{s1} &= \varepsilon_{cu} (d_1 - c)/c = \frac{0.003}{272}(40 - 272) = -0.0025, && f_{s1} = -413.68 \ N/mm^2 \\
\varepsilon_{s2} &= \varepsilon_{cu} (d_2 - c)/c = \frac{0.003}{272}(185 - 272) = -0.00096, && f_{s2} = \text{-}192 \ N/mm^2 \\
\varepsilon_{s3} &= \varepsilon_{cu} (d_3 - c)/c = \frac{0.003}{272}(315 - 272) = 0.0004, && f_{s3} = 94 \ N/mm^2 \\
\varepsilon_{s4} &= 0.002, && f_{s4} = 413.68 \ N/mm^2
\end{aligned}
$$

The resultants in the concrete and steel reinforcement are

$$
\begin{aligned}
C &= 0.85 f'_c b \beta c = 0.85 \times 27.57 \times 500 \times 0.85 \times 272 = 2.71 \times 10^6 \ N \\
R_{s1} &= f_{s1} A_{s1} = -413.68 \times 1520 = -6.28 \times 10^5 \ N \\
R_{s2} &= f_{s2} A_{s2} = -192 \times 760 = -1.46 \times 10^5 \ N \\
R_{s3} &= f_{s3} A_{s3} = 94 \times 760 = 71,440 \ N \\
R_{s4} &= f_{s4} A_{s4} = 413.68 \times 1,520 = 6.28 \times 10^5 \ N
\end{aligned}
$$

The load and moment capacities are

$$
\begin{aligned}
P_n &= C - \sum_{i=}^{n_s} R_{si} = 2.78 \times 10^6 \ N \\
M_n &= P_n e = C\left(\frac{h}{2} - c_g\right) + \sum_{i=}^{n_s} R_s \left(d_i - \frac{h}{2}\right) = 6.42 \times 10^8 \ N \ mm
\end{aligned}
$$

with

$$
e = M_n / P_n = 230 \ mm
$$

Finally, the factored capacity is

$$
\phi (P_n, M_n) = 0.65 \times \left(2.78 \times 10^6, 6.42 \times 10^8\right) = \left(1.80 \times 10^6 \ N, 4.17 \times 10^8 \ N \ mm\right)
$$

Strengthened case (Point C) Due to the presence of the FRP, the compressive strain is equal to the confined one, i.e., $\varepsilon_c = \varepsilon_{ccu} = 0.0043$. *The neutral axis and the strains in the steel reinforcement bars are equal to*

$$c = \frac{\varepsilon_{ccu}}{(\varepsilon_{ccu} + \varepsilon_y)} d = \frac{0.0043}{(0.0043 + 0.00207)} 460 = 311\ mm$$

$$\varepsilon_{s1} = \varepsilon_{ccu}(d_1 - c)/c = 0.0043(40 - 311)/311 = -0.0038, \qquad f_{s1} = -413.68\ N/mm^2$$

$$\varepsilon_{s2} = \varepsilon_{ccu}(d_2 - c)/c = 0.0043(185 - 311)/311 = -0.0017, \qquad f_{s2} = -340\ N/mm^2$$

$$\varepsilon_{s3} = \varepsilon_{ccu}(d_3 - c)/c = 0.0043(315 - 311)/311 = 0.000055, \qquad f_{s3} = 11\ N/mm^2$$

$$\varepsilon_{s4} = 0.00207, f_{s4} = 413.68\ N/mm^2$$

The stress resultants in the concrete and steel reinforcement are

$$C = \alpha_c f'_{cc} b \beta_c c = 0.947 \times 31.20 \times 500 \times 0.829 \times 311 = 3.83 \times 10^6\ N$$

$$R_{s1} = f_{s1} A_{s1} = -413.68 \times 1520 = -6.28 \times 10^5\ N$$

$$R_{s2} = f_{s2} A_{s2} = -340 \times 760 = -2.58 \times 10^5\ N$$

$$R_{s3} = f_{s3} A_{s3} = 11 \times 760 = 8360\ N$$

$$R_{s4} = f_{s4} A_{s4} = 413.68 \times 1520 = 6.28 \times 10^5\ N$$

The load and movement capacity are

$$P_n = C - \sum_{i=}^{n_s} R_{si} = 4.05 \times 10^6\ N$$

$$M_n = P_n e = C\left(\frac{h}{2} - \beta_c c/2\right) + \sum_{i=}^{n_s} R_s \left(d_i - \frac{h}{2}\right) = 7.42 \times 10^8\ N\ mm$$

with

$$e = M_n/P_n = 183\ mm$$

Finally, the factored capacity is

$$\phi(P_n, M_n) = 0.65 \times \left(4.05 \times 10^6, 7.42 \times 10^8\right) = \left(2.63 \times 10^6\ N, 4.82 \times 10^8\ N\ mm\right)$$

Step 4. Fixed value of the eccentricity (Point E) *It is required to calculate the point in the interaction diagrams for a value of eccentricity "e", equal to the intersection with the interaction diagram of a line equal to 45°, i.e.,* $e = M_n/P_n = 100$ *mm. Since, the eccentricity is lower than the one corresponding to the case of balanced strain configuration, i.e.,* $e = 152$ *mm, an initial value of the ratio* c/d *greater than* 0.6 *is assumed. The point* P_n, M_n, *in the interaction diagram is determined by means of the following iterative procedure.*

Unstrengthened case: *A trial value of the ratio* c/d *is chosen to be* 0.7. *The position of the neutral axis and the strain distribution on the section are*

$$c = 0.7d = 322 \ mm$$

$$\varepsilon_{s1} = \frac{\varepsilon_{cu}}{c}(d_1 - c) = \frac{0.003}{322}(40 - 322) = -0.0026, \qquad f_{s1} = -413.68 \ N/mm^2$$

$$\varepsilon_{s2} = \frac{\varepsilon_{cu}}{c}(d_2 - c) = \frac{0.003}{322}(185 - 322) = -0.0012, \qquad f_{s2} = -255 \ N/mm^2$$

$$\varepsilon_{s3} = \frac{\varepsilon_{cu}}{c}(d_3 - c) = \frac{0.003}{322}(315 - 322) = -0.65 \times 10^{-4}, \qquad f_{s3} = -13 \ N/mm^2$$

$$\varepsilon_{s4} = \frac{\varepsilon_{cu}}{c}(d_4 - c) = \frac{0.003}{322}(460 - 322) = 0.00128, \qquad f_{s4} = 257 \ N/mm^2$$

Stress resultants in concrete and steel reinforcements are

$$C = 0.85 f'_c b \beta c = 0.85 \times 27.57 \times 500 \times 0.85 \times 322 = 3.20 \times 10^6 \ N$$

$$R_{s1} = f_{s1} A_{s1} = -413.68 \times 1520 = -6.28 \times 10^5 \ N$$

$$R_{s2} = f_{s2} A_{s2} = -255 \times 760 = -1.93 \times 10^5 \ N$$

$$R_{s3} = f_{s3} A_{s3} = -13 \times 760 = -9,880 \ N$$

$$R_{s4} = f_{s4} A_{s4} = 257 \times 1,520 = 3.90 \times 10^5 \ N$$

The load and moment capacities

$$P_n = C - \sum_{i=}^{n_s} R_{si} = 3.65 \times 10^6 \ N$$

$$M_n = P_n e = C\left(\frac{h}{2} - c_g\right) + \sum_{i=}^{n_s} R_s \left(d_i - \frac{h}{2}\right) = 5.89 \times 10^8 \ N\,mm$$

with $e = M_n/P_n = 161 \ mm$. *The percentage error is equal to* $ERR = (161 - 100)/100 = 61\,\%$, *and thus another value of the ratio* c/d *should be used. The pair* P_n, M_n, *corresponding to* $e = 100 \ mm$, *is determined iteratively increasing the value of* c/d *until the ratio* M_n/P_n *is equal to* $100 \ mm$. *The solution is given by a value* c/d *equal to* 0.875, *corresponding to the following values*
Neutral axis position

$$c = 0.875d \cong 402 \ mm$$

Strain distribution

$$\varepsilon_{s1} = \varepsilon_{cu}/c \times (d_1 - c) = 0.003/402\,(40 - 402) = -0.0027, \qquad f_{s1} = -413.68 \ N/mm^2$$

$$\varepsilon_{s2} = \varepsilon_{cu}/c \times (d_2 - c) = 0.003/402\,(185 - 402) = -0.0016, \qquad f_{s2} = -324 \ N/mm^2$$

$$\varepsilon_{s3} = \varepsilon_{cu}/c \times (d_3 - c) = 0.003/402\,(315 - 402) = -0.652 \times 10^{-4}, \qquad f_{s3} = -130 \ N/mm^2$$

$$\varepsilon_{s4} = \varepsilon_{cu}/c \times (d_4 - c) = 0.003/402\,(460 - 402) = 0.000428, \qquad f_{s4} = 85 \ N/mm^2$$

Stress resultants

$$C_c = 0.85 f_c' b\beta c = 0.85 \times 27.57 \times 500 \times 0.85 \times 402 = 4.01 \times 10^6\ N$$
$$R_{s1} = f_{s1}A_{s1} = -413.68 \times 1520 = -6.28 \times 10^5\ N$$
$$R_{s2} = f_{s2}A_{s2} = -324 \times 760 = -2.46 \times 10^5\ N$$
$$R_{s3} = f_{s3}A_{s3} = -130 \times 760 = -9.88 \times 10^4\ N$$
$$R_{s4} = f_{s4}A_{s4} = 87 \times 1520 = 1.32 \times 10^5\ N$$

Load and moment capacities

$$P_n = 4.85 \times 10^6\ N \quad M_n = 4.85 \times 10^8\ N\ mm \quad e = \frac{M_n}{P_n} \approx 100\ mm$$

Factored capacity

$$\phi(P_n, M_n) = 0.65 \times \left(4.85 \times 10^6, 4.85 \times 10^8\right) = \left(3.15 \times 10^6\ N, 3.15 \times 10^8\ N\ mm\right)$$

Strengthened case *A trial value of the ratio c/d is chosen, equal to c/d=0.865. The position of the neutral axis and the strain distribution of the section are*

$$c = 0.865 \times d = 398\ mm$$
$$\varepsilon_{s1} = \varepsilon_{ccu}(d_1 - c)/c = 0.0043(40 - 398)/398 = -0.0039, \qquad f_{s1} = -413.68\ N/mm^2$$
$$\varepsilon_{s2} = \varepsilon_{ccu}(d_2 - c)/c = 0.0043(185 - 398)/398 = -0.0023, \qquad f_{s2} = -413.68\ N/mm^2$$
$$\varepsilon_{s3} = \varepsilon_{ccu}(d_3 - c)/c = 0.0043(315 - 398)/398 = -0.0009, \qquad f_{s3} = -182\ N/mm^2$$
$$\varepsilon_{s4} = \varepsilon_{ccu}(d_4 - c)/c = 0.0043(460 - 398)/398 = 0.00067, \qquad f_{s3} = 134\ N/mm^2$$

Stress resultants

$$C = \alpha_c f_{cc}' b\beta_c c = 0.947 \times 31.20 \times 500 \times 0.829 \times 312 = 4.88 \times 10^6\ N$$
$$R_{s1} = f_{s1}A_{s1} = -413.68 \times 1520 = -6.28 \times 10^5\ N$$
$$R_{s2} = f_{s2}A_{s2} = -413.68 \times 760 = -3.14 \times 10^5\ N$$
$$R_{s3} = f_{s3}A_{s3} = 182 \times 760 = 1.38 \times 10^5\ N$$
$$R_{s4} = f_{s4}A_{s4} = 134 \times 1520 = 2.04 \times 10^5\ N$$

Load and moment capacities

$$P_n = C - \sum_{i=}^{n_s} R_{si} = 5.75 \times 10^6\ N$$
$$M_n = P_n e = C\left(\frac{h}{2} - \beta_c c/2\right) + \sum_{i=}^{n_s} R_s\left(d_i - \frac{h}{2}\right) = 6 \times 10^8\ N\ mm$$

with

$$e = M_n/P_n = 104\ mm$$

Finally, the factored capacity is

$$\phi(P_n, M_n) = 0.65 \times \left(5.76 \times 10^6, 6 \times 10^8\right) = \left(3.74 \times 10^6\ N, 3.90 \times 10^8\ N\ mm\right)$$

Step 5. Pure flexure (Point F) *The analysis is similar to Example 1, with flexural area is equal to zero ($A_f = 0$). According to ACI guidelines only FRP strengthening is not allowed for confined concrete section, thus only unstrengthened case is discussed. Since A_{s3} and A_{s2} are close to the neutral axis position, their contribution is neglected and thus only outer steel bars are taken into account.*

First, a strain distribution with tensile steel bars yielded, compressive steel bars not yielded. The tensile and compressive steel bars are denoted $A_s = A_{s4}$, and $A'_s = A_{s1}$, respectively, which are located at distances d, e, with respect to the extreme compression fiber. The position of the neutral axis is derived by (13.42) for $A_f = 0$ and taking into account a new term corresponding to the compressive steel reinforcement as follows

$$0.85 f'_c \beta_1 bc + A'_s f'_s - A_s f_y = 0$$

with

$$f'_s = E_s \varepsilon_{cu} \left(\frac{c - e}{c} \right)$$

which can be arranged as

$$A_1 c^2 + A_2 c + A_3 = 0 \rightarrow c \approx 48 \ mm$$

with

$$A_1 = 0.85 f'_c \beta_1 b = 9963 \ N/mm$$
$$A_2 = (A'_s E_s \varepsilon_{cu} - A_s f_y) = 2.83 \times 10^5 \ N$$
$$A_3 = -A'_s E_s \varepsilon_{cu} e = -3.65 \times 10^7 \ N \ mm$$

The solution of the previous equations gives the position of the neutral axis $c = 48$ mm. Once the position of the neutral axis is determined, the strains on the member should be verified by checking the following inequalities

$$\varepsilon_s = \varepsilon_{cu} \frac{d - c}{c} = 0.003 \frac{460 - 48}{48} = 0.0257 \geq 0.002 = \varepsilon_{sy} \qquad \textit{(verified)}$$
$$\varepsilon'_s = \varepsilon_{cu} \frac{c - e}{c} = 0.003 \frac{48 - 40}{48} = 0.00050 < 0.002 = \varepsilon_{sy} \qquad \textit{(verified)}$$

The nominal moment capacity is obtained by using (13.47) with $A_f = 0$ and a new term due to compressive steel reinforcement as follows

$$M_n = A_s f_y \left(d - \beta_1 c/2 \right) + A'_s f'_s \left(\beta_1 c/2 - e \right)$$
$$M_n = 1520 \times 413.68 \times (460 - 0.85 \times 48/2) + 1520 \times 2.0 \times 10^5 \times 0.0005 \times (0.85 \times 48/2 - 40)$$
$$M_n = 2.73 \times 10^8 \ Nmm$$

Finally, the factored capacity is equal to

$$\phi \left(P_n, M_n \right) = 0.9 \left(0, 2.73 \times 10^8 \right) = \left(0, 2.46 \times 10^8 \ N \ mm \right)$$

Exercises

Exercise 13.1 *Determine the factored moment capacity ϕM_n of the rectangular section subjected to interior environmental exposure with an* initial *deformation in the concrete substrate $\varepsilon_{bi} = 0.0060$. The material and geometric data of the section are: concrete, $f_c' = 20.68\ MPa$, width $b = 300\ mm$, height $h = 600\ mm$, reinforcement depth $d = 560\ mm$; steel reinforcement: $A_s = 602\ mm^2$, $f_y = 413\ MPa$; FRP, $f_{fu}^* = 2800\ MPa$, $\varepsilon_{fu}^* = 0.017$, $E_f = 164000\ MPa$, $b_f = 250\ mm$, $t_f = 0.333\ mm$, $n = 1$.*

Exercise 13.2 *Determine the FRP area and the bending strength corresponding to the balanced strengthening configuration, for the section reported in Exercise 13.1.*

Exercise 13.3 *Calculate the maximum distributed load intensity "w" that a simply supported beam of length $L = 5\ m$ is able to carry. The geometric and material data of the cross-section are: concrete, $f_c' = 27.57\ MPa$, width $b = 300\ mm$, height $h = 500\ mm$, depth of the tension reinforcement $d = 460\ mm$; steel reinforcement, $A_s = 602\ mm^2$, $f_y = 413\ MPa$; FRP, $f_{fu}^* = 2800\ MPa$, $\varepsilon_{fu}^* = 0.017$, $Ef = 164000\ MPa$, $b_f = 200\ mm$, $t_f = 1.0\ mm$, $n = 1$.*

Exercise 13.4 *Design the FRP for a simply supported beam with length $L = 7\ m$. The external uniform dead and live loads are: $w_{DL} = 11.25\ N/mm$ (Dead), and $w_{LL} = 13.25\ N/mm$ (Live). The geometric and material data of the RC structure and FRP are: concrete, $f_c' = 34.5\ MPa$, width $b = 300\ mm$, height $h = 500\ mm$, $d = 460\ mm$; longitudinal steel reinforcement, $A_s = 1,000\ mm^2$, $f_y = 413\ MPa$; FRP, $f_{fu}^* = 2,800\ MPa$, $\varepsilon_{fu}^* = 0.016$, $t_f = 0.33\ mm$.*

Exercise 13.5 *A cantilever beam with an interior exposure condition is subjected to a concentrated end force $V_u = 2.88 \times 10^5\ N$. The RC data are: concrete, $f_c' = 20.68\ MPa$, width $b = 300\ mm$, height $h = 600\ mm$, reinforcement depth $d = 560\ mm$; transverse steel reinforcement, $A_{sv} = 127.1\ mm^2$, $f_y = 413\ MPa$, $s_s = 200\ mm$. It is required to design the FRP shear strengthening for both continuous and discontinuous distribution utilizing the following FRP material: $f_{fu}^* = 3640\ MPa$, $\varepsilon_{fu}^* = 0.016$, $t_f = 0.2\ mm$.*

Exercise 13.6 *Design a nonslender circular column with six vertical rebars and circumferential rebar cage, subjected to a factored ultimate loading $V_u = 3.78 \times 10^6\ N$. The section properties are: concrete, $f_c' = 27.57\ MPa$, diameter $D = 500\ mm$; longitudinal steel reinforcement, $A_s = 1,406.72\ mm^2$, $f_y = 413\ MPa$; FRP, $f_{fu}^* = 3,800\ MPa$, $\varepsilon_{fu}^* = 0.0167$, $t_f = 0.30\ mm$.*

Exercise 13.7 *Consider a nonslender column with cross-section as in Figure 13.20 and $d_1 = 40\ mm$, $A_{s1} = 803.8\ mm^2$, $d_2 = 225\ mm$, $A_{s2} = 401.9\ mm^2$, $d_3 = 375\ mm$, $A_{s3} = 401.9\ mm^2$, $d_4 = 40\ mm$, $A_{s4} = 803.8\ mm^2$, $h = b = 600\ mm$. The materials of the RC column are $f_c' = 27.5\ MPa$, $f_y = 413\ MPa$. Calculate the interaction diagram for the following cases: (1) unstrengthened member, (2) strengthened member with $n = 3$ layers of carbon–epoxy FRP with properties $f_{fu}^* = 3,800\ MPa$, $\varepsilon_{fu}^* = 0.0161$, $t_f = 0.5\ mm$.*

Appendix A

Gauss Distribution

The probability density function (PDF) is

$$p(\alpha; \overline{\overline{\alpha}}, \alpha_\sigma) = \frac{1}{\alpha_\sigma \sqrt{2\pi}} \exp\left[-\frac{(\alpha - \overline{\overline{\alpha}})^2}{2\alpha_\sigma^2}\right] \tag{A.1}$$

where $\overline{\overline{\alpha}}$, α_σ, are the mean and standard deviation of the random variable α, respectively.

The cumulative density function (CFD) is

$$F(\alpha; \overline{\overline{\alpha}}, \alpha_\sigma) = \int_{-\infty}^{\alpha} p(\alpha')\, d\alpha' = \frac{1}{2}\left[1 + erf(z)\right] \tag{A.2}$$

where $erf(z)$ is the error function [161],which is readily available in tabular and approximate numerical form (e.g., in MATLAB®), and

$$z = \frac{\alpha - \overline{\overline{\alpha}}}{\alpha_\sigma \sqrt{2}} \tag{A.3}$$

When $\overline{\overline{\alpha}} = 0$, and taking into account that $erf(-z) = -erf(z)$, we have

$$A(\alpha) = F(\alpha) - F(-\alpha) = erf(z) \tag{A.4}$$

which is used in Chapters 4 and 8.

Appendix B

Weibull Distribution

The probability density function [268, 269] is zero for stress $\sigma < 0$ and

$$f(\sigma; m, \sigma_0) = \frac{m}{\sigma_0}\left(\frac{\sigma}{\sigma_0}\right)^{m-1} exp\left(-\left(\frac{\sigma}{\sigma_0}\right)^m\right) \quad ; \quad \sigma \geq 0 \tag{B.1}$$

where $m > 0$ is the *shape parameter*, also called *Weibull modulus*, and $\sigma_0 > 0$ is the *scale parameter*, also called *characteristic strength,* not to be confused with the average strength.

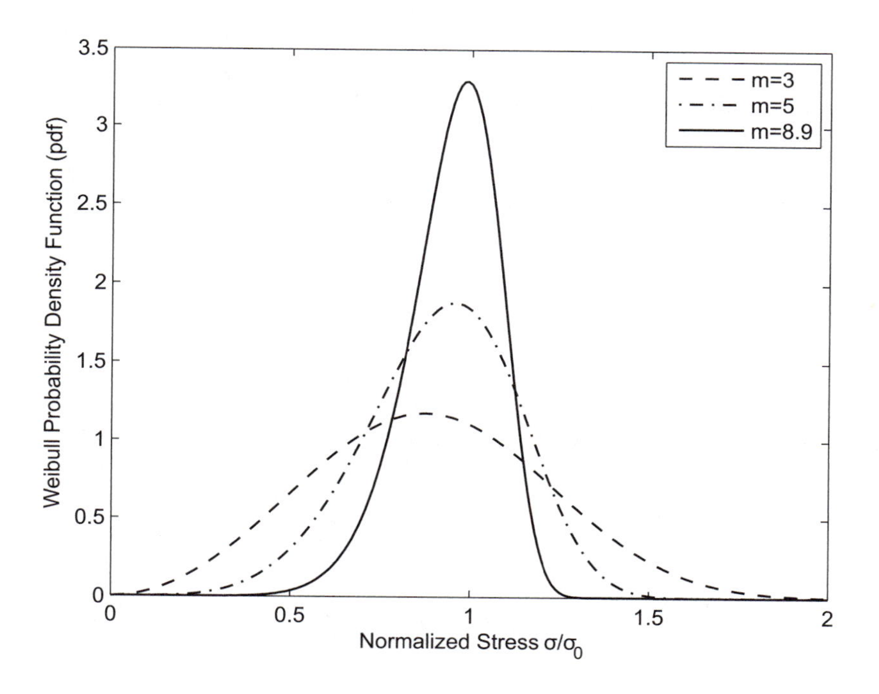

Figure B.1: Values of the Weibull probability density function (pdf) for shape $m = 3, 5, 8.9$ and scale $\sigma_0 = 1$.

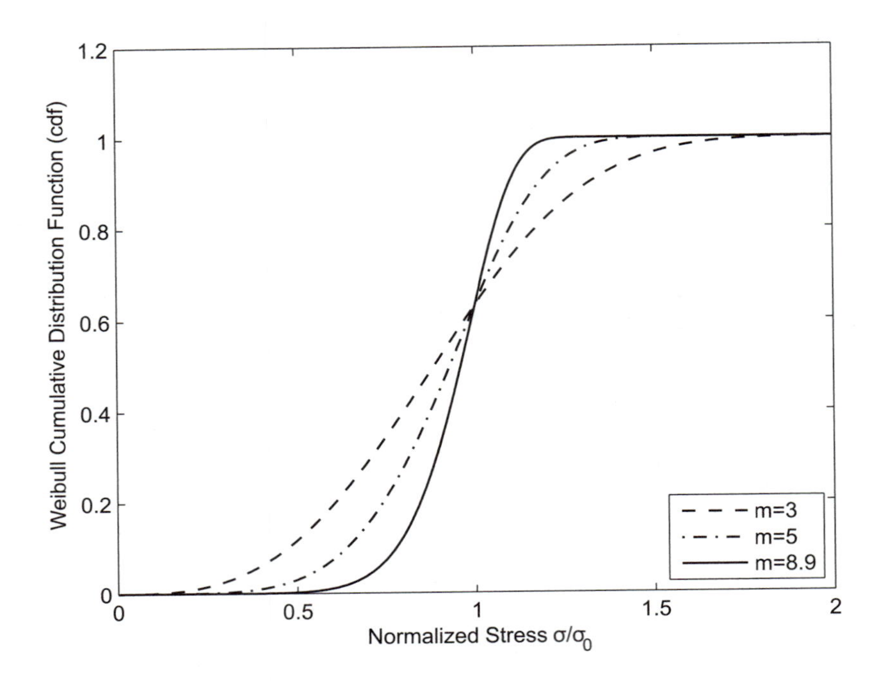

Figure B.2: Values of the Weibull cumulative distribution function (cdf) for shape $m = 3, 5, 8.9$ and scale $\sigma_0 = 1$.

When used to describe the probability of failure of fibers in a dry (not impregnated) tow, σ is the strength of individually tested fibers. The expected or mean value σ_{av} is defined as the sum of the probability of each outcome (fiber strength) multiplied by the value of the outcome, yielding

$$\sigma_{av} = \sigma_0 \Gamma \left(1 + \frac{1}{m}\right) \tag{B.2}$$

where Γ is the Gamma function (B.6). Since the average strength σ_{av} of testing a set (population) of hundreds of individual fibers is usually reported along with the shape parameter m, one can use (B.2) to calculate the scale parameter σ_0.

The cumulative distribution function (CDF) is

$$F(\sigma; m, \sigma_0) = 1 - exp\left[-\left(\frac{\sigma}{\sigma_0}\right)^m\right]$$
$$\lim_{\sigma \to \infty} F(\sigma; m, \sigma_0) = 1. \tag{B.3}$$

Since the parameters m, σ_0 are determined by breaking fibers of length L_0, but failure in a unidirectional composite occurs over a different length δ, called the *ineffective length* [155], the CDF [2, (8.36)][1] is scaled by δ/L_0 as

$$F(\sigma_f; m, \sigma_0) = 1 - exp\left[-\frac{\delta}{L_0}\left(\frac{\sigma_f}{\sigma_0}\right)^m\right]$$
$$= 1 - exp\left[-\delta\alpha{\sigma_f}^m\right] \tag{B.4}$$

with

$$\alpha = \frac{1}{L_0\sigma_0^m} = \frac{1}{L_0}\left[\frac{\Gamma(1+1/m)}{\sigma_{av}}\right]^m \tag{B.5}$$

For each value of fiber stress σ_f, 100 times the CDF represents the percentage of broken fibers, and $100(1 - F)$ represents the percentage of unbroken fibers.

The Gamma function [270] is an extension of the factorial function $n!$ to real and complex numbers. For a positive real number r it is defined by

$$\Gamma(r) = \int_0^{inf} t^{r-1}e^{-t}dt \tag{B.6}$$

and for positive integers n, it reduces to

$$\Gamma(n) = (n-1)! \tag{B.7}$$

[1] Note that [2, (8.36)] uses $\overline{\sigma}$ to denote fiber stress, while in this textbook the same symbol is used to denote composite effective stress.

Appendix C

Conversion of Units

The conversions presented in this section are used as follows. Given a quantity in the U.S. customary system (units on the left of the conversions), write down the value and replace the customary units by the conversion factor and the SI units. For example, a continuous strand mat weighs $w = 1.5$ oz/ft^2. In SI units the weight is

$$w = 1.5\frac{oz}{ft^2} = 1.5\frac{28.350\ g}{0.0929\ m^2} = 457.75\frac{g}{m^2}$$

or, using Table C.1, the conversion factor for oz/ft^2 is 305 g/m^2, so

$$w = 1.5\frac{oz}{ft^2} = 1.5 \times 305\ \frac{g}{m^2} = 457.5\ \frac{g}{m^2}$$

which are not identical due to rounding.

The conversion from SI to customary units is illustrated by the following example. The density of E-glass is $\rho = 2.5$ g/cm^3. In the U.S. customary system, ρ is

$$\rho = 2.5\frac{g}{cm^3} = 2.5 \times \left(\frac{1}{27.713}\frac{lb}{in^3}\right) = 0.090\frac{lb}{in^3}$$

or, using Table C.1, the conversion factor for lb/in^3 is 27.713 g/cm^3 (also labeled g/cc), so

$$\rho = 2.5\frac{g}{cm^3} = \frac{2.5}{27.713} = 0.090\frac{lb}{in^3}$$

Any conversion factor can be derived by writing the unit in one system (say lb/in^2), then replacing each individual unit by the equivalent in the other system. For example

$$\sigma = \frac{lb}{in^2} = \frac{4.4482\ N}{(0.0254\ m)^2} = 6895\frac{N}{m^2} = 6.895\ kPa = \frac{1}{145}\ MPa$$

The conversion factors most frequently used in this book are listed in Table C.1.

Table C.1: Conversion factors.

	Given a quantity in	**Multiply by**	**To obtain**
Length	inches (in)	25.400	millimeters (mm)
	inches (in)	2.5400	centimeters (cm)
	feet (ft)	0.3048	meters (m)
	yards (yd)	0.9144	meters (m)
	miles (mi)	1.6093	kilometers (km)
Area	square inches (in^2)	645.16	sq. millimeters (mm^2)
	square feet (ft^2)	0.0929	square meters (m^2)
	square yards (yd^2)	0.8361	square meters (m^2)
	square miles (mi^2)	2.5900	sq. kilometers (km^2)
Volume	cubic inches (in^3)	16387	cu. millimeters (mm^3)
	cubic feet (ft^3)	0.02832	cubic meters (m^3)
	cubic yards (yd^3)	0.76456	cubic meters (m^3)
Mass	ounces (oz)	28.350	grams (g)
	pounds (lb)	0.4536	kilograms (kg)
	kips (1000 lb)	0.4536	tonnes (metric ton)
	tons (Tn)	0.9072	tonnes (metric ton)
	tonnes (metric ton)	1000	kilograms (kg)
Density	pounds/cu. in. (lb/in^3)	27.713	grams/cc (g/cm^3)
Force	pound (lb) (force)	4.4482	Newton (N)
	kip (1000 lb)	4.4482	kilo Newton (kN)
	$kg\ m\ s^{-2}$	1.0	Newton (N)
	kg (force)	1/9.81	Newton (N)
Stress or Pressure	pounds in^{-2} (psi)	6.8948	kilo Pascal (kPa)
	kip in^{-2} (ksi)	6.8948	mega Pascal (MPa)
	$N\ m^{-2}$	1.0	Pascal (Pa)
Moment of Inertia	in^4	416231	mm^4
	ft^4	0.00863	m^4
Moment	kip foot (1000 lb ft)	1.35582	kilo Newton meter (kNm)
	pound inch (lb in)	0.11298	Newton meter (Nm)
Yield	yard/pound (yd/lb)	2.01587	meters/kilograms (m/kg)
Viscosity	centipoises	10^{-2}	poises (g/cm s)
	$lb\ ft^{-1}\ sec^{-1}$	$1.4882\ 10^3$	centipoises
Energy	Btu	$1.0550\ 10^3$	Joules (Nm)
Fracture Energy	$in\ lb\ in^{-2}$	0.175126	$kJ\ m^{-2}$
1/Yield	lb/yd	496,238	TEX (g/km)
Denier	g/(9,000 m)	1/9	TEX (g/km)
CTE	(in/in)/°F	1.8	(mm/mm)/°C
Fabric weight	oz/ft^2	305	g/m^2
	oz/yd^2	34	g/m^2

Bibliography

[1] Barbero, E. J., Web resource, www.mae.wvu.edu/barbero/icmd/.

[2] Barbero, E. J., *Finite Element Analysis of Composite Materials*, CRC Press, Boca Raton, FL, 2007.

[3] The Basics of Fiberglass. Fiber Glast Development Co. Brookville, OH. Web resource, http://www.fiberglast.com.

[4] A Step by Step Guide to Molding Fiberglass. Fiber Glast Development Co. Brookville, OH. Web resource, http://www.fiberglast.com.

[5] Advanced Moldmaking and Plug Construction. Fiber Glast Development Co. Brookville, OH. Web resource, http://www.fiberglast.com.

[6] Vacuum Bagging and Sandwich Core Construction. Fiber Glast Development Co. Brookville, OH. Web resource, http://www.fiberglast.com.

[7] The Art of Moldless Composites. Fiber Glast Development Co. Brookville, OH. Web resource, http://www.fiberglast.com.

[8] Cosmetic Gel Coat and Fiberglass Repair. Fiber Glast Development Co. Brookville, OH. Web resource, http://www.fiberglast.com.

[9] Fiberglass Repairs Made Easy. Fiber Glast Development Co. Brookville, OH. Web resource, http://www.fiberglast.com.

[10] Composite Materials and Manufacturing. Society of Manufacturing Engineers (SME), Dearborn, MI. Web Resource, http://www.sme.org.

[11] Composites Manufacturing Series DVD Package. Society of Manufacturing Engineers (SME), Dearborn, MI. Web resource, http://www.sme.org.

[12] Peters, S., editor, *Applications: Construction, Encyclopedia of Composites*, Chapman Hall, London, 1998.

[13] Jones, R. M., Reproduced from *Mechanics of Composite Materials*, Figure 1.15, p. 23, Copyright 1975, with permission from Taylor and Francis.

[14] Budynas, R. G. and Nisbett, J. K., *Shigley's Mechanical Engineering Design*, McGraw-Hill, New York, NY, 8th edition, 2008.

[15] Schwartz, M. M., *Composite Materials Handbook*, McGraw-Hill, New York, NY, 2nd edition, 1992.

[16] Harper, C. A., *Handbook of Plastics, Elastomers and Composites*, McGraw-Hill, New York, NY, 2nd edition, 1991.

[17] Bunsell, A. R., *Fibre Reinforcements for Composite Materials, Composite Materials Series 2*, Elsevier, Amsterdam, 1988.

[18] Vasiliev, V. V., *Mechanics of Composite Structures*, Taylor and Francis, Bristol, PA, 1993.

[19] Lee, S. M., editor, *International Encyclopedia of Composites*, volume 4, VHC, 1991, Figure 296 and Table 9.

[20] Swanson, S. R. and Qian, Y., Compos. Sci. Technol. **43** (1992) 197.

[21] Colvin, G. E. and Swanson, S. R., J. Eng. Mater. Technol. **112** (1990) 61.

[22] Wang, H. and Vu-Khanh, T., J. Compos. Mater. **28(8)** (1994) 684.

[23] Fiberite APC-2 Thermoplastic Laminates. Material Data Sheet. ICI Advanced Materials, Winona, MN.

[24] Camanho, P. P., Maimi, P., and Dávila, C. G., Compos. Sci. Technol. **67** (2007) 2715.

[25] Soden, P. D., Hinton, M. J., and Kadour, A. S., *Chapter 2: Failure Criteria in Fibre Reinforced Polymer Composites*, Elsevier, Amsterdam, 2004.

[26] Brunswick Technologies Inc. Product Catalog, Brunswick, ME.

[27] Clarke, J. L., editor, *Structural Design of Polymer Composites: EUROCOMP Design Code and Handbook*, Chapman and Hall, London, UK, 1996.

[28] National Research Council of Canada, Ontario, Canada, *National Building Code of Canada*, 2005.

[29] Salmon, C. G., Johnson, E., J., and Malhas, F. A., *Steel Structures–Design and Behavior*, Pearson Prentice-Hall, Upper Saddle River, NJ, 5th edition, 2009.

[30] Anderson, T. L., *Fracture Mechanics*, CRC, Boca Raton, FL, 1995.

[31] Rikards, R. et al., Eng. Fract. Mech. **61** (1998) 325.

[32] Varna, J., Joffe, R., Akshantala, N., and Talreja, R., Compos. Sci. Technol. **59** (1999) 2139.

[33] Compiled from Brunswick Technology Product Catalog, Brunswick Technology Inc., Brunswick, ME, with permission.

[34] Katz, H. and Mileski, J., editors, *Handbook of Fillers for Plastics*, Van Nostrand Reinhold/Wiley, Hoboken, NJ, 1987.

[35] Schwartz, M. M., *Composite Materials, Volume 1: Properties, Nondestructive Testing and Repair*, Prentice-Hall, New York, 1997.

[36] Warner, S. B., *Fiber Science*, Prentice-Hall, New York, 1975.

[37] Charrier, J.-M., *Polymeric Materials and Processing*, Hanser, Munich, 1990.

[38] Lee, S. M., editor, *International Encyclopedia of Composites: Reinforcements*, volume 5, VCH, New York, 1991.

[39] Moss, T., *Composites: An Insider's Technical Guide to Corporate America's Activities*, New York, 2nd edition, 1994.

[40] The Society of Plastic Industry (SPI). Composites Institute, New York, NY.

[41] SACMA, Arlington, VA, *SRM 1-94, Recommended Test Method for Compressive Properties of Oriented Fiber-Resin Composites.*

[42] Tarnolpolsky, Y. M. and Kincis, T., *Static Test Methods for Composites*, Van Nostrand Reinhold Company, New York, 1981.

[43] Gauthier, M. M., *Engineered Materials Handbook*, ASM International, Materials Park, OH, 1995.

[44] Lissart, N. and Lamon, J., J. Materials Sci. **32** (1997) 6107.

[45] Loidl, D., Paris, O., Rennhofer, H., Muller, M., and Peterlik, H., Carbon **45** (2007) 2801.

[46] Zinck, P., Gaerard, J. F., and Wagner, H. D., Eng. Fract. Mech. **69** (2002) 1049.

[47] Owens Corning Fiberglass Co., Granville, OH.

[48] Lee, S. M., editor, *International Encyclopedia of Composites: Filament Winding*, volume 2, VCH, New York, 1990.

[49] Barbero, E. J. and Damiani, T., ASCE J. Compos. Constr. **7** (2003) 3.

[50] Barbero, E. J. and Damiani, T., J. Reinf. Plastics Compos. **22** (2003) 373.

[51] U. S. Department of Transportation, NHTSA 49 CFR Part 571, Federal Register, Vol. 59, No. 248, 1994.

[52] Lupescu, M. B., *Fiber Reinforcements for Composite Materials*, Editura Tehnica, Bucuresti, 2004.

[53] Hubca, G. et al., *Composite Materials*, Editura Tehnica, Bucuresti, 1999.

[54] Lee, S. M., editor, *International Encyclopedia of Composites: Carbon Fiber, Evolving Technology*, volume 1, VCH, New York, 1990.

[55] Toray, Kirkland, WA.

[56] Hercules, Wilmington, DE.

[57] Talreja, R., *Fatigue of Composite Materials*, Technomic, Lancaster, PA, 1987.

[58] Iijima, S., Nature **354 (6348)** (1991) 56.

[59] Iijima, S. and Ichihashi, T., Nature **363, (6430)** (1993) 603.

[60] Bethune, D. S. et al., Nature **363, (6430)** (1993) 605.

[61] Treacy, M. M. J., Ebbesen, T. W., and Gibson, J. M., Nature **381(6584)** (1996) 678.

[62] Poncharal, P., Wang, Z. L., Ugarte, D., and de Heer, W. A., Science **283(5407)** (1999) 1513.

[63] Falvo, M. R. et al., Nature **389 (6651)** (1997) 582.

[64] Wong, E. W., Sheehan, P. E., and Lieber, C. M., Science **277** (1997) 1971.

[65] Salvetat, J. P. et al., Adv. Mater. **11(2)** (1999) 161.

[66] Yu, M., Lourie, O., Dyer, M. J., Kelly, T. F., and Ruoff, R. S., Science **287** (2000) 637.

[67] Yu, M. F., Files, B. S., Arepalli, S., and Ruoff, R. S., Phys. Rev. Lett. **84** (2000) 5552.

[68] Yakobson, B. I., Brabec, C. J., and Bernholic, J., Phys. Rev. Lett. **76 (14)** (1996) 2511.

[69] Lu, J. P., J. Phys. and Chem. Solids **58 (11)** (1997) 1649.

[70] Krishnan, A., Dujardin, E., Ebbesen, T., Yianilos, P., and Treacy, M. M. J., Phys. Rev. B. **58** (1998) 14013.

[71] Manchado, M. A. L., Valentini, L., Biagiotti, J., and Kenny, J. M., Carbon **43(7)** (2005) 1499.

[72] McIntosh, D., Khabashesku, V., and Barrera, E., J. Phys. Chem. C. **111(2)** (2007) 1592.

[73] McIntosh, D., Khabashesku, V., and Barrera, E., Chem. Mater. **18** (2006) 4561.

[74] Ericson, L. M. et al., Science **305** (2004) 1447.

[75] Li, W. Z. et al., Science **274 (5293)** (1996) 1701.

[76] Potschke, P., Bhattacharyya, A., Janke, A., and Goering, H., Compos. Interf. **10 (4–5)** (2003) 389.

[77] Haggenmueller, R., Gommans, H., Rinzler, A., Fischer, J., and Winey, K., Chem. Phys. Lett. **330(3–4)** (2000) 219.

[78] Coleman, J. N. et al., Appl. Phys. Lett. **82(11)** (2003) 1682.

[79] Vigolo, B. et al., Science **290(5495)** (2000) 1331.

[80] Stevens, J. L. et al., Nano Letters **3(3)** (2003) 331.

[81] DuPont Elastomers, Stow, OH.

[82] Teijin Ltd., New York, NY.

[83] Akzo Nobel, Knoxville, TN.

[84] Zylon Fiber. Technical Information, Toyobo Co.

[85] Spectra Fiber 900, 1000, 2000. Product Information Sheet, Honeywell Inc.

[86] Vectran Fiber. Product Information Sheet, Kuraray Inc.

[87] Textron Systems Division, Wilmington, MA.

[88] Nextel Ceramic Fiber, 3M Center, St. Paul, MN.

[89] Lager, J. R. and June, R. R., J. Compos. Mater. (1969) 48.

[90] Lee, S. M., editor, *International Encyclopedia of Composites: Metal Matrix Composites, Aluminum*, volume 3, VCH, New York, 1990.

[91] Lee, S. M., editor, *International Encyclopedia of Composites: Ceramic Matrix Composites for Advanced Rocket Engines Turbomachinery*, volume 6, VCH, New York, 1991.

[92] Lee, S. M., editor, *International Encyclopedia of Composites: Construction, Structural Applications*, volume 6, VCH, New York, 1991.

[93] Ashland Chemical Co., Columbus, OH, *Hetron and Aropol Resin Selection Guide for Corrosion Resistant FRP Applications.*

[94] Schweitzer, P. A., *Corrosion Resistance Tables*, Dekker, New York, NY, 1995.

[95] Adams, R. C. and Bogner, B. R., **1-C** (1993) 1, SPI Composites Institute 48th Conference.

[96] Derakane Epoxy Vinyl Ester Resins Technical Data. Dow Chemical Co., TX.

[97] Lee, H. and Neville, K., *Handbook of Epoxy Resins*, McGraw-Hill, Inc., CA, 1967.

[98] Shell Chem. Tech. Bulletin SC-712-88, SC-856-88R, SC-1011-89, etc. Shell Chem. Co., Oak Brook, IL.

[99] Corrosion Resistance Guide, Indspec Chem. Co., Pittsburgh, PA, 1990.

[100] Fire PFR2 Test Results and Approvals, Indspec Chem. Co., Pittsburgh, PA, 1990.

[101] Lee, S. M., editor, *International Encyclopedia of Composites: Phenolic Resins and Composites*, volume 4, VCH, New York, 1991.

[102] Lee, S. M., editor, *International Encyclopedia of Composites: Flammability*, volume 2, VCH, New York, 1990.

[103] Plastics Design Library-William Andrew Publishing, Norwich, NY, *Plastics Design Library: The effect of Creep and Other Time Related Factors on Plastics.*

[104] Plastics Design Library-William Andrew Publishing, Norwich, NY, *Plastics Design Library: The Effect of Temperature and Other Factors on Plastics and Elastomers*, 2nd edition.

[105] Ferry, J. D., *Viscoelastic Properties of Polymers*, Wiley, New York, NY, 3rd edition, 1980.

[106] Nielsen, L. E. and Landel, R. F., *Mechanical Properties of Polymer and Composites*, Dekker, New York, NY, 2nd edition, 1994.

[107] Kachanov, L. M., *Introduction to Continuum Damage Mechanics*, Kluwer, Dordretch, 1990.

[108] Luciano, R. and Barbero, E. J., ASME J. Appl. Mech. **62** (1995) 786.

[109] Barbero, E. J. and Luciano, R., Int. J. Solids Struct. **32** (1995) 1859.

[110] Harris, J. S. and Barbero, E. J., J. Reinf. Plas. Compos. **17** (1998) 361.

[111] Qiao, P., Barbero, E. J., and Davalos, J. F., J. Compos. Mater. **34** (2000) 39.

[112] Struik, L. C. E., *Physical Aging in Amorphous Polymers and Other Materials*, Elsevier, Amsterdam, 1978.

[113] Barbero, E. J. and Ford, K. J., ASME J. Eng. Mat. Tech. **126** (2006) 413.

[114] Barbero, E. J. and Rangarajan, S., ASTM J. Testing Eval. **33** (2005) 377.

[115] Barbero, E. J. and Ford, K. J., SAMPE J. Adv. Mater. **38** (2006) 7.

[116] Arridge, R. G. C., *An Introduction to Polymer Mechanics*, Taylor and Francis, London, 1985.

[117] Kumar, A. and Gupta, R. K., *Fundamentals of Polymers*, McGraw-Hill, New York, NY, 1998.

[118] ASTM International Web Resource., `www.astm.org`.

[119] Shen, C. H. and Springer, G. S., J. Compos. Mater. **10** (1976) 2.

[120] Springer, G. S., editor, *Environmental Effects of Composite Materials. vol. I–III*, Technomic, PA, 1981.

[121] Barbero, E. J., J. Compos. Mater. **43(19)** (2008) 2109.

[122] Weeton, J. W., Peters, D. M., and Thomas, K. L., editors, *Engineer's Guide to Composite Materials*, ASM International, OH, 1986.

[123] Hollaway, L., editor, *Handbook of Polymer Composites for Engineering*, Woodhead, Cambridge, England, 1994.

[124] Naito, K., Tanaka, Y., Yang, J.-M., and Kagawa, Y., Carbon **46** (2008) 189.

[125] Montes-Moran, M. A., Martinez-Alonso, A., Tascon, J. M. D., Paiva, M. C., and Bernardo, C. A., Carbon **39** (2001) 1057.

[126] Paiva, M. C., Bernardo, C. A., and Nardin, M., Carbon **38** (2000) 1323.

[127] Piquero, T., Vincent, H., Vincent, C., and Bouix, J., Carbon **33** (1995) 455.

[128] Okoroafor, E. U. and Hill, R., Ultrasonics **33** (1995) 123.

[129] Zairi, F., Nait-Abdelaziz, M., Gloaguen, J. M., Bouaziz, A., and Lefebvre, J. M., Int. J. Solids Struct. **45** (2008) 5220.

[130] Ni, Q.-Q. and Jinen, E., Eng. Fract. Mech. **56** (1997) 779.

[131] Cabral-Fonseca, S., Paiva, M. C., Nunes, J. P., and Bernardo, C. A., Polymer Testing **22** (2003) 907.

[132] Post, N. L., Cain, J., McDonald, K. J., Case, S. W., and Lesko, J. J., Eng. Frac Mech. **75** (2008) 2707.

[133] Kanari, M., Tanaka, K., Baba, S., and Eto, M., Carbon **35** (1997) 1429.

[134] Iwashita, N., Sawada, Y., Shimizu, K., Shinke, S., and Shioyama, H., Carbon **33** (1995) 405.

[135] Lee, S. M., Adapted from *International Encyclopedia of Composites: Filament Winding,* vol. 2, pp. 167–182, S. M. Lee editor, VCH Publishers, New York, 1990.

[136] Reinhart, T., editor, *Engineering Materials Handbook. Vol. 1: Composites,* ASM International, Metals Park, OH, 1987.

[137] Tang, J., Lee, W. I., and Springer, G. S., J. Compos. Mater. **21** (May 1987) 421.

[138] Foley, M. F. and Gutowski, T., 23rd Int. SAMPE Technical Conference, October 21–24 (1991) 326.

[139] Liu, B., Bruschke, M. V., and Advani, S. G., *Liquid Injection Molding Simulation — LIMS User's Manual — Version 3.0*, University of Delaware, DE, 1994.

[140] Lodge, C., *RTM Process Invades High-Volume Markets*, Plastics World, Delhi, 1985.

[141] Stark, E. B. and Breitigam, W. V., *Engineered Materials Handbook: Composites, vol. 1*, ASM International, Novelty, OH, 1993.

[142] Galli, E., Plastic Design Forum, March/April (1982) 47.

[143] Hard Core Dupont SCRIMP^{TM} Systems, New Castle, DE.

[144] Liquid Injection Molding Simulation (LIMS), Composite TechBriefs, Center for Composite Materials, University of Delaware, Newark, DE.

[145] Mase, G. E. and Mase, G. T., *Continuum Mechanics for Engineers*, CRC Press, Boca Raton, FL, 2nd edition, 1992.

[146] Luciano, R. and Barbero, E. J., Int. J. Solids Struct. **31** (1994) 2933.

[147] Halpin, J. and Tsai, S. W., *Effects of Environmental Factors on Composite Materials*, AFML-TR, 1969.

[148] Jones, R. M., *Mechanics of Composite Materials*, Taylor and Francis, Washington, D.C., 1975.

[149] Hashin, Z. and Rosen, B. W., J. Appl. Mech. (1964) 223.

[150] Tsai, S. W. and Hahn, H. T., *Introduction to Composite Materials*, Technomic, Lancaster, PA, 1980.

[151] Halpin, J. and Pagano, N. J., J. Compos. Mater. **3** (1969) 720.

[152] Agarwal, B. D. and Broutman, L. J., *Analysis and Performance of Fiber Composites*, Wiley, New York, NY, 2nd edition, 1990.

[153] Barbero, E. J. and Kelly, K., J. Compos. Mater. **27** (1993) 1214.

[154] Kelly, K. and Barbero, E. J., Int. J. Solids Struct. **30** (1993) 3417.

[155] Rosen, B. W., AIAA J **2** (1964) 1985.

[156] Barbero, E. J., J. Compos. Mater. **32** (1998) 483.

[157] Yurgartis, S. W., Compos. Sci. Technol. **30** (1987) 279.

[158] Barbero, E. J. and Tomblin, J. S., Int. J. Solids Struct. **33** (1996) 4379.

[159] Rosen, B. W., *Fiber Composite Materials (Chapter 3)*, ASM International, Metals Park, OH, 1965.

[160] Godoy, L. A., *Theory of Stability: Analysis and Sensitivity*, McGraw-Hill, 1999.

[161] Erdelyi, A., *Higher Transcendental Functions*, volume 2, McGraw-Hill, New York, NY, 1953.

[162] Häberle, J. G., *Strength and Failure Mechanisms of Unidirectional Carbon Fibre-Reinforced Plastics*, PhD thesis, Imperial College, London, 1991.

[163] Daniels, J. A. and Sandhu, R. S., *Composite Materials: Testing and Design, STP 1206 vol. 11*, ASTM International, Philadelphia, PA, 1993.

[164] Stellbrink, K. K., *Micromechanics of Composites*, Hanser Publishers, Munich, 1996.

[165] Nielsen, L. E., J. Compos. Mater. **1** (1967) 100.

[166] Tada, H., Paris, P., and Irwin, G., *The Stress Analysis of Cracks Handbook*, ASME, New York, 3rd edition, 2000.

[167] Dávila, C. G., Camnanho, P. P., , and Rose, C. A., J. Compos. Mater. **39** (2005) 323.

[168] Dvorak, G. J. and Laws, N., J. Compos. Mater. **21** (1987) 309.

[169] Wang, A. S. D., Compos. Tech. Review **6** (1984) 45.

[170] Davies, P., *Protocols for Interlaminar Fracture Testing of Composites*, Polymer and Composites Task Group. European Structural Integrity Society (ESIS), Plouzané, France, 1992.

[171] Puck, A. and Schurmann, H., Compos. Sci. Technol. **58** (1998) 1045.

[172] Sonti, S. S., Barbero, E. J., and Winegardner, T., **10-C** (1995) 1, SPI Composites Institute 50th Annual Conference.

[173] Kim, Y., Davalos, J. F., and Barbero, E. J., J. Compos. Mater. **30** (1996) 536.

[174] Vinson, J. R. and Sierakowski, R. L., *The Behavior of Structures Composed of Composite Materials*, Martinus Nijhoff, The Netherlands, 1987.

[175] Reddy, J. N., *Mechanics of Laminated Composite Plates-Theory and Analysis*, CRC Press, Boca Raton, FL, 1997.

[176] Young, W. C., *Roark's Formulas for Stress and Strain*, McGraw-Hill, New York, NY, 6th edition, 1989.

[177] Ugural, A. C., *Stresses in Plates and Shells*, CRC Press, Boca Raton, FL, 3rd edition, 2009.

[178] Gere, J. M. and Timoshenko, S. P., *Mechanics of Materials*, PWS-Kent, Boston, MA, 3rd edition, 1990.

[179] McCormac, J. and Elling, R. E., *Structural Analysis*, Harper & Row, New York, NY, 1988.

[180] ASME, New York, NY, *ASME Boiler and Pressure Vessel Code, Section XIII.*

[181] Nahas, M. N., J. Compos. Technol. Research **8** (1986) 138.

[182] Hull, D., *An Introduction to Composite Materials*, Cambridge University Press, Cambridge, 2nd edition, 1996.

[183] Waddoups, M. E., *Characterization and Design of Composite Materials Workshop*, Technomic, Bristol, PA, 1968.

[184] Barbero, E. J., *Introduction to Composite Materials Design*, Taylor and Francis, Philadelphia, PA, 1st edition, 1999.

[185] Pipes, R. B. and Cole, B. W., J. Compos. Mater. **7** (1973) 246.

[186] Camanho, P. P., Dávila, C. G., Pinho, S. T., Iannucci, L., and Robinson, P., Composites Part A **37** (2006) 165.

[187] Cortes, D. H. and Barbero, E. J., Annals of Solid and Structural Mechanics (2009).

[188] Hart-Smith, L. J., The Institution of Mechanical Engineers, Part G: J. Aerospace Eng. **208** (1994) 9.

[189] Wang, J. Z. and Soche, D. F., J. Compos. Mater. **27** (1993) 40.

[190] Tan, S. C., *Stress Concentrations in Laminated Composites*, Technomic, Lancaster, PA, 1994.

[191] Savin, G. N., *Stress Distribution Around Holes*, NASA TT-F-607, 1970.

[192] Weeton, J. W., editor, *Engineered Materials Handbook, Vol. 1: Composites*, ASM International, Metals Park, OH, 1987.

[193] Lekhnitskii, S. G., *Theory of Elasticity of an Anisotropic Body*, Holden-Day, San Francisco, CA, 1963.

[194] Whitney, J. M. and Nuismer, R. J., J. Compos. Mater. **8** (1974) 253.

[195] Nuismer, R. J. and Whitney, J. M., *Uniaxial Failure of Composite Laminates Containing Stress Concentrations, Fracture Mechanics of Composites, STP 593*, ASTM International, Philadelphia, PA, 1975.

[196] Peterson, R. E., *Stress Concentration Factors*, Wiley, New York, NY, 1974.

[197] Varellis, I. and Norman, T. L., Compos. Sci. Technol. **51** (1994) 367.

[198] Friedrich, K., editor, *Application of Fracture Mechanics to Composite Materials*, Elsevier, Amsterdam, 1989.

[199] Barbero, E. J., Reproduced from *Finite Element Analysis of Composite Materials,* Figure 8.1, p. 192, Copyright 2008, with permission from Taylor and Francis.

[200] Hahn, H. T., Erikson, J. B., and Tsai, S. W., *Characterization of Matrix/Interface-Controlled Strength of Composites*, Martinus Nijhoff, 1982, In *Fracture of Composite Materials,* G. Sih and V. P. Tamuzs, ed.

[201] Barbero, E. J. and Cortes, D. H., Composites: Part B (2009).

[202] Barbero, E. J., Sgambitterra, G., Adumitroiae, A., and Martinez, J., (submitted).

[203] Barbero, E. J., Sgambitterra, G., and Tessler, A., (submitted).

[204] Chou, T.-W., *Microstructural Design of Fiber Composites*, Cambridge University Press, Cambridge, MA., 1992.

[205] Ito, M. and Chou, T.-W., Compos. Sci. Technol. **57** (1997) 787.

[206] Ito, M. and Chou, T.-W., J. Compos. Mater. **32** (1998) 2.

[207] Scida, D. and Aboura, Z., Compos. Sci. Technol. **59** (1997) 1727.

[208] Scida, D. and Aboura, Z., Compos. Sci. Technol. **59** (1999) 505.

[209] Naik, N., *Woven Fabric Composites*, Technomic, Lancaster, PA, 1994.

[210] Naik, N. and Ganesh, V., Compos. Sci. Technol. **45** (1992) 135.

[211] Naik, N. and Ganesh, V., J. Compos. Mater. **30** (1996) 1779.

[212] Naik, N. and Ganesh, V., Reprinted from *J. Materials Science,* 32, Thermo-Mechanical Behavior of Plain Weave Fabric Composites: Experimental Investigations, Figure 1–2, p. 271, Copyright 1997, with permission from Springer.

[213] Barbero, E. J., Damiani, T. M., and Trovillion, J., Int. J. Solids Struct. **42** (2005) 2489.

[214] Barbero, E. J., Trovillion, J., Mayugo, J., and Sikkil, K., Composite Struct. **73** (2006) 41.

[215] Barbero, E. J., Lonetti, P., and Sikkil, K., Composites: Part B **37** (2006) 137.

[216] Falzon, J. and Herszberg, I., J. Compos. Mater. **30** (1996) 1210.

[217] Sheng, S. Z. and Hoa, S. V., J. Compos. Mater. **35** (2001) 1701.

[218] Naik, N. and Ganesh, V., J. Materials Sci. **32** (1997) 267.

[219] Hahn, H. T., J. Compos. Mater. **9** (1975) 316.

[220] Ketter, R., Lee, G., and Sherwood, P., *Structural Analysis and Design*, McGraw-Hill, NY, 1979.

[221] Cook, R. D. and Young, W. C., *Advanced Mechanics of Materials*, Macmillan, New York, NY, 1st edition, 1985.

[222] Cook, R. D. and Young, W. C., *Advanced Mechanics of Materials*, Prentice-Hall, Upper Saddle River, NJ, 2nd edition, 1999.

[223] Budinas, R. G. and Nisbett, J. K., *Shigley's Mechanical Engineering Design*, McGraw-Hill, New York, NY, 8th edition, 2008.

[224] Boresi, A. P., Schmidt, R. J., and Sidebottom, O. M., *Advanced Strength of Materials*, Wiley, New York, NY, 5th edition, 1993.

[225] Godoy, L. A., Barbero, E. J., and Raftoyiannis, I., Computers and Struct. **56** (1995) 1009.

[226] Barbero, E. J., Godoy, L. A., and Raftoyiannis, I., Int. J. Numer. Methods Eng. **39** (1996) 469.

[227] Barbero, E. J. and Tomblin, J., Thin-Walled Struct. **18** (1994) 117.

[228] Barbero, E. J. and Raftoyiannis, I., Composite Struct. **27** (1994) 261.

[229] Smith, J. C., *Structural Analysis*, Harper & Row, New York, NY, 1988.

[230] Godoy, L. A., Barbero, E. J., and Raftoyiannis, I., J. Compos. Mater. **29** (1995) 591.

[231] Massa, J. and Barbero, E. J., J. Compos. Mater. **1** (1998).

[232] Hong, C.-H. and Chopra, I., J. Amer. Helicopter Soc. **30** (1985) 57.

[233] Librescu, L. and Khedir, A. A., AIAA J. **26** (1988) 1373.

[234] Chandra, R., Stemple, A. D., and Chopra, I., J. Aircraft **27** (1990) 619.

[235] Smith, E. C. and Chopra, I., J. Amer. Helicopter Soc. **36** (1991) 23.

[236] Whitney, J. M., *Structural Analysis of Laminated Anisotropic Plates*, Technomic, Lancaster, PA, 1987.

[237] U. S. Department of Defense, Washington, DC, *Military Handbook MIL-HDBK-17-3F Polymer Matrix Composites, Vol. III, Utilization of Data*, 2002.

[238] Niu, M. C. Y., *Composite Airframe Struct.*, Comilit Press, Hong Kong, 1992.

[239] Niu, M. C. Y., *Airframe Structural Design*, Comilit Press, Hong Kong, 1988.

[240] Megson, T. H. G., *Aircraft Structures*, John Wiley, New York, NY, 2nd edition, 1990.

[241] Gibbs and Cox Inc., New York, NY, *Marine Design Manual for Fiberglass Reinforced Plastics*, 1960, Published by McGraw-Hill.

[242] Marchant, A., *Design of Bottom Shell Laminate for a GFRP Vessel*, Woodhead, Cambridge, MA, 1994, In *Handbook of Polymer Composites for Engineers, L.* Hollaway, ed.

[243] PASCO. Structural Panel Analysis and Sizing Code, NASA LaRC, Langley, VA, `rcc.uga.edu/`.

[244] VICONOPT. Vibration and Instability of Plate Assemblies Including Shear and Anisotropy (VIPASA) with Constraints and Optimization, Old Dominion Univ, Williamsburg, VA, `rcc.uga.edu/`.

[245] Dickson, J. N. and Biggers, S. B., *POSTOP: Post Buckled Open-Stiffener Optimum Panels-Theory and Capability, NASA CR-172259*, 1983.

[246] Bushnell, D., Computer and Struct. **25** (1987) 469, Available from Lockheed Palo Alto Research Laboratory.

[247] Troitsky, M. S., *Stiffened Plates-Bending, Stability, and Vibrations*, Elsevier, 1976.

[248] Arnold, R. R. and Mayers, J., Int. J. Solids Struct. **20** (1984) 863.

[249] Deo, R. B., Kan, H. P., and Bhatia, N. M., *Design Development and Durability Validation of Post Buckled Composite and Metal Panels*, U.S. Air Force Flight Dynamics Lab, 1989, WRDC-TR-89-3030, 4 Volumes.

[250] Reproduced from *Military Handbook MIL-HDBK-17-3F, Polymer Matrix Composites, Vol. III, Utilization of Data,* U.S. Department of Defense, 2002.

[251] Spier, E. E., Experimental Mech. **18** (1978) 401.

[252] Deo, R. B., Agarwal, B. L., and Madenci, E., *Design Methodology and Life Analysis of Post Buckled Metal and Composite Panels*, U.S. Air Force Flight Dynamics Lab, 1985, AFWAL-TR-85-3096, Volume I.

[253] Deo, R. B., Agarwal, B. L., and Madenci, E., Reproduced from *AFWAL-TR-85-3096,* vol. 1, 1985 with permission.

[254] Peters, S. T., editor, *Handbook of Composites*, Chapman and Hall, London, UK, 1998.

[255] Uniform Building Code. Int. Conf. Building Officials, Whittier, CA, 1991.

[256] American Concrete Institute, Farmington Hills, MI, *ACI 440.2R-08, Guide for the Design and Construction of Externally Bonded FRP Strengthening System for Strengthening Concrete Structures*, 2008.

[257] American Concrete Institute, Farmington Hills, MI, *ACI 318-08, Building Code Requirements of Structural Concrete, and Commentary*, 2008.

[258] Nawy, E. G., *Reinforced Concrete: A Fundamental Approach*, Pearson Prentice-Hall, Upper Saddle River, NJ, 2005.

[259] Hollaway, L. C. and Leeming, M. B., *Strengthening of Reinforced Concrete Structures: Using Externally-Bonded FRP Composites in Structural and Civil Engineering*, CRC Press, Boca Raton, FL, 2000.

[260] Park and Paulay, *Reinforced Concrete Structures*, John Wiley & Sons, Inc., New York, NY, 1975.

[261] Bruno, D., Carpino, R., and Greco, F., Compos. Sci. Technol. **67** (2007) 1459.

[262] Greco, F., Lonetti, P., and Nevone, P., Eng. Fract. Mech. **74** (2007) 346.

[263] Greco, F. and Lonetti, P., Composites: Part B **40** (2009) 379.

[264] Nanni, A. and Bradford, N. M., Construct. Build. Mater. **9** (1995) 115.

[265] Teng, J. G., Chen, J. F., Smith, S. T., and Lam, L., *FRP Strengthened RC Struct.*, John Wiley & Sons, Inc., West Sussex, UK, 2002.

[266] De Lorenzis, L. and Tepfers, R., J. Compos. Constr. **7** (2003) 219.

[267] Lam, L. and Teng, J., J. Reinf. Plastics Compos. **22** (2003) 1149.

[268] Weibull distribution, `en.wikipedia.org/wiki/Weibull\_distribution`.

[269] Hahn, G. G. and Shapiro, S. S., *Statistical Models in Engineering*, John Wiley, New York, NY, 1967.

[270] Gamma function, `en.wikipedia.org/wiki/Gamma\_function`.

Index